*Gertz I. Likhtenshtein,
Jun Yamauchi, Shin'ichi Nakatsuji,
Alex I. Smirnov, and Rui Tamura*

Nitroxides

Further Reading

M. Kaupp, M. Bühl, V. G. Malkin (Eds.)

Calculation of NMR and EPR Parameters

Theory and Applications

2004
ISBN 978-3-527-30779-1

P. G. Wang, T. B. Cai, N. Taniguchi (Eds.)

Nitric Oxide Donors

For Pharmaceutical and Biological Applications

2005
ISBN 978-3-527-31015-9

R.-A. Eichel, S. Weber

Introduction to EPR Spectroscopy

Applications from Life to Materials Science

2009
ISBN 978-3-527-31849-0

*Gertz I. Likhtenshtein, Jun Yamauchi,
Shin'ichi Nakatsuji, Alex I. Smirnov,
and Rui Tamura*

Nitroxides

Applications in Chemistry, Biomedicine,
and Materials Science

WILEY-VCH Verlag GmbH & Co. KGaA

The Authors

Prof. Dr. Gertz I. Likhtenshtein
Ben-Gurion University of the Negev
Department of Chemistry
P.O. Box 653
Beer-Sheva 84105
Israel

Prof. Dr. Jun Yamauchi
Kyoto University
Yoshida Nihonmatsu-cho
Sakyo-ku
Kyoto 606-8501
Japan

Prof. Dr. Shin'ichi Nakatsuji
University of Hyogo
Graduate School of Materials Science
Department of Materials Science
3-2-1 Kouto, Kamigori, Ako-gun
Hyogo 678-1297
Japan

Prof. Dr. Alex I. Smirnov
North Carolina State University
Department of Chemistry
2620 Yarbrough Drive
Campus Box 8204
Raleigh, NC 27695-8204
USA

Prof. Dr. Rui Tamura
Kyoto University
Graduate School of Human and
Environmental Studies
Yoshida Nihonmatsu-cho, Sakyo
Kyoto 606-8501
Japan

All books published by Wiley-VCH are carefully produced. Nevertheless, authors, editors, and publisher do not warrant the information contained in these books, including this book, to be free of errors. Readers are advised to keep in mind that statements, data, illustrations, procedural details or other items may inadvertently be inaccurate.

Library of Congress Card No.: applied for

British Library Cataloguing-in-Publication Data
A catalogue record for this book is available from the British Library.

Bibliographic information published by the Deutsche Nationalbibliothek
Die Deutsche Nationalbibliothek lists this publication in the Deutsche Nationalbibliografie; detailed bibliographic data are available in the Internet at http://dnb.d-nb.de

© 2008 WILEY-VCH Verlag GmbH & Co. KGaA, Weinheim

All rights reserved (including those of translation into other languages). No part of this book may be reproduced in any form – by photoprinting, microfilm, or any other means – nor transmitted or translated into a machine language without written permission from the publishers. Registered names, trademarks, etc. used in this book, even when not specifically marked as such, are not to be considered unprotected by law.

Printed in the Federal Republic of German
Printed on acid-free paper

Cover design Adam Design, Weinheim
Typesetting SNP Best-set Typesetter Ltd., Hong Kong
Printing Strauss GmbH, Mörlenbach
Bookbinding Litges & Dopf GmbH, Heppenheim

ISBN: 978-3-527-31889-6

Contents

Preface *XIII*
Symbols and Abbreviations *XV*

1 Fundamentals of Magnetism *1*
Jun Yamauchi
1.1 Magnetism of Materials *1*
1.1.1 Historical Background *1*
1.1.2 Magnetic Moment and its Energy in a Magnetic Field *3*
1.1.3 Definitions of Magnetization and Magnetic Susceptibility *4*
1.1.4 Diamagnetism and Paramagnetism *5*
1.1.5 Classification of Magnetic Materials *6*
1.1.6 Important Variables, Units, and Relations *7*
1.2 Origins of Magnetism *8*
1.2.1 Origins of Diamagnetism *8*
1.2.2 Origins of Paramagnetism *10*
1.2.3 Magnetic Moments *13*
1.2.4 Specific Rules for Many Electrons *15*
1.2.5 Magnetic Moments in General Cases *17*
1.2.6 Zeeman Effect *18*
1.2.7 Orbital Quenching *19*
1.3 Temperature Dependence of Magnetic Susceptibility *21*
1.3.1 The Langevin Function of Magnetization and the Curie Law *21*
1.3.2 The Brillouin Function of Magnetization and the Curie Law *22*
1.3.3 The Curie–Weiss Law *24*
1.3.4 Magnetic Ordered State *26*
1.3.5 Magnetic Interactions *30*
1.3.5.1 Exchange Interaction *30*
1.3.5.2 Dipolar Interaction *31*
1.3.6 Spin Hamiltonian *34*
1.3.7 Van Vleck Formula for Susceptibility *35*
1.3.8 Some Examples of the van Vleck Formula *36*
1.3.8.1 The Curie Law *36*
1.3.8.2 Zero-Filed Splitting Case *37*

Nitroxides: Applications in Chemistry, Biomedicine, and Materials Science
Gertz I. Likhtenshtein, Jun Yamauchi, Shin'ichi Nakatsuji, Alex I. Smirnov, and Rui Tamura
Copyright © 2008 WILEY-VCH Verlag GmbH & Co. KGaA, Weinheim
ISBN: 978-3-527-31889-6

1.3.8.3	Spin Cluster Case—The Dimer Model 37
1.3.8.4	Multiple-spin Cluster Case—The Triangle or Others 39
1.3.8.5	Temperature-Independent Paramagnetism 39
1.3.9	Low-Dimensional Interaction Network 40
1.4	Experimental Magnetic Data Acquisition 43
1.4.1	Methods 43
1.4.2	Evaluations of Magnetic Susceptibility and Magnetic Moment 44
	References 45

2	**Molecular Magnetism** 47
	Jun Yamauchi
2.1	Magnetic Origins from Atoms and Molecules 47
2.1.1	Historical Background 47
2.1.2	Spin States Derived from Chemical Bonds 48
2.1.3	Organic Free Radicals 50
2.1.4	Coordinate Compounds 50
2.2	Characteristics of Molecular Magnetism 51
2.2.1	Molecular Paramagnetism 51
2.2.2	Magnetic Properties of Organic Free Radicals 52
2.3	Nitroxide as a Building Block 53
2.3.1	Stability of the N–O Bond 53
2.3.2	Structural Resonance of the N–O Bond 54
2.3.3	Molecular and Magnetic Interactions between Nitroxides 55
2.3.4	Nitroxides as Building Block 56
2.4	Low-Dimensional Properties of Nitroxides 57
2.4.1	One-Dimensional Magnetism 57
2.4.1.1	TANOL (TEMPOL) 57
2.4.1.2	F_5PNN 59
2.4.2	Interchain Interaction and Spin Long-Range Ordering 60
2.4.3	Two-Dimensional Magnetism 63
2.4.3.1	DANO 63
2.4.3.2	*p*-NPNN 64
2.4.4	Coordination of Nitroxide with Metal Ions 65
2.4.4.1	Cu^{2+}, Mn^{2+}-TANOL (TEMPOL) 65
2.4.4.2	Mn^{2+}–IPNN 66
	References 68

3	**Fundamentals of Electron Spin Resonance (ESR)** 71
	Jun Yamauchi
3.1	Magnetic Resonance of Electron and Nuclear Spins 71
3.1.1	Historical Background 71
3.1.2	Classification of Magnetic Resonance 72
3.2	Principle of Electron Spin Resonance (ESR) 72
3.2.1	Principle of ESR from Spectroscopic Interpretation 72
3.2.2	Principle of ESR from Resonance Interpretation 74
3.2.3	Bloch Equation 77

3.2.3.1	Solutions of the Bloch Equation	*77*
3.2.3.2	Absorption Line Shape	*78*
3.2.3.3	Relaxation Times	*81*
3.2.4	Modified Bloch Equation	*82*
3.2.5	Hyperfine Interaction	*84*
3.2.5.1	Interaction of the Electron Spin with Nuclear Spins	*84*
3.2.5.2	Hyperfine Splitting	*85*
3.2.5.3	Hydrogen Atom ($S = 1/2$ and $I = 1/2$)	*87*
3.2.5.4	Spin Polarization Mechanism	*88*
3.2.6	Fine Structure	*90*
3.2.7	Dynamical Phenomena	*91*
3.2.7.1	Correlation Time	*91*
3.2.7.2	Rotational Correlation Time	*92*
3.2.7.3	Chemical Exchange	*94*
3.3	Anisotropic Parameters in Crystal	*94*
3.3.1	g-Anisotropy	*94*
3.3.2	A-Anisotropy	*96*
3.3.3	D-Anisotropy	*99*
3.3.4	Anisotropic Parameters from ESR Powder Pattern	*99*
3.4	Pulsed ESR	*102*
3.4.1	Fundamental Concept of FT-ESR	*102*
3.4.2	Electron Spin Echo (ESE)	*104*
3.4.2.1	Two-Pulse Method	*104*
3.4.2.2	Three-Pulse Method	*105*
3.4.2.3	Inversion Recovery Method	*106*
3.4.2.4	Echo-Detected ESR (ED-ESR)	*106*
3.4.2.5	Nutation Spectroscopy	*107*
3.4.3	ESEEM	*108*
3.5	Double Resonance	*108*
3.5.1	ENDOR	*108*
3.5.2	TRIPLE	*110*
3.5.3	ELDOR	*111*
3.5.4	Pulsed Methods for Double Resonance	*113*
3.6	ESR of Magnetic Materials	*114*
3.6.1	Low-Dimensional Magnetic Materials	*114*
3.6.2	Ferromagnetic Resonance (FMR)	*115*
3.6.3	Antiferromagnetic Resonance (AFMR)	*116*
3.6.4	Ferrimagnetic Resonance	*118*
	References	*118*
4	**Recent Advantages in ESR Techniques Used in Nitroxide Applications**	*121*
	Alex I. Smirnov	
4.1	Introduction	*121*
4.2	Macromolecular Distance Constraints from Spin-Labeling Magnetic Resonance Experiments	*122*
4.2.1	Continuous Wave ESR of Nitroxide–Nitroxide Pairs	*123*

4.2.2	Continuous Wave ESR of Nitroxide–Metal Ion Pairs	125
4.2.3	Time Domain Magnetic Resonance of Spin Pairs	128
4.2.3.1	Nitroxide Spin Labels in Protein Structure Determination by NMR	128
4.2.3.2	Electronic Relaxation Enhancement in Spin Pairs	130
4.2.3.3	Pulsed Double Electron-Electron Resonance	131
4.2.3.4	ESR Double Quantum Coherence Experiments	133
4.2.4	Distance and Angular Constraints by ESR of Spin Pairs at High Magnetic Fields	134
4.3	Multiquantum ESR	135
4.4	Spin-Labeling ESR of Macroscopically Aligned Lipid Bilayers and Membrane Proteins	136
4.4.1	Mechanical Alignment of Lipid Bilayers on Planar Surfaces	136
4.4.2	Alignment of Discoidal Bilayered Micelles by Magnetic Forces	138
4.4.3	Nanopore-Confined Cylindrical Bilayers	139
4.5	Spin-labeling ESR at High Magnetic Fields	141
4.5.1	Spin-Labeling HF and Multifrequency ESR in Studying Molecular Dynamics	142
4.5.1.1	Stochastic Liouville Theory of Slow Motion ESR Spectra Simulations	143
4.5.1.2	Molecular Dynamics Simulation Methods	145
4.5.2	High Field ESR in Studying Nitroxide Microenvironment	146
4.5.2.1	Probing Local Polarity and Proticity of Membrane and Proteins	146
4.5.2.2	Site-Directed pH-Sensitive Spin-labeling: Differentiating Local pK and Polarity Effects by High-Field ESR	150
4.6	Perspectives	152
	Acknowledgements	154
	References	154
5	**Preparations, Reactions, and Properties of Functional Nitroxide Radicals**	**161**
	Shin'ichi Nakatsuji	
5.1	Short Historical Survey and General Preparative Methods of NRs	161
5.2	Early Progress toward FNRs for Organic Magnetic Materials	164
5.3	Organic Ferromagnets Based on NRs	165
5.4	Charge-Transfer Complexes/Radical Ion Salts Based on NR	168
5.5	Donors and Acceptors Carrying NRs and the Derived CT Complexes/Radical Salts	172
5.6	Suprampolecular Spin Systems Carrying NRs	175
5.7	Photochromic Spin Systems Carrying NRs	184
5.8	FNRs for Biomedicinal Applications	191
5.9	Functional Nitrones	196
5.10	Conclusion	197
	References	198

6	**Nitroxide Spin Probes for Studies of Molecular Dynamics and Microstructure** *205*
	Gertz I. Likhtenshtein
6.1	Nitroxide Molecular Dynamics *205*
6.1.1	Introduction *205*
6.1.2	Molecular Dynamics of Surrounding Molecules *206*
6.1.2.1	Microviscosity and Fluidity *206*
6.1.2.2	Motion of Macromoles as a Whole and Segmental Dynamics *207*
6.1.2.3	Low-Amplitude High-Frequency Motion and Phonon Dynamics *208*
6.2	The Spin Label–Spin Probe Methods *212*
6.3	Spin Oximetry *215*
6.4	Determination of the Immersion Depth of Radical and Flourescent Centers *217*
6.4.1	Analysis of Power Saturation Curves in Solids by CW ESR *217*
6.4.2	Determination of Depth of Immersion of a Luminescent Chromophore and a Radical Using Dynamic Exchange Interactions *218*
6.5	Nitroxide as Polarity Probes *220*
6.6	Electrostatic Effects in Molecules in Solutions *221*
6.6.1	Effect of Charge on Dipolar Interactions Between Protons and a Paramagnetic Species *221*
6.6.2	Impact of Charge on Spin Exchange Interactions Between Radicals and Paramagnetic Complexes *223*
6.7	Spin-Triplet–Fluorescence–Photochrome Method *224*
6.8	Dual Fluorophore–Nitroxide as Molecular Dynamics Probes *227*
6.9	Nitroxide Spin pH Probes *230*
6.10	Nitroxides as Spin Probes for SH Groups *231*
	References *232*

7	**Nitroxide Redox Probes and Traps, Nitron Spin Traps** *239*
	Gertz I. Likhtenshtein
7.1	Nitroxide Redox Probes *239*
7.1.1	Introduction *239*
7.1.2	Quantitative Characterization of Antioxidant Status *240*
7.1.3	Antioxidant Activity *242*
7.1.3.1	SOD Mimetic Activity *242*
7.1.3.2	Spin Trapping by Nitroxides *244*
7.2	Nitron Spin Trapping *247*
7.2.1	Introduction *247*
7.2.2	Chemical Structure and Reactions of Nitrones with Radicals *248*
7.2.3	Non-Radical Reactions of Nitrones *253*
7.2.4	Nitric Oxide Trapping *254*
7.2.5	Thermodynamics and Kinetics of Nitron Reactions *254*
7.3	Dual Fluorophore–Nitroxides (FNRO·) as Redox Sensors and Spin Traps *256*

7.3.1	Introduction 256
7.3.2	Analysis of Antioxidant Status 257
7.3.3	Analysis of Superoxide and Nitric Oxide by Pyren-Nitronyl 261
7.3.3.1	Superoxide Analysis 261
7.3.3.2	Nitric Oxide Analysis 262
7.3.4	Dual Molecules as Spin Traps 263
	References 264

8 Nitroxides in Physicochemistry 269
Gertz I. Likhtenshtein

8.1	Polymers 269
8.1.1	Introduction 269
8.1.2	Polymerization: Nitroxides Mediated Living Polymerization (NMLP) 269
8.1.2.1	Phenomenon of and Chemistry of Nitroxide Mediated Living Polymerization 270
8.1.2.2	Thermodynamic and Kinetics 272
8.1.3	Molecular Dynamics and Microstructure of Polymers 274
8.1.3.1	Introduction 274
8.1.3.2	Polymers Segmental Dynamics 276
8.1.3.3	Spatial and Orientational Distribution of Nitroxides 279
8.2	Nitroxides in Photochemistry and Photophysics 279
8.2.1	Fluoresence Quanching, Photoelectron Transfer and Photoreduction 279
8.2.1.1	Duel Fluorophore–Nitroxide Compounds 279
8.2.1.2	Nitroxides in Multispin Systems 284
8.2.1.3	Spin Trapping in Photochemical Reactions 285
8.3	Complexes Transition Metals with Nitroxide Ligands 287
8.4	Nitroxides in Inorganic Chemistry 292
8.4.1	Langmuir–Blodgett (LB) Films on Inorganic Substrates 293
8.4.2	Surface Microstructure and Dynamics 294
8.4.3	Nanoparticles 295
8.4.4	Local Acidity 296
	References 297

9 Organic Functional Materials Containing Chiral Nitroxide Radical Units 303
Rui Tamura

9.1	Introduction 303
9.2	Synthesis and Structure of Chiral NRs 304
9.2.1	Chiral Five-Membered Cyclic NRs 304
9.2.1.1	Chiral α-NNs 304
9.2.1.2	Chiral PROXYLs 305
9.2.1.3	Chiral DOXYLs 313
9.2.2	Chiral Six-Membered Cyclic NRs 314

9.2.3	Miscellaneous Examples *314*	
9.3	Magnetic Properties of Chiral NRs in the Solid State *315*	
9.3.1	Chiral Nitoxide-Mn^{2+} Complex Magnets *315*	
9.3.2	Chiral Multispin System *316*	
9.4	Properties of Chiral NRs in the Liquid Crystalline State *318*	
9.4.1	DOXYL and TEMPO Liquid Crystals *318*	
9.4.2	PROXYL Liquid Crystals *320*	
9.4.2.1	Phase-Transition Behavior *320*	
9.4.2.2	Ferroelectric Properties *322*	
9.4.2.3	Nonlinear Mesoscopical–Ferromagnetic Interactions *323*	
9.5	Application of Redox Properties of NRs *323*	
9.5.1	Oxidation Catalyst *323*	
9.5.1.1	Achiral Catalyst *323*	
9.5.1.2	Chiral Catalyst *325*	
9.5.2	Radical Battery *325*	
9.6	Conclusion *326*	
	References *327*	

10 Spin Labeling in Biochemistry and Biophysics *331*
Gertz I. Likhtenshtein

10.1	Proteins and Enzymes *331*	
10.1.1	Intramolecular Dynamics and Conformational Transition in Enzymes *332*	
10.1.1.1	Introduction *332*	
10.1.1.2	Low-Temperature Molecular Dynamics *332*	
10.1.1.3	Protein Dynamics at Ambient Temperature *334*	
10.1.2	Conformational Changes in Proteins and Enzymes, and Mechanism of Intramolecular Dynamics *336*	
10.1.3	Structure of the Enzymes' Active Centers *338*	
10.1.4	Site-Directed Spin-Labeling (SDSL) *340*	
10.1.4.1	Introduction *340*	
10.1.4.2	Soluble Proteins *341*	
10.1.4.3	Rhodopsin and Bacteriorhodopsin *344*	
10.1.4.4	Muscle Proteins *344*	
10.1.4.5	Membrane Proteins *345*	
10.2	Biomembranes *346*	
10.2.1	Structure and Dynamics *347*	
10.2.1.1	Location of Labels, Water, and Oxygen in Membranes *347*	
10.2.1.2	Membrane Microstructure *348*	
10.2.2	Membrane Dynamics *349*	
10.2.3	Proteins and Peptides in Membranes *350*	
10.3	Nucleic Acids *354*	
10.3.1	Introduction *354*	
10.3.2	DNA *355*	

10.3.3	RNA 358
10.4	Polysacchrides and Dextrins 360
10.4.1	Cotton and Cellulose 360
10.4.2	Cyclodextrins 362
	References 363

11 Biomedical and Medical Applications of Nitroxides 371
Gertz I. Likhtenshtein

11.1	Cells and Tissues. Biomedical Aspects 371
11.1.1	Cell Membrane Fluidity 371
11.1.2	Cells Redox Status 372
11.1.3	Nitroxides as Cell Protectors 373
11.2	Nitroxides *In Vivo* 374
11.2.1	Introduction 374
11.2.2	Nitroxide *In Vivo* Biochemistry. Biomedical Aspects 375
11.2.2.1	Antioxidant Activity of Nitroxides 375
11.2.2.2	Detection of Reactive Radicals: Spin-Trapping 376
11.2.2.3	Spin Farmokinetics *In Vivo* 378
11.2.2.4	Spin pH Probing 380
11.2.2.5	Spin Imaging 381
11.2.2.6	Nitroxide Spin Probe Oximetry 383
11.3	Medical Application of Nitroxides 385
11.3.1	Nitroxides and Nitrons as Drugs 385
11.3.2	Protection in Animal Model Diseases 386
11.3.3	Human Diseases. Therapeutic Aspects 388
11.3.4	Nitroxides in Clinics 389
11.4	Areas Related to Future Development of Nitroxide Applications in Biomedicine 391
	References 393

12 Conclusion 401

Index 405

Preface

Stable nitroxide radicals have proved to be effective tools in solving many problems in chemistry, physics, biology, and biomedicine at the molecular level. The nitroxide labels are used as "molecular rulers" to measure the distances between chosen groups and to measure the size, form, and micro-relief of objects of interest. The labels provide information that helps the scientist to understand the structure and molecular dynamics of individual molecules, polymers, liquid crystals, enzymes, proteins, membranes, and nucleic acids and how they function. Recently, new important developments of nitroxides as redox-probes, spin-traps, imaging and pharmokinetic reagents, and magnetic materials have been reported. Nitroxide derivatives are successfully used for the investigation of chemical kinetics, photophysical, and photochemical processes. Therapeutic and clinical applications of nitroxides appear to be a new advance in medicine.

This volume covers all aspects of this field. It also critically discusses recent results obtained with the use of nitroxides and gives an analysis of developments in the field in the future.

This book is a view of the area by a group of scientists with long-term experience in the investigation of chemistry, physicochemistry, biochemistry, and biophysics of nitroxides. It is not intended to provide an exhaustive survey of each topic but rather a discussion of their theoretical and experimental background, and recent developments. The literature of nitroxides is vast and many scientists have made important contributions in the area so that it is impossible in the space allowed in this book to give a representative set of references. The authors apologize to those they have not been able to include. More than 1100 references are given, which should provide a key to essential relevant literature.

In Chapters 1 to 3 (J. Yamauchi) of the present monograph, the general theoretical, and experimental background is expounded for magnetic properties and ESR techniques. Chapter 4 (A. Smirnov) describes recent advances in modern ESR technique and related areas, which to a considerable extent were stimulated by the growing requirements of nitroxide application in biology and biomedicine. Fundamentals, recent results in preparation, and basic chemical properties of nitroxides are the main subject of Chapter 5 (S. Nakatsuji)

Chapters 6 and 7 (G.I. Likhtenshtein) offer a brief outline of principles and current results in nitroxide application as spin-labels and spin-probes in the investigation of the molecular dynamics and microstructure of biological and nonbiological objects, and as redox-probes and spin-traps. An important role of nitroxides as spin pH-meters, spin oximeters, and reactive radical–nitron adducts is elucidated. These chapters form the basis for subsequent profound studies of molecular properties of various objects such as polymers, inorganic materials, complexes of nitroxides with paramagnetic metals, photochemical systems, and so forth (Chapter 8 by G.I. Likhtenshtein). Advantages in a new area, the construction and investigation of magnetic materials on the basis of nitroxides are the subjects of Chapter 9 (R. Tamura). Chapter 10 (G.I. Likhtenshtein) discusses recent results from the utilization of the spin-labeling method for the investigation of molecular structure, dynamics, and functional activity of proteins, enzymes, biomembranes, nucleic acids, and polysaccharides. The concluding Chapter 11 (G.I. Likhtenshtein) considers biomedical, therapeutic, and clinical applications of nitroxides in areas which appear to be of great importance for human wellbeing.

Chemical, biochemical, biomedical, and material researchers may find in this book knowledge about fundamentals, instrumentation, data interpretation, capacity, and recent advances in nitroxide applications. It will help them to understand how nitroxides can help to solve their own problems. Physicists and experts in ESR instrumentation may learn about current problems and achievements in various areas of chemistry and molecular biology, and in the rapidly developing field of the application of nitroxides in biomedicine and medicine in particular. The book is also suited as a subsidiary text for instructors, graduate, and undergraduate students of university biochemical and chemical departments.

The authors are very grateful to Drs A. Rockenbauer, V. Khramtsov, F. Villamena, and A. Wasserman for their valuable advice and fruitful discussions. Finally, the authors are deeply indebted to Dr H. Tsue for his generous help in the preparation of the manuscript.

January 2008 *The authors*

Symbols and Abbreviations

Symbols

α	electron spin quantum number $m_s = 1/2$
	angle, alternating parameter
	anisotropic exchange parameter
α_n	nuclear spin quantum number $m_I = 1/2$
β	spin quantum number $m_s = -1/2$
	angle
β_n	nuclear spin quantum number $m_I = -1/2$
δ_{ij}	Kronecker δ
θ	angle
	Curie-Weiss constant
ΔH	line-width
$\Delta H_{1/2}$	half height line-width
ΔH_{pp}	peak-to-peak line-width
ΔH_{msl}	maximum-slope line-width
$\Delta H_{\omega 1/2}$, $H_{\omega pp}$	line-width in frequency
ε_0	dierectric constant
ε_F	Fermi energy
η	$|J'/J|$
Γ	molecular field coefficient
γ	gyromagnetic ratio
	exchange interaction parameter
	anisotropic exchange parameter
γ_n	nuclear gyromagnetic ratio
ϕ	molecular orbital
	relaxation function
Ψ	molecular orbital
ψ	atomic orbital
ν	frequency
μ	magnetic permeability
μ_0	magnetic permeability of free space

Symbols and Abbreviations

μ_B	Bohr magneton
μ_n	nuclear Bohr magneton
$\mathcal{H}$	Hamiltonian
ρ	spin density, life-time probability
Λ	$\Lambda_{\mu\nu}$-tensor
$\Lambda_{\mu\nu}$	second order perturbation of spin–orbit interaction
λ	spin–orbit coupling constant
τ, τ_c	correlation time, life time
τ_C	correlation time
ω	(angular) frequency
ω_C	cyclotron frequency
ω_L	Larmorfrequency
χ	magnetic susceptibility
$\chi_{//}$	parallel magnetic susceptibility
$\chi_\perp$	perpendicular magnetic susceptibility
$\chi_{dia}, \chi_{para}, \chi_{TIP}$	diamagnetic, paramagnetic, and temperature-independent paramagnetic magnetic susceptibility
A	hyperfine coupling constant
A_{aniso}	anisoptpic hyperfine splitting A_x, A_y, A_z or A_{xx}, A_{yy}, A_{zz}
$\mathbf{A}_{dip}$	dipolar hyperfine tensor
A_{iso}	average of **A**-tensor
A_{sol}	hyperfine coupling constant in solution
$\mathbf{A_0}$	hyperfine tensor of Fermi contact term
A_0	Fermi contact hyperfine coupling constant
$\mathbf{B}$	magnetic induction
B_J	Brillouin function
C	Curie constant
	Coulomb integral
c	light velocity
$\mathbf{D}$	**D**-tensor
D	zero-field splitting parameter
E	energy, zero-field splitting parameter
F	force
FT	Fourier transformation
G	Gaussian (function) complex magnetization autocorrelation function
g	g-value
g_e	free electron g-value
g_n	nuclear g-value
g_{iso}	average of g-tensor
g_{sol}	g-value in solution
h	Planck constant

Symbols and Abbreviations

$\mathbf{H}, H$	magnetic field (strength)
H_1	oscillating magnetic fiel
H_a	anisotropic field
H_e	exchange field
H_C	coercivity (coercive force)
H_{cr}	critical field
H_{eff}	effective field
H_{dip}	dipolar field
H_{ex}	exchange field
$\mathbf{H}_{mol}, H_{mol}$	Wiss molecular field
$\mathbf{I}$	nuclear spin operator
I	nuclear spin angular momentum
	light intensity
I	nuclear spin quantum number
	function of η
$\mathbf{J}$	total angular momentum operator
J	total angular momentum
J	quantum number of total angular momentum
	exchange integral
	spectral density
	power spectrum
J	the resonance integral or coupling factor
K	anisotropy constant
k	Boltzmann constant
	rate constant
$\mathbf{L}$	orbital angular momentum operator
	nitary matrix
L	L-band
$\mathbf{L}$	orbital angular momentum
L	quantum number of orbital angular momentum
L	Lorentzian (function)
	Langevin function
l	azimutal quantum number
$\mathbf{M}$	magnetization
M	magnetization (M_x, M_y, M_z, M_0)
M_S	saturation magnetization
M_r	residual magnetization
$\mathbf{m}$	magnetic moment
$\mathbf{m}_{eff}, m_{eff}$	effective magnetic momnt
m_e	mass of electron
m_p	mass of proton
m_i ($i = l, S, I$)	quantum number
n	principal quantum number
Q_{ij}^i	McConnell proprotionality

R	relaxation term
S	spin operator (electron)
S	spin angular momentum (electron)
S	spin quantum number (electron)
s	spin quantum number $s = 1/2$ (electron)
	saturation factor
Tr	trace
T_i ($i = x,y,z$)	triplet basis functions
T_1	spin–lattice relaxation time
T_2	spin–spin relaxation time
T_C	Curie temperature
T_N	Neel temperature
t	time
u	magnetization
v	magnetization
X	X-ray, X-band
z	number of the nearest neighbor or the interchain paths

Abbreviations

CT	charge-transfer
CW-ESR	continuous wave-electron magnetic resonance
DANO	di-p-anisyl nitroxide
DDQ	2,3-dichloro-5,6-dicyano-p-benzoquinone
DEER	double electrone electron resonance
DMPO	5,5-dimethylpyrroline-N-oxide
DPPH	diphenyl picryl hydrazyl
DQC	double quantum coherence
DTBNO	di-t-butyl nitroxide
ED-ESR	echo detected ESR
ELDOR	electron double resonance
ESR	electron spin resonance
ENDOR	electron nuclear double resonance
ESEEM	electron spin echo envelope modulation
FT-ESR	Fourier transform-electron spin resonance
hfcc	hyperfine coupling constant
hfs	hyperfine splitting
HOMO	highest occupied molecular orbital
HTS	high temperature-series (expansion)
IN	imine nitroxide
LCAO	linear combination of atomic orbital
LUMO	lowest unoccupied molecular orbital
MD	molecular dynamics
MM	molecular mechanics

MO	molecular orbital
NMR	nuclear magnetic resonance
NMLP	nitroxide mediated leavingpolymerization
NR, NRO·	nitroxide radical
NNR, NNRO·	nitronyl nitroxide radical
PBN	phenyl *t*-butyl nitrone
POBN	α–(4-pyridyl-1-oxide)-*N*-*t*-butylnitrone
SDSL	site-directed spin-labeling
SO	super oxide
SOD	super oxide dismutase
SOMO	singly occupied molecular orbital
SQUID	superconducting quantum interference device
PEDRI	proton electron double resonance imaging instrumentation
PELDOR	pulse electron-electron double resonance
ROS	reactive oxygen species
SIFTER	single-frequency technique for refocusing
TANOL	2,2,6,6-tetramethylpiperidin-4-ol-1-oxyl
TEMPOL	2,2,6,6-tetramethylpiperidinyl-1-oxyl
TEMPO	2,2,6,6-tetramethylpiperidinyl-1-oxy
TPV	1,3,5-triphenylverdazyl
PROXYL	2,2,5,5-tetramethylpyrrolidinyl-1-oxy
TCNE	tetracyanoethylene
TCNQ	7,7,8,8-tetracyanoquinodimethane
TCNQF$_4$	2,3,5,6-tetrafluoro-7,7,8,8-tetracyanoquinodimethane
TMM	trimethylenemethane
TTF	tetrathiafulvalene
	bis(ethylenedithio)tetrathiafulvalene
TMTSF	tetramethyltetrathiafulvalene
TRESR	time-resolved electron spin resonance

1
Fundamentals of Magnetism

Jun Yamauchi

1.1
Magnetism of Materials

1.1.1
Historical Background

Magnets play a crucial role in a modern life; as we know, a vast number of devices are employed in the electromagnetic industry. In ancient times human beings experienced magnetic phenomena by utilizing natural iron minerals, especially magnetite. It was not until modern times that magnetic phenomena were appreciated from the standpoint of electromagnetics, to which many physicists such as Oersted and Faraday made a great contribution. In particular, Ampère explained magnetic materials in 1822, based on a small circular electric current. This was the first explanation of a molecular magnet. Furthermore, Ampère's circuital law introduced the concept of a magnetic moment or magnetic dipoles, similar to electric dipoles. Macroscopic electromagnetic phenomena are depicted in Figure 1.1, in which a bar magnet and a circuital current in a wire are physically equivalent. Microscopic similarity is shown in Figure 1.2, in which a magnetic moment or dipole and a microscopic electron rotational motion are comparable but not

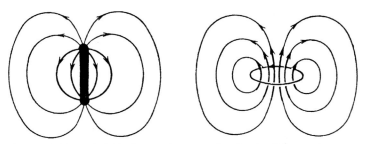

Figure 1.1 Magnetic fields due to a bar magnet and a circuital current.

Nitroxides: Applications in Chemistry, Biomedicine, and Materials Science
Gertz I. Likhtenshtein, Jun Yamauchi, Shin'ichi Nakatsuji, Alex I. Smirnov, and Rui Tamura
Copyright © 2008 WILEY-VCH Verlag GmbH & Co. KGaA, Weinheim
ISBN: 978-3-527-31889-6

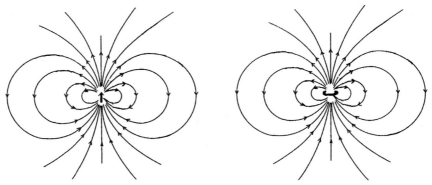

Figure 1.2 Magnetic fields due to a magnetic moment and a small circular current.

discriminated at all. The true understanding of the origin of magnetism, however, has come with quantum mechanics, newly born in the twentieth century.

Before the birth of quantum mechanics vast amounts of data concerning the magnetic properties of materials were accumulated, and a thoroughly logical classification was achieved by observing the response of every material to a magnetic field. These experiments were undertaken using magnetic balances invented by Gouy and Faraday. The principle of magnetic measurement is depicted in Figure 1.3, in which the balance measures the force exerted on the materials in a magnetic field. In general, all materials are classified into two categories, diamagnetic and paramagnetic substances, depending on the directions of the force. The former tend to exclude the magnetic field from their interior, thus being expelled effect in the experiments of Figure 1.3. On the other hand, some materials are attracted

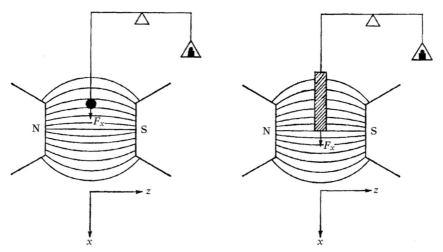

Figure 1.3 Faraday and Gouy balances for magnetic measurements. Force (F_x) is measured.

by the magnetic field. This difference between diamagnetic and paramagnetic substances is caused by the absence or presence of the magnetic moments that some materials possess in atoms, ions, or molecules. Curie made a notable contribution to experiments, and was honored with Curie's law (1895). Our understanding of magnetism was further extended by Weiss, leading to antiferromagnetism and ferromagnetism, which imply different magnetic interactions of magnetic moments with antiparallel and parallel configurations. These characteristics are involved in the Curie–Weiss law. The details will be described one after another in the following sections.

1.1.2
Magnetic Moment and its Energy in a Magnetic Field

The magnetic field generated by an electrical circuit is given as

$$\oint H \cdot dl = I \tag{1.1}$$

That is, the total current, I, is equal to the line integral of the magnetic field, H, around a closed path containing the current. This expression is called "Ampère's circuital law". The magnetic field generated by a current loop is equivalent to a magnetic moment placed in the center of the current. The magnetic moment is the moment of the couple exerted on either a bar magnet or a current loop when it is in an applied magnetic field [1]. If a current loop has an area of A and carries a current I, then its magnetic moment is defined as

$$|m| = IA \tag{1.2}$$

The cgs unit of the magnetic moment is the "emu", and, in SI units, magnetic moment is measured in Am^2. The latter unit is equivalent to JT^{-1}. The magnetic field lines around the magnetic moment are shown above in Figure 1.2. In materials the origins of the magnetic moment and its magnetic field are the electrons in atoms and molecules comprising the materials. The response of materials to an external magnetic field is relevant to magnetic energy, as follows:

$$E = -m \cdot H \tag{1.3}$$

This expression for energy is in cgs units, and in SI units the magnetic permeability of free space, μ_0, is added.

$$E = -\mu_0 m \cdot H \tag{1.4}$$

This expression in SI units is also represented using the magnetic induction, B, as defined in the next section. Therefore, the following expression is convenient in SI units:

$$E = -\boldsymbol{m} \cdot \boldsymbol{B} \tag{1.5}$$

The SI unit of magnetic induction is T (tesla).

1.1.3
Definitions of Magnetization and Magnetic Susceptibility

Each magnetic moment of a molecular magnet, including atoms or ions, is accounted for as a whole by vector summation. This physical parameter needs a counting base, such as unit volume, unit weight, or, more generally, unit quantity of substance. The last one is the mol (mole), which is widely used in chemistry. This is used in the definition of magnetization, $\boldsymbol{M}$, of materials. The units of magnetization, therefore, are emu cm^{-3}, emu g^{-1}, and emu mol^{-1}, or in SI units, A m^{-1}, A m^2 kg^{-1}, and A m^2 mol^{-1}, in which A m^2 may be replaced by J T^{-1}.

$\boldsymbol{M}$ is a property of the material, depending on the individual magnetic moments of its constituent magnetic origins. Considering the vector sum of each magnetic moment, the magnetization reflects the magnetic interaction modes at a microscopic molecular level, resulting in remarkable experimental behaviors with respect to external parameters such as temperature and magnetic field. Magnetic induction, $\boldsymbol{B}$, is a response of the material when it is placed in a magnetic field, $\boldsymbol{H}$. The general relationship between $\boldsymbol{B}$ and $\boldsymbol{H}$ may be complicated, but it is regarded as a consequence of the magnetic field, $\boldsymbol{H}$, and the magnetization of the material, $\boldsymbol{M}$:

$$\boldsymbol{B} = \boldsymbol{H} + 4\pi\boldsymbol{M} \tag{1.6}$$

This is an expression in cgs units. In SI units the relationship between $\boldsymbol{B}$, $\boldsymbol{H}$, and $\boldsymbol{M}$ is given using the permeability of free space, μ_0, as

$$\boldsymbol{B} = \mu_0(\boldsymbol{H} + \boldsymbol{M}) \tag{1.7}$$

The unit of magnetic induction, in cgs and SI units, is G (gauss) and T (tesla), respectively, and the conversion between them is $1\,G = 10^{-4}\,T$.

Since the magnetic properties of the materials should be measured as a direct magnetization response to the applied magnetic field, the ratio of $\boldsymbol{M}$ to $\boldsymbol{H}$ is important:

$$\chi = M/H \tag{1.8}$$

This quantity, χ, is called "magnetic susceptibility". The magnetization of ordinary materials exhibits a linear function with $\boldsymbol{H}$. Strictly speaking, however, magnetization also involves higher terms of $\boldsymbol{H}$, and is manifested in the $\boldsymbol{M}$ vs. $\boldsymbol{H}$ plot (a magnetization curve). Ordinary weak magnetic substances follow $\boldsymbol{M} = \chi \boldsymbol{H}$. The unit of susceptibility is emu cm^{-3} Oe^{-1} in cgs units, and because of the equality of $1\,G = 1\,Oe$, the unit emu cm^{-3} G^{-1} is also allowed. In some literature, especially in chemistry, χ is given in units of emu mol^{-1}. It should be noted that, in SI units, susceptibility is dimensionless.

The relation between M and H is the susceptibility: the ratio of B to H is called "magnetic permeability"

$$\mu = B/H \tag{1.9}$$

Two equations relating B with H and M (1.6 and 1.7) and the definitions of χ and μ lead to the following relations:

$$\mu = 1 + 4\pi\chi \quad \text{(in cgs units)} \tag{1.10}$$

$$\mu/\mu_0 = 1 + \chi \quad \text{(in SI units)} \tag{1.11}$$

Here, Equation 1.11 indicates the dimensionless relation, and the magnetic permeability of free space, μ_0, appears again. The permeability of a material measures how permeable the material is to the magnetic field. In the next section the physical explanation will be given after the introduction of magnetic flux.

1.1.4
Diamagnetism and Paramagnetism

Every material shows either positive or negative magnetic susceptibility, that is, $\chi > 0$ or $\chi < 0$. In magnetophysics or magnetochemistry this nature is referred to as "paramagnetism" (displayed by a "paramagnetic material") in the case of $\chi > 0$ and as "diamagnetism" (displayed by a "diamagnetic material") in the case of $\chi < 0$. In the M–H curve this behavior is discriminated as a positive or negative slope, as shown in Figure 1.4. Usually, a diamagnetic response toward an external magnetic field is so minor that its slope is very small compared to the paramagnetic case. The difference between paramagnetism and diamagnetism is solely

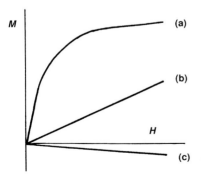

Figure 1.4 Schematic field dependencies of magnetization of (a) ferromagnetic, (b) paramagnetic, and (c) diamagnetic materials.

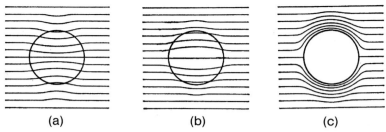

Figure 1.5 Magnetic flux in (a) paramagnetic, (b) diamagnetic, and (c) superconductive materials.

attributed to whether or not the material possesses magnetic moments in atomic, ionic, and molecular states.

Paramagnetic materials sometimes experience magnetic phase transitions at low temperatures. This means cooperative orderings of magnetic moments occur through exchange and dipolar interactions between them. There exist several ordering patterns which specify the vector arrangement of magnetic moments. Ferromagnetic and antiferromagnetic types are typical with parallel and antiparallel orientations, respectively. These magnetisms are called "ferromagnetism" and "antiferromagnetism". Phenomena concerning the cooperative ordering of magnetic moments are very attractive targets for investigation not only experimentally but also theoretically.

In view of the relationship between χ and μ, positive or the negative magnetic susceptibility corresponds to an increase or decrease in permeability, respectively, in comparison with the applied magnetic field. In order to gain more insight, the concept of "magnetic flux" or "flux density" is discussed here. Magnetic induction, **B**, is the same idea as the density of flux, Φ/A, inside the medium, by analogy with $H = \Phi/A$ in free space. Here, A is the cross-section. This indicates the difference between the external and internal flux, implying the degree of permeability of the magnetic field within a medium. This is illustrated in Figure 1.5, in which the lines indicate the magnetic flux. Perfect diamagnetism, see Figure 1.5c, is specified by **B** = 0 and is manifested inside superconductors (the "Meissner effect"). From the standpoint of magnetic flux, materials are characterized as "diamagnetic" and either "paramagnetic" or "antiferromagnetic" when magnetic flux inside is less than outside, and the reverse, respectively. In the case of ferromagnetic materials magnetic flux inside is very much greater than that outside. Ferromagnetic materials tend to concentrate magnetic flux within the medium and are characterized by a net overall magnetic moment, which is referred to as "spontaneous magnetization".

1.1.5
Classification of Magnetic Materials

The basic concept of the magnetic materials is summarized diagrammatically with the help of magnetic moments represented by arrows. No magnetic moment exists

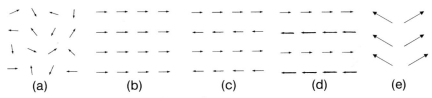

Figure 1.6 Disordered and ordered states of magnetic moments: (a) paramagnetic; (b) ferromagnetic; (c) antiferromagnetic; (d) ferrimagnetic; and (e) canting antiferromagnetic states.

in diamagnetic materials and the magnetic field applied induces a magnetic flux opposite to it. In substances possessing any magnetic moments, each magnetic moment is randomly orientated by thermal agitations, as shown in Figure 1.6a. A decrease of temperature, however, causes magnetic interactions between each magnetic moment to predominate over the thermal energy in the surroundings, thus some ordering of magnetic moments is brought about below the phase transition temperatures. Two typical ordering modes are depicted in Figure 1.6b (a ferromagnetic case) and in Figure 1.6c (an antiferromagnetic case). Materials possessing any magnetic moments respond positively to the magnetic field applied, resulting in both increases of χ and μ.

Here, a comment on the ordering in antiferromagnetism is appropriate. The antiparallel configuration of the magnetic moments has some orientational varieties. One variation concerns a different magnitude in the magnetic moments of each antiferromagnetically interacting pair. This case is termed "ferrimagnetism" (as applied to "ferrimagnetic materials") and is shown in Figure 1.6d. Such a ferrimagnetic material possesses a net magnetic moment even when the antiparallel array of each moment occurs. Because of this net magnetic moment, although the magnitude itself is far less compared with the ferromagnetic case, the magnetic susceptibility becomes far greater than for a paramagnetic material. The other typical antiparallel arrangement occurs in the case of deviation of co-linearity of magnetic moments (see, Figure 1.6e), which is called "canting antiferromagnetism". If the canting direction is not countervailed as a whole, then the net magnetic moment survives. Both ferrimagnetism and canting antiferromagnetism are sometimes termed "weak ferromagnetism" on the basis of their spontaneous magnetizations, though these are very small compared to genuine ferromagnetic materials.

1.1.6
Important Variables, Units, and Relations

In consideration of the difference of the unit systems, cgs and SI, the important variables and relations in magnetic study which we have introduced so far are summarized here [1].

	Variables	cgs	SI	Conversion
Energy	E	erg	J (joule)	$1\,\text{erg} = 10^{-7}\,\text{J}$
Magnetic field	H	Oe (oersted)	Am^{-1}	$1\,\text{Oe} = 79.58\,\text{Am}^{-1}$
Magnetic induction	B	G (gauss)	T (tesla)	$1\,\text{G} = 10^{-4}\,\text{T}$
Magnetic flux	Φ	Mx (maxwell)	Wb (weber)	$1\,\text{Mx} = 10^{-8}\,\text{Wb}$
Magnetization	M	emu cm^{-3}	Wb m^2	$1\,\text{emu cm}^{-3} = 12.57\,\text{Wb m}^{-2}$

	Relations	cgs units	Relations	SI units
Magnetic energy	$E = -\mathbf{m}\cdot\mathbf{H}$	erg	$E = -\mu_0\mathbf{m}\cdot\mathbf{H} = -\mathbf{m}\cdot\mathbf{B}$	J
Magnetic susceptibility	$\chi = M/H$	$\text{emu cm}^{-3}\,\text{Oe}^{-1}$	$\chi = M/H$	dimensionless
Magnetic permeability	$\mu = B/H$ $= 1 + 4\chi$	G Oe^{-1}	$\mu = B/H = \mu_0(1+\chi)$	$\text{T A}^{-1}\text{m} = \text{H m}^{-1}$

SI units represented by SI fundamental constituents, kg, m, s, and A.

SI symbol	SI unit	Fundamental constituent
N	newton	kg m s^{-2}
J	joule	$\text{kg m}^2\text{s}^{-2}$
T	tesla	$\text{kg s}^{-2}\text{A}^{-1}$
Wb	weber	$\text{kg m}^2\text{s}^{-2}\text{A}^{-1}$
H	henry	$\text{kg m}^2\text{s}^{-2}\text{A}^{-2}$

1.2
Origins of Magnetism

1.2.1
Origins of Diamagnetism

Diamagnetic materials innately possess no magnetic moments in the atoms, ions, or molecules which are their constituents, with the exception that magnetic moments interact with each other most strongly as an "antiparallel pair" so that at ambient temperatures they behave in a diamagnetic ways. Keeping this exception in mind, therefore, a rare origin of diamagnetism is strong twin coupling of magnetic moments in an antiferromagnetic manner. It may be pointed out that, in this case, paramagnetism turns up at an elevated temperature region. Apart from this exception, what is the general origin of diamagnetism? This may be understood on the basis of Lentz law, which states that, when a magnetic field is applied to a circuit, the current is induced so as to reduce the increased magnetic flux caused by the magnetic field. This means that the circuit is accompanied by a magnetic moment opposite to the applied magnetic field. This is equivalent to the diamagnetism caused by the Larmor precession of electrons. As a simple example we consider a spherical electron distribution around a nucleus and an electron on the sphere at a distance, r (Figure 1.7a).

The electric induction attributed to the applied magnetic field along the z-axis occurs in a plane normal to the magnetic field. The radius, a, of this circuit is related as $a^2 = x^2 + y^2$, where the coordinate of the electron is (x, y, z). The induced magnetic moment on this loop is expressed in emu using the electron mass, m_e, as

$$\delta m = -(e^2/4m_e c^2)<x^2+y^2>H \qquad (1.12)$$

Here, the symbol $<>$ indicates an average in Figure 1.7a. Spherical symmetry assumes $<x^2> = <y^2> = <z^2> = <r^2>/3$, giving

$$\delta m = -(e^2/6m_e c^2)<r^2>H \qquad (1.13)$$

This is summed up for all electrons in the atom, and a molar magnetic susceptibility, χ_M, is given in emu, using the Avogadro constant, N_A, as follows:

$$\chi_M = -(N_A e^2/6m_e c^2)\sum <r_k^2> \qquad (1.14)$$

This formulae is endorsed by quantum mechanics and the temperature-independency of this value may be understood from the evaluation of $<r_k^2>$ using the wave functions. The energy difference between the wave functions with different

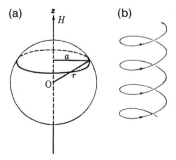

Figure 1.7 (a) Electron rotation in the radius, a; (b) cyclotron motion caused by the magnetic field.

radial parts is approximates 10 eV. The thermal energy kT at room temperature amounts to approximately 1/40 eV. Thus, diamagnetism exhibits no dependence on temperature. Atomic diamagnetism usually is of the order of $\sim 10^{-6}$ emu, and increases in absolute magnitude for larger atoms with a bigger atomic number because they have a wider radial distribution function.

Diamagnetism for the free electron model is commented on here. It is well known that free electrons are moved by the applied magnetic field, showing a helical motion along the direction of the field. This is called a "cyclotron motion", which is a counterclockwise circular locus with respect to the magnetic field (see, Figure 1.7b). The frequency of the cyclotron is twice the Larmor frequency and it produces magnetic moments again in opposition to the magnetic field. This magnetic moment is given in emu

$$\delta m = -(ea^2/2c)\omega_C \qquad (1.15)$$

where ω_C is the cyclotron frequency and given by $\omega_C = eH/mc$. This is a diamagnetic contribution. However, when these diamagnetic contributions caused by the cyclotron motion are averaged classically for the electron assembly, then macroscopic diamagnetism vanishes. This is the theorem of Miss van Leeuwen. Landau brought this problem to a settlement considering quantization of the helical motions. To cut the matter short, diamagnetic susceptibility is given in emu cm^{-3} as

$$\chi = -(n/2E_F)\mu_B^2 \qquad (1.16)$$

Here, n is a density of free electrons and E_F the Fermi energy. The new symbol, μ_B, is called the "Bohr magneton", which is defined as a unit of magnetic moment. The detail will be described in the following section, referring to the discussion of the origin of paramagnetism.

To summarize this section, diamagnetism is a counteraction of electrons against the magnetic field so as to reduce the increment of magnetic flux caused by the applied field. This is a universal action of electrons. In the presence of magnetic moments originating from electrons such as paramagnetic materials, the opposite action – that is, a cooperative increase of the magnetic flux – takes place very readily, as if a magnetic bar is aligned along the direction of magnetic field. Note that these reactions, diamagnetic and paramagnetic, are combined additively. Now we are ready to comprehend the origins of paramagnetism.

1.2.2
Origins of Paramagnetism

Here we seek the origins of magnetic moments that give rise to paramagnetism. Briefly, magnetic moments are attributable to the angular momenta of the electrons in the atom. The image of the angular momentum of an electron corresponds to Ampère's circulating circuit, leading to the magnetic moment at the

1.2 Origins of Magnetism

atomic level, even in the absence of a magnetic field. As we see from the wave functions of the electrons in the hydrogen atom based on the Schrödinger equation and the introduction of the electron spin, there exist two types of angular momenta. One is orbital angular momentum and the other is spin angular momentum. Spin angular momentum is an intrinsic part of an electron itself, regardless of location inside or outside the atom. Finally, the orbital and the spin angular momenta are combined, giving the total angular momentum, which produces the magnetic moments.

The quantum theory of a hydrogen atom is derived from the Schrödinger equation, as follows.

$$\mathcal{H}\Psi = E\Psi$$

$$\mathcal{H} = -\frac{\hbar^2}{2m_e}\left[\frac{1}{r}\frac{\partial^2}{\partial r^2}r + \frac{1}{r^2}\left\{\frac{1}{\sin^2\theta}\frac{\partial^2}{\partial\varphi^2} + \frac{1}{\sin\theta}\frac{\partial}{\partial\theta}\sin\theta\frac{\partial}{\partial\theta}\right\}\right] - \frac{e^2}{4\pi\varepsilon_0 r} \quad (1.17)$$

The first term is the kinetic energy, and the potential energy term, $-(1/4\pi\varepsilon_0)(e^2/r)$, is the Coulomb interaction between the electron and the nucleus (proton). As a result of the spherical symmetry of the Coulomb potential, the wave function, Ψ, is separated into the product of the functions of each variable in the spherical coordinates.

$$\Psi nlm_l(r, \theta, \varphi) = Rnl(r)\Theta lm_l(\theta)\Phi m_l(\varphi) \quad (1.18)$$

The $Rnl(r)$ is called the radial part of the wave function and comprises the associated Laguerre functions. This function contains two types of quantum numbers, n, the principal quantum number and l, the azimutal quantum number. The angular parts of the spherical coordinates are combined into $Ylm_l(\theta, \varphi) = \Theta lm_l(\theta)\Phi m_l(\varphi)$, which are known as "spherical harmonics". The angular parts are labeled by two quantum numbers, l and m_l. The latter is named the "magnetic quantum number", and plays an important role when the magnetic field is applied. The characteristics of the orbital motion of the electron around the nucleus may be described by wave functions with a particular set of quantum numbers, n, l, m_l. These quantum numbers vary under some limitations over integer numbers as

$$\begin{aligned} n &= 1, 2, 3, \ldots \\ l &= 0, 1, 2, 3, \ldots, (n-1) \\ m_l &= 0, \pm 1, \pm 2, \pm 3, \ldots, \pm l \end{aligned} \quad (1.19)$$

The quantum number, l, is replaced by the conventional terminology, s, p, d, ..., corresponding to $l = 0, 1, 2, \ldots$, respectively. These are a natural consequence of physical meanings of the wave function so that the probability of finding an electron in radial and angular motions described by $\Psi nlm_l(r, \theta, \varphi)$ is given by $|\Psi nlm_l(r, \theta, \varphi)|^2$. Thus, the wave function must be a finite, continuous, and one-valued function, and furthermore it is normalized to 1.

The essential point in this situation is that the angular motion which is specified by $Ylm_l(\theta, \varphi)$ is concerned with the orbital angular momentum as long as l is not equal to zero. The magnitude of the orbital angular momentum of an individual electron with the quantum numbers l and m_l is calculated by operating the relevant operators $\mathbf{L}^2$ and $\mathbf{Lz}$ for the angular momentum $\mathbf{L}$,

$$|L| = \sqrt{l(l+1)}\hbar, \quad Lz = m_l\hbar \quad (1.20)$$

From the nature of quantum numbers (1.19), we see that $\mathbf{L} \neq 0$ unless $l = 0$. Consequently, the s electrons occupying the s orbitals ($l = 0$) have zero orbital angular momentum, and therefore, no contribution to the magnetic moments. In order to attain an orbital angular momentum pictorially, the case of d-orbitals ($l = 2$) is illustrated in Figure 1.8a, where the five components, $m_l = 2, 1, 0, -1, -2$ are differentiated with respect to the direction of the magnetic field. The magnitude of the orbital angular momentum of the d-orbital is $\sqrt{6}\hbar$ and a little larger than the projected value of the moment to the magnetic field direction. This means that the orbital angular momentum vector can never align along the direction of the magnetic field but makes a precession and forms a cone around the magnetic field direction. This is a quantization image for the angular momentum by the applied magnetic field.

Next we consider the spin angular momentum. For the spin motion of the electron around its own axis, the spin quantum numbers have to be introduced, analogous to the quantum numbers, l and m_l, for the orbital angular momentum.

$$s = 1/2, \quad m_s = \pm 1/2 \quad (1.21)$$

The spin angular momentum, $\mathbf{S}$, and its components are given similarly from the general character of angular momentum

$$|S| = \sqrt{s(s+1)}\hbar = (\sqrt{3}/2)\hbar, \quad Sz = m_s\hbar = (\pm 1/2)\hbar \quad (1.22)$$

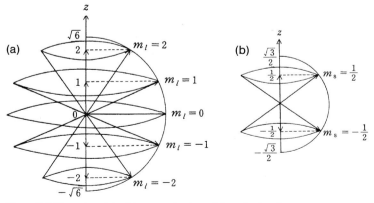

Figure 1.8 Vector model of the quantization of the orbial and spin angular momenta: (a) $l = 2$, (b) $s = 1/2$.

1.2 Origins of Magnetism

The vector model of the spin angular momentum is also schematically shown in Figure 1.8b. The application of the magnetic field produces two kinds of magnetic moments making precession in a cone. The projection value of the magnetic moments is given by the components of the spin quantum number, m_s, similar to the magnetic quantum number, m_l, of the orbital angular momentum. The spin quantum number is never derived from the Schrödinger Equation 1.17. Taking the relativistic effects into account, Dirac modified the Schrödinger equation, giving rise to the freedom of spin in the electron. Instead of solving the relativistic Dirac equation, we review several experimental matters which led to evidence for the intrinsic presence of electron spin. It was in 1922 that Stern and Gerlach reported to Bohr the atomic beam experiment through the magnetic field gradient, implying that an Ag beam split into two lines when a magnetic field was applied. According to quantum mechanics Ag has no orbital angular momentum. Nevertheless, an Ag atom is classified into two types, one attracted and the other repelled by the magnetic field. This means that even the s electron (5s electron) behaves magnetically and any magnetic moment must be caused by any kind of motion. Thus, two kinds of spin motions, in a clockwise and an anticlockwise manner, were postulated and the angular moment of self-rotations can produce magnetic moments in directions parallel and antiparallel to the magnetic field. The other evidence of the electron spin concerns why the atomic spectrum of Na splits into two D-lines even in the absence of the magnetic field. This phenomenon cannot be explained without "spin" of the electron.

Now we have two kinds of angular momenta in the atom as an origin of the magnetic moment. The orbital and the spin angular momenta are combined vectorialy and we define the total angular momentum, using new quantum numbers, j and m_j. In one electron system (l and $s = 1/2$), the total quantum numbers are simple, like $j = l + 1/2$ and $j = l - 1/2$, and m_j is given for each j value as

$$m_j = j, j-1, ,, -j+1, -j \tag{1.23}$$

From these quantum numbers, the total angular momentum, J, and its components, J_z, is give by

$$|J| = \sqrt{j(j+1)}\hbar, \quad J_z = m_j \hbar \tag{1.24}$$

In this case the state multiplicity is $2j + 1$ and the degeneracy is lifted by the application of the magnetic field, which can explain any Zeeman effect in the materials. The excited state of 3p orbital ($l = 1$) occupied by one electron ($s = 1/2$) for Na is characterized by the quantum numbers, $j = 3/2$ and $1/2$, giving rise to D-line splitting even without the magnetic field.

1.2.3
Magnetic Moments

We have learned that the electrons have two kinds of angular momenta and behave like charged particles forming a loop, giving rise to magnetic moments. Ampère's law makes it possible to formulate the relationship between angular momentum and the magnetic moment. Considering the relation of (1.2), $|m| = I\,A$, a simple treatment of a circular orbit concludes the magnetic moment relating to the angular momentum, m_l.

$$m_z = -\mu_B m_l \quad (m_l = l, l-1, \,,\, -l+1, -l)$$
$$\mu_B = e\hbar/2m_e \tag{1.25}$$

This means that the magnetic moment can be measured in the unit of μ_B, which is called the "Bohr magneton". Note that the magnetic moment vector is opposite to the direction of the angular momentum because of the negative charge of the electron and that this magnetic moment is a projection value to the quantization axis (magnetic field direction). The relationship between the magnetic moment and the angular momentum operator is written as

$$m = -\mu_B L \tag{1.26}$$

and the magnitude of the magnetic moment becomes $\mu_B \sqrt{l(l+1)}$.

The magnitude of the Bohr magneton is very important in magnetic science and its value is given in SI units as

$$\mu_B = 9.274 \times 10^{-24}\,\mathrm{J T^{-1}} \tag{1.27}$$

In cgs units it is given as

$$\mu_B = e\hbar/2m_e c = 0.927 \times 10^{-20}\,\mathrm{erg\,G^{-1}} \tag{1.28}$$

For the spin angular momentum the deduction of the relationship is not simple because we cannot modify the relation classically but instead resort to the theory of quantum electrodynamics. However, the relation itself seems simple, almost analogous to the orbital angular momentum

$$m = -g_e \mu_B S, \quad m_z = -g_e \mu_B m_s, \quad (m_s = 1/2, -1/2) \tag{1.29}$$

in which the newly introduced proportional constant g_e, which is called the g-factor of the electron, is given to be $g_e = 2.002319$ after the relativistic correction. Otherwise $g_e = 2$ is frequently used. Considering the spin quantum numbers are $s = 1/2$ and $m_s = \pm 1/2$, the magnetic moment of the electron is counted as one Bohr magneton.

In summary, the combined magnetic moments are given as a result of the two contributions from each angular momentum, as follows:

$$m = -\mu_B(L + g_e S) \tag{1.30}$$

In this section it is instructive also to describe the nuclear magnetic moment in comparison with the above-mentioned electron case. The nuclear magnetic moment originating from the nuclear spin quantum number, I, is given by a similar relation to (1.29) for the electron spin as follows:

$$m_n = g_n \mu_n \mathbf{I}, \quad m_{nz} = g_n \mu_n m_I, \quad (m_I = I, I-1, , , -I+1, -I) \qquad (1.31)$$

Here, $\mathbf{I}$ is a nuclear spin operator and g_n is a proportionality constant called the "nuclear g-value". Each nucleus possesses its original I and g_n-value. The nuclear Bohr magneton, μ_n, is a unit of the nuclear magnet and is defined in SI units, analogous to the Bohr magneton (1.25), by

$$\mu_n = e\hbar/2m_p = 5.05824 \times 10^{-27} \, \text{J T}^{-1}. \qquad (1.32)$$

Here m_p represents a mass of proton. Therefore, the ratio of $|\mu_B/\mu_n|$, which is equal to m_p/m_e, is in the order of 10^3, indicating the dominant contribution of the electron to the magnetic moments of materials and the far stronger magnetic interactions between the electron magnetic moments. For ^{1}H (proton) $I = 1/2$ and $g_n = 5.585$, yielding $m_n = 2.7927 \, \mu_n$. ^{14}N with $I = 1$ possesses $m_n = 0.4036 \, \mu_n$. Finally, it may be understood that the relations (1.31) have a positive sign, differing from the electron case with a negative sign, owing to the positive charge of the nucleus.

1.2.4
Specific Rules for Many Electrons

In general atoms or ions there exist many electrons (except the hydrogen-like atoms) so we have to take into account additional rules concerning electron configurations; these are called the "Aufbau principle". It is easy to understand that electrons occupy wave functions with lower energy first. The second rule tells us that double occupation with the same quantum numbers (n, l, m_l, m_s) is prohibited; this is the well-known "Pauli exclusion principle". Because the orbital wave function is designated by (n, l, m_l), a maximum of two electrons can occupy each orbital wave function and their spin quantum numbers $m_s = \pm 1/2$ should be different. This means that an electron is likely to make a pair with the opposite spin. In the occupation process we need an additional law for the degenerate orbitals such as $l \neq 0$. This is "Hund's rule". First, the electrons maximize their total spin, which is realized when each electron occupies an individual orbital separately with parallel spins. After the one-electron occupations are completed within the degenerate orbitals, antiparallel spins start to reside. Second, for a given spin arrangement the electron configuration for the lowest energy results in the largest total orbital angular momentum.

For a many-electron atom we have to consider the orbit–orbit, spin–orbit, and spin–spin interactions between the angular momenta of the individual electron specified by the quantum numbers, l and s. The orbital angular momentum induces a magnetic moment at the nucleus, and hence exerts a magnetic field at the electron, which interacts with an electron magnetic moment originated from its spin. This magnetic interaction mechanism is called the spin–orbit coupling, which is the most important interaction in magnetism and magnetic resonance because it actually couples between the orbital wave functions and electron spins. The magnitude of the spin–orbit coupling is determined by presuming the orbiting motion of the nucleus around the electron specified by the electron wave function. Therefore, spin–orbit interaction is proportional to the nuclear charge and thus nuclear number, Z, as is expressed by the Hamiltonian between the orbital and spin operators, $\mathbf{l}_i$ and $\mathbf{s}_i$ for an i electron

$$\mathcal{H} = \zeta \mathbf{l}_i \cdot \mathbf{s}_i, \quad \zeta = 2Z\mu_B^2 <1/r^3> \tag{1.33}$$

where $<1/r^3>$ means an orbital average. For many-electron cases this coupling has to be summed up. In this process, when the spin–orbit coupling is weak for the light atoms, the couplings between the individual orbital angular momenta and the individual spin angular momenta become predominant. Consequently, the summation of (1.33) is transformed into the next Hamiltonian as a result of $\mathbf{L} = \Sigma \mathbf{l}_i$ and $\mathbf{S} = \Sigma \mathbf{s}_i$.

$$\mathcal{H} = \lambda \mathbf{L} \cdot \mathbf{S} \tag{1.34}$$

This is an important Hamiltonian called the spin–orbit coupling. In summing up the individual orbital and spin angular momenta, special rules are concluded for the characteristic configurations of electrons in a certain shell. For a fully filled shell, $\mathbf{L}$ and $\mathbf{S}$, therefore the spin–orbit coupling vanishes. Besides, a half-filled shell gives $L = 0$ and $S = (2l + 1)/2$; again, no spin–orbit interaction. We classify the remaining configurations by "less than half" and "more than half". The coefficient λ of the shell-electron number, n, is given as $\lambda = \zeta/n$ for the former case and as $\lambda = -\zeta/(4l + 2 - n)$ for the latter case, and a positive or negative λ is deduced for the "less than half" or "more than half" configurations, respectively. For example, therefore, a "less than half" atom is likely to couple the total orbital and total spin momenta with an antiparallel configuration

Next we proceed to Russell–Saunders coupling. As we see from the above discussion, a weak spin–orbit coupling is assumed in this derivation. The allowed values of the added angular momenta are explained for simplicity in the two-electron case, the orbital quantum numbers, l_1 and l_2 and the spin quantum numbers, s_1 and s_2. The allowed total orbital and spin quantum numbers are as follows:

$$L = l_1 + l_2, l_1 + l_2 - 1, \,,\, |l_1 - l_2|, \quad S = s_1 + s_2, s_1 - s_2 \tag{1.35}$$

Actually $s_1 = s_2 = 1/2$, then we obtain $S = 1$ and $S = 0$. This method is repeated for more than three electrons. The components of the combined angular momenta are specified using m_L and m_S, analogous to the m_l and m_s, as follows:

$$m_L = L, L-1, \,,\, -L, \quad m_s = S, S-1, \,,\, -S \tag{1.36}$$

1.2 Origins of Magnetism

The values $2L + 1$ or $2S + 1$ are the number of the components which belong to L or S quantum numbers, respectively. These are called the "orbital multiplicity" and the "spin multiplicity" indicating the number of the degenerate states in free atoms. The corresponding angular momenta are given using these quantum numbers as

$$|L| = \sqrt{L(L+1)}\hbar, \quad L_z = m_L\hbar$$
$$|S| = \sqrt{S(S+1)}\hbar, \quad S_z = m_S\hbar \quad (1.37)$$

The total angular momentum is then determined by the same vector summation of the total orbital and spin angular momenta.

$$J = L+S, L+S-1,\,,\,|L-S|, \quad m_J = J, J-1,\,,\,-J \quad (1.38)$$

The multiplicity is $2J + 1$, and the magnitudes of the total angular momentum are

$$|J| = \sqrt{J(J+1)}\hbar, \quad J_z = m_J\hbar \quad (1.39)$$

Now the description of the atomic state can be characterized using the three angular momenta, L, S, and J, and their quantum numbers, L, S, and J. In general, the atomic or its ionic states are specified by the atomic "term", like $^2P_{3/2}$ or $^2P_{1/2}$ for 3p-excited Na. The central capital alphabets, S, P, D, F,,, mean $L = 0, 1, 2, 3,$,,, respectively, and the superscript indicates the multiplicity of $2S + 1$ and the subscript J quantum numbers. Incidentally, the Na ground state is represented by $^2S_{1/2}$ and its excited states by a lower $^2P_{1/2}$ and a higher $^2P_{3/2}$. This is the origin of the two D-lines.

In closing this section, we remark on the case in which the Russell–Saunders coupling fails. For the heavier atoms the spin–orbit coupling becomes so strong that, in atoms such as the actinides like U, the spin and orbital angular momenta of the individual electrons couple first, and then the combined quantum number, $j_i = l_i + s_i$, becomes a good quantum number. The resultant angular momentum, j_i, interacts with another, giving the total angular momentum,

$$j_i = l_i + s_i, \quad J = \sum j_i \quad (1.40)$$

This is the j–j coupling and the quantum numbers, L and S, are meaningless. Nevertheless, the Russell–Saunders coupling can be applied effectively even for the rare earth elements (lanthanoids). Usually, therefore, there is no need to take account of the j–j coupling.

1.2.5
Magnetic Moments in General Cases

As we summarized above, in Section 1.2.3, the combined magnetic moment is given by 1.30. This relation is derived for one electron having l and s quantum

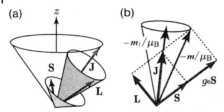

Figure 1.9 (a) Angular momenta, L, S, and J, and (b) magnetic moments, m and m_j, related with L, S, and J.

numbers. For many electrons this is also the case in which the meanings, **L** and **S**, are modified as the operators referring to the combined orbital and spin angular momentum, $\mathbf{L} = \Sigma \mathbf{l}_i$ and $\mathbf{S} = \Sigma \mathbf{s}_i$. In the case of the effective Russell–Saunders coupling, the total quantum number, J, as a result of the combined contribution between the total orbital and spin angular momenta, becomes a good quantum number. This means that the two physically significant parameters, the magnetic moment, **m**, and the total angular momentum, **J**, are not collinear. This situation is depicted in Figure 1.9, assuming $g_e = 2$. The projected magnitude of the magnetic moment, **m**, along the **J** axis is given as a following relation of the total angular momentum operator, **J**.

$$m_J = -g_J \mu_B \mathbf{J} \tag{1.41}$$

$$g_J = 1 + \frac{J(J+1) + S(S+1) - L(L+1)}{2J(J+1)}$$

Here, g_J is called the Landè g-factor. The magnitude of the total magnetic moment and its component are given as follows

$$|m_J| = g_J \mu_B \sqrt{J(J+1)}, \quad m_{Jz} = g_J \mu_B m_J \tag{1.42}$$

where $m_J = J, J-1, , , -J$. The special important cases are $S = 0$ or $L = 0$, that is, either contribution of the orbital or the spin angular momentum to the magnetic moment. Then the above equation implies $g_J = 1$ for $S = 0$ and $g_J = 2$ for $L = 0$. The latter case is "spin only" contribution to the magnetic moment,

$$|m_S| = g_e \mu_B \sqrt{S(S+1)}, \quad m_{Sz} = g_e \mu_B m_S \tag{1.43}$$

where $m_S = S, S-1, , , -S$.

1.2.6
Zeeman Effect

The Zeeman effect was observed in the spectroscopy of the emitted light from the atoms under the influence of the magnetic field. Compared to the atomic spectra without the magnetic field, the additional splittings of the spectra were detected, which are ascribed to the interaction of the atomic magnetic moments with the magnetic field. The energy of this interaction is given, in cgs units, by

$$E = -\mathbf{m}\cdot\mathbf{H} \tag{1.44}$$

In SI units this relation is modified by μ_0 as $E = -\mu_0 \mathbf{m}\cdot\mathbf{H}$, thus,

$$E = -\mathbf{m}\cdot\mathbf{B} \tag{1.45}$$

Replacing $\mathbf{m}$ with the angular momentum operator, $\mathbf{J}$, the Hamiltonian becomes

$$\mathcal{H} = g_J \mu_B \mathbf{J}\cdot\mathbf{H}, \quad \mathcal{H} = g_J \mu_B \mathbf{J}\cdot\mathbf{B} \tag{1.46}$$

This is called the "Zeeman Hamiltonian" (the Zeeman term) or the "Zeeman energy". From these energy representations, it is clear that the Zeeman energy depends not only on the quantum number, J, but also on the quantum numbers, L and S, because g_J includes L, S, and J. In "spin only" case, it is given by

$$\mathcal{H} = g_e \mu_B \mathbf{S}\cdot\mathbf{H}, \quad \mathcal{H} = g_e \mu_B \mathbf{S}\cdot\mathbf{B} \tag{1.47}$$

Considering the components of J, the Zeeman splitting can be explained. For example, the D-lines of Na are observed as $^2P_{3/2} \Leftrightarrow {}^2S_{1/2}$ and $^2P_{1/2} \Leftrightarrow {}^2S_{1/2}$. Under the application of the external magnetic field the line number increases depending on each component of the sublevels. Historically, the Zeeman effects were discriminated as the normal and anomalous Zeeman effects. The reason of this situation is attributable to incomplete understanding in the periods of no idea of the electron spin. The "anomalous" term is no more anomalous after the introduction of the assured existence of the electron spin and hence of the spin–orbit coupling.

1.2.7
Orbital Quenching

The expectation value of the orbital angular momentum, L, for a certain orbital, Ψ, is obtained from the integral, $<L> = \hbar \int \Psi^* \mathbf{L} \Psi d\tau$. Here $\mathbf{L}$ is the operator of the orbital angular momentum. The quenching of the orbital angular momentum means $<L> = 0$, that is, the expectation value of any component of the orbital angular momentum vanishes. Under what circumstances is the angular momentum quenched? Let us see how it comes for L_z, which is represented as $L_z = (1/i)(x\partial/\partial y - y\partial/\partial x) = (1/i)(\partial/\partial \varphi)$. The important point is that L_z is a pure imaginary operator (this is also the case for the other components). In addition, the operator, L_z, is a "Hermitian operator", so the diagonal element must be real. As long as the wave function, Ψ, is real, the matrix element must be zero from the requirement of "Hermiticity". This is called the "orbital quenching". The real wave function can be brought about when the electronic orbital motion interacts strongly with the crystalline electric fields. This has something to do with the lifting of the degenerate orbital energies. Non-degenerate atomic or molecular orbitals must be real. This theorem is easily comprehensible if one considers that, for an assumed complex wave function, the complex conjugate of the wave function also satisfies the original Schödinger equation with a same eigenvalue. We conclude that the crystal field produced by a symmetric environment can, at least partially, quench the orbital angular momentum of the atom. In this situation, $\mathcal{H} = \lambda \mathbf{L} \cdot \mathbf{S}$ should be treated as a perturbation for the discussion of the spin system. In the complete quenching cases, the quantum number, J, and its operator, $\mathbf{J}$, can be replaced by the quantum number, S, and its operator, $\mathbf{S}$, respectively, in concert with the replacement of g_J with g_e and, therefore, the magnetic moment includes only spin origin, $\mathbf{m} = -g_e \mu_B \mathbf{S}$. The partial or incomplete quenching implies remaining orbital angular momentum to some extent, resulting in some contribution of the orbital angular momentum to the magnetic moments. In the perturbation of $\lambda \mathbf{L} \cdot \mathbf{S}$, the g-factor deviates from g_e, the magnetic moment being $\mathbf{m} = -g \mu_B \mathbf{S}$. In this context observation of the g-value in the electron spin resonance (ESR), spectroscopy is of considerable significance.

As an example of orbital quenching, the magnetic data are summarized for the transition metal ions in comparison with the data of the rare-earth ions (Table 1.1). The experimental magnetic moment, m, is listed in the unit of μ_B and the theoretically estimated values correspond to the quenching and non-quenching cases based on the formulae, $g_e\sqrt{S(S+1)}$ and $g_J\sqrt{J(J+1)}$. The data of the transition ions are in good agreement with the value, $g_e\sqrt{S(S+1)}$, rather than the data from the total angular momentum, J. This means almost complete quenching of the orbital angular momentum, and accordingly the magnetic origin is exclusively attributed to spin, so that this is a so-called "spin only" case or magnetism. In some examples, such as Fe^{2+}, Co^{2+}, or Ni^{2+}, a little deviation is noticeable. These belong to the incomplete or partial quenching case and sometimes the orbital wave

Table 1.1 Magnetic moments of $3d^n$ and $4f^n$ ions.

n	Ions	μ/μ_B	S	J
1	Ti^{3+}, V^{4+}	1.8	1.73	1.55
2	V^{3+}	2.8	2.83	1.63
3	V^{2+}, Cr^{3+}	3.8	3.87	0.77
4	Cr^{2+}, Mn^{3+}	4.9	4.90	0
5	Mn^{2+}, Fe^{3+}	5.9	5.92	5.92
6	Fe^{2+}	5.4	4.90	6.70
7	Co^{2+}	4.8	3.87	6.63
8	Ni^{2+}	3.2	2.83	5.59
9	Cu^{2+}	1.9	1.73	3.55
10	Zn^{2+}	0	0	0

$S = 2\sqrt{S(S+1)}$, $J = g_J\sqrt{J(J+1)}$

functions are nearly degenerate. Consequently, remarkable g-factors apart from the free electron, g_e, and its anisotopy are observed for these materials. For the 4f^n ions from Ce^{3+} and Yb^{3+}, on the other hand, the agreement between the experiment and the theory is considered good except Eu^{3+}. The rare-earth ions containing f-electrons as an origin of magnetism are good examples for the Russell–Saunders coupling. The exceptional discrepancy in the data of Eu^{3+} is explained that the quantum numbers are $L = 3$, $S = 3$, and $J = 0$~6 in the Russell–Saunders category, and $3.4\mu_B$ is due to the excited states above the ground $^7F_0$ ($J = 0$) populated at room temperature.

1.3
Temperature Dependence of Magnetic Susceptibility

1.3.1
The Langevin Function of Magnetization and the Curie Law

We discuss an ensemble of non-interacting magnetic moments with the same origin in the applied field, H, at the temperature, T. The probability of occupying an energy state, $E = -\mathbf{m}\cdot\mathbf{H}$, is given by Boltzmann statistics, that is, $\exp(-E/kT) = \exp(mH\cos\theta/kT)$, where θ is an angle of the magnetic moment, m, to the applied field, H, and m and H indicate the magnitude of each vector. One has to know the number of the magnetic moments lying between the angles, θ and $\theta + d\theta$, with respect to the magnetic field. Its probability, $P(\theta)$, is related to the fractional area, dA, of the surface of the sphere covering the angles between θ and $\theta + d\theta$ at a constant radius, r. In view of $dA = 2\pi r^2 \sin\theta d\theta$, the overall probability, including the above-mentioned Boltzmann factor, is given by

$$P(\theta) = \frac{e^{mH\cos\theta/kT} \sin\theta d\theta}{\int_0^\pi e^{mH\cos\theta/kT} \sin\theta d\theta} \tag{1.48}$$

The magnetization, M, parallel to the applied field is a total vector sum of each component, $m\cos\theta$, and therefore, the magnetization of the whole system amounts to

$$M = Nm \frac{\int_0^\pi e^{mH\cos\theta/kT} \cos\theta \sin\theta d\theta}{\int_0^\pi e^{mH\cos\theta/kT} \sin\theta d\theta} \tag{1.49}$$

Here, N is the number of the magnetic moment, m, in the whole system. This equation can be represented by the following formulae after the integrals are carried out mathematically.

$$M = Nm[\coth(mH/kT) - kT/mH] = NmL(\alpha) \tag{1.50}$$

The function $L(\alpha) = \coth(\alpha) - 1/\alpha$ as a function of $\alpha = mH/kT$, is called the "Lengiven function", which is shown in Figure 1.10. We check the features of the

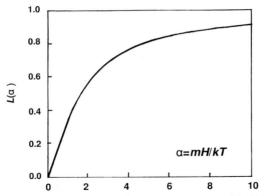

Figure 1.10 The Lengevin function L(α), expressed in M/Nm vers. α = mH/kT.

Lengiven function in the specific areas of $\alpha \gg 1$ and $\alpha \ll 1$. For $\alpha \gg 1$ this is the case either in a very large magnetic field, H, or at very low temperature, T, near zero kelvin. Then $L(\alpha) \to 1$, and M approaches Nm. The largest value, $M = Nm$, is equivalent to the complete alignment of the magnetic moments along the magnetic field, H. What about $\alpha \ll 1$, which may be achieved by the opposite parameters setting to the $\alpha \gg 1$? In this case the Lengiven function can be expanded as a Taylor series. Keeping only the prominent term, we have

$$M = Nm^2(H/3kT) \tag{1.51}$$

This relation indicates that the magnetization is proportional to the applied field and inversely proportional to the temperature. Thus, the magnetic susceptibility $\chi = M/H$ is obtained as

$$\chi = C/T, \quad C = Nm^2/3kT \tag{1.52}$$

This relation was experimentally obtained by Curie, and is called the "Curie law", where the constant, C, is a Curie constant. In conclusion, the magnetic susceptibility of paramagnetic materials without particular magnetic interactions obeys this law, and the characteristics of this behavior are ascertained by a simple formula; that is, an inverse proportionality to the temperature.

1.3.2
The Brillouin Function of Magnetization and the Curie Law

As we have discussed, each magnetic moment is expressed by $m = -g_j\mu_B \mathbf{J}$, and its Zeeman energy is $E = g_j\mu_B \mathbf{J} \cdot \mathbf{H} = g_j\mu_B m_j H$, where m_j takes $J, J-1, \, , \, -J$. Thus, we calculate $<m_j>$ instead of $<\cos\theta>$ by replacing the integral in the average with Σ of m_j. Eventually we obtain the Brillouin function, $B_j(\alpha)$.

1.3 Temperature Dependence of Magnetic Susceptibility

$$M = Ng_j\mu_B J\left\{\frac{2J+1}{2J}\coth\frac{2J+1}{2J}\alpha - \frac{1}{2J}\coth\frac{1}{2J}\alpha\right\} = Ng_j\mu_B J B_J(\alpha) \quad (1.53)$$

The function $B_J(\alpha) = \{(2J + 1)/2J\}\coth\{(2J + 1)/2J\}\alpha - (1/2J)\coth\alpha/2J$ is called the "Brillouin function" as a function of $\alpha = g_j\mu_B JH/kT$, which is equal to the Lengiven function in the limit of $J \to \infty$. The Brillouin functions for some typical quantum numbers are depicted in comparison with the experimental data (Figure 1.11) [2], where the transition ions, Fe^{3+} and Cr^{3+}, are the case of the orbital quenching (see Table 1.1) so that J should be replaced by S in the Brillouin function, and the data for Gd^{3+} is also the case of $L = 0$, therefore $J = S = 7/2$, $g_j = 2$. We examine the specific areas of $\alpha \gg 1$ and $\alpha \ll 1$. These conditions are expected in a similar manner regarding the parameters, H and T. In the conditions of $\alpha \gg 1$, we get $B_J(\alpha) \to 1$. Consequently, the magnetization approaches $M = Ng_j\mu_B J$, the saturated values in Figure 1.11. For $\alpha \ll 1$ the Brillouin function can be also expanded in a Taylor series. Keeping the first meaningful term, the magnetic susceptibility is represented by the Curie law, similar in form to the Lengevin case:

$$\chi = C/T, \quad C = Ng_j^2\mu_B^2 J(J+1)/3k \quad (1.54)$$

Comparing the two derived forms of the Curie law from the Langiven and the Brillouin functions, one sees that the Curie constant sheds light on the quantum mechanical meaning of the microscopic magnetic moment, that is,

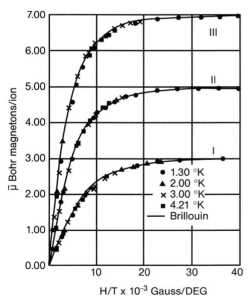

Figure 1.11 Brillouin function for $J = 3/2$ (I), $J = 5/2$ (II), and $J = 7/2$ (III) and magnetic data of Cr^{3+} ($S = 3/2$), Fe^{3+} ($S = 5/2$), and Gd^{3+} ($S = 7/2$).

$$m^2 = g_J^2 \mu_B^2 J(J+1) \tag{1.55}$$

This is the previous conclusion from the operator representation of m_J (1.41) and its magnitude (1.42). In this context the effective Bohr magneton, m_{eff}, is defined as

$$m_{eff} = g_J \mu_B \sqrt{J(J+1)} \tag{1.56}$$

Again, in the case of orbital quenching, m_{eff} is equal to $g_e \mu_B \sqrt{S(S+1)}$. Finally, it may be added that the magnetic susceptibility, χ, in SI units should be multiplied by μ_0 in the Curie law, and, thus, the Curie constant becomes $\mu_0 N g_J^2 \mu_B^2 J(J+1)/3k$.

1.3.3
The Curie–Weiss Law

In reality the observed magnetic susceptibilities do not obey the Curie law. This is because, in the above derivation of the Curie law, we have assumed isolated magnetic moments and thus no magnetic interactions are included. Many magnetic materials possess various magnetic interactions, more or less, between the individual magnetic moments, leading to the Curie–Weiss law

$$\chi = C/(T - \theta) \tag{1.57}$$

where the correction term, θ, has the unit of temperature, and is called the "Weiss constant", which is empirically evaluated from a plot of $1/\chi$ vs T. These techniques are shown schematically in Figure 1.12 in comparison with the Curie law. The intercepts of the data with the abscissa take place away from the origin, whereas the straight line crosses at the origin for the Curie law. Here we derive the Curie–Weiss law on the assumption of the existence of magnetic interactions. Although Weiss did not explain the details of the interactions between the mag-

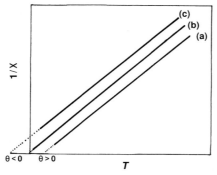

Figure 1.12 Curie–Weiss laws with (a) $\theta > 0$ and (c) $\theta < 0$ compared with (b) Curie law.

netic moments, the fundamental concept is "a molecular field" arising from the magnetization and acting on the magnetic moments in addition to the external magnetic field. The molecular field is directly proportional to the magnetization, M, and the effective magnetic field, H_{eff}, is expressed as combined with the applied magnetic field, H.

$$H_{\text{eff}} = H + \Gamma M \tag{1.58}$$

Where a term, ΓM, is a molecular field and Γ is called the "molecular field coefficient". In the Curie law relation $M/H = C/T$, and H_{eff} of (1.58) is inserted into this magnetic field, H, then we have the following relation:

$$M = CH/(T - C\Gamma) \tag{1.59}$$

In the expression of $\chi = M/H$, the Curie–Weiss law is obtained with the Weiss constant, $\theta = C\Gamma$. The Curie–Weiss law predicts anomalous behaviors at the temperature, $T_C = \theta$. The divergence of the magnetic susceptibility corresponds to the phase transition to the spontaneously magnetic ordered phase. The phase transition temperature is called the "Curie temperature" (T_C). Below this temperature, the material exhibits ferromagnetism with a spontaneous magnetization. A positive value of θ indicates that a molecular field is acting in the same direction as an applied field, so the magnetic moments are likely to align in parallel with each other, in the same direction as the magnetic field.

On the other hand, we sometimes observe a negative value of θ, in which the arrangement of the magnetic moment seems opposite, like antiferromagnetism. We see Nèel's interpretation on the antiferromagnetic formalism. In the simplest alignment of magnetic moments, one can presume two sublattices, each of which comprises the same magnetic moments in the same orientation. These structurally identical sublattices are labeled A and B, and have magnetic interactions with each other, A–A, A–B, and B–B. Ignoring the A–A and B–B interactions, the magnetic moments in the A sublattice see the molecular field generated by the magnetic moments in the sublattice B, and vice versa. In comparison with the ferromagnetic case, the molecular field is apparently opposite in direction; thus, we assume

$$H_{\text{eff}}^A = H - \Gamma M_B, \quad H_{\text{eff}}^B = H - \Gamma M_A \tag{1.60}$$

Here, M_A and M_B are the magnetizations of the sublattices A and B, respectively. Following the same procedure in the ferromagnetic case, we have the sublattice magnetizations, M_A and M_B.

$$M_A = C'(H - \Gamma M_B)/T, \quad M_B = C'(H - \Gamma M_A)/T \tag{1.61}$$

The total magnetization, M, is given by $M = M_A + M_B$, resulting in

$$M = 2C'H/(T+C'T) \tag{1.62}$$

In the expression of $\chi = M/H$, the Curie–Weiss law is obtained with the negative Weiss constant, $\theta = -C'T$. The Curie–Weiss law predicts anomalous behavior at the temperature, $T_N = -\theta$. Although the divergence of the magnetic susceptibility, like a ferromagnetic case, is not concomitant, the phase transition from a paramagnetic to an antiferromagnetic state takes place. In the antiferromagnetic ordered state each sublattice is spontaneously magnetized just like the spontaneous magnetization of the ferromagnets. This phase transition temperature is called the "Nèel temperature" (T_N).

In conclusion, the Curie–Weiss law is compatible with the existence of ferromagnets and antiferromagnets. The characteristics of the magnetic ordered states have to be described more, and the mechanisms of the interactions of the magnetic moments must be scrutinized for further comprehension of the magnetism from a quantum mechanical point of view. This is a physical research subject on magnetic cooperative phenomena.

1.3.4
Magnetic Ordered State

We focus our attention to how the magnetic ordered states come out. According to the Curie–Weiss law, magnetic susceptibility at temperatures crossing the phase transition, T_C, is discontinuous and it diverges at $T = T_C$, then what happens in between the paramagnetic and ferromagnetic phase transition? Let's consider again the Brillouin function as a function of $\alpha = g_J\mu_B JH/kT$. In the Weiss molecular field, the external magnetic field, H, is replaced by H_{eff}, including $\Gamma M(T)$. In the absence of the external magnetic field, H, the following two relations are worked out:

$$M(T)/M(0) = B_J(\alpha), \quad M(T)/M(0) = (kT/Ng_J^2\mu_B^2 J\Gamma)\alpha \tag{1.63}$$

Here, $M(0) = Ng_J\mu_B J$ is the maximum magnetization at $T = 0$. The Brillouin function varies as a function of α, as is shown above in Figure 1.11, whereas the latter is a linear function of α. The significant physical solutions are those where the two curves intersect. The unquestioned solution that occurs at the origin is devoid of meaning. Lowering the temperature the slope of the linear function gradually decreases, so that we have another intersection in addition to the origin, revealing the presence of the spontaneous magnetization. Figure 1.13 illustrates this mathematical meaning in three temperature regions, $T > T_C$, $T = T_C$, and $T < T_C$. It is essential that, at T_C, the linear function is a tangent line to the Brillouin function, and that, below T_C, spontaneous magnetization starts to grow. The spontaneous magnetization, $M(T)/M(0)$ is plotted in Figure 1.14 as a function of T/T_C for $J = 1/2$, $J = 1$, and $J = \infty$.

It is added as a summary that, in the approximation of the Weiss molecular field, the Curie temperature, T_C, is expressed with $m_{\text{eff}} = g_J\mu_B\sqrt{J(J+1)}$ as

$$T_C = Nm_{\text{eff}}^2 \Gamma/3k \tag{1.64}$$

This relation reveals that the larger the quantum number, J, and molecular field coefficient, Γ, the higher T_C is expected to be implying the large magnetic moments and the strong interactions between them are effective to obtain ferromagnetic materials at high temperatures.

Here we describe the distinguishable response of ferromagnetic materials: it is a magnetization curve under a magnetic field cycle in which, after the magnetic field is applied to reach a certain high value, the field is reduced to zero, and then it is reversed in direction, making a loop. The magnetization, M, is traced out versus H, as shown in Figure 1.15, and is called a "hysteresis curve". The initial increase of the magnetization starts at the origin (the unmagnetized state), O, and it reaches a maximum value (the "saturation magnetization"), $Ms = Ng_J\mu_B J$. In the reducing process of the field the magnetization does not conform to the origi-

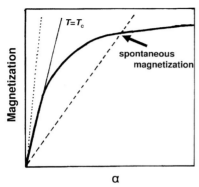

Figure 1.13 Graphical illustration of Brillouin function and spontaneous magnetization. At $T > T_c$ the dotted line crosses only at the origin ($\alpha = 0$) and at $T < T_c$ the dashed line hits at $\alpha \neq 0$. At $T = T_c$ a solid line becomes a tangent.

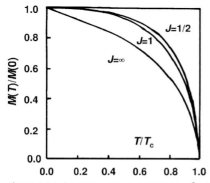

Figure 1.14 Spontaneous magnetization for $J = 1/2$, $J = 1$, and $J = \infty$.

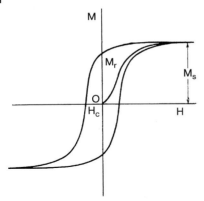

Figure 1.15 Hysteresis curve. Magnetization initially starts at the origin (O) and reaches its saturation magnetization (M_s). During the process of reducing magnetic fields, magnetization remains at $H = 0$ (M_r) and, for an opposite magnetic field, H_c, magnetization vanishes.

nal increasing curve, but remains at a certain value at $H = 0$. This is called the "residual magnetization" that corresponds to a genuine spontaneous magnetization. The reversed magnetic field gradually decreases the residual magnetization and finally makes the magnetization vanish at the field, $H = H_C$, which is named the "coercivity" or "coercive force". The hysteresis loop is completed after a cyclic application of the magnetic field. The important parameters in the evaluation of the ferromagnetic materials consist of these three values, M_S, M_r, and H_C, and every combination of these parameters is useful for practical applications depending on the various targets. In particular, large M_r means a strong magnet, and the coercivity, H_C, discriminates the materials as either soft or hard magnets. A soft magnet is likely to be magnetized easily and is also easily demagnetized.

The initial unmagnetized state in ferromagnetic materials may need some elucidation. The magnetic domain model explains that, although each domain has spontaneous magnetization, the domains are arranged in such a manner as to cancel out the net magnetization. Once the magnetic field is applied, the domains move to the direction of the magnetic field. The magnetic domains are small regions, but may be observed by several methods. Fine magnetic particles are attracted onto the surface and image up to the domain boundary where the direction of the magnetic moments changes. Another method utilizes the magneto-optic effect using polarized light.

In the case of the antiferromagnetic ordered state, the two sublattices possess their own magnetizations, which are oriented in the opposite direction and at half the magnitude compared to the ferromagnetic case. However, each magnetization obeys the ferromagnetic spontaneous magnetization curve (see above, Figure 1.14) at half magnitude. Consequently, the net magnetization is almost zero, giving the same order of magnetic susceptibility as the paramagnetic state. The most important difference observed in this ordered state is the anisotropic susceptibilities, $\chi_\parallel$

and $\chi_\perp$, the parallel axis being defined along the magnetization direction, which is called an "easy axis". Therefore, the external magnetic field can be applied along the easy axis or perpendicular to it and the magnetic susceptibilities, $\chi_\parallel$ and $\chi_\perp$, are illustrated schematically in Figure 1.16, in which $\chi_\parallel$ decreases linearly toward zero on lowering the temperature, whereas $\chi_\perp$ stays constant. A powdered sample usually exhibits their averaged value,

$$\chi = (\chi_\parallel + 2\chi_\perp)/3 \qquad (1.65)$$

as plotted by the dashed line.

The most important magnetic behaviors of the antiferromagnets are the magnetic phase change with increasing magnetic field, especially a field applied along the easy axis; that is, the direction of the magnetic moments. The magnetic energy in the parallel arrangement (see Figure 1.16a) exceeds the assumed energy in the perpendicular arrangement (see Figure 1.16b) at a certain magnitude of the magnetic field, and the parallel orientation abruptly changes into the perpendicular one, conserving the antiparallel orientation. This magnetic phase change induced by the magnetic field is called the "flopping" of the magnetic moments or spins. Thus, the state is called the "spin-flopped" state, and the magnetic field which induces the transition is named the "critical field", H_{cr}, or "spin-flop field", H_{sf}. This magnetic phase transition depends on the anisotropic energy of the magnetic moment orientation. Theoretical consideration concludes the relation of H_{cr} with $\chi_\parallel$ and $\chi_\perp$

$$H_{cr} = \sqrt{2K/(\chi_\perp - \chi_\parallel)} \qquad (1.66)$$

where K indicates the anisotoropy constant, which causes the magnetic moments to align toward the easy axis in the absence of the magnetic field. Similarly, the

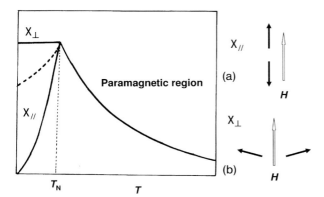

Figure 1.16 Magnetic susceptibility of an antiferromagnet. The dashed line indicates a powder susceptibility at the antiferromagnetic region, $T < T_N$.

antiferromagnetic arrangement of the magnetic moments is eventually unstable under high magnetic fields, and exhibits, to some extent, a tendency to approach the ferromagnetic state. Therefore, magnetization behavior in antiferromagnetic materials provide interesting magnetic properties with respect to the applied magnetic field. These phenomena are also important targets of magnetic investigations, and newly classified magnetism called metamagnetism is gathering much attention. The spin-flopped state can be referred to as antiferromagnetic resonance (AFMR), which will be dealt with later (see Section 3.6.3).

1.3.5
Magnetic Interactions

1.3.5.1 Exchange Interaction

Now we have prepared ourselves to have a deeper insight into various magnetic interactions between the magnetic moments. Let us consider two-electron systems, such as an He atom or a hydrogen molecule. As a result of spin multiplicity there exist singlet and triplet states and, therefore, four wave functions, $^1\Psi_0$, $^3\Psi_1$, $^3\Psi_0$, and $^3\Psi_{-1}$

$$^1\Psi_0 = \frac{1}{\sqrt{2}}(|\phi_a \alpha \phi_b \beta| - |\phi_a \beta \phi_b \alpha|)$$

$$^3\Psi_1 = |\phi_a \alpha \phi_b \alpha|$$

$$^3\Psi_0 = \frac{1}{\sqrt{2}}(|\phi_a \alpha \phi_b \beta| + |\phi_a \beta \phi_b \alpha|)$$

$$^3\Psi_{-1} = |\phi_a \beta \phi_b \beta|$$

(1.67)

where ϕ_a and ϕ_b are orbital functions and α and β the spin function indicating $m_S = 1/2$ and $-1/2$, respectively. The total wave functions, including the orbital and spin functions, are represented by "Slater's determinants". We focus our attention on the energy difference of the singlet and triplet states, evaluating the repulsion energy between the two electrons given by $\mathcal{H}' = (1/4\pi\varepsilon_0)(e^2/r)$, where r is the distance between the two electrons. The kinetic and other potential energies of the two electrons with the nucleus constitute the Hamiltonian, $\mathcal{H}_0$. The expectation values of $\mathcal{H}_0$ are the same for the two electrons, and we examine the expectation values of $\mathcal{H}$ for the above-mentioned wave functions. Then we obtain the energies for $^1\Psi_0$ (E_S) and for $^3\Psi_1$, $^3\Psi_0$ and $^3\Psi_{-1}$ (E_T) as follows.

$$E_S = C + J, \quad E_T = C - J$$

$$C = \frac{1}{4\pi\varepsilon_0}\int \phi_a(1)\phi_b(2)\frac{e^2}{r}\phi_a(1)\phi_b(2)d\tau$$

(1.68)

$$J = \frac{1}{4\pi\varepsilon_0}\int \phi_a(1)\phi_b(2)\frac{e^2}{r}\phi_b(1)\phi_a(2)d\tau$$

The integrals, C and J, are called the "Coulomb" and "exchange" integrals, respectively. The energy separation between the triplet and singlet states is $2J$, as

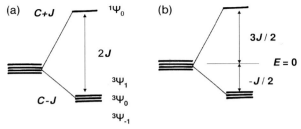

Figure 1.17 Energy shift due to Coloumb and exchange integrals and attributed to spin Hamiltonian, $-2JS_1 \cdot S_2$.

shown in Figure 1.17a. This quantum mechanical result is important because, when J is positive, the triplet states are more stable than the singlet state, meaning the ferromagnetic parallel spin is favorable. Considering that this energy-splitting originates from a difference in the spin multiplicity, we may put up the effective spin Hamiltonian as a phenomenological equivalence:

$$\mathcal{H}\text{ex} = -2JS_1 \cdot S_2 \tag{1.69}$$

The eigenvalues of this Hamiltonian may be obtained by the operation of $S_1 \cdot S_2$ as $(3/2)J$ for the singlet state and $(-1/2)J$ for the triplet state, the separation being $2J$ (Figure 1.17b). Hex is called the "Heisenberg Hamiltonian" or a magnetic interaction of the Heisenberg type, expressing the exchange interaction between the electron spins. It is noted that the spin interaction of Heisenberg type is isotropic, as indicated by $S_1 \cdot S_2$.

In ferromagnetic or antiferromagnetic materials the following general expression is usual as an exchange interaction:

$$\mathcal{H}\text{ex} = -2\sum_{<ij>} J_{ij} S_i \cdot S_j \tag{1.70}$$

As one can see from the r-dependence of the exchange integral, J, the magnitude of J_{ij} decreases quite rapidly depending on the distance r_{ij} between the electrons, i and j, and the nearest neighboring pairs are to be taken into account. Now the meaning of the Weiss molecular field, $H_{mol} = \Gamma M$, is comprehensible to be $H_{mol} = -2zJ<Sz>/g_e\mu_B$, in which z is the number of the nearest neighboring spins and $<Sz>$ the average value of Sz.

1.3.5.2 Dipolar Interaction

One more important magnetic interactions is called a "magnetic dipolar interaction", given by the two magnetic dipoles or moments, m_1 and m_2, as

$$U(m_1, m_2, r) = \frac{m_1 \cdot m_2}{r^3} - \frac{3(m_1 \cdot r)(m_2 \cdot r)}{r^5} \tag{1.71}$$

Here **r** is a connecting vector between the two magnetic moments. This formula is a classical analog, but we replace m_1 and m_2 with explicit forms of the spin operators, $\mathbf{S}_1$ and $\mathbf{S}_2$, leading to

$$\mathcal{H}(\mathbf{S}_1, \mathbf{S}_2, \mathbf{r}) = g_e^2 \mu_B^2 \left\{ \frac{\mathbf{S}_1 \cdot \mathbf{S}_2}{r^3} - \frac{3(\mathbf{S}_1 \cdot \mathbf{r})(\mathbf{S}_2 \cdot \mathbf{r})}{r^5} \right\} \tag{1.72}$$

Seeking the spin Hamiltonian of the magnetic dipolar interaction, the terms relevant to the orbital wave functions are integrated for an average, and then the above Hamiltonian is expressed in a tensor form as

$$\mathcal{H}_{SS} = g_e^2 \mu_B^2 [\mathbf{S}_{1x}\ \mathbf{S}_{1y}\ \mathbf{S}_{1z}] \cdot \begin{bmatrix} \left\langle \frac{r^2-3x^2}{r^5} \right\rangle & \left\langle -\frac{3xy}{r^5} \right\rangle & \left\langle -\frac{3xz}{r^5} \right\rangle \\ \left\langle -\frac{3xy}{r^5} \right\rangle & \left\langle \frac{r^2-3y^2}{r^5} \right\rangle & \left\langle -\frac{3yz}{r^5} \right\rangle \\ \left\langle -\frac{3xz}{r^5} \right\rangle & \left\langle -\frac{3yz}{r^5} \right\rangle & \left\langle \frac{r^2-3z^2}{r^5} \right\rangle \end{bmatrix} \cdot \begin{bmatrix} \mathbf{S}_{2x} \\ \mathbf{S}_{2y} \\ \mathbf{S}_{2z} \end{bmatrix} \tag{1.73}$$

Next we define the operator $\mathbf{S} = \mathbf{S}_1 + \mathbf{S}_2$, this expression is made up to

$$\mathcal{H}_{SS} = \frac{1}{2} g_e^2 \mu_B^2 [\mathbf{S}_x\ \mathbf{S}_y\ \mathbf{S}_z] \cdot \begin{bmatrix} \left\langle \frac{r^2-3x^2}{r^5} \right\rangle & \left\langle -\frac{3xy}{r^5} \right\rangle & \left\langle -\frac{3xz}{r^5} \right\rangle \\ \left\langle -\frac{3xy}{r^5} \right\rangle & \left\langle \frac{r^2-3y^2}{r^5} \right\rangle & \left\langle -\frac{3yz}{r^5} \right\rangle \\ \left\langle -\frac{3xz}{r^5} \right\rangle & \left\langle -\frac{3yz}{r^5} \right\rangle & \left\langle \frac{r^2-3z^2}{r^5} \right\rangle \end{bmatrix} \cdot \begin{bmatrix} \mathbf{S}_x \\ \mathbf{S}_y \\ \mathbf{S}_z \end{bmatrix} \tag{1.74}$$

This is finally expressed by the following spin Hamiltonian.

$$\mathcal{H}_{SS} = \mathbf{S} \cdot \mathbf{D} \cdot \mathbf{S} \tag{1.75}$$

Here **D** is a 3×3 matrix and is called **D**-tensor, satisfying Tr[**D**] = 0. In the SI system, all relevant equations in the derivation must be multiplied by $\mu_0/4\pi$. This is a quantum mechanical expression of the magnetic dipolar interaction.

It is usually convenient to discuss the **D**-tensor as a diagonalized form, and thus we have

$$\mathcal{H}_{SS} = D_x \mathbf{S}_x^2 + D_y \mathbf{S}_y^2 + D_z \mathbf{S}_z^2, \quad D_x + D_y + D_z = 0 \tag{1.76}$$

Here D_x, D_y, and D_z are the principal values, and owing to Tr[**D**] = 0, we define two independent parameters, D and E, as

$$D = \frac{3}{2} D_z, \quad E = \frac{1}{2}(D_x - D_y) \tag{1.77}$$

The transformation of (1.77) results in

$$\mathcal{H}_{SS} = D\{Sz^2 - \frac{1}{3}S(S+1)\} + E(Sx^2 - Sy^2) \qquad (1.78)$$

The introduced parameters, D and E, are named the "zero-field splitting constants" after the fact that this interaction can split the spin states even in the absence of the external magnetic field. Eventually both parameters are given by the orbital integrations below:

$$D = \frac{\mu_0}{4\pi}\frac{3}{4}g^2\mu_B^2 \left\langle \frac{r^2 - 3z^2}{r^5} \right\rangle$$

$$E = \frac{\mu_0}{4\pi}\frac{3}{4}g^2\mu_B^2 \left\langle \frac{x^2 - y^2}{r^5} \right\rangle \qquad (1.79)$$

When we apply these relations to the delocalized system, the spin densities, ρ_i and ρ_j, are used:

$$D = \frac{\mu_0}{4\pi}\frac{3}{4}g^2\mu_B^2 \sum_{<ij>} \frac{r_{ij}^2 - 3z_{ij}^2}{r_{ij}^5}\rho_i\rho_j$$

$$E = \frac{\mu_0}{4\pi}\frac{3}{4}g^2\mu_B^2 \sum_{<ij>} \frac{x_{ij}^2 - y_{ij}^2}{r_{ij}^5}\rho_i\rho_j \qquad (1.80)$$

Then knowledge of the spin density and the distance vector **r** between the two electrons makes it possible to evaluate the zero-field splitting constants, D and E.

Three important remarks are presented here: (1) this spin Hamiltonian is effective for $S \geq 1$; (2) this interaction produces a so-called "anisotropic energy", which governs the preferential orientation of the spins or magnetic moments; (3) this is a point-dipole approximation. Comment (1) means this interaction vanishes in the case of $S = 1/2$. Concerning the second comment, the magnetic dipolar interaction works out uniaxially in symmetry for $E = 0$ and an orthorhombicity in symmetry becomes essential for $E \neq 0$. The final comment (3) is often utilized for an approximate evaluation of the distance, r, between the spins. Neglecting the spread of the electron wave functions, the representations of D and E are deduced as $E = 0$ and D, as follows:

$$D = \frac{3}{2}g^2\mu_B^2/r^3 \qquad (1.81)$$

If the useful relation is expressed using $D/g\mu_B$ in mT and r in nm, then we have

$$D/g\mu_B = 1.391 g/r^3 \qquad (1.82)$$

When $g = 2.0023$ and $r = 1$ nm, then this formula yields $D = 2.785$ mT.

1.3.6
Spin Hamiltonian

In the previous section we have discussed the magnetic interactions on the basis of electron spins and retained a deep insight into the magnetic interactions. The concept of a spin Hamiltonian was partially introduced and seems useful in pursuing investigations which are essentially associated with the electron spins, especially magnetism and electron spin resonance. Let us summarize here the fundamental Hamiltonians which determine the electronic states and their energies from the quantum mechanical point of view.

$$\mathcal{H} = \mathcal{H}_{kin} + \mathcal{H}_{pot} + \mathcal{H}_{cr} + \mathcal{H}_{LS} + \mathcal{H}_{Ze} + \mathcal{H}_{spin} \tag{1.83}$$

The first two terms concern the kinetic and the potential energies important for determining electronic orbital wave functions, and the third one is an energy derived from the crystal or ligand field that causes a shift of the electronic energy. The following two terms, $\mathcal{H}_{LS}$ and $\mathcal{H}_{Ze}$, have been introduced as the spin–orbit coupling and the Zeeman energy, respectively. The final term includes every interaction regarding the electron spins. The exchange interaction, $\mathcal{H}_{ex}$, and the magnetic dipolar interaction, $\mathcal{H}_{ss}$, are its members, and in later chapters about magnetic resonance nuclear spin will be incorporated. In view of the orbital quenching we try to find an effective Hamiltonian by dealing with the two terms, $\mathcal{H}_{LS}$ and $\mathcal{H}_{Ze}$, as a perturbation [3]

$$\mathcal{H}' = \lambda \mathbf{L} \cdot \mathbf{S} + \mu_B (\mathbf{L} + g_e \mathbf{S}) \cdot \mathbf{H} \tag{1.84}$$

The orbital wave functions are expressed as $|0\rangle, \ldots, |n\rangle$, and their energies as $E_0, \ldots, E_n$, are determined from $\mathcal{H}_0 = \mathcal{H}_{kin} + \mathcal{H}_{pot} + \mathcal{H}_{cr}$. The second-order perturbation results in

$$\mathcal{H} = \mu_B \mathbf{S} \cdot (g_e \mathbf{1} - 2\lambda \Lambda) \cdot \mathbf{H} - \lambda^2 \mathbf{S} \cdot \Lambda \cdot \mathbf{S} + \mu_B^2 \mathbf{H} \cdot \Lambda \cdot \mathbf{H} \tag{1.85}$$

Here Λ is a tensor which is composed of the matrix elements given by

$$\Lambda_{\mu\nu} = \sum_{n \neq 0} \frac{\langle 0|L_\mu|n\rangle \langle n|L_\nu|0\rangle}{E_n - E_0}, \quad \mu, \nu = x, y, z \tag{1.86}$$

Concerning the Zeeman term, **1** is a unit matrix and the **g**-tensor representation is convenient:

$$\mathbf{g} = g_e \mathbf{1} - 2\lambda \Lambda \tag{1.87}$$

1.3 Temperature Dependence of Magnetic Susceptibility

Then the Zeeman Hamiltonian and the above result are written as follows:

$$\mathcal{H}_{Ze} = \mu_B \mathbf{S} \cdot \mathbf{g} \cdot \mathbf{H}$$
$$\mathcal{H} = \mu_B \mathbf{S} \cdot \mathbf{g} \cdot \mathbf{H} - \lambda^2 \mathbf{S} \cdot \Lambda \cdot \mathbf{S} + \mu_B^2 \mathbf{H} \cdot \Lambda \cdot \mathbf{H} \tag{1.88}$$

Several points about these terms arise. Hereafter we call the **g**-tensor a g-value, which is anisotropic, and is given by

$$g_{ij} = g_e \delta_{ij} - 2\Lambda_{ij} \tag{1.89}$$

Here δ_{ij} is a Kronecker δ. It is manifest that the anisotropy of g-value results from spin–orbit coupling, more precisely from the orbital contribution. The deviation of the g-value from the free electron and its anisotropy are undoubtedly evident in the ESR observations. It is interesting that the second term, $-\lambda^2 \mathbf{S} \cdot \Lambda \cdot \mathbf{S}$, has a similar form to the magnetic dipolar interaction, $\mathcal{H}_{SS} = \mathbf{S} \cdot \mathbf{D} \cdot \mathbf{S}$. Considering the diagonalized Λ_{ij}, the formalism becomes the same so that this effect is included representatively in $\mathcal{H}_{SS} = \mathbf{S} \cdot \mathbf{D} \cdot \mathbf{S}$ (1.75). Therefore, the zero-field splitting parameters, D and E, are deduced, as given in (1.78). The second term, $-\lambda^2 \mathbf{S} \cdot \Lambda \cdot \mathbf{S}$, is considered to be the result of a two-electron interaction of higher order as is seen from the fact that the validity of this relation is $S \geq 1$ and the term vanishes when $S = 1/2$. In addition, the nature of the anisotropy is also the same, and therefore a specific terminology is conferred as "one-ion anisotropy" or "one-ion anisotropy constant". The last term (1.85) is independent of electron spins and indicates the higher order of orbital magnetic moments induced by the magnetic field. This is an origin of the van Vleck paramagnetism, which is independent of temperature. Summarizing all spin Hamiltonians we have

$$\mathcal{H} = \mu_B \mathbf{S} \cdot \mathbf{g} \cdot \mathbf{H} + \mathbf{S} \cdot \mathbf{D} \cdot \mathbf{S} - 2\sum_{<ij>} J_{ij} \mathbf{S}_i \cdot \mathbf{S}_j \tag{1.90}$$

1.3.7
Van Vleck Formula for Susceptibility

Here we deal with a general method for calculating magnetic susceptibility. When a substance is placed in an external field, its magnetization is given by its energy variation with respect to the field, $M = -\partial E/\partial H$. This magnetization is obtained by summing the microscopic magnetizations, m_n, weighted by the Boltzmann factors. Because an analogous relation for m_n also holds as

$$m_n = -\partial E_n / \partial H \tag{1.91}$$

it leads to the final formula

$$M = \frac{N \sum_n (-\partial E_n/\partial H) \exp(-E_n/kT)}{\sum_n \exp(-E_n/kT)} \tag{1.92}$$

where E_n ($n = 1, 2, 3, \ldots$) is a quantum mechanical energy in the presence of the magnetic field, which will be fundamentally evaluated once the Hamiltonian of the system is assumed. In this context the spin Hamiltonian discussed above for the magnetic interactions plays an important role. Thus, we require knowledge of E_n as a function of H. Van Vleck proposed some legitimate approximations, assuming the energy is an expansion in a series of the applied field [4, 5]:

$$E_n = E_n^{(0)} + E_n^{(1)} H + E_n^{(2)} H^2 + \cdots \quad (1.93)$$

Where $E_n^{(0)}$ is the energy in zero field, the term linear to H is called the first-order Zeeman term, and $E_n^{(2)}$ the second-order Zeeman, that is

$$E_n^{(1)} = <n|\mathcal{H}_{\text{Ze}}|n>, \quad E_n^{(2)} = \sum_{m \neq n} <n|\mathcal{H}_{\text{Ze}}|m>^2/(E_n^{(0)} - E_m^{(0)}) \quad (1.94)$$

where $\mathcal{H}_{\text{Ze}}$ is as described in the previous section. The other nomenclatures are standard in perturbation theory. From (1.91) and (1.93) we have

$$m_n = -E_n^{(1)} - 2E_n^{(2)} H - \cdots \quad (1.95)$$

The second approximation is the expansion of the exponential under $H/kT \ll 1$. This is fulfilled under the conditions of H not being too large and T not too low. The final form will be

$$M = \frac{N \sum_n (-E_n^{(1)} - 2E_n^{(2)} H)(1 - E_n^{(1)} H/kT) \exp(-E_n^{(0)}/kT)}{\sum_n (1 - E_n^{(1)} H/kT) \exp(-E_n^{(0)}/kT)} \quad (1.96)$$

The absence of the magnetization at zero magnetic field, $M = 0$, requires

$$\sum_n E_n^{(1)} \exp(-E_n^{(0)}/kT) = 0 \quad (1.97)$$

Substituting this relation into (1.96) and retaining only terms linear in H, finally we obtain the susceptibility given by

$$\chi = \frac{N \sum_n (E_n^{(1)2}/kT - 2E_n^{(2)}) \exp(-E_n^{(0)}/kT)}{\sum_n \exp(-E_n^{(0)}/kT)} \quad (1.98)$$

This is the "van Vleck formula". When the states are degenerate, the summation is repeated n times, where n is a degeneracy. Let us explain how to utilize this formula and derive the analytical formulae of the magnetic susceptibility for several examples.

1.3.8
Some Examples of the van Vleck Formula

1.3.8.1 The Curie Law
Consider the spin degeneracy, S, with an orbital singlet, $L = 0$, the energy levels corresponding to the energy expansion in the derivation of the van Vleck formula are

$$E_n^{(0)} = 0, \quad E_n^{(1)} = g\mu_B m_S, \quad E_n^{(2)} = 0 \tag{1.99}$$

Then the result of χ (1.98) coincides with the Curie law of $\chi = Ng^2\mu_B^2 S(S+1)/3kT$.

1.3.8.2 Zero-Filed Splitting Case
In this case the following Hamiltonian is exemplified.

$$\mathcal{H} = \mu_B \mathbf{S} \cdot \mathbf{g} \cdot \mathbf{H} + \mathbf{S} \cdot \mathbf{D} \cdot \mathbf{S} \tag{1.100}$$

Here we assume the **g**- and **D**-tensors have the same principal axis and the rhombic parameter, E, is zero. The energies compared to the zero-field energy, $E_0 = 0$, will be given for the axial (parallel) direction of H as

$$E_1 = g_{\parallel}\mu_B H + D, \quad E_2 = -g_{\parallel}\mu_B H + D \tag{1.101}$$

Then the van Vleck formula for the parallel magnetic susceptibility is given by

$$\chi_{\parallel} = \frac{2Ng_{\parallel}^2 \mu_B^2}{kT} \frac{\exp(-D/kT)}{1 + 2\exp(-D/kT)} \tag{1.102}$$

When the magnetic field is perpendicular to the axial direction, the energies for the much larger D compared to the Zeeman energy become

$$E_1 = D, \quad E_2 = -g_{\perp}^2 \mu_B^2 H^2/D, \quad E_3 = g_{\perp}^2 \mu_B^2 H^2/D + D \tag{1.103}$$

As one can see from the relation, $E_n^{(1)} = 0$, for $n = 1, 2,$ and 3. The final result is

$$\chi_{\perp} = \frac{2Ng_{\perp}^2 \mu_B^2}{D} \frac{1 - \exp(-D/kT)}{1 + 2\exp(-D/kT)} \tag{1.104}$$

The temperature dependences of (1.102) and (1.104) and their inverse susceptibilities are drawn for certain parameters (Figure 1.18).

1.3.8.3 Spin Cluster Case – The Dimer Model
This model is expressed in the following Hamiltonian.

$$\mathcal{H} = -2J\mathbf{S}_1 \cdot \mathbf{S}_2 + g\mu_B(S_{1z} + S_{2z})H \tag{1.105}$$

The first term indicates the exchange-coupled two spins, $S_1 = S_2 = 1/2$, and the second term the Zeeman Hamiltonian in the applied magnetic field (this direction is specified by the z-axis). Consequently the spin system forms the singlet ($S = 0$) and triplet ($S = 1$) sublevels, the latter includes the energy terms dependent on the magnetic field. When the sign of J is negative, the energy diagram for the ground singlet and excited triplet energy is realized, as is often observed in actual cases

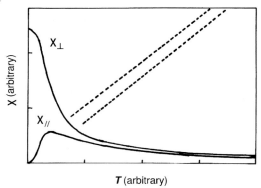

Figure 1.18 $\chi_\|$ (1.102) and $\chi_\perp$ (1.104) in the case of D > 0. The dotted lines show that each inverse $1/\chi_\|$ and $1/\chi_\perp$ follows a Curie–Weiss law above a certain temperature.

such as molecular magnetic materials or metal ion pairs. Thus, this model is well known as the "dimer" model or the "singlet–triplet" (ST) model. The energy of the four states is given as

$$E_1 = (3/2)J$$
$$E_2 = -(1/2)J + g\mu_B H, \quad E_3 = -(1/2)J, \quad E_4 = -(1/2)J - g\mu_B H \tag{1.106}$$

The state, E_1, is the singlet and E_2, E_3, and E_4 belong to the triplet. Applying the van Vleck formula, we obtain the magnetic susceptibility for the dimer model or ST model.

$$\chi = \frac{2Ng^2\mu_B^2}{3kT} \frac{\exp(-2|J|/kT)}{1 + 3\exp(-2|J|/kT)} \tag{1.107}$$

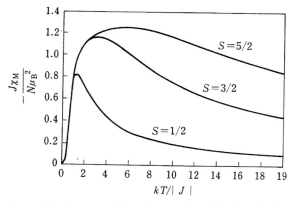

Figure 1.19 Dimer models in the case of J < 0 for S = 1/2 (ST model), S = 3/2, and S = 5/2.

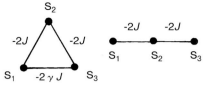

Figure 1.20 Triangle three-spin cluster and linear three-spin cluster.

Several temperature variations for $S = 1/2$, $S = 3/2$, and $S = 5/2$ pairs are shown in Figure 1.19, which are actually observed for the paired samples consisting of Cu^{2+}, Cr^{3+}, and Mn^{2+} ions.

1.3.8.4 Multiple-spin Cluster Case – The Triangle or Others

A group of interacting magnetic moments form a magnetic cluster and exhibit a prominent temperature-dependent magnetic susceptibility. For instance, triangle and linear trimer, square and linear tetramer, or star-burst spin networks are typical magnetic clusters. Even for these complicated spin systems the van Vleck formula works out efficiently, as long as appropriate Hamiltonians for the systems are deduced or presumed. Here we examine two cases consisting of three spins. The spin–exchange coupling structure is shown in Figure 1.20.

The spin–exchange interaction in an isosceles triangular three-spin system is expressed by the spin Hamiltonian:

$$\mathcal{H} = -2J(\mathbf{S}_1 \cdot \mathbf{S}_2 + \mathbf{S}_2 \cdot \mathbf{S}_3 + \gamma \mathbf{S}_3 \cdot \mathbf{S}_1) \tag{1.108}$$

In the case of $\gamma = 1$, it reduces to a regular triangle, whereas the linear trimer case corresponds to $\gamma = 0$. With the help of the Kambe formula [6] the magnetic susceptibility for $\gamma = 1$ leads to

$$\chi = \frac{Ng^2\mu_B^2}{4kT} \frac{5 + \exp(-3J/kT)}{1 + \exp(-3J/kT)} \tag{1.109}$$

In the regular triangle ($\gamma = 1$) with $J < 0$, it provides a strange situation, that is, one spin remains as a spin-frustrated state and this position is not designated in the ground state, thus presenting a spin frustration problem, which is an interesting target of research not only in theory but also in experiment. Space limitations mean we cannot discuss spin-clusters further here and the reader is referred to more magnetism-oriented books [4, 5].

1.3.8.5 Temperature-Independent Paramagnetism

Here we consider the ground state with $E_0^{(0)} = 0$ as an energy origin. In addition, it has no angular momentum and is therefore diamagnetic. Using $E_0^{(1)} = 0$ and $\chi = -2NE_0^{(2)}$ from the van Vleck formula, magnetic susceptibility is concluded to be given by

$$\chi = -2N\sum_{m\neq 0}<0|\mathcal{H}_{\text{Ze}}|m>^2/(E_0^{(0)} - E_m^{(0)}) \tag{1.110}$$

This relation indicates that the diamagnetic ground state may be coupled with the excited states by the second-order perturbation of the Zeeman Hamiltonian, eventually yielding paramagnetic susceptibility. All denominators in the equation are negative and no Boltzmann statistics are included so that the derived susceptibility features temperature-independent paramagnetism. This temperature-independent susceptibility, χ_{TIP}, is almost of the same order of magnitude of diamagnetism, although the sign is opposite, and it is actually observed in many materials. Materials containing the transition metal ions, in particular, exhibit a relatively large contribution, for instance $1\sim 2 \times 10^{-4}$ emu mol^{-1} for Ni^{2+} or Co^{3+}, having a singlet ground state, $^1A_{1g}$. The large contribution of χ_{TIP} is accounted for by the low-lying excited states coming from the factor, $1/(E_0^{(0)} - E_m^{(0)})$. It may be necessary to remark that temperature-independent paramagnetism is potentially generated from coupling between the ground state and the excited states, regardless of the ground state being diamagnetic or paramagnetic. In this context, for every paramagnetic material, χ_{TIP} is superimposed on the usual paramagnetism.

1.3.9
Low-Dimensional Interaction Network

When we consider magnetic interactions in the crystal, the magnetic networks of the interacting magnetic moments are generally three-dimensional. The exchange interaction shows dominant dependences on the distance between the two magnetic origins as well as on the electron distribution of the wave function in which the electron spins reside. Thus, the exchange interaction parameter, J, possesses the most remarkable value in a certain crystal direction. With this fact in mind, the magnetic interaction network in general is likely to become low-dimensional. These magnetic properties are categorized as low-dimensional problems, which attract much interest not only experimentally but also theoretically. In this sense one-dimensional magnetism, called "magnetic linear chains", is most important. But sometimes we encounter layered magnetic systems as a two-dimensional model.

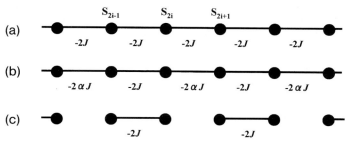

Figure 1.21 Magnetic interaction networks: (a) a regular linear chain; (b) an alternating linear chain; (c) dimer.

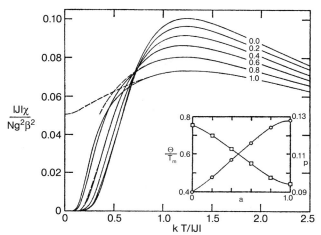

Figure 1.22 Magnetic susceptibilities for $\alpha = 1$, 0.2, 0.4, 0.6, 0.8, and 1.0 in the one-dimensional magnetic chain. The calculations were made for a 10-spin chain in the Hamiltonian (1.111). The Bonner and Fisher case is $N \to \infty$ (the dashed line for $\alpha = 1.0$). The dimer (S–T) model is given by $\alpha = 0$. The alternating chains correspond to $\alpha = 0.2$–0.8. The insert shows the Weiss constant divided by the temperature of the susceptibility maximum (circles) and the product of maximum susceptibility times the corresponding temperature, $p = \chi_{max} kT_{max}/Ng^2\mu_B^2$ (squares), versus α.

The expression "linear chain" refers to a magnetic chain within which each magnetic origin interacts with its two nearest neighbors only. The Hamiltonian of such a system will be given by using the exchange interaction of Heisenberg type, $-2J\mathbf{S}_1 \cdot \mathbf{S}_2$

$$\mathcal{H} = -2J \sum_{i=1}^{N/2} (\mathbf{S}_{2i-1} \cdot \mathbf{S}_{2i} + \alpha \mathbf{S}_{2i} \cdot \mathbf{S}_{2i+1}) \qquad (1.111)$$

Depending on the sign of J, ferromagnetic and antiferromagnetic chains are possible as far as $\alpha > 0$. As is easily recognized, a variety of magnetic linear chains are available (Figure 1.21). The simplest case is $\alpha = 0$, the system being reduced to the dimer model, which of course is not characteristic of the magnetic chain. For $\alpha = 1$ the system conforms to an $S = 1/2$ uniform chain or "regular chain". The magnetic susceptibility for antiferromagnetic chains was first calculated by Bonner and Fisher [7]. Data for $\alpha = 1$ (Figure 1.22) indicate unique behavior due to no energy gap above the ground state. The magnetic susceptibility reaches a maximum value, χ_{max}, at the temperature, T_{max}, given by

$$\chi_{max}/(Ng^2\mu_B^2/|J|) = 0.07346, \quad kT_{max}/|J| = 1.282 \qquad (1.112)$$

Below T_{max} the susceptibility decreases gradually but it seems to remain constant towards the temperature of zero kelvin. According to Bonner and Fisher's calculation the asymptotic susceptibility is expressed by the following relation:

$$\chi_{T=0}/(Ng^2\mu_B^2/|J|) = 0.05066 \tag{1.113}$$

The uniform ferromagnetic chain with $S = 1/2$ is expressed by [8]

$$\chi = (Ng^2\mu_B^2/4kT)\{1+(J/kT)^a\} \tag{1.114}$$

with an exponent, a, depending on the temperature. For a high temperature limit like $kT/J > 1$, $a = 1$ and for $T \to 0$ a approaches 4/5. A numerical expression is also proposed using $K = J/2kT$ as

$$\begin{aligned}\chi = (Ng^2\mu_B^2/4kT)\{&(1+5.79599K+16.902653K^2+29.376885K^3\\&+29.832959K^4+14.036918K^5)/(1+2.79799K+7.0086780K^2\\&+8.6538644K^3+4.5743114K^4)\}^{2/3}\end{aligned} \tag{1.115}$$

On the other hand, the intermediate cases of $0 < \alpha < 1$ for $J < 0$ are classified as an alternating chain with strong and weak antiferromagnetic interactions, when it is known that the system has an energy gap. The calculations for these chains [9] are also depicted in Figure 1.22 above, together with the result of a regular linear chain. Due to the energy gap the magnetic susceptibilities tend to zero on lowering the temperature. Thus, we specify the linear chain models as the "regular" Heisenberg linear chain for $\alpha = 1$ and as the "alternating" Heisenberg chain for $0 < \alpha < 1$.

Some peculiar cases are described. One is the case of $\alpha < 0$, in which the alternating chain has both antiferromagnetic and ferromagnetic interactions. The magnetic susceptibility is numerically evaluated under some conditions. The other case is for $\alpha \to -\infty$, and then we have a regular chain with $S = 1$. This system is theoretically predicted to have a unique ground state with a gapped excitation spectrum. This gap is called the Haldane gap, and the magnetization curve at zero kelvin draws much attention from theoreticians and experimentalists as well.

Furthermore, the present magnetic chains interact with each other, forming a two-dimensional network with antiferromagnetic or ferromagnetic weak interactions. Among them, of course for special cases, a so-called uniform two-dimensional network is possible. These interchain interactions or two-dimensionality problems provide very interesting reseach subjects regarding the magnetic behavior of materials in comparison to the three-dimensional magnetic materials.

The following Hamiltonian is also noted:

$$\mathcal{H} = -2J\sum_{i=1}^{N}\{\alpha S_{zi}S_{zi+1}+\gamma(S_{xi}S_{xi+1}+S_{yi}S_{yi+1})\} \tag{1.116}$$

This Hamiltonian includes an anisotropic exchange effect except in the case of $\alpha = 1$ and $\gamma = 1$ of the Heisenberg exchange model. The most usual cases are of $\alpha = 1$ and $\gamma = 0$ or $\alpha = 0$ and $\gamma = 1$. These models are referred to as the "Ising" model and the "X–Y" model, respectively. The magnetic properties, including thermal behaviours, may be analytically solved for the Ising model [10], and thus qualitative understandings on magnetism become possible. However, the Ising or X–Y model is applicable only for large anisotropic materials, such as Ni^{2+} and Co^{2+}, and, therefore, it is rare to analyse magnetic data from the standpoint of the Ising or X–Y model.

1.4
Experimental Magnetic Data Acquisition

1.4.1
Methods

Several methods (magnetometers) are utilized for the measurement of magnetic susceptibility. Historically, methods such as Gouy and Faraday methods, are classified as a "force method". The force exerted on a sample for the Faraday method is given in the coordinates of Figure 1.1, shown earlier:

$$F_x = M_z \partial H_z / \partial x = v\chi H_z \partial H_z / \partial x \tag{1.117}$$

The determination of $H_z \partial H_z / \partial x$ is carried out by using known standard samples. The force direction coincides with gravity, so that a modified Faraday method, using a "torsion balance", was invented, in which the field gradient is generated along the horizontal direction (see the y-axis in Figure 1.1). The horizontal torsion is canceled using an inductive coil set on the balance, where the feedback current is a measure of the force. The limit of the measurement amounts to $10^{-10}\,\mathrm{emu\,g^{-1}}$. Dynamic methods are also applicable; one is a vibrating sample magnetometer (VSM) and another is a magnetic induction method which involves applying an oscillating magnetic field. The detection systems consist of a detection coil by which the generated electromotive force, V(t), or the mutual induction coefficient of the secondary coil (a Hartshorn bridge circuit is utilized) are measured, respectively. The latter method is important for determining the complex susceptibility, $\chi(\omega)$, which is given by

$$\chi(\omega) = \chi'(\omega) - i\chi''(\omega) \tag{1.118}$$

AC susceptibilities, in general, emphasize the magnetic loss or relaxation phenomena, and far more important is that $\chi''(\omega)$ is related to magnetic resonance phenomena, as an absorption of energy. The third method is based on the superconducting quantum interference effect. This magnetometer is called a SQUID (superconducting quantum interference device), which utilizes a superconducting ring with a weak-point junction, counting a quantized magnetic flux as a generated current (the Josephson effect). The SQUID magnetometer is commercially available and is the most prevalent method, and besides, data acquisitions in cgs units is very easy.

1.4.2
Evaluations of Magnetic Susceptibility and Magnetic Moment

The observed magnetic susceptibility comprises three main contributions

$$\chi_{obs} = \chi_{para} + \chi_{TIP} + \chi_{dia} \tag{1.119}$$

The first two terms are paramagnetic and the last one is diamagnetic, and the last two terms exhibit no temperature dependence. In order to evaluate the temperature-dependent χ_{para} and compare it with theoretical calculations, we have to subtract the other two contributions from χ_{obs}. The first thing to do is correction of diamagnetism. The term χ_{dia} may be evaluated by using appropriate substances which have a similar composition and molecular structure. However, this is not necessarily universal. Pascal allotted semiempirically diamagnetic susceptibilities of the composition atoms or atomic groups from the total diamagnetic observation (recent recommended values were summarized earlier in Table 1.2 [4, 5]). This validity is based on the additive law for each diamagnetic contribution. The next procedure is how to remove χ_{TIP}. For organic free radicals it may be negligible

Table 1.2a Diamagnetic susceptibilities of atoms and ions (10^{-6} emu mol^{-1}).

H	−2.93	F	−6.3	Li$^+$	−1.0	F$^-$	−9.1
C	−6.00	Cl	−20.1	Na$^+$	−6.8	Cl$^-$	−23.4
N	−5.57	Br	−30.6	K$^+$	−14.9	Br$^-$	−34.6
N (in ring)	−4.61	I	−44.6	Rb$^+$	−22.5	I$^-$	−50.9
N (amide)	−1.54	S	−15.0	Cs$^+$	−35.0	CN$^-$	−13.0
N (diamide)	−2.11	Se	−23	NH^{4+}	−13.3	CNS$^-$	−31.0
O (alcohol, ether)	−4.61	P	−10	Mg^{2+}	−5.0	ClO$_4^-$	−32.0
O (carbonyl)	−1.73	As	−21	Ca^{2+}	−10.4	OH$^-$	−12.0
O (carboxyl)	−7.93	Si	−13	Zn^{2+}	−15.0	O^{2-}	−7.0
		B	−7	Hg^{2+}	−40.0	SO$_4^{2-}$	−40.1

Table 1.2b Diamagnetic susceptibilities of ligands and constitutive correction.

H$_2$O	−13	C=C	+5.5	CHCl$_2$	+6.43
NH$_3$	−18	C≡C	+0.8	CBr	+4.1
CO	−10	C=C–C=C	+10.6	Benzene	−1.4
CH$_3$COO$^-$	−30	C=C–C	+4.5	Cyclohexane	+3.0
Oxalate	−25	N=N	+1.85	Piperidine	+3.0
Pyridine	−49	C≡N	+0.8	Imidazole	+8.0
Bipyridine	−105	C=N	+8.15	C (tertiary)	−1.29
Pyrazine	−50	C=N–N=C	+10.2	C (quaternary)	−1.54
Ethylenediamine	−46	N=O	+1.7	C (one aromatic ring)	−0.24
Acetylanetonato	−52	CCl	+3.1	C (two aromatic rings)	−3.10
o-Phenanthroline	−128	CCl$_2$	+1.44	C (three aromatic rings)	−4.0

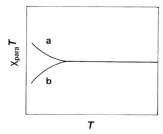

Figure 1.23 $\chi_{para} T$–T plot. At low temperatures it deviates from (a) a constant line upward for ferromagnetic interaction and (b) downward for antiferromagnetic interaction. The constant value at higher temperature regions gives the effective magnetic moment, m_{eff}, and, accordingly, the spin quantum number of the sample.

because the orbital excited states lie far above the ground state. When we can assume or approximate the temperature dependence in the higher temperature region, a tactful method is applicable for deducing temperature-independent terms. Considering thermal agitation, the only constant contribution remains at $T \to \infty$. This asymptotic value corresponds to the remaining constant terms $(\chi_{TIP} + \chi_{dia})$.

The plot of the effective magnetic moment, m_{eff}, versus T from the obtained temperature-dependent χ_{para} is useful in knowing whether the magnetic interaction is ferromagnetic or antiferromagnetic and also what the magnetic moment or J (or S) value is. Considering the Curie law (1.54), m_{eff} can be evaluated from the asymptotic constant value at higher temperatures in the $\chi_{para} T$–T plot (Figure 1.23). In lower temperature regions this plot deviates upward or downward, suggesting the ferromagnetic or antiferromanetic interactions of the magnetic moments, respectively.

References

1 Spaldin, N.A. (2003) *Magnetic and Materials, Fundamentals and Device Applications.* Cambridge University Press, pp. 7, 16–17.
2 Henry, W.E. (1952) *Physical Review,* **88**, 559–62.
3 Pryce, M.H.L. (1950) *Proceedings of the Physical Society,* **A63**, 25–29.
4 Carlin, R.L. (1986) *Magnetochemistry,* Springer-Verlag, pp. 3, 20–21.
5 Kahn, O. (1993) *Molecular Magnetism.* VCH, Weinheim, pp. 3–7.
6 Kambe, K. (1950) *Progress of Theoretical Physics,* **5**, 48–51.
7 Bonner, J.C. and Fisher, M.E. (1964) *Physical Review,* **135**, A640–58.
8 Baker, G.A., Jr, Rushbrooke, G.S. and Gilbert, H.E. (1964) *Physical Review,* **135**, A1272–7.
9 Duffy, W. Jr and Barr, K. (1968) *Physical Review,* **165**, 647–54.
10 Onsager, L. (1944) *Physical Review,* **65**, 117–49.

2
Molecular Magnetism

Jun Yamauchi

2.1
Magnetic Origins from Atoms and Molecules

2.1.1
Historical Background

Traditional magnets are made of iron- or nickel-based materials and many magnetic devices of industrial importance consist of transition or rare-earth metal ions. It was not until three decades ago that metal-free magnetic materials were made from light elements such as carbon and nitrogen. This is because of material science making rapid progress recently. In the time of Faraday or Curie, most molecular or organic compounds without metal ions as a constituent were found to be diamagnetic. Of course, a magnetic study of the oxygen molecule (O_2) was carried out by Curie, and this work, combined with work on nitric oxide (NO), may be counted as an origin of molecular magnetism. It is well-recognized that the history of free radicals started from Gomberg's famous experiment in 1900 using diphenylmethyl chloride, producing triphenylmethyl (**1** in Figure 2.1) [1, 2]. However, it was only recently that this new species proved to be a radical and a paramagnetic molecule. Moreover, the magnetic evaluations of organic free radicals in a solid state were reported much later.

Traditionally, in magnetic materials containing transition or rare-earth metal ions, the magnetism originates from electron spins or orbital angular momentum of those atoms. In this sense, it may be called "atom-based magnetism". On the other hand, magnetism is generated by spin angular momentum of light elements, such as carbon, nitrogen or oxygen, in which the magnetic origin comes from mainly unpaired electrons in p-orbitals. Thus, we can set up a new class of magnetism known as "molecule-based magnetism" or "molecular magnetism". As a matter of fact, molecule-based magnetism is typically embodied in organic free radicals. In addition, organic molecules, which are usually diamagnetic, are likely to form complex compounds with magnetic metals or ions. In these materials magnetism is ascribed to atomic ions similar to the atom-based compounds. Such

---

Nitroxides: Applications in Chemistry, Biomedicine, and Materials Science
Gertz I. Likhtenshtein, Jun Yamauchi, Shin'ichi Nakatsuji, Alex I. Smirnov, and Rui Tamura
Copyright © 2008 WILEY-VCH Verlag GmbH & Co. KGaA, Weinheim
ISBN: 978-3-527-31889-6

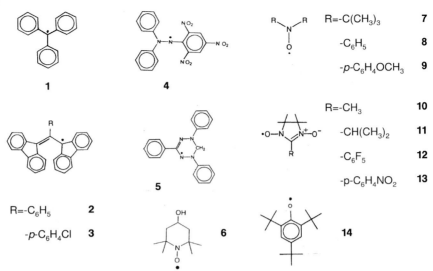

Figure 2.1 Representative neutral organic radicals: (**1**) (TPM: triphenyl methyl); (**2**) (BDPA: 1,3-bisdiphenylene-2-phenylally); (**3**) (p-Cl-BDPA: 1,3-bisdiphenylene-2-p-chlorophenylally); (**4**) (DPPH: diphenylpicrylhidrazyl); (**5**) (TPV: 2,4,6-triphenylverdazyl); (**6**) (TANOL,TEMPOL: 2,2,6,6-tetramethyl-4-piperidinol-1-oxyl); (**7**) (DTBNO: di-t-butyl nitroxide; (**8**) (DPNO: diphenyl nitoxide); (**9**) (DANO: di-p-anisyl nitroxide); (**10**) (MeNN: methyl nitronyl nitoxide); (**11**) (IPNN: isopropyl nitronyl nitroxide); (**12**) (F_5NN: perfluorophenyl nitronyl nitroxide); (**13**) (p-NPNN: p-nitro phenyl nitronyl nitroxide); (**14**) (TTBP: 2,4,6-tri-t-butylphenoxyl).

a coordinate compound is also one part of molecular magnetism in contrast to atom-based inorganic magnetic compounds.

Molecular magnetism has advantages in as much as it is freely utilized and is useful in practical applications, compared with the atom-based magnetism. This statement is compatible with the vast possibilities of organic free radicals, and recent developments in nano-technology or molecular manipulation may permit various combinations of organic molecules and metal ions. In this context the future of the molecular magnetism is highly promising.

2.1.2
Spin States Derived from Chemical Bonds

In the usual covalent-bond formation, each of the two atoms participating in the bond supplies one electron, giving rise to the electron pair in the molecular orbital (MO). As is seen from Figure 2.2a, an unpaired electron in each atom joins and disappears, and as a result the formed molecule gains a stabilized energy coming from the chemical bond. From the magnetic point of view this indicates that the paramagnetic nature of atoms disappears due to the chemical bond and diamag-

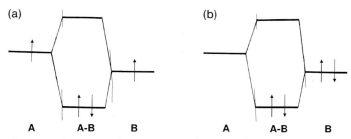

Figure 2.2 Electron-pair formation in covalent (a) and coordinate bonds (b).

netism is brought forth. This means that atom-aggregating molecules are mostly diamagnetic through the chemical bonds. Thus, we have to expect singular situations for the magnetic molecules. In other words, we have to produce odd-electron molecules in which at least one unpaired electron remains without forming a chemical bond. It is usually said that the unpaired electron resides in a SOMO (singly occupied molecular orbital). This particular situation is exemplified in NO and NO_2 molecules and is likely in radical molecules, as in the case of triphenyl-methyl. The Gomberg experiment can be interpreted such that the cleavage of the chemical bond has a possibility of giving rise to molecular paramagnetism under specific conditions of impeding further chemical formations once the radical is produced. These conditions will be satisfied by either a large amount of radical stabilization or a great molecular hindrance protecting the radical site. The former explanation concerns the large delocalization energy of the molecule and the latter the substitution of large bulky groups in the atoms neighboring the radical site. Regarding paramagnetic oxygen (O_2), which is an even-electron molecule, the molecular orbitals consisting of the 2p atomic orbitals are partly degenerate and happen to be occupied by the two electrons as HOMO (highest occupied molecular orbital). Therefore, the Hund rule works out for producing an $S = 1$ (triplet) ground state of O_2.

Another type of covalent bond is a "coordinate bond", shown in Figure 2.2b. The bond-forming mechanism is a HOMO–LUMO interaction, where HOMO is given by a ligand possessing the electron pair and LUMO (lowest unoccupied molecular orbital) accepts those electrons in the formed molecular orbital with HOMO. LUMO mostly corresponds with the atomic orbital of metal ions. These molecules are called "coordinate" or "complex" compounds. In these chemical bonds the magnetism is clearly diamagnetic. The magnetism arises from the metal ion itself, such as transition or rare earth ions. At this point it may be atom-based magnetism. But, consider the molecular character of the ligands – they have the aforementioned characteristic properties of the molecular magnetic materials and provide unprecedented magnetic behaviors which are different from inorganic compounds.

2.1.3
Organic Free Radicals

Since Gomberg's finding many organic free radicals have been synthesized. Their fundamental structures come from C·, N·, >N–O· and O·, where · indicates an unpaired electron as an origin of paramagnetism. Several neutral radicals referred in Chapters 2 and 3 are listed above in Figure 2.1. There are also many ion-radical molecules derived from oxidation or reduction of the diamagnetic molecules. These chemical processes produce the odd-electron molecules through one-electron withdrawing or donating, resulting in cation or anion radical salts, respectively. Besides, if this reaction takes place between electron-donating and -accepting molecules, there is a possibility of complex molecules consisting of cation and anion radicals. These are called charge-transfer (CT) complexes, which are not only magnetically interesting materials but also important electron-conductive materials.

As for the system of MO degeneracy and even for electron arrangements like O_2, some molecules have a higher symmetry than C_3 and the MOs become degenerate, causing such diamagnetic molecules to be paramagnetic by one- or two-electron reduction and oxidation. The latter especially yields an $S = 1$ (triplet) ground state. One more comment regarding molecular paramagnetsm is instructive. This concerns the orbital orthogonality in organic molecules. For instance, carbene or nitrene molecules exhibit a ground triplet state ($S = 1$), which is one of the high spin states target molecules.

2.1.4
Coordinate Compounds

It is well established that the magnetism of transition metal ions varies with their ligands. This is something like controlled magnetism found in organic molecules. These ideas have been developed very recently to further combinations of molecule-based and atom-based magnetism. This notion is gaining momentum and is contributing to newly born magnetic phenomena. The key point is that the electron is likely to delocalize within the molecular framework, so that magnetic interactions are easily conveyed from one to the other, being quite different from atom-based magnetism.

Although in principle any organic free radical can interact with metal ions, the most widely studied systems incorporate nitroxide radicals. Consequently there exist two magnetic origins, one metal-based and one molecular-based. The interaction networks necessarily become complicated and generate many interesting phenomena. There are several types of interactions between the metal ion and the nitroxide. First, the nitroxide interacts with the metal ions directly through the oxygen atom, forming a kind of covalent bond by spin pairing. In this situation, however, if the interaction is weak, the through-space exchange mechanism becomes operative, yielding either antiferromagnetic or ferromagnetic spin configurations. Second, weak coupling between the metal and the radical spins often happens when the NO group does not directly bond to the metal ion. In spite of the distance, the exchange interaction, although it may be small, is not negligible either through a super-exchange (through-bond) interaction or a through-space spin exchange.

When the nitroxide binds through its oxygen atom to a paramagnetic metal ion, two cases can occur: (1) spin-antiparallel (antiferromagnetic) and (2) spin-parallel (ferromagnetic). The sign of the coupling can be predicted based on orbital overlap considerations. A significant overlap between the metal and radical orbitals usually results in (1) and situation where the orbitals at orthogonal to each other gives rise to (2). Regarding case (1), an interesting magnetism takes place when the metal ions possess a spin quantum number larger than a half ($S \geq 1$). For instance, Mn^{2+} has $S = 5/2$ and, interacting with $S = 1/2$ of nitroxide, the combined spin levels become $S = 3/2$ and $S = 7/2$. When we have a repeated interaction network consisting of such an interaction unit, ferrimagnetism would be realized. The experimental examples are described in Section 2.4.4.

2.2
Characteristics of Molecular Magnetism

2.2.1
Molecular Paramagnetism

The characteristic origin of molecular magnetism is derived from p-orbitals, especially for organic free radicals. The wave functions of the p-orbitals extend their electron density toward a uniaxial direction in space. The exchange interaction given by $-2J\mathbf{S}_1 \cdot \mathbf{S}_2$, between the two unpaired electrons which occupy one of the p-orbitals is inevitably influenced in magnitude by the spatial orientation of the two molecules. Uniaxial overlapping of the p-orbitals is essential for a strong magnetic interaction. Thus, this type of orientation dependence is the most outstanding feature of molecular magnetism.

In addition to the above through-space molecular interaction, the through-bond interaction plays a crucial role in enhanced magnetic properties. This is quite natural in the intramolecular interactions, but even in the intermolecular interactions the through-bond interaction becomes important via intermolecular forces or hydrogen bonds. In discussing this in details molecular orbital treatments are indispensable tools. Thus, the prominent approach in the molecular magnetism is characterized by these two terms, the through-space and through-bond interactions. In particular, since the intramolecular interaction is largely governed by SOMO, the topological symmetry of the constituent molecules has to be taken into consideration. This approach will be semi-empirically established by accumulating and analysing various molecular data and will become a guiding principle for the molecular design of new magnetic materials.

As for the through-bond interaction and topological effect, analogous phenomena have been pointed out for atom-based materials. The former is equivalent to the superexchange interaction as was observed in MnO or MnF$_2$. The Mn^{2+}–Mn^{2+} interaction is mediated through the 2p-orbitals of the O^{2-} or F$^-$ ions. This is literally a through-bond interaction. Besides, these mediated ions are sometimes arranged linearly or perpendicularly, giving rise to 180° or 90° exchange interactions. Depending on the topological configuration, an antiferromagnetic or ferromagnetic exchange interaction can be generated. However, freedom for new magnetic designs is rather restricted in atom-based magnetic materials, compared with molecule-based ones.

One more significant feature of molecular paramagnetism is very small anisotropy in the exchange interaction, which is expressed by the Heisenberg type, $-2J\mathbf{S}_1 \cdot \mathbf{S}_2$. This corresponds to almost perfect quenching of the orbital angular momentum, and consequently, from the relations (1.87) or (1.89), to quite small deviation of g-values from the free electron one, as is evidenced by ESR observation. For instance, nitroxide radicals usually possess g-values ranging from 2.010 to 2.0023 (average value 2.006), depending on the molecular axes (see Table 3.1). Therefore, organic free radicals are considered to be a quite good exemplification of quantum spin, $S = 1/2$ and so embody remarkable quantum effects. Since magnetic susceptibility in the paramagnetic region reflects the anisotropy of these g-values in the order of g^2 there is no prominent difference between the susceptibilities determined from a powder and a single crystal.

On the other hand, for the coordinate compounds the situation is not universal. Even for the orbital quenching systems a certain amount of contribution from the spin-orbit interaction is noticeable and therefore the g- and χ-anisotropies become more significant. Nevertheless, the exchange interaction can be treated as an Heisenberg Hamiltonian. This is the case in coordinate compounds containing Cu^{2+} or Mn^{2+}. Among the coordinate compounds, however, there exist many cases in which the quenching of the orbital angular momentum is not perfect. The situation varies depending on the crystalline fields; that is, the type of ligands surrounding the metal ions. These situations are realized in the magnetic data of Table 1.2, especially for Fe^{2+}, Co^{2+} or Ni^{2+}. Then the exchange interaction is described in the form of the Ising or XY model. As a result of a large g-anisotropy the magnetic susceptibilities are differentiated by the crystal orientation with respect to the magnetic field direction.

2.2.2
Magnetic Properties of Organic Free Radicals

Here, the characteristics of molecular magnetism are summarized for the organic free radicals. The anisotropy of the g-values is relevant with the spin-orbit interaction coefficient, λ, which increases in the heavier element. Thus, from the carbon-based radical to oxygen-based one, $\Delta g = g - g_e$ increases proportionally to the atomic weight, but is still less than 0.01. Therefore, the isotropic exchange Hamiltonian of the Heisenberg type is usually dominant and there is no need to consider an anisotropic exchange. Then the anisotropic part for determining the spin orientation in space originates from the magnetic dipolar interactions. This anisotropic energy determines the preferential orientation of spins in the crystal. Nevertheless, the intermolecular dipolar energy is estimated to be of the order of 10 mT in the corresponding local magnetic field, which is negligibly small compared with the exchange interaction term. Even in the triplet state ($S = 1$) the zero-field splitting parameter is only one order higher than that of the intermolecular case. Thus, organic free radicals comprise an ideal spin quantum system.

The through-bond interaction is significant through the SOMO, especially of π type. The spin-polarization effect with regard to whether the spin-up or spin-down is polarized is influenced by the atom number constituting the MO between the interaction sites. In SOMO, this spin density shows a tendency to alternate spin-up and spin-down on the neighboring atoms, and therefore, whether the atom number between the interacting sites is odd or even is a key for the through-bond interaction. This concept may lead to a topological consideration. These points are useful for a new challenge of molecular design. From the point of view of the through-space interaction, the interaction network is likely to be linear due to the uniaxial robe of a p-orbital, and in some special cases, depending on the molecular packing, it becomes planar. As a result, the organic free radicals supply a good quantum model system for low-dimensional magnetism. Readers are recommended to consult several instructive books about molecular magnetism [3–5].

2.3
Nitroxide as a Building Block

2.3.1
Stability of the N–O Bond

Hereafter we focus our attention on the magnetic properties of one specific organic free radical, nitroxide, which is the main subject of this book in the later chapters. Similar to the paramagnetic gas, NO, the nitroxide radical has one unpaired electron ($S = 1/2$) and possesses two substituents on the nitrogen site. The paramagnetism originates from the one-electron occupancy of the 2pz-orbital, that is, a π-orbital which is delocalized over at least two atoms, N and O. If the substituent groups on both sides of the nitrogen atom contain π-orbitals, the delocalization of the π-electron spreads over the whole molecular framework. Then the radical gains considerable stabilization energy. In contrast with these nitroxides of π-type, there is a typical nitroxide in which the unpaired electron is localized exclusively on N–O, which has been proved to be remarkably useful for applications in both the biology and material science fields. Suppose that there are two N–C bonds in addition to the N–O bond and that the carbons have no π-orbital and make chemical bonds using sp^3-hybridized orbitals, causing the exclusive localization of the π-electron on the N–O group. Because of the low delocalization energy, its stability must be attained to prevent the radical site from making up a chemical bond. This steric hindrance of the substituent groups on the N site plays a crucial role in the stability of nitroxides. For example, di-*t*-butyl nitroxide (DTBNO, 7) is a very stable liquid at room temperature. The bulky substituents, two *t*-butyls, cover the N–O radical part, resulting in no more reaction. These localized nitroxides are excellent candidates for spin-proving or spin-labeling methods, which will be treated in detail.

2.3.2
Structural Resonance of the N–O Bond

One of the most interesting points is the structural resonance phenomenon as described by limiting structures in the valence-bond theory (Figure 2.3).

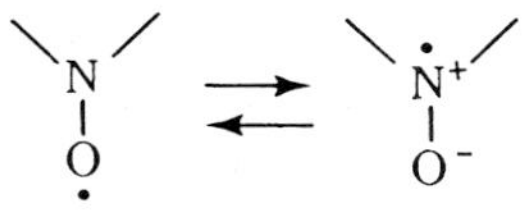

$$\begin{array}{c}\diagdown\diagup\\ N\\ |\\ \overset{\bullet}{O}\end{array} \quad \rightleftarrows \quad \begin{array}{c}\diagdown\diagup\\ \overset{\bullet}{N}{}^{+}\\ |\\ O^{-}\end{array}$$

Figure 2.3 Resonance structures of nitroxide.

This structural resonance in nitroxide radicals becomes possible because the bond-forming atoms, N and O, are neighboring elements and possess odd and even valence-electrons, respectively. The phenomenon is of great significance in the following points. First, the ionic structure is hydrophilic and soluble in water and polar solvents as well, whereas typical organic solvents may dissolve more easily in the neutral form. This ambivalent character makes nitroxide radicals unique and useful, and their potential use is enormous, in particular in biology. Second, this resonance equilibrium may change depending on the surroundings and whether the nitroxides are in polar or non-polar environments. This character is important when the radical is utilized as a spin-proving or spin-labeling reagent. As a measure of this indicator, the unpaired electron density (called the "spin density") on the nitrogen atom is a key. This point is understandable from the limiting structures of each side of the resonance equilibrium. The unpaired electron is nominally localized on the oxygen atom in the non-polar structure and increases its probability of staying on the nitrogen atom as the equilibrium transfers to the right side. The more the polar structure contributes, the more spin density on the nitrogen atom is observed. In fact, this effect may be ascertained from ESR spectra measuring the nitrogen hyperfine coupling constant. This spin-density variation and the nitrogen hyperfine splitting in the ESR spectra, is the third point. According to molecular orbital calculations, the spin density on the oxygen atom is a little larger than that on the nitrogen atom, which may be a reflection of the larger electronegativity of the oxygen atom. However, almost equal amounts of the unpaired electrons stay on the nitrogen. In this sense, the magnetism is molecule-based even in the localized N–O radicals, which is basically different from atom-based magnetism. Therefore, the interaction schemes in the nitroxide radicals have diversity, variety and singularity.

2.3.3
Molecular and Magnetic Interactions between Nitroxides

The functional group, N–O, of nitroxides radicals shows a marked tendency towards molecular interactions with either parallel or antiparallel configurations (Figure 2.4a, b). Both forms are depicted as a dimeric structure, but the parallel forms are repeatedly connected, forming a magnetic linear chain. On the other hand, in spite of the repeated structure of antiparallel form in the crystal, dimer interaction is favored so that the magnetism obeys a dimer (S–T) model or strongly alternating linear chain model. Provided the dimer interaction is very powerful, the magnetism as a whole becomes diamagnetic. Let us look at the crystal structure of Fremy's salt (Figure 2.4c), which is known as an inorganic nitroxide [6]. Because of the strong pairing interaction, the ground state becomes a singlet, the excited triplet state being far above the ground state. As a whole, the crystal displays no paramagnetic behaviors at ambient temperature. But its paramagnetism comes into existence when it is dissolved in alkaline water. The dimer model is the simplest magnetic interaction which governs the magnetism of organic free radicals.

Nitroxides provide many interesting low-dimensionality problems in magnetism. One-dimensional molecular arrays and, as a result, one-dimensional magnetic properties have gained much attention to date, and, moreover, the interchain magnetic interactions cause magnetic phase transitions so that the nitroxides are treated as a research target of cooperative phenomena in the solid state. Two-dimensional molecular arrays are also possible for nitroxides in a specific molecular stacking and then they supply sufficient data for theoretical discussions on the two-dimensional problems of isotropic Heisenberg spin systems. It is in this molecular packing of nitroxides that the first organic ferromagnetic material was discovered amongst the nitronyl nitroxides.

The molecular exchange interaction in spin-delocalized systems like nitroxides has to be modified. As the exchange parameter, J, between the electrons, i and j, is given by (1.70), the wave functions, Ψ_i and Ψ_j of each unpaired electron, i and

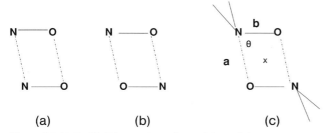

Figure 2.4 N–O···N–O interaction scheme (a) parallel; (b) antiparallel; and (c) Fremy's salt interaction in which $a = 2.86 \text{ Å}$, $b = 1.28 \text{ Å}$, $\theta = 88°$. Fremy's salt is diamagnetic owing to a strong exchange interaction between the N–Os.

j, should be involved as delocalized molecular orbitals of each unpaired electron, such as LCAO-MO (linear combination of atomic orbitals-MO). In place of φ_i and φ_j we insert $\Phi_A = \Sigma_i C_i \Psi_i$ and $\Phi_B = \Sigma_i C_i \Psi_i$ where A and B indicate each unpaired electron of molecules, A and B, and the suffix i means the atomic site on which the unpaired electron is delocalized. Intuitive inspection will yield the following expression the molecular exchange Hamiltonian [7]

$$\mathcal{H} = -2\sum_{<AB>} S_A \cdot S_B \sum_{<ij>} J_{ij}^{AB} \rho_i^A \rho_j^B \qquad (2.1)$$

where S_A and S_B are the spin operator of molecules, A and B, J_{ij}^{AB} is the exchange integral between the orbitals on atoms *i* and *j* belonging to A and B, respectively, and ρ_i^A and ρ_j^B are spin densities on atoms *i* and *j* of each orbital, A and B, respectively. The second sum is taken for all the pairs of interacting atoms that can be formed in molecules, A and B, and the first sum extends over all the interacting molecules in the solid. The greatest contribution will be given in practice by the nearest neighbors. This is McConnell's formulation, which emphasizes the importance of the product of the spin densities of different atoms of neighboring molecules. Furthermore, it gives important information about spin alignments, that is to say, even for an antiferromagnetic interaction, $J_{ij}^{AB} < 0$, if the spin densities, ρ_i^A and ρ_j^B having opposite sign are dominant, the ferromagnetic spin alignment will be established. The fact that the sign of the spin density is a crucial key for the spin alignment, parallel or antiparallel, is one of the most fascinating properties of molecular magnetic materials and will help the design of new substances possessing unique and peculiar magnetic properties.

2.3.4
Nitroxides as Building Block

Nitroxides, as we have seen from their molecular characters, possess a unique atomic group, N–O, and because of this characteristic bond, they provide a diversity and variety in molecular packing in crystals. From a magnetic point of view a nitroxide molecule with various kinds of substitutions at the nitrogen atom works as if it is a building block for making up functional magnetic materials. Much attention has been paid to these molecular designs and crystal formations recently. The magnetic network is basically produced through the chemical or electrostatic interaction of the N–O group with another N–O itself or other atomic groups. This tendency seems to be enhanced by the electronegative oxygen atom of N–O, which usually spreads out from the molecular center. One specific category may be a hydrogen bond, through which any kind of intermolecular interaction can be produced. This hydrogen bridge promises a possibility of complicated magnetic networks and high utility for application devices.

Related to this N–O network, there exist unique nitroxides with two N–O groups within a molecule, which may potentially form other different interaction networks. These nitroxides are called nitronyl nitroxides (NN) and some examples are shown in Figure 2.1 **(10–13)**. As Figure 2.5 indicates the N–O radical site can

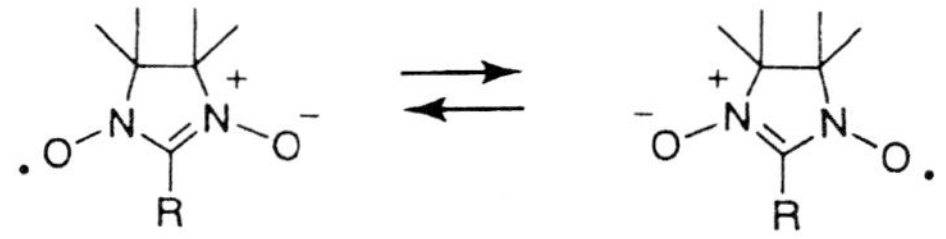

Figure 2.5 Switching of the N–O radical site.

be switched so fast that these two structures are not differentiated. As a matter of fact, the spin density distribution on both N–O bonds is symmetrical, as is concluded from ESR measurement. This molecule can spread out its bonding hand, N–O, to both sides of molecules, which makes it possible to form a two-dimensional magnetic network. These nitronyl nitroxides will surely present other promising magnetic properties.

Finally, we mention the combination of nitroxide molecules with an atom-based magnet as ligands, forming a coordinate compound. This possibility is reasonable because we can synthesize a nitroxide radical with substituents capable of ligation to metal ions. In addition, the direct coordination ability of the N–O group is high, so that we know some examples of direct legation by the nitroxide radicals. For these materials, the magnetic behaviors observed experimentally need additional analysis and understanding of the new mechanism. The combined magnetic materials formed by nitroxides and the atom-based magnetism of the metal ions will extend possibilities in magnetic science in the future.

2.4
Low-Dimensional Properties of Nitroxides

2.4.1
One-Dimensional Magnetism

2.4.1.1 TANOL (TEMPOL)

The formal name of this nitroxide radical is 2,2,6,6-tetramethyl-4-piperidinol-1-oxyl (6), which may be the first example that obeys the linear-chain susceptibility with an exchange interaction of the Heisenberg type, as calculated by Bonner and Fisher. The data is shown as a $\log\chi$ vs $\log T$ plot in Figure 2.6 [8, 9]. In the higher temperature region it follows the Curie–Weiss law with the Weiss constant, $\theta = -6.0$ K. As the temperature is lowered, however, it starts to deviate from this law and reaches a broad maximum, $\chi_{max} = 226 \times 10^{-4}$ emu mol^{-1} at $T_{max} = 6.5$ K, below which χ decreases comparatively slowly as the temperature decreases further. This behavior indicates the short-range magnetic ordering of the unpaired electrons among adjacent molecules owing to negative exchange interaction. It is worth remarking that χ can approach a finite non-zero value, $\chi_{T=0}$ at $T = 0$ K. It was evaluated to be $\chi_{T=0} = 180 \times 10^{-4}$ emu mol^{-1} by extrapolation. This finding implies no energy gap between the ground state and the first excited state, just like the calculation by Bonner and Fisher for $\alpha = 1$ and $N \to 8$ in the Hamiltonian (1.111). Theoretically $\chi_{T=0}$ is given by $\chi_{T=0}/(Ng^2\mu_B^2/|J|) = 0.05066$, which will be

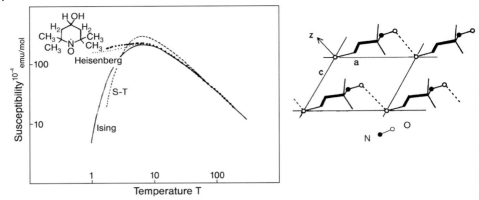

Figure 2.6 Magnetic susceptibility and molecular network of TANOL (TEMPOL). The molecules form molecular chains along an a-axis through hydrogen bonds but magnetic chains may be along the z-direction.

compared with that obtained experimentally. To do so, first we evaluate the exchange parameter, J, from the T_{max} value. Using the relation of (1.112), $|J|/k = 5.0$ K is obtained. Inserting $g = 2.005$ determined from the powder ESR spectrum, the maximum susceptibility, χ_{max} and the extrapolated susceptibility at 0 K, $\chi_{T=0}$, are evaluated to be 218×10^{-4} emu mol^{-1} and 174×10^{-4} emu mol^{-1}, respectively, both of which are in fairly good agreement with those of the observed values, $\chi_{max} = 226 \times 10^{-4}$ emu mol^{-1} and $\chi_{T=0} = 180 \times 10^{-4}$ emu mol^{-1}. Without the ambiguity of the determined parameter, J, the ratio, $\chi_{max}/\chi_{T=0}$, may be useful: $\chi_{max}/\chi_{T=0} = 1.3$ (experiment) while $\chi_{max}/\chi_{T=0} = 1.45$ (theory).

The above conclusion is supported by crystallographic analysis [10]. The crystal structure of TANOL is monoclinic: $\mathbf{a} = 7.052 \pm 0.010$ Å, $\mathbf{b} = 14.081 \pm 0.018$ Å, $\mathbf{c} = 5.780 \pm 0.010$ Å, $\beta = 118°40' \pm 10'$. The nitroxide molecules are aggregated by the hydrogen bond and form chains parallel to the a-axis (Figure 2.6). Although the molecules line up along the a-axis through the N–O···H–O, it is difficult to conclude whether the direction of the linear exchange interaction is in the the direction of the hydrogen bonding, rather it is highly plausible that it is along the c-axis, where the lattice constant is the shortest and the through-space exchange interaction directly between the N–O moieties seems more significant than the through-bond one. Another possibility is seen along the z-axis in Figure 2.6. This is an oxygen-mediated interaction like N–O···O···N–O, which may be the most plausible linear chain. At any rate, we need experimental evidence using a single crystal by which the correlation of the magnetism with the structural properties will be elucidated. One method will be described in the ESR chapter on low-dimensional materials, that is, the measurements of temperature-dependent g-values (Section 3.6.1). The measured specific heat of the magnetic contribution in TANOL also shows a broad maximum corresponding to the short-range magnetic ordering around 4 K [11]. The analysis indicates the exchange interaction parameter, $|J|/kT$

= 4.16 K based on the Bonner and Fisher theory, is in excellent agreement with the susceptibility results.

In summary, the magnetism of TANOL is described by the one-dimensional magnetic chain model with an antiferromagnetic exchange interaction of the Heisenberg type. The interchain magnetic interaction will be referred to later together with a discussion of its magnetic phase transition occurring at 0.49 K.

2.4.1.2 F$_5$PNN

The formal name of this nitroxide radical is pentafluorophenyl nitronyl nitroxide (12). The magnetism of nitronyl nitroxides has been extensively studied and gained much attention. The materials are very stable and afford a good crystallinity, probably due to their structural features mentioned above. The crystal structure of F$_5$PNN and its magnetic properties above 1.5 K have been reported [12, 13]. The compound has a uniform chain structure at room temperature as is shown by the dotted lines in Figure 2.7 representing the N–O···N–O contacts. Additional features are that the N–O and N$^+$–O$^-$ in the molecule have no structural discrimination and spread their intermolecular hands on both sides. Magnetic susceptibility (Figure 2.7) has a maximum at 3.4 K and, contrary to the behavior of TANOL, it decreases continuously and rapidly after reaching the maximum. The final extrapolated prediction seems to give $\chi = 0$ as T approaches 0 K. This suggests the existence of an energy gap above the singlet ground state. The temperature dependence is explained not by a regular linear chain model, as was successful for TANOL, but rather by an alternating Hamiltonian. Quantitative analysis confirms an alternating antiferromagnetic chain, (1.111), with the exchange parameter,

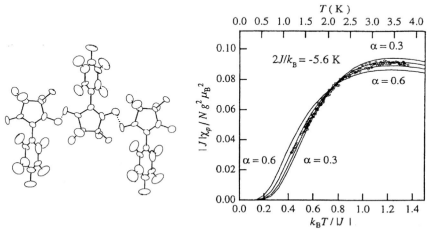

Figure 2.7 Molecular network of F$_5$PNN and magnetic susceptibility. The dotted lines indicate N–O and N–O interactions. The solid curves in the right figure are obtained for α = 0.3, 0.4, 0.5 and 0.6; g = 2.0 is assumed.

$J/k = -2.8$ K, and the alternating parameter, $\alpha = 0.4$. In order to confirm the existence of an energy gap, the magnetization processes were also examined below 2 K. Corresponding to the weak exchange interaction ($\alpha = 0.4$ pairs), the first step of the magnetization curve was observed, and the magnetization rises further, reaching the second step, which is equivalent to the saturation magnetization. Like the present experiment, the alternating Heisenberg antiferromagnetic case is undoubtedly evidenced by the magnetization measurements at low temperature. Regarding the magnetism of F_5PNN, the crucial problem remains unsolved, that is, the magnetic alternation ascertained by the quantitative susceptibility and magnetization analyses is contradictory to the regular molecular array assured by crystal analysis. Considering the large deviation of $\alpha = 0.4$ from $\alpha = 1$ (regular case), some elucidations may be necessary, such as the structural variation of alternating pair interaction at low temperature or the additional interaction path which makes the magnetic interaction irregular.

An example of the alternating chain structure having two types of N–O contacts has also been reported [13]. MeNN (methyl nitronyl nitroxide, 10) holds the N–O···Me and N–O···N–O contacts repeatedly within the structural chain; that is, $\alpha < 0$ in (1.111). The former magnetic interaction is antiferromagnetic, whereas the latter becomes ferromagnetic. This example indicates the diversity and variety of magnetic building schemes in nitroxide radicals.

2.4.2
Interchain Interaction and Spin Long-Range Ordering

Magnetic long-range ordering throughout a crystal can be induced by a three-dimensional network of magnetic interactions. The simple Ising models supply analytical solutions implying anomalous spikes in the magnetic susceptibilities or λ-type anomalies in the heat capacity data [14]. It was reported that genuine Heisenberg exchange interactions in the one- or two-dimensional arrays cannot induce any magnetic phase transitions, and besides the latter case holds special solutions in the spin statistics [15, 16]. In the actual materials, however, the molecules are arranged three-dimensionally even in the one-dimensional magnetic chain or two-dimensional magnetic plane systems. Thus it is reasonable to assume, more or less, interchain magnetic interactions or interplane magnetic interaction. These weak exchange interactions, even though they may be negligibly small compared with the major magnetic interaction, are attributable to the three-dimensional spin-array in the crystal. Theoretical predictions relate the transition temperature T_{tr} with the main and interchain exchange interaction parameters, J and J'. For examples, in the mean molecular field theory these quantities are related by

$$kT_{tr} \approx S^2 \sqrt{2z|J \cdot J'|} \tag{2.2}$$

Here, z is the number of the interchain interaction paths. Oguchi used the Green function method to estimate J' [17]; T_{tr} is related to $\eta = |J'/J|$, as follows:

$$kT_{tr}/|J| = 4S(S+1)/3I(\eta),$$

$$I(\eta) = \frac{1}{\pi^3}\int\int_0^\pi\int \frac{dq_x dq_y dq_z}{\eta(1-\cos q_x) + \eta(1-\cos q_y) + (1-\cos q_z)} \quad (2.3)$$

For several one-dimensional radicals, the antiferromagnetic long-range ordering was shown by magnetic heat capacity measurements. In addition to TANOL, the magnetic data of carbon-centered radicals (BDPA-Bz, **2** with benzene and p-Cl-BDPA, **3** and the nitrogen-centered radical (TPV, **5**) are listed in Table 2.1 [18]. The low-dimensionality is measured by the ratio, $\eta = |J'/J|$, with TANOL showing the best linear chain model. Comparison of the data of BDPA-Bz and p-Cl-BDPA may indicate the effect of chlorine substitution on the interchain interaction. The low-dimensionality may be reduced in p-Cl-BDPA and instead the phase transition temperature was increased, which is a plausible occurrence in view of the theoretical predictions.

Table 2.1 One-dimensional antiferromagnets of organic radicals

Radicals	T_N/K	J/k/K	|J|/kT_N|	|J'/J|	|J'/k|/K	Comments and references
TANOL	0.49	−5.0	10.2	0.004	0.02	One-dimensional chain [8, 9, 19]
BDPA-Bz	1.695	−4.4	2.6	0.04	0.18	One-dimensional chain [9, 20]
p-Cl-BDPA	3.25	−4.4	1.4	0.2	0.88	One-dimensional chain [9, 21]
TPV	1.7	−6.0	3.5	0.2	1.2	Railroad chain [22, 23]

The hydrostatic pressure dependences of the magnetic transition temperature and the magnetic interactions in TANOL have been investigated by heat capacity measurements at low temperatures [24]. The presumed magnetic interaction network, $\eta = |J'/J| = 0.01$, would be influenced by the applied pressure and its effect can be monitored by the changes in the three-dimensional magnetic ordering, which was observed at $T_N = 0.49$ T at the ambient pressure, $P_0 = 1$ kbar. According to magnetic heat capacity data, the broad maximum indicating the magnetic short-range ordering and the λ anomaly indicating the magnetic transition temperature were moved to higher temperature regions in linear proportion to the pressure. The transition temperature, $T_N(P)$, at the pressure, P, is approximated by the following relation in comparison with its value $P = 1$ kbar (P_0):

$$T_N(P)/T_N(P_0) = 1 + 0.15P \quad (2.4)$$

As is shown in Figure 2.8, $T_N(P)/T_N(P_0)$ is almost doubled at $P = 6$ kbar. This enhancement is remarkable among many magnetic substances. The slope in Figure 2.8, which is defined as a coefficient γ in (2.4) by $\gamma = d\{T_N(P)/T_N(P_0)\}/dP$ is compared for several other low-dimensional magnetic materials in Table 2.2 [24]. Another organic antiferromagnetic material, p-Cl-BDPA, was found to be

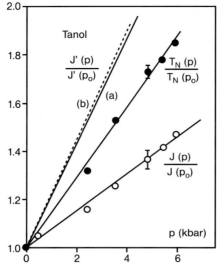

Figure 2.8 Pressure dependence of magnetic interactions and phase transition temperature in TANOL. T_N, J and J' for P are compared with those for ambient pressure, P_0.

Table 2.2 Pressure dependence of magnetic transition temperature of magnetic compounds

Compounds	γ/kbar^{-1}	Compounds	γ/kbar^{-1}
$(NH_4)2CuBr_4 \cdot 2H_2O$	0.019	$K_2CuCl_4 \cdot 2H_2O$	−0.016
$Mn(HCOO)2 \cdot 2H_2O$	0.022	$MnBr_2 \cdot 4H_2O$	0.014
$(CH_3)4NMnCl_3 (P < 3\,kbar)$	0.019	$CuCl_2 \cdot 2H_2O$	0.048
$(CH_3)4NMnCl_3 (P > 3\,kbar)$	0.058	$CoCl_2 \cdot 6H_2O$	0.038
TANOL	0.15	p-Cl-BDPA	0.083
p-NPNN (cf. Section 2.4.3.2)	−0.08		

$\gamma = 0.083\,\mathrm{kbar}^{-1}$. The relative changes of $J(P)/J(P_0)$ and $J'(P)/J'(P_0)$ for the intrachain interaction, $J(P)$, and the interchain interaction, $J'(P)$, respectively, are also depicted in Figure 2.8:

$$J(P)/J(P_0) = 1 + 0.076P, \quad J'(P)/J'(P_0) = 1 + 0.22P \tag{2.5}$$

It is clearly concluded that the enhancement effect in TANOL is much greater in $J'(P)$ than in $J(P)$. These findings surely suggest the exchange interaction mechanisms in the intra- and interchain interactions based on the crystal analysis (Figure 2.8b) and are indicative of so-called "softness" of organic magnetic materials.

2.4.3
Two-Dimensional Magnetism

2.4.3.1 DANO

The formal name of this nitroxide radical is di-p-anisyl nitroxide (**9**), for which the crystal structure are shown in Figure 2.9 [25]. The crystal belongs to the orthorhombic space group Aba2 and the unit cell with four molecules has dimensions of $a = 7.33$ Å, $b = 26.8$ Å and $c = 6.25$ Å. Ignoring the twist of the two phenyl rings, it can be said that all the molecules in the unit cell occupy crystallographic equivalent sites and make the plane formed by the –C–NO–C– group parallel to the bc-plane and the axis of the N–O bond to the c-axis, resulting in the $2p\pi$ orbital of the N–O group being stretched along the a-axis. The molecular layer parallel to the ac-plane is separated from the adjacent one by 13.4 Å. As a consequence, the DANO crystal may be regarded as forming a two-dimensional magnetic lattice with a nearly quadratic network.

The magnetic susceptibility exhibited in Figure 2.9 has a broad maximum at 4.6 K with $\chi_{max} = 304 \times 10^{-4}$ emu mol^{-1} [26]. Besides a vague anomaly near 2.7 K, reported by Duffy et al. [27], a distinct minimum of the magnetic susceptibility was observed at 1.67 K. The insert in Figure 2.9 shows the magnetic susceptibility of a single crystal along three crystallographic axes. Takizawa applied the high-temperature-series expansion (HTS) method for the two-dimensional Heisenberg spin system, in which the magnetic susceptibility is expanded as a function of $x = kT/2JS(S+1)$ as

$$Ng^2\mu_B^2/\chi\cdot|J| = 3x + \sum_n C_n/x^{n-1} \quad (n = 1, 2, \ldots) \tag{2.6}$$

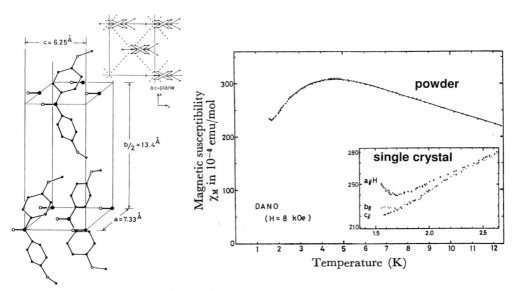

Figure 2.9 Crystal structure of DANO and magnetic susceptibility.

where C_n is the HTS coefficient. By utilizing the first 10 terms for $S = 1/2$, the broad maximum can be reproduced using $J/k = -2.45$ K as a solid line in Figure 2.9 indicates. This J/k value is in excellent agreement with the value determined from the molecular field theory with the Weiss constant, $\theta = -4.9$ K and $z = 4$. On the basis of the phase transition temperature, $T_N = 1.67$ K, the interplane exchange interaction, J', can be estimated to be $|J'/J| = 10^{-3}$.

Finally, a comment on the heat capacity measurement is given. The magnetic phase transition at $T_N = 1.67$ K is fully supported by the anisotropic (single crystal) and powder magnetic susceptibilities and also by the ESR and NMR shifts indicating the long-range-ordering of the spins. Nevertheless, no anomaly was observed either at T_N in the specific heat measurements [27]. This unusual result may be attributable to the well-developed short-range magnetic ordering near T_N. It was pointed out that, in the two-dimensional system, the specific heat anomaly is expected to be extremely small even if magnetic ordering occurs. Thus, it may be concluded that DANO crystal is a good example of the nearly ideal two-dimensional magnet.

2.4.3.2 p-NPNN

The formal name of this nitroxide radical is p-nitrophenyl nitronyl nitroxide (13). There are four polymorphic forms, α-, β-, γ- and δ-phases, known for p-NPNN. All these phases have ferromagnetic interactions, and among them the orthorhombic β-phase is the most stable and has been shown to undergo a magnetic phase transition to a ferromagnetic ordered state at 0.60 K. The molecular arrangement on the ac-plane is shown in Figure 2.10 [28]. The magnetic susceptibility of a β-phase crystal was first measured, indicating the Curie–Weiss law with the positive Weiss constant, $\theta = 1$ K. The transition to the ferromagnetic ordered state was discovered in the β-phase by the measurements of ac susceptibility and heat capacity [29]. The ac susceptibility diverges at around 0.6 K and the heat capacity displays a sharp peak of λ-type at $T_C = 0.6$ K. This ferromagnetism of p-NPNN in the β-phase is also shown by the magnetization curve below T_C, characteristic of the hysteresis loop (Figure 2.10) with an easy saturation and a small coercive force [29, 30]. The spontaneous magnetization obeys

$$M(T) \propto M(0)\{1-(T/T_C)^\alpha\}^\beta \tag{2.7}$$

with $\alpha = 1.86$, $\beta = 0.32$. The critical exponent of the magnetization as a form, $M(T) \propto (T_C - T)^\beta$, is $\beta = 0.32$, in agreement with a value of $1/3$ expected for a three-dimensional Heisenberg system. It is remarked that, when pressure is applied to the crystal, the ac susceptibility showed remarkable change. A decrease in the transition temperature was reported in the initial gradient as $dT_C/dP = -0.048$ K/kbar [31]. The slope, γ, is comparable to those of TANOL or p-Cl-BDPA (Table 2.2), although its sign is opposite. In addition, the phase diagram seems complicated in the higher pressure region, implying a delicate and fine magnetic interaction system in the first ferromagnet, p-NPNN in the β-phase. Following the finding of the first example of an organic ferromagnet in the β-phase of p-NPNN, the highest T_C is documented to be 1.48 K for diazaadamantane dinitroxide [32].

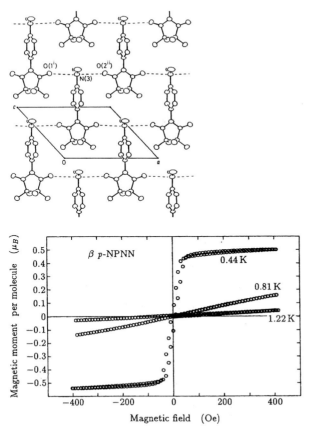

Figure 2.10 Crystal structure of p-NPNN and hysteresis curve.

2.4.4
Coordination of Nitroxide with Metal Ions

2.4.4.1 Cu^{2+}, Mn^{2+}-TANOL (TEMPOL)

Extensive work on the complexes of transition-metal ions with nitroxide ligands has been performed by Italian groups, by whom hexafluoroacetyl acetonate (hfac) was chosen as planar ligands for the metal ion and on the perpendicular axis of the hfac ligands TANOL is located. The crystallographic analysis disclosed a Cu^{2+}–TANOL structure as is shown in Figure 2.11a, which indicates the axial coordination by the two oxygen atoms of the N–O and O–H of the nitroxide [33, 34]. The system seems to form an alternating linear chain. However, it suggests that the alternating parameter, α, in (1.111) is quite small so that the magnetism is at least approximated by a simple dimer model, leading to the exchange parameter, $J/kT = 6.5 \text{ cm}^{-1}$ (ferromagnetically coupled) [35]. In the case of the nitroxide ligand in an axial coordination on the copper ion Figure 2.11b, as seen in the crystal structure, the overlap between the metal and ligand magnetic orbitals must be

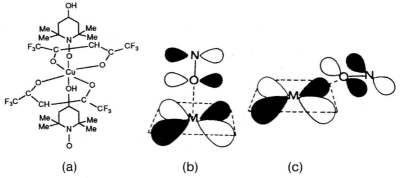

Figure 2.11 (a) $Cu^{2+}(hfac)_2$–TANOL complex; (b) axial coordination of N–O; (c) equatorial coordination of N–O.

zero. Therefore, no antiferromagnetic contribution would be expected. On the other hand, in an equatorial coordination like Figure 2.11c, there can be direct overlap with the magnetic orbital of Cu^{2+}, and consequently a large antiferromagnetic pairing of the spins [34, 36]. In the $Mn(hfac)_2$-$(nitroxide)_2$, in which the nitroxides coordinates on both sides of the Mn^{2+}, the spin quantum number, S, is different; $S = 5/2$ for Mn^{2+} and $S = 1/2$ for the two nitroxides, then the antiferromagnetic coupling produces one $S = 3/2$, two $S = 5/2$ and one $S = 7/2$ states. The magnetic data showed that the $S = 3/2$ is a ground state with the $S = 5/2$ states at 525 and 735 cm^{-1} [37].

2.4.4.2 Mn^{2+}–IPNN

The magnetic properties of magnetic linear chain materials formed by manganese (II) hexafluoroacetylacetonate ($Mn(hfac)_2$) and the nitronyl nitoxide (NN) have been investigated for various types of NN derivatives. We take 2-isopropyl nitronyl nitroxide (IPNN, **11**) as an example [38, 39]. X-ray analysis of $Mn(hfac)_2$–IPNN confirmed the one-dimensionality of the interacting network (Figure 2.12), in which NN is bridging with $Mn(hfac)_2$ on the N–O groups at both sides, forming the hexacoordinated structure of Mn^{2+} by the six oxygen atoms. Magnetic susceptibility measurements in the range 6~300 K show a divergence at low temperature that is compatible with an infinite chain structure. The single crystal magnetic susceptibilities (Figure 2.12) clearly indicate a magnetic phase transition at 7.6 K. Considering Mn^{2+} ($S = 5/2$) and IPNN ($S = 1/2$) with antiferromagntic interactions between them, a ferrimagnetic chain would be expected. In view of the lack of any reasonable exchange pathway between the chains it is suggested that the origin of the ferromagnetic transition is essentially the dipolar interaction, from which the anisotropy fields may be estimated. The spin alignment in the ordered phase was determined to be orthogonal to the chain axis. Similar orderings have been ascertained in the complexes with different substituents in place of iso-propyl, in which the transition temperatures varied from 5.3 K to 8.6 K. Interestingly, furthermore,

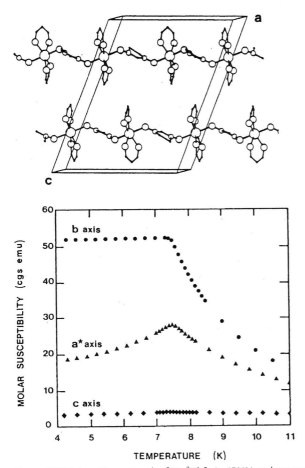

Figure 2.12 Interacting network of $Mn^{2+}(hfac)_2$–IPNN and magnetic susceptibility.

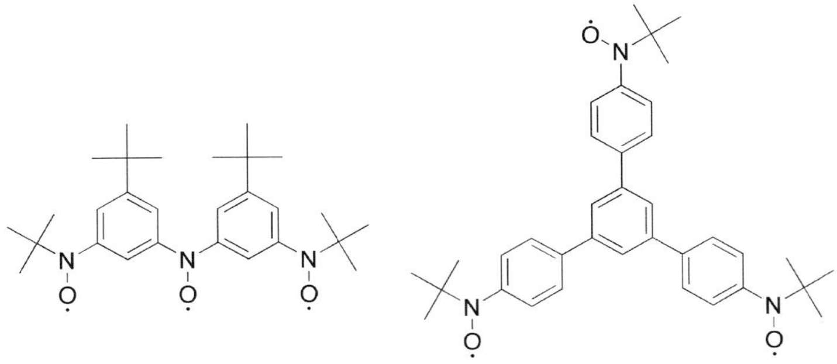

15

16

Figure 2.13 Nitroxide polyradicals, $S = 3/2$.

ESR characteristics of low-dimensional substances (see Section 3.6.1) were also displayed for Mn(hfac)$_2$–IPNN, showing the direction of the magnetic linear chain.

It has been reported that Mn(hfac)$_2$ is also an excellent spin source coupled with nitroxide polyradicals, such as shown in Figure 2.13 [40]. The interaction network was found to be fundamentally two-dimensional and to undergo three-dimensional magnetic orders at comparatively higher temperatures, indicating relatively strong exchange interaction. The triradicals were mixed with Mn(hfac)$_2$, yielding 2:3 complexes, {Mn(hfac)$_2$}$_3$·**15**$_2$ and {Mn(hfac)$_2$}$_3$·**16**$_2$. It turned out that a transition temperature of the former complex was 46 K and its magnetization amounted to 9 Bohrs, which can be justified by the structural data and assuming the manganese ions $3 \times S = 15/2$ and the triradical $2 \times S = 3$ with an antiparallel configuration. Thus, {Mn(hfac)$_2$}$_3$·**15**$_2$ is a ferrimagnet with ferromagnetic ($J_{intra} > 0$) and antiferromagnetic ($J_{inter} < 0$) networks. In general, Mn(hfac)–nitroxide complexes show ferrimagnetism.

References

1 Gomberg, M. (1900) *Berichte der Deutschen Chemischen Gesellschaft*, **33**, 3150–63.
2 Gomberg, M. (1900) *Journal of the American Chemical Society*, **22**, 757–71.
3 Gatteschi, D., Kahn, O., Millar, J.S. and Palacio, F. (eds) (1991) *Magnetic Molecular Materials*, Kluwer Academic Publishers.
4 Millar, J.S. and Drillon, M. (2001–2005) *Magnetism: Molecules to Materials, I–V*, Wiley–VCH, Weinheim.
5 Makarova, T.L. and Palacio, F. (eds) (2006) *Carbon Based Magnetism*, Elsevier.
6 Howie, R.A., Glasser, L.S.D. and Moser, W. (1968) *Journal of the Chemical Society A*, 3043–7.
7 Mconnell, H.M. (1963) *The Journal of Chemical Physics*, **39**, 1910.
8 Yamauchi, J., Fujito, T., Ando, E. Nishiguchi, H., Deguchi, Y. (1968) *Journal of the Physical Society of Japan*, **25**, 1558–61.
9 Yamauchi, J. (1971) *Bulletin of the Chemical Society of Japan*, **44**, 2301–8.
10 Lajzerowicz-Bonneteau, J. (1968) *Acta Crystallographica*, **B24**, 196–9.
11 Lemaire, H., Rey, P., Rassat, A., de Combarieu, A. and Michel, J.C. (1968) *Molecular Physics*, **14**, 201–8.
12 Hosokoshi, Y., Tamura, M., Shiomi, D., Iwasaki, N., Nozawa, K., Kinoshita, M., Katori, H.A. and Goto, T. (1994) *Physica B*, **201**, 497–9.
13 Takahashi, M., Hosokoshi, Y., Nakano, H., Goto, T., Takahashi, M. and Kinoshita, M. (1997) *Molecular Crystals and Liquid Crystals*, **306**, 111–18.
14 Onsager, L. (1944) *Physical Review*, **65**, 117–49.
15 Mermin, N.D. and Wagner, H. (1966) *Physical Review Letters*, **17**, 1133–6.
16 Stanley, H.E. and Kaplan, T.A. (1966) *Physical Review Letters*, **17**, 913–15.
17 Oguchi, T. (1964) *Physical Review*, **133**, A1098–9.
18 Yamauchi, J. and Deguchi, Y. (1977) *Bulletin of the Chemical Society of Japan*, **50**, 2803–4.
19 Boucher, J.P., Nechtschein, M. and Saint-Paul, M. (1973) *Physics Letters*, **42A**, 397–8.
20 Duffy, W., Jr., Dubach, J.F., Pianetta, P.A., Deck, J.F., Strandburg, D.L. and Miedema, A.R. (1972) *The Journal of Chemical Physics*, **56**, 2555–61.
21 Yamauchi, J., Adachi, K. and Deguchi, Y. (1973) *Journal of the Physical Society of Japan*, **35**, 443–7.
22 Azuma, N., Yamauchi, J., Mukai, K., Ohya-Nishiguchi, H. and Deguchi, Y. (1973) *Bulletin of the Chemical Society of Japan*, **46**, 2728–34.

23 Takeda, K., Deguchi, H., Hoshiko, T., Konishi, K., Takahashi, K. and Yamauchi, J. (1989) *Journal of the Physical Society of Japan*, **58**, 3361–70.
24 Takeda, K., Uryu, N., Inoue, M. and Yamuchi, J. (1987) *Journal of the Physical Society of Japan*, **56**, 736–41.
25 Hanson, A.W. (1953) *Acta Crystallographica*, **6**, 32–4.
26 Takizawa, O. (1976) *Bulletin of the Chemical Society of Japan*, **49**, 583–8.
27 Duffy, Jr, W., Strandburg, D.L. and Deck, J.P. (1969) *Physical Review*, **183**, 567–72.
28 Awaga, K., Inabe, T., Nagashima, U. and Maruyama, Y. (1989, 1990) *Journal of the Chemical Society, Chemical Communications*, **520**, 1617–89.
29 Tamura, M., Nakazawa, Y., Shiomi, D., Nozawa, K., Hosokoshi, Y., Ishikawa, M., Takahashi, M. and Kinoshita, M. (1991) *Chemical Physics Letters*, **186**, 401–4.
30 Nakazawa, Y., Tamura, M., Shirakawa, N., Shiomi, M., Takahashi, M., Kinoshita, M., and Ishikawa, M. (1992) *Physical Review B, Condensed Matter*, **46**, 8906–14.
31 Mito, M., Kawae, T., Takumi, M., Nagata, K., Tamura, M., Kinoshita, M. and Takeda, K. (1997) *Physical Review B, Condensed Matter*, **56**, R14255–8.
32 Chiarelli, R., Novek, A., Rassat, A. and Tholence, L.J. (1993) *Nature*, **363**, 147–9.
33 Anderson, O.P. and Kuechler, T. (1980) *Inorganic Chemistry*, **19**, 1417–22.
34 Caneschi, A., Gatteschi, D. and Sessoli, R. (1989) *Accounts of Chemical Research*, **22**, 392–8.
35 Bencini, A., Benelli, C., Gatteschi, D. and Zanchini, C. (1984) *Journal of the American Chemical Society*, **106**, 5813–18.
36 Bencini, A. and Gatteschi, D. (1990) *EPR of Exchange Coupled Systems*, Springer-Verlag, p. 195.
37 Drickman, M.H., Porter, L.C. and Doedens, R.J. (1986) *Inorganic Chemistry*, **25**, 2595–9.
38 Caneschi, A., Gatteschi, D., Rey, P. and Sessoli, R. (1988) *Inorganic Chemistry*, **27**, 1756–61.
39 Caneschi, A., Gatteschi, D., Renard, P., Rey, P. and Sessoli, R. (1989) *Inorganic Chemistry*, **28**, 1976–80.
40 Inoue, K., Hayamizu, T., Iwamura, H. Hashizume, D. and Ohashi, Y. (1996) *Journal of the American Chemical Society*, **118**, 1803–4.

3
Fundamentals of Electron Spin Resonance (ESR)

Jun Yamauchi

3.1
Magnetic Resonance of Electron and Nuclear Spins

3.1.1
Historical Background

Zavoisky published the first article on paramagnetic resonance in 1945, which was followed by two more articles in the next year [1–3]. The first paper was entitled "Spin-magnetic resonance in paramagnetics", dealing with hydrated cupric chloride under the conditions of frequency, $v = 1.33 \times 10^5$ kHz, and magnetic field, $H = 47.6$ Oe. He described the ratio, $v/H = 2.791 \times 10^6$, which gives the spin precession frequency in the field of 10 Oe and from this result it follows that the spin quantum number of cupric chloride is equal to $S = 1/2$. It was only one year later that the initial nuclear magnetic resonance (NMR) work appeared [4, 5]. These magnetic resonances are phenomena relevant with magnetic moments derived either from the electrons or the nuclei. It is clear that the origin of the nuclear magnetic moment is exclusively the nuclear spin, but as for the electronic magnetic moment, the orbital angular momentum of the electron contributes, in addition to the electron spin origin, giving the total magnetic moments. In this description it is perhaps inaccurate to say the magnetic resonance of electron spin or electron spin resonance (ESR), but rather electron paramagnetic resonance (EPR) may be appropriate. In this text we use ESR, considering the quenching of the orbital angular momentum for organic free radicals and sometimes for transition metal ions of the iron group.

ESR and NMR applied to a certain objective share a common Hamiltonian representing their energy terms as an operator, and both include physical phenomena of the Larmor precession, spin–spin interactions, and similar relaxation processes. Instrumentally, however, they are quite different, mainly owing to the frequencies of the oscillation field operated in both experiments. That is to say, ESR spectroscopy needs microwave techniques rather than radiofrequency in the NMR spectroscopy. The following fact may be another significant difference: that NMR magnets need higher magnetic field strength with much better homogeneity than those required for ESR. These two factors made the ESR and NMR techniques progress independently, at least in technical devices. But it may be appropriate to emphasize that both magnetic resonances share many theoretical equations in the essential meanings, which can be translated between by the replacement of the nuclear magnetic moment, m_n, and the electron magnetic moment, m, under appropriate approximations.

The Fourier transform (FT) operation of data acquisition in the time domain is now exclusively utilized in NMR spectroscopy and, recently, an advanced ESR spectroscopy using pulsed techniques has progressively become dominant. The observations of Hahn's echo [6] have added novel concepts in magnetic resonance. Spectrum acquisition in the frequency domain is referred to as continuous-wave ESR (CW-ESR), which is a main subject of this text, but recent advances in pulsed ESR, including the echo techniques in the time-domain, will also be mentioned in this chapter.

3.1.2
Classification of Magnetic Resonance

Magnetic resonance can be observed for materials containing permanent magnetic moments, which originate from either electrons or nuclei (Section 1.2.3). The latter terminology is straightforward: nuclear magnetic resonance (NMR). The electron-originated magnetic moments behave in various fashions, so that there are many terminologies specifying the characteristics of the magnetic properties. Electron paramagnetic resonance (EPR) derives from signal studies entitled magnetic resonance in paramagnets and extensive contributions from metal ion groups. Here, however, we specify "paramagnetic" as an electron origin, thus from the classification of its magnetism, paramagnetic, antiferromagnetic, or ferromagnetic states are clearly differentiated, referring to electron paramagnetic resonance (EPR), antiferromagnetic resonance (AFMR), or ferromagnetic resonance (FMR). The special antiferromagnetic situation of ferrimagnetism, called ferrimagnetic resonance, may be relative to AFMR, but the resonance behaviors are similar to FMR. Theoretical analyses were made for FMR by Kittel [7] and for AFMR by Nagamiya *et al.* [8]. On the other hand, the naming of ESR is more specific, with particular emphasis on electron-spin oriented resonance. Some scientists recommend us to use electron magnetic resonance (EMR) as a whole, in contrast to nuclear magnetic resonance (NMR).

3.2
Principle of Electron Spin Resonance (ESR)

3.2.1
Principle of ESR from Spectroscopic Interpretation

Let us consider the electron magnetic moment $m = -g\mu_B S$, instead of $m = -g_j\mu_B J$, in an external magnetic field, H. The Zeeman Hamiltonian with an isotropic g-value is represented by

$$\mathcal{H} = g\mu_B S \cdot H \tag{3.1}$$

This magnetic energy is given for the electron spin up ($m_S = 1/2$, αspin) and electron spin down ($m_S = 1/2$, βspin), respectively, by

$$E_\alpha = g\mu_B H/2, \quad E_\beta = -g\mu_B H/2 \tag{3.2}$$

The energy splitting is the Zeeman effect, and is called Zeeman splitting. The energy difference becomes $\Delta E = g\mu_B H$. Here we apply the spectroscopic energy relation of light absorption or emission, $E_2 - E_1 = h\nu$, where ν is the frequency of light and h Planck's constant, then the following relation must hold as a result of the law of energy conservation:

$$h\nu = g\mu_B H \tag{3.3}$$

This relation is usually said to be the fundamental equation of the resonance condition for ESR. Quick evaluation using $g = 2$ indicates that, at the magnetic field of a conventional magnet, the frequency of light (electromagnetic wave) amounts to the microwave region. For the measurements of ESR spectra the microwave frequency, ν, is kept constant and the external magnetic field, H, is swept, the spectral chart being given by the magnetic field strength in the abscissa and the absorbed microwave power in the ordinate scales. These images are depicted in Figure 3.1. For a given microwave frequency, ν, the resonance absorption takes place at the magnetic field strength,

$$H_0 = h\nu/g\mu_B \tag{3.4}$$

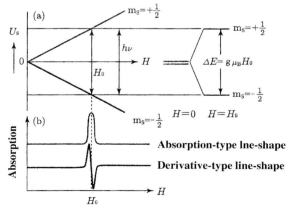

Figure 3.1 Zeeman effect of (a) $S = 1/2$ and (b) ESR spectra. The energy separation at $H = H_0$ is $\Delta E = g\mu_B H_o$, which is equal to $h\nu$ at resonance.

Actual ESR absorptions are given as a spectrum having a line-shape function and are recorded in its derivative shape (Figure 3.1b). This is a spectroscopic image of the ESR phenomena.

From this ESR measurement, the most important data is a g-value, which is calculated from $g = h\nu/\mu_B H_0$. One resonance position, H_0, corresponds to one g-value in a given steady frequency, ν, in the experiment, but it should be noted that the relation between these parameters that specifies the resonance position is in an inverse proportion. That is to say, the lower resonance position (smaller H_0) corresponds to the larger g-value. It is also important that the resonance position, H_0, is not necessarily indicative of the nature of the materials, but, rather, the g-value reflects the individuality of the materials and includes much information: the species of the atom on which the unpaired electron sits; the molecular orbitals in which the unpaired electron resides; the nature of the chemical bonds to which the unpaired electron belongs; and, finally, the surrounding molecular environments from which the unpaired electron feels interactions. The materials possess their own proper magnitude of g-values, which makes it possible to specify or predict the various situations of the unpaired electron.

Here is a comment on the expression of a magnetic field strength in ESR spectra. In cgs units, the magnetic field, *H*, and magnetic induction, *B*, take the same value such as $1\,\text{Oe} = 1\,\text{G}$. In SI units, however, both quantities differ by a factor, μ_0. Considering the conversion factor of *B* between the cgs and SI units is 10^{-4} so that $1\,\text{G} = 10^{-4}\,\text{T}$, an abscissa scale or a magnitude of the magnetic field is customarily expressed in a unit of magnetic induction (T or mT). This may be rationalized by referring to Sectoin 1.1.2.

3.2.2
Principle of ESR from Resonance Interpretation

In contrast to the interpretation of ESR as matching the photon energy of the applied microwave to the Zeeman energy splitting, another essential interpretation is "resonance"; that is, what resonates with each other in the ESR? The answer will be inherent in the methodological features of ESR and NMR, which utilize the oscillating magnetic field perpendicular to the static and sweeping magnetic field. The oscillating microwave can be adsorbed in ESR as a consequence of the "resonance". In addition, magnetic resonance, especially in the pulsed measurements, detects the perpendicular components of the magnetizations influenced by the perpendicular oscillating (microwave) magnetic field. Let us contemplate magnetic resonance from the aspect of the Larmor precession.

The magnetic moment in the static magnetic field feels torque $H \times M$. As a result, it rotates around the static magnetic field at the frequency given below:

$$\omega_L = \gamma H \qquad (3.5)$$

Here, γ is called the "gyromagnetic ratio", which means the ratio between the magnetic moment and the angular momentum. For electron spin $\gamma = -g\mu_B/\hbar$

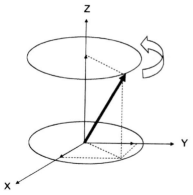

Figure 3.2 Larmor precession of magnetization or spin. It precesses at a frequency of γH. M_x and M_y are oscillating whereas M_z is kept constant.

and this is negative because of the opposite direction of the magnetic moment and the angular momentum vectors. On the other hand, for a nuclear spin, $\gamma_n = g_n\mu_n/\hbar$ it is positive. The positive or negative sign only determines the rotating direction around the static field so that throughout this book we use positive γ value for the electron as $\gamma = g\mu_B/\hbar$. The classical image is drawn in Figure 3.2, which is a Larmor precession motion. For free electrons $\nu_L = \omega_L/2\pi$ is given by $\nu_L = 28.0247\,\text{MHz}\,\text{mT}^{-1}$. In this situation, suppose that the rotating magnetic field, H_1, is applied on the xy plane perpendicular to the magnetic field (z-axis), such as

$$H_1 = H_1(\cos\omega t, \sin\omega t, 0) \tag{3.6}$$

This rotating magnetic field is actually produced from the linearly polarized oscillating magnetic field, $2H_1\cos\omega t$. What kind of relation will be set forth if the Larmor precession of the magnetic moment and the rotating magnetic field coincide always in phase? This is easily from (3.5), leading to $h\nu = g\mu_B H$, which is the equation of the resonance condition (3.3). This workout can be interpreted classically that the Larmor precession motion is enhanced by the in-phase application of the external magnetic field in the xy-plane, and through this resonance the magnetic system absorbs the energy from the radiation field. When the frequency of the rotating external field does not coincide with the Larmor frequency owing to the external static field, no resonance and no energy transfer takes place. In the ESR experiments, the external magnetic field is swept and then the Larmor frequency increases, up to the steady rotating (microwave) frequency set in the spectrometer for observation. This is literally a resonance situation. In order to obtain more visible insight, we derive several mathematical relations.

Now we have two coordinate systems, one the static or experimental coordinate system, the xyz-system, and the other the rotating coordinate system, the $x'y'z'$-

system. Of course we take $z = z'$. The total magnetic fields in the ESR experiment are represented by

$$\bm{H} = (H_1 \cos \omega t, H_1 \sin \omega t, H) \quad (3.7)$$

The equation of motion of the magnetization, $\bm{M}$, as viewed from the frame rotating at the frequency, ω, is given from classical mechanics as

$$d'\bm{M}/dt = \gamma (\bm{H} - \bm{\omega}/\gamma) \times \bm{M} \quad (3.8)$$

where the prime specifies differentiation with respect to the rotating frame. The effective field is defined as $\bm{H}_{eff} = \bm{H} - \bm{\omega}/\gamma$ and then

$$d'\bm{M}/dt = \gamma \bm{H}_{eff} \times \bm{M} \quad (3.9)$$

The magnetization along the z direction will precess about H_{eff} in the rotating frame, as shown in Figure 3.3, in which the angles, θ and α, are defined as an angle of H_{eff} from the z-axis and the magnetization from the z-axis after a precession time, t, respectively. Those parameters are given by the following equations.

$$\sin\theta = H_1/\sqrt{(H-\omega/\gamma)^2 + H_1^2}, \quad \cos\alpha = \cos^2\theta + \sin^2\theta \cos\omega' t \quad (3.10)$$

Next, the frequency of the rotating frame, ω, is set equal to the Larmor frequency owing to the external magnetic field, H, then we can see what happens with the above two relations. The magnetic z-component vanishes, resulting in $H_{eff} = H_1$ and in this situation the cone of the magnetization motion becomes plane perpendicular to the constant H_1 in the rotating frame, leading to a repeated up-and-down change of the magnetization. The turnaround time of this motion is given by $t = 2\pi/\gamma H_1$. The macroscopic magnetizations up and down may correspond to the microscopic spin down and up, or α and β spins. This spin change due to the rotating magnetic field is understood as a resonance phenomenon from the specific resonance condition of $\omega = \omega_L$ or $\omega = \gamma H$.

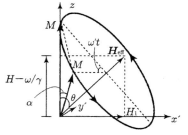

Figure 3.3 Magnetization precesses around H_{eff} at the frequency γH_{eff} in the rotating frame.

Pulsed ESR techniques utilize microwave irradiation during a certain short time as a pulse. At resonance the magnetization vector turns around up and down during the time of the pulse width. There exist specific angles of rotation such as 90° or 180°. These are called 90° ($\pi/2$) or 180° (π) pulses. Each pulse width is given by

$$\Delta_{90} = \pi/2\gamma H_1, \quad \Delta_{180} = \pi/\gamma H_1 \tag{3.11}$$

After these pulse irradiations, the magnetic (spin) system is set in a thermally non-equilibrium state, starting to return to the original thermal situation through any relaxation mechanisms. The principle of pulsed ESR concerns the observations of this recovery in time-domain as a free induction decay (FID) or spin echoes.

We must be careful not to misunderstand that the resonance phenomenon mentioned above is straightforwardly equivalent to the resonance observation. The above explanation indicates a continuous or cyclic absorption and emission of light (electromagnetic wave or microwave in the ESR) in the period over a certain time, $t = 2\pi/\gamma H_1$. What can be observed in the ESR experiment is a net absorption of the electromagnetic wave applied as an oscillating field perpendicular to the static magnetic field. It is added here that, in order to observe the net absorption, we need an important mechanism called "relaxation", which will be divided into two kinds, spin–lattice and spin–spin relaxation mechanisms, and these will be the next theme.

3.2.3
Bloch Equation

3.2.3.1 Solutions of the Bloch Equation

Bloch proposed a set of phenomenological equations to describe the dynamic behavior of interacting nuclear spins [9]. These equations, which include relaxation times, are also applicable to the electrons. The underlying concepts are so useful that we describe the magnetic resonance phenomenologically in Sections 3.2.3 and 3.2.4, and take advantage of the results. What is called the Bloch equation is basically the torque equation of motion with the relaxation term, R:

$$dM/dt = \gamma(H \times M) + R \tag{3.12}$$

The relaxation term, R, is represented as

$$R = \left(-\frac{M_x}{T_2}, -\frac{M_y}{T_2}, -\frac{M_z - M_0}{T_1}\right) \tag{3.13}$$

Here, M_0 is a magnetization along the z-axis at the thermal equilibrium and $M = (M_x, M_y, M_z)$. T_1 and T_2 are the relaxation times specifying the rates to return to the equilibrium for the z-component and for x- or y-components, respectively.

In order to obtain the steady-state solutions, we consider the rotating coordinates with frequency of ω and the magnetizations are translated to the in-phase and out-of-phase components for the oscillating field, H_1, as

$$M_x = u\cos\omega t + v\sin\omega t, \quad M_y = u\sin\omega t - v\cos\omega t \tag{3.14}$$

Then we can derive three equations for u, v, M_z;

$$\frac{du}{dt} = \gamma H_0 v - v\omega - \frac{u}{T_2}$$

$$\frac{dv}{dt} = -\gamma H_0 u + u\omega + \gamma H_1 M_z - \frac{v}{T_2} \tag{3.15}$$

$$\frac{dM_z}{dt} = -\gamma H_1 v - \frac{M_z - M_0}{T_1}$$

Presuming the steady state conditions, $du/dt = dv/dt = dM_z/dt = 0$, one obtains the solutions

$$u = \frac{\gamma H_1 M_0 T_2^2(\omega_0 - \omega)}{1 + T_2^2(\omega_0 - \omega)^2 + \gamma^2 H_1^2 T_1 T_2}$$

$$v = \frac{\gamma H_1 M_0 T_2}{1 + T_2^2(\omega_0 - \omega)^2 + \gamma^2 H_1^2 T_1 T_2} \tag{3.16}$$

$$M_z = \frac{M_0\{1 + T_2^2(\omega_0 - \omega)^2\}}{1 + T_2^2(\omega_0 - \omega)^2 + \gamma^2 H_1^2 T_1 T_2}$$

Here, $\omega_0 = \gamma H_0$ and H_0 satisfies the equation of the resonance condition (3.3). These magnetizations, u, v and M_z, are the solutions of the Bloch equation (3.12).

3.2.3.2 Absorption Line Shape

Several useful characteristics are picked out for elucidation. We focus our attention on the v-mode, which is called the "absorption" mode, whereas the u-mode is called the "dispersion" mode. The name absorption comes from the fact that the average rate at which energy is absorbed per unit volume by the sample from the H_1 field depends on the out-of-phase component, v. The absorption and dispersion modes are depicted as a function of $\omega - \omega_0$ in Figure 3.4.

The function of the v-mode is a maximum at $\omega = \omega_0$, $v(\omega_0)$ being given by

$$v(\omega_0) = \gamma H_1 M_0 T_2 / (1 + \gamma^2 H_1^2 T_1 T_2) \tag{3.17}$$

This absorption intensity increases on increasing H_1 and reaches the maximum value, $(1/2)M_0\sqrt{T_2/T_1}$, when $\gamma^2 H_1^2 T_1 T_2 = 1$. The H_1 dependence of the absorption implies saturation phenomenon in the magnetic resonance. In this context $s = \gamma^2 H_1^2 T_1 T_2$ is defined as a saturation factor. In usual experiments we must be

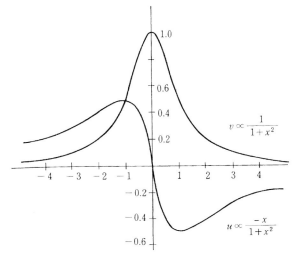

Figure 3.4 Dispersion (u-mode) and absorption (v-mode) as a function of $x = \omega - \omega_0$.

careful about this saturation factor of the sample so as to avoid this effect by keeping the conditions, $s = \gamma^2 H_1^2 T_1 T_2 \ll 1$. Then the inherent line shape for the magnetic resonance will be obtained. This shape function is called the Lorentzian (line-shape), and is given by

$$L(\omega - \omega_0) = \frac{1}{\pi} \frac{T_2}{1 + T_2^2(\omega - \omega_0)^2} \tag{3.18}$$

This shape function is normalized as is seen from the integration over $-\infty < \omega - \omega_0 < +\infty$. By simple mathematical calculations, the actual line shape obtained from the experiment has a form

$$L(H) = \frac{I_m}{1 + [(H - H_0)/(\Delta H_{1/2}/2)]^2} \tag{3.19}$$

using the absorption line intensity, I_m, and the line-width, $\Delta H_{1/2}$ (Figure 3.5). The line-width, $\Delta H_{1/2}$, is defined as the width between the resonance points of half-height, $I_m/2$, corresponding to $\Delta\omega_{1/2}$. These parameters are related to T_2 as follows.

$$\Delta\omega_{1/2} = 2/T_2, \quad \Delta H_{1/2} = 2/\gamma T_2 \tag{3.20}$$

As a matter of fact, usual ESR spectra are recorded in a derivative line shape, and it is not easy to calculate $\Delta H_{1/2}$ from spectra, but rather it is simple to obtain so-called "peak-to-peak" or "maximum-slope" line-width, ΔH_{pp} or ΔH_{msl}. This parameter is also related to T_2 in the Lorentzian line shape:

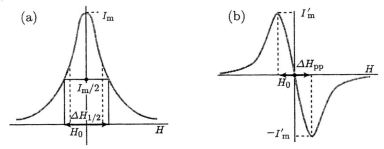

Figure 3.5 Line-shape parameters characterizing absorption intensity and line-width: (a) absorption mode and (b) its derivative curve.

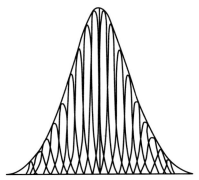

Figure 3.6 Inhomogeneous line exhibiting a spin-packet. The line comprises each homogeneous line with a statistical intensity distribution. Homogeneous lines are expressed as a Lorentzian line-shape, and inhomogeneous lines as a Gaussian line-shape.

$$\Delta H_{pp} = 2/\sqrt{3}\gamma T_2 \tag{3.21}$$

The line-width is an important quantity in the ESR data, but we must be careful about saturation of the resonance absorptions.

Here, the Gaussian line-shape in comparison to the Lorentzian line-shape is explained. For a diluted spin system the Lorenzian line-shape is usually observed, and we say the system is "homogeneous". In a crystal, however, the spin system is an ensemble of many spins which are subject to slightly different local fields and their fluctuations. This system is called an "inhomogeneous" system, and exhibits a different line-shape, called the Gaussian line-shape, convoluted by each Lorentzian line-shape. Assuming a statistical distribution from the central absorption line, the line-shape may be illustrated by Figure 3.6, which is called a "spin-packet". The Gaussian line-shape is given by

$$G(\omega-\omega_0) = \frac{1}{\sqrt{2\pi}\sigma} \exp\left[-\frac{(\omega-\omega_0)^2}{2\sigma^2}\right] \qquad (3.22)$$

The distribution function is related to a statistical normal distribution (Gaussion distribution) function, and σ in the formula is a dispersion parameter for the distribution. Two line-widths of the Gaussian are given from the line-shape function:

$$\Delta\omega_{1/2} = 2\sqrt{2\ln 2}\,\sigma, \qquad \Delta\omega_{pp} = 2\sigma \qquad (3.23)$$

Analogous to the Lorentzian expressed by a variable H, $L(H)$, the Gaussian line-shape is

$$G(H) = I_m \exp\left[-(\ln 2)\left(\frac{H-H_0}{\Delta H_{1/2}/2}\right)^2\right] \qquad (3.24)$$

I_m and $\Delta H_{1/2}$ have the same meaning, shown in Figure 3.5. The line-widths expressed by the magnetic field, $\Delta H_{1/2}$ and ΔH_{pp}, are obtained by dividing the $\Delta\omega_{1/2}$ and $\Delta\omega_{pp}$ relations by γ. A facile judgment of whether the absorption line is Lorentzian or Gaussian is possible by comparing the relation between $\Delta H_{1/2}$ and ΔH_{pp}, $\Delta H_{1/2}/\Delta H_{pp} = \sqrt{3}(=1.732)$ for Lorentzian and $\Delta H_{1/2}/\Delta H_{pp} = \sqrt{2\ln 2}(=1.1774)$ for Gaussian.

Interpreting the line-width in another way, it may result from spin–spin magnetic interactions of dipolar and exchange types. That is to say, we consider the resultant local magnetic fields acting on the objective spin. Combining the dipolar local field (H_{dip}) and an exchange local field (H_{ex}), the line-width is given by

$$\Delta H_{1/2} = H_{dip}^2 / H_{ex} \qquad (3.25)$$

In other words, the exchange phenomenon works to average the dipolar local field. Thus, we have terminology regarding the line-width, such as "dipolar broadening" or "exchange narrowing". In addition, we sometimes observe line-width narrowing at high temperatures. In view of the rapid motion at high temperatures, the local fields may be reduced by averaging: we call this "motional narrowing". The line-width and line-shape are inherently interesting problems in magnetic resonance.

3.2.3.3 Relaxation Times

The relaxation term, R, in the Bloch equation includes the coefficients, $1/T_1$ and $1/T_2$, expressing the rates at which the non-equilibrium magnetization, $M = (M_x, M_y, M_z)$, approaches its thermal equilibrium value, $M = (0, 0, M_0)$. The relaxation times, T_1 and T_2, are called the "spin–lattice" and "spin–spin" relaxation times, respectively. T_1 is also called the longitudinal relaxation time because it relates to the rate of change of the magnetization parallel to the applied field. This process obeys the exponential rate law characterized by the time constant, T_1. The spin–lattice relaxation is named after its relaxation process, in which the excess energy in the spin system absorbed from the radiation field is dissipated to the non-spin system, mainly to the thermal energy of the surroundings of the spins. The latter energy system, in general, is termed "lattice".

As readers can understand from the line-width relations, (3.20) and (3.21), the line-width is governed by a spin–spin relaxation, which does not involve the transfer of energy between the spin and the lattice. The non-equilibrium M_x and M_y values come from an ensemble of the spins precessing in an applied field. We note that if all spins are in an in-phase precession, the x- and y-components of the magnetization remain constant, but actually a gradual dephasing occurs, resulting in their magnitude vanishing. This relaxation process can be fulfilled through the fluctuating magnetic interactions between the spins, producing slightly different local fields in the z-direction at each site. Thus there will be a spread of the Larmor frequencies, indicating the dephasing phenomenon and the thermal equilibrium situation. This dephasing process also follows an exponential rate law characterized by the time constant, T_2. The loss of the components of the magnetization rotating in the x–y plane is concomitant with the loss of the precessing phases. Thus, T_2 is also called the "transverse relaxation time".

Although we mentioned that the line-width is characterized by the spin–spin relaxation time, T_2, it should involve the contribution from the spin–lattice relaxation time, T_1. Because of the energy fluctuation caused by this relaxation process, the uncertainty principle gives us the relation

$$\Delta E\, \Delta t \geqq \hbar \tag{3.26}$$

This principle states that if we wish to make accurate measurements of the energy of a quantum system we should wait for a long time between the measurements [10], giving us the width of the spectral line, $\Delta\omega$,

$$\Delta\omega = 1/\tau \tag{3.27}$$

where τ is the life-time of the states involved in the transition and is comparable to T_1. When T_1 is very short, the line-width becomes broad, in addition to the T_2 contribution. In solids the relation $T_1 \gg T_2$ usually holds, leading to (3.20) and (3.21). A remarkable fluctuation case, such as non-viscous liquid, may produce the circumstance of $T_1 \sim T_2$. The above-mentioned contribution to the line-width is always effective when the energy levels have an appreciably short life-time compared to the inverse of the resonance frequency of the transition. This is the case which we frequently encounter when dealing with dynamic processes in chemistry.

3.2.4
Modified Bloch Equation

The adaptation of the Bloch equation to take account of exchange phenomena was first made by Gutowsky et al. [11], in the context of NMR. An alternative procedure was later proposed by McConnell [12] and we will use this method here for ESR [13]. We start from the equations, (3.15), for the x- and y-components of the magnetization in the rotating frame, and introduce complex magnetization, defined by $G = u + iv$, then we have

$$dG/dt + \alpha G = i\gamma H_1 M_0 \qquad (3.28)$$

where $\alpha = 1/T_2 - i(\omega_0 - \omega)$. Now we consider some spin exchange or chemical exchange phenomena in equilibrium as follows (Figure 3.7).

$$A \rightleftarrows B$$

Figure 3.7 Spin exchange or chemical exchange phenomena in equilibrium.

We characterize this equilibrium in terms of the life-times, τ_A and τ_B, and the fractions occupying each sites, ρ_A and ρ_B. It follows that $\rho_A = \tau_A/(\tau_A + \tau_B)$ and $\rho_B = \tau_B/(\tau_A + \tau_B)$ in addition to $\rho_A + \rho_B = 1$. Thus, we write down a similar equation to (3.28) for G_A and G_B with M_0 replaced by $M_0\rho_A$ and $M_0\rho_B$, respectively. Now we add the transfer of the spins between the two sites. Then we obtain the modified Bloch equations:

$$dG_A/dt + \alpha_A G_A = i\gamma H_1 M_0 \rho_A + G_B/\tau_B - G_A/\tau_A$$
$$dG_B/dt + \alpha_B G_B = i\gamma H_1 M_0 \rho_B + G_A/\tau_A - G_B/\tau_B \qquad (3.29)$$

On the basis of the steady-state approximation, $dG_A/dt = 0$ and $dG_B/dt = 0$, the complex magnetization is eventually given by

$$G = G_A + G_B = i\gamma H_1 M_0 \frac{(\tau_A + \tau_B) + \tau_A \tau_B(\rho_A \alpha_A + \rho_B \alpha_B)}{(\alpha_A \tau_A + 1)(\alpha_B \tau_B + 1) - 1} \qquad (3.30)$$

The imaginary part of G, relating to the out-of-phase magnetization, v, gives an absorption line shape.

The simplest case is well worth careful perusal, that is, (1) $1/T_{2A} = 1/T_{2B} = 0$, (2) $\tau_A = \tau_B = \tau$. In these approximations, the imaginary part of G is given by

$$v = \frac{1}{2}\gamma H_1 M_0 \frac{\tau(\omega_A - \omega_B)^2}{(\omega - \omega_A)^2(\omega - \omega_B)^2 + 4\tau^2(\omega - \bar{\omega})^2} \qquad (3.31)$$

Here, $\bar{\omega} = (\omega_A + \omega_B)/2$. The line-shape variations are drawn for $\tau(\omega_A - \omega_B) = 0.5 \sim 10$ in Figure 3.8. In the slow exchange case we observe two lines at ω_A and ω_B (Figure 3.8a) and as the life-times decrease, these two lines broaden and merge eventually into a single line at $\bar{\omega}$ (Figure 3.8f). In the slow- and rapid-exchange limits, each line-shape is represented by the Lorentzian. The results obtained here are very useful when we analyze the ESR spectra reflecting the spin dynamic processes in many organic radicals.

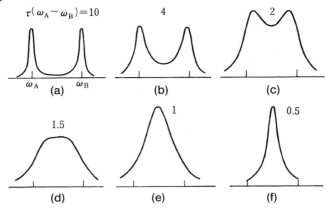

Figure 3.8 Line-shape variation depending on $\tau(\omega_B - \omega_A)$.

3.2.5
Hyperfine Interaction

3.2.5.1 Interaction of the Electron Spin with Nuclear Spins

An ESR spectrum of an isolated electron with the magnetic moment due to $S = 1/2$, as is derived from the Bloch equation, consists of one absorption line with a certain line-width, but experimentally we often observe a number of lines with a simple intensity ratio and a central symmetry pattern, especially in solution. When the spectral structure arises from the interaction of the electron spin with nuclear spins, it is called the "hyperfine structure" owing to the hyperfine interaction, which is a basic magnetic interaction between the electron and nuclear magnetic moments. We obtain the hyperfine Hamiltonian for the magnetic interaction between the electron and nuclear spins by replacing $m_1 = -g\mu_B S$ and $m_2 = g_n\mu_n I$ in (1.71):

$$\mathcal{H}(S, I, r) = -g_e\mu_B g_n\mu_n \left\{ \frac{S \cdot I}{r^3} - \frac{3(S \cdot r)(I \cdot r)}{r^5} \right\} \tag{3.32}$$

Now we evaluate the spatial parts by integration using the electron wave function, $\psi nlm_l(r, \theta, \phi)$. The Hamiltonian includes the term, $1/r^3$, which becomes infinite unless the wave function goes to zero very rapidly as $r \to 0$. The difficulty occurs when an electron in an s-orbitals or $\psi(0) \neq 0$. The detail of this treatment is far from the content of this book and we summarize only the result. Eventually we obtain two terms for the hyperfine interaction. One is called the "dipolar term" and the other is the "isotropic term". The former hyperfine anisotropy is given by

$$\mathcal{H}_{dip} = -g\mu_B g_n\mu_n [S_x \ S_y \ S_z] \cdot \begin{bmatrix} \left\langle \frac{r^2-3x^2}{r^5} \right\rangle & -\left\langle \frac{3xy}{r^5} \right\rangle & -\left\langle \frac{3xz}{r^5} \right\rangle \\ -\left\langle \frac{3xy}{r^5} \right\rangle & \left\langle \frac{r^2-3y^2}{r^5} \right\rangle & -\left\langle \frac{3yz}{r^5} \right\rangle \\ -\left\langle \frac{3xz}{r^5} \right\rangle & -\left\langle \frac{3yz}{r^5} \right\rangle & \left\langle \frac{r^2-3z^2}{r^5} \right\rangle \end{bmatrix} \cdot \begin{bmatrix} I_x \\ I_y \\ I_z \end{bmatrix} \tag{3.33}$$

$$= S \cdot A_{dip} \cdot I$$

where < > means integration over the spatial coordinates, and we simply denote $\mathcal{H}_{dip} = \mathbf{S} \cdot \mathbf{A}_{dip} \cdot \mathbf{I}$. The matrix representation of $\mathbf{A}_{dip}$ is a tensor and, from its definition, it is easily seen that its diagonal sum (trace) vanishes:

$$\text{Tr}[\mathbf{A}_{dip}] = 0 \tag{3.34}$$

The isotropic term, on the other hand, becomes

$$\mathcal{H}_{iso} = (8\pi/3) g\mu_B g_n \mu_n |\psi(0)|^2 \mathbf{S} \cdot \mathbf{I} \tag{3.35}$$

Thus, we have $\mathcal{H}_{iso} = A_0 \mathbf{S} \cdot \mathbf{I}$,

$$A_0 = (8\pi/3) g\mu_B g_n \mu_n |\psi(0)|^2 \tag{3.36}$$

Here A_0 is an isotropic hyperfine (coupling) constant and this term is also called the "Fermi contact interaction" or simply "Fermi interaction". Finally, the hyperfine spin Hamiltonian is summarized as combined

$$\mathcal{H}_{hf} = A_0 \mathbf{S} \cdot \mathbf{I} + \mathbf{S} \cdot \mathbf{A}_{dip} \cdot \mathbf{I} = \mathbf{S} \cdot \mathbf{A} \cdot \mathbf{I}$$
$$\mathbf{A} = \mathbf{A}_0 + \mathbf{A}_{dip} \tag{3.37}$$

Using the unit matrix, $\mathbf{1}$, A_0 is combined in the $\mathbf{A}$ tensor; that is, $\mathbf{A}_0 = A_0 \mathbf{1}$. The mathematical unitary transformation of the tensor $\mathbf{A}$ leads to a principal axis system in which the off-diagonal elements are all zero and we know the direction cosine of these axis directions. In the principal axis systems, the hyperfine coupling constant, $\mathbf{A}$, is given by

$$\mathbf{A} = A_0 \begin{bmatrix} 1 & 0 & 0 \\ 0 & 1 & 0 \\ 0 & 0 & 1 \end{bmatrix} + \begin{bmatrix} A'_x & 0 & 0 \\ 0 & A'_y & 0 \\ 0 & 0 & A'_z \end{bmatrix} \tag{3.38}$$

Here the relations, $A_i = A_0 + A'_i$ ($i = x, y, z$) and $\text{Tr}[\mathbf{A}] = A_i = 3A_0$, holds. Therefore, the average of the A anisotropy leads to A_0, which is observed in the ESR spectrum in solution.

$$A_{iso} = (A_x + A_y + A_z)/3 = A_0 \tag{3.39}$$

3.2.5.2 Hyperfine Splitting
We consider the spin Hamiltonian involving the hyperfine interaction.

$$\mathcal{H} = \mu_B \mathbf{S} \cdot \mathbf{g} \cdot \mathbf{H} + \mathbf{S} \cdot \mathbf{A} \cdot \mathbf{I} \tag{3.40}$$

For simplicity we examine the hyperfine splitting of the ESR spectrum due to the hyperfine interaction for an isotropic spin Hamiltonian, then

$$\mathcal{H} = g\mu_B \mathbf{S}\cdot\mathbf{H} + A_0 \mathbf{S}\cdot\mathbf{I} \tag{3.41}$$

In a small magnetic field the angular momenta of $\mathbf{S}$ and $\mathbf{I}$ are combined, giving rise to new angular momentum, $\mathbf{F} = \mathbf{S} + \mathbf{I}$, and in a high magnetic field, however, $\mathbf{S}$ and $\mathbf{I}$ behave independently, in which the energies of the $S = 1/2$ levels are given by

$$E_\pm = \pm(1/2)g\mu_B H \pm (1/2)m_I A_0 \tag{3.42}$$

where m_I is a component of I. For instance, the energy diagrams for $I = 1/2$ and $3/2$ are depicted in Figure 3.9. The low field behaviors must be calculated more rigorously using secular determinants or the perturbation method. The resonance condition, including the nuclear components, will be estimated from the ESR transitions, $\Delta m_S = \pm 1$ and $\Delta m_I = 0$, then we have

$$h\nu = g\mu_B H_0 + A_0 m_I \tag{3.43}$$

At a constant ν, we have several resonance lines depending on m_I, as is shown in Figure 3.9. The separation for each absorption line is calculated to be equal to $A_0/g\mu_B$, and the line number of the splitting coincides with the number of the m_I components; that is, $2I + 1$. From the observed ESR spectra we document the line separation measured in a magnetic field strength, and this is a value of hyperfine coupling constant (hfcc), A_0, although truly it is a divided value by $g\mu_B$. We usually list the hyperfine data in the unit of G (or Oe) or T (mT). Of course, we may use the energy dimension for A_0, such as cm^{-1}.

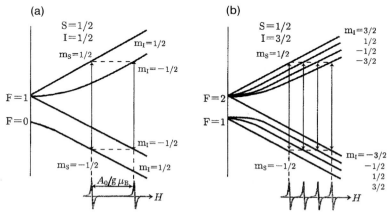

Figure 3.9 Energy levels attributed to hyperfine (hf) interaction and ESR spectra with hf structure: (a) $S = 1/2$, $I = 1/2$ and (b) $S = 1/2$, $I = 3/2$. Each hf splitting is separated by $A_0/g\mu_B$.

3.2.5.3 Hydrogen Atom (S = 1/2 and I = 1/2)

As the simplest example for the hyperfine interaction, we study here more rigorously a hydrogen atom as an objective of the ESR phenomenon. The Hamiltonian of the system, including the nuclear Zeeman term, is given by

$$\mathcal{H} = g_e \mu_B S \cdot H + A_0 S \cdot I - g_n \mu_n I \cdot H \tag{3.44}$$

As a zero-order spin function, we set the combinations α and β for $S = 1/2$ and α_n and β_n for $I = 1/2$, such as $\psi_1 = |\alpha\alpha_n\rangle$, $\psi_2 = |\alpha\beta_n\rangle$, $\psi_3 = |\beta\alpha_n\rangle$, and $\psi_4 = |\beta\beta_n\rangle$. The matrix elements for the above Hamiltonian on this basis set are evaluated as follows

$$\begin{bmatrix} Z_e + Z_n + A_0/4 & 0 & 0 & 0 \\ 0 & Z_e - Z_n - A_0/4 & A_0/2 & 0 \\ 0 & A_0/2 & -Z_e + Z_n - A_0/4 & 0 \\ 0 & 0 & 0 & -Z_e - Z_n + A_0/4 \end{bmatrix}$$

Here Z_e and Z_n indicate the Zeeman energies of the electron and the nucleus, respectively; that is, $Z_e = g_e \mu_B H/2$ and $Z_n = -g_n \mu_n H/2$. The energy eigenvalues and the wave functions may be obtained from solving the secular determinant comprising this matrix. The four energies are given by

$$\begin{aligned} E_1 &= \frac{1}{2} g_e \mu_B H + \frac{1}{4} A_0 - \frac{1}{2} g_n \mu_n H \\ E_2 &= -\frac{1}{2} g_e \mu_B H + \frac{1}{4} A_0 + \frac{1}{2} g_n \mu_n H \\ E_3 &= -\frac{1}{4} A_0 + \frac{1}{2} \sqrt{(g_e \mu_B + g_n \mu_n)^2 H^2 + A_0^2} \\ E_4 &= -\frac{1}{4} A_0 - \frac{1}{2} \sqrt{(g_e \mu_B + g_n \mu_n)^2 H^2 + A_0^2} \end{aligned} \tag{3.45}$$

The states of E_3 and E_4 are mixed states of ψ_2 and ψ_3, which are ascribable to the presence of the off-diagonal elements between ψ_2 and ψ_3. Now we had a strict solution for the Figure 3.9a. Actually, we have triplet degeneracy for $F = 1$ with an energy $A_0/4$ and singlet one for $F = 0$ with $-3A_0/4$. The low and high field line positions, H_l and H_h, are calculated by ignoring some trivial higher terms in the E_3 and E_4 equations.

$$\begin{aligned} H_l &= \frac{2h\nu - A_0 + \sqrt{4h^2\nu^2 - 4h\nu A_0 - 3A_0^2}}{4 g_e \mu_B} \\ H_h &= \frac{2h\nu + A_0 + \sqrt{4h^2\nu^2 + 4h\nu A_0 - 3A_0^2}}{4 g_e \mu_B} \end{aligned} \tag{3.46}$$

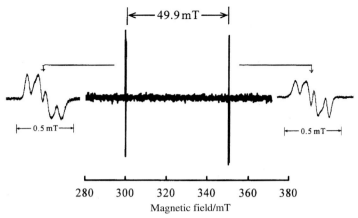

Figure 3.10 ESR spectrum of a trapped hydrogen atom in β − Ca$_3$(PO$_4$)$_2$. The hyperfine components, $m_i = \pm 1/2$, are overlapped with ^{31}P hyperfine (superhyperfine) splitting.

These results contain two important implications:

1. The separation of the two lines, $\delta H = H_h − H_l$ is not equal to $A_0/g_e\mu_B$.
2. The resonance center, $(H_h + H_l)/2$, does not coincide with $H_0 = h\nu/g_e\mu_B$, the resonance position without the hyperfine interaction term.

Normal data acquisition for hyperfine splitting is performed through reading the line separation, δH, but this is not necessarily equal to A_0 in the Hamiltonian. The second fact (2) concerns g-value accuracy, indicating the nominal resonance position is not suitable for determining the true g-value. The larger the hyperfine interaction becomes compared to the Zeeman term, the more careful we must be when documenting the A_0 and g-values. In general, rigorous solutions are rarely obtained for actual cases, and we resort to a perturbation method.

Figure 3.10 is a spectrum of a hydrogen atom trapped in β-tricalcium phosphate after X-ray irradiation [14], which includes mainly its hyperfine separation of 49.9 mT. The hyperfine splitting is so large that the separation of $\delta H = H_h − H_l$ does not indicate the true A_0 value. From an exact solution, or taking into account the perturbation method, the actual hyperfine coupling constant was determined to be 49.6 mT. In addition, the resonance center of the splitting never indicates the true $g = 2.00219$.

3.2.5.4 Spin Polarization Mechanism

We generally encounter isotropic hyperfine splittings (hfs) in the ESR observations of transition metal ions. For instance, Mn^{2+} ions ($I = 5/2$) show splitting of ~8 mT. Considering the wave function in which the unpaired electron resides, d-orbitals are characterized as $|\psi(0)|^2 = 0$, which means no isotropic hfs. Then how does this large splitting result? Here we predict a polarization of s-electrons, such as those in 1s-, 2s- and 3s-orbitals, which are usually shared by the paired electrons of α and β spins in an equal probability. If we take into account the exchange interaction between those s-electrons and the unpaired electrons in the d-orbitals, the opposite spin to that of the unpaired electron may be induced on the s-orbitals, hence $|\psi(0)|^2 \neq 0$. This is one spin polarization mechanism and is assured by the negative sign of A_{Mn}.

Another important mechanism of spin polarization is generated through a σ-bond of the C–H fragment for most free radicals. Consider the carbon atom incorporated in the conjugated system of organic radicals. The unpaired electron occupies a molecular π-orbital delocalized over the carbon atom framework where the hydrogen atoms are attached. How does the 1s orbital of the hydrogen atom acquire an unpaired electron? The model shown in Figure 3.11 is a C–H fragment containing a $2p_z$ π-orbital, in which the unpaired electron resides, and a σ-bond with the hydrogen atom, which is one of the sp² hybrid orbitals. One would consider structures (a) and (b) to be equally weighted. However, if we consider the interaction between the σ and π systems, structure (a) is slightly preferable. The reason is that the exchange interaction between the π and the σ electrons of the carbon atom is favorable for the spin parallel configuration like (a). A little weighted contribution may cause a slight polarization of an α spin on the carbon and hence of the β spin on the hydrogen. The latter gives rise to a hydrogen hfcc with a negative sign. Consequently, the relation between the hfcc of the hydrogen atom, A_H, and the spin density on the carbon atom, ρ_C, is expressed as

$$A_H = Q_{CH}^H \rho_C \tag{3.47}$$

This formula was first proposed by McConnell [15], and the general formula also applicable for other atoms, $A_i = Q_{ji}^i \rho_j$, is called "McConnell's relation". The constant, Q_{CH}^H, is the hyperfine splitting for the unit spin density on the carbon atom and has a negative sign. Experimentally this absolute value could be determined if the ESR spectrum of the ·CH radical such as planar ·CH₃ or its analogs were observed. The documented data for many π-conjugated radicals predict the most probable value for Q_{CH}^H with the combination of the theoretical spin

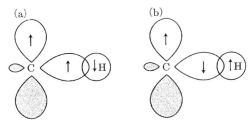

Figure 3.11 C–H fragment with $2p_z$π-orbital and sp²σ-orbital. Configuration (a) is a little stabilized compared with (b), causing β-spin polarization on H.

density calculations. The value of Q_{CH}^H is determined semi-empirically to be $-2.3 \sim -2.7\,\text{mT}$.

Regarding nitroxide radicals, the nitrogen atom ($I = 1$) usually has an isotropic hf of $1\sim1.5\,\text{mT}$, depending on the substituents in the molecule. First, the spin polarization on the nitrogen 1s and 2s orbitals may be plausible from the nitrogen π-orbital, and the next is the $\sigma - \pi$ polarization between the sp^2 σ-orbital and the $2p_z$ π-orbital of the nitrogen atom and, finally, the $\sigma - \pi$ polarization from the oxygen π-orbital through the N–O σ-bond is also effective. These are summed up in the nitrogen isotropic hfcc. The dipolar term of the nitrogen hyperfine interaction is also remarkable, but is mainly influenced from the nitrogen $2p_z$-orbital, a bit from the oxygen because of the far distance. In general, the McConnell relation will be written down as the total contributions of the spin polarization in the following form

$$A_n = Q_n \rho_n + \Sigma_i Q_{ni} \rho_{ni} \tag{3.48}$$

where n indicates the atom in question and n_i the adjacent atoms.

3.2.6
Fine Structure

We have already discussed the two-electron spin system, S_1 and S_2, in Section 1.3.5 and obtained the spin Hamiltonian, $\mathcal{H}_{SS} = \mathbf{S}\cdot\mathbf{D}\cdot\mathbf{S}$ or $\mathcal{H}_{SS} = DxSx^2 + DySy^2 + DzSz^2$, where $\mathbf{S} = \mathbf{S}_1 + \mathbf{S}_2$ and $\mathbf{D}$ is D-tensor. This spin–spin interaction yields an ESR pattern called the fine structure, which is different from the hyperfine structure. First, we examine the fine structure of the triplet state, $S = 1$. We have a basis set of four spin states coming from S_1 and S_2, which are written $|\alpha\alpha\rangle$, $|\alpha\beta\rangle$, $|\beta\alpha\rangle$, and $|\beta\beta\rangle$. From these spin functions we make up triplet ($S = 1$) and singlet ($S = 0$) eigenfunctions, each being characterized by the operators, $\mathbf{S}^2$ and Sz:

$$\begin{array}{lll}
|+1\rangle: & |\alpha\alpha\rangle & \\
|0\rangle: & \frac{1}{\sqrt{2}}(\alpha\beta + \beta\alpha)\\
|-1\rangle: & |\beta\beta\rangle &
\end{array} \Bigg\} S=1 \quad \begin{array}{l} m_S = +1 \\ m_S = 0 \\ m_S = -1 \end{array} \tag{3.49}$$

$$\left|\frac{1}{\sqrt{2}}(\alpha\beta - \beta\alpha)\right\rangle \quad S=0 \quad m_S = 0$$

However, this set is not suitable as an eigenfunction of the Hamiltonian, $\mathcal{H}_{SS} = \mathbf{S}\cdot\mathbf{D}\cdot\mathbf{S}$, at zero magnetic field. Thus, we define new basis functions as below:

$$|T_x\rangle = \frac{1}{\sqrt{2}}(|-1\rangle - |+1\rangle)$$

$$|T_y\rangle = \frac{i}{\sqrt{2}}(|-1\rangle + |+1\rangle) \tag{3.50}$$

$$|T_z\rangle = |0\rangle$$

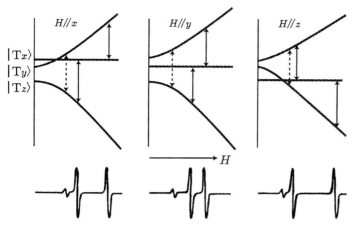

Figure 3.12 Energy levels of $S = 1$ (triplet state) and ESR spectra with fine structure depending on the magnetic field orientations. The dotted lines indicate $\Delta m_s = \pm 2$ (forbidden transition).

Using $|T_x\rangle$, $|T_y\rangle$, and $|T_z\rangle$, we evaluate the matrix elements for the Hamiltonian, $\mathcal{H}_{Ze} + \mathcal{H}_{SS}$.

$$\begin{bmatrix} -D_x & ig\mu_B H_z & ig\mu_B H_y \\ -ig\mu_B H_z & -D_y & -ig\mu_B H_x \\ -ig\mu_B H_y & ig\mu_B H_x & -D_z \end{bmatrix}$$

The secular determinant obtained from the above matrix determines the energies of the triplet states for $H//x$, $H//y$, and $H//z$, shown in Figure 3.12. In view of the transition, $\Delta m_s = \pm 1$, schematic ESR spectra in the derivative form are predicted, yielding two lines with an equal intensity. The separations in the unit of magnetic field are given as $D + 3E$, $D - 3E$, and $2D$. The dotted arrows in Figure 3.12 indicate the transition, $\Delta m = \pm 2$, which is usually forbidden, but partially arrowed in the higher order of calculation. This resonance is named the forbidden transition or "half-field resonance" after its resonance position, half of the $g = 2$ position. The triplet spectrum cannot be observed in solution because of the rapid tumbling motion of the molecules (cf. **Tr[D]** = 0), so that the solution must be frozen to get the fine structure. The frozen or polycrystalline sample of the triplet species exhibits so-called powder-patterns, which will be discussed at Section 3.3.4.

3.2.7
Dynamical Phenomena

3.2.7.1 Correlation Time
Dynamical phenomena can be treated based on fluctuation in local fields due to irregular Brownian motions, molecular rotations, vibrations or chemical exchanges.

These time-dependent terms are represented by the perturbing Hamiltonian, $\mathcal{H}'(t)$, to the Zeeman term, $\mathcal{H}_0 = g\mu_B \mathbf{S} \cdot \mathbf{H}_0$. Now the time-dependent fluctuating field involves many frequency components as is given by its FT:

$$F(\omega) = \int f(t) e^{i\omega t} dt \tag{3.51}$$

Using $f(t)$ at a time of t and $f(t + \tau)$ at a retarded time τ, we define an autocorrelation function, $G(\tau)$ as

$$G(\tau) = \overline{f^*(t+\tau)f(t)} \tag{3.52}$$

This is an average over all times, t. This autocorrelation function, depending on τ, takes large values for small τ and vanishes for large τ. Assuming this τ dependency to be exponential, we introduce a correlation time, τ_C, which measures a sustaining duration of the fluctuation field or a losing time of the memory of the fluctuation field. Thus, we have

$$G(\tau) = \overline{f^*(t)f(t)} \exp(-|\tau|/\tau_C) \tag{3.53}$$

Then we define the power spectrum or spectral density, $J(\omega)$, by the FT of $G(\tau)$ and the above $G(\tau)$ leads to the following $J(\omega)$:

$$J(\omega) = \frac{2\tau_C}{1+\omega^2\tau_C^2} \overline{f^*(t)f(t)} \tag{3.54}$$

This spectral density shows a rapid decay in the larger frequency region above $\omega \sim \tau_C$.

One useful example is used to deduce the relaxation times, T_1 and T_2, as the local fluctuating field, $\mathbf{H}(t)$ for $\mathcal{H}'(t) = g\mu_B \mathbf{S} \cdot \mathbf{H}(t)$ [16]

$$\frac{1}{T_1} = \frac{g^2\mu_B^2}{\hbar^2}\overline{(H_x^2 + H_y^2)}\frac{\tau_C}{1+\omega_0^2\tau_C^2}, \quad \frac{1}{T_2} = \frac{g^2\mu_B^2}{\hbar^2}\left\{\tau_C\overline{H_z^2} + \frac{1}{2}\overline{(H_x^2 + H_y^2)}\frac{\tau_C}{1+\omega_0^2\tau_C^2}\right\} \tag{3.55}$$

In addition, if $\overline{H^2(t)} = \overline{H_x^2(t)} + \overline{H_y^2(t)} + \overline{H_z^2(t)}$ and $\overline{H_x^2(t)} = \overline{H_y^2(t)} = \overline{H_z^2(t)}$ are reasonable, then we have

$$\frac{1}{T_1} = \frac{g^2\mu_B^2}{3\hbar^2}\overline{H^2}\frac{2\tau_C}{1+\omega_0^2\tau_C^2}, \quad \frac{1}{T_2} = \frac{1}{T_2'} + \frac{1}{2T_1} \tag{3.56}$$

$1/T_2' = (g^2\mu_B^2/3\hbar^2)\overline{H^2(t)}\tau_C$ is often called the "secular broadening", whereas $1/2T_1$ is the "non-secular" or "life-time" broadening.

3.2.7.2 Rotational Correlation Time

The ESR spectra of VO^{2+} in solution consist of a hyperfine splitting from the vanadium nucleus ($I = 7/2$), and appreciable spectral variations dependent on the microwave frequency [17]. Each line-width for the hyperfine splitting varies with the component of the nuclear spin quantum number, m_I, given by a quadratic equation

$$\Delta H_{pp} = a + bm_I + cm_I^2 \qquad (3.57)$$

Kivelson related these coefficients with the molecular rotational effect due to the anisotropic g- and A-values [18]. The coefficients, a, b, and c, of course, include the rotational correlation time, τ_C, of the paramagnetic species in solution and may be related to the solvent viscosity, η, and temperature, T, in the Stokes–Einstein model

$$\tau_C = 4\pi r^3 \eta / 3kT \qquad (3.58)$$

where r indicates the molecular radius in a molecular shape approximation using a sphere. The precise analyses of the VO^{2+} spectra supply information on the rotational correlation time and the anisotropy of the g- and A-values [19].

A similar effect is witnessed for nitroxide radicals in solution. The nitrogen hyperfine splitting does not have equal intensity, but usually the broadest component is in the highest field $m_I = 1$). In addition, in a viscous solution or at low temperature, spectra of DTBNO (7 in Figure 2.1) start to deform, showing an anisotropic effect attributed to incomplete molecular rotations (Figure 3.13). Several parameters that characterize the nitroxide g- and A-anisotropies also determine the spectral features, from which the rotational correlation time, τ_C, may be estimated in the spin-labeling and proving studies. Mathematical processes have

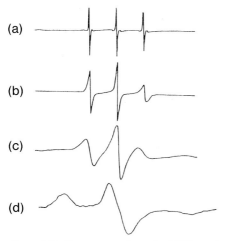

Figure 3.13 Temperature-dependent ESR spectra of DTBNO in solution. The three nitrogen hf lines do not show an equal intensity owing to the nitrogen g- and hf-anisotropy (a), and on lowering the temperature this effect is emphasized (b) and (c), leading to the almost rigid line-shape that includes every anisotropy (d).

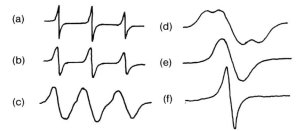

Figure 3.14 Concentration-dependent ESR spectra of DTBNO in solution. These are due to a spin–spin exchange effect in addition to a dipolar interaction. (a) Less than 10^{-4} mol/L^{-1}, (b) ~10^{-2} mol/L^{-1}, (c) ~2×10^{-2} mol/L^{-1}, (d) ~8×10^{-2} mol/L^{-1}, (e) ~10^{-1} mol/L^{-1} and (f) pure DTBNO.

been developed for spectral simulations by Freed [20]. Dynamical phenomena in nitroxide will be examined in Chapter 6.

3.2.7.3 Chemical Exchange

The dynamical line-shape effect was described by a phenomenological treatment of the modified Bloch equation in Section 3.2.4. A number of phenomena have been documented in solution dynamics. The most fundamental reaction is an electron or a proton transfer reaction and a lot of chemical exchange reactions or conformational changes have been pointed out [21–24]. When a certain ESR active species is concerned in these chemical processes, ESR spectra surely supply unique information. In this context, the concentration dependence of the hyperfine ESR spectra of nitroxide radicals in solution is a typical example of spin exchange, as is shown in Figure 3.14. This phenomenon has to be kept in mind when observing high-resolution ESR. It is well-known that dissolved oxygen molecules have broadened line-width because of the spin exchange between the radicals and unpaired electrons ($S = 1$) of the oxygen molecules.

3.3
Anisotropic Parameters in Crystal

3.3.1
g-Anisotropy

In ESR spectroscopy the fundamental parameters, g-values, hyperfine coupling constants, and zero-field splitting constants are given as a tensor of a physical quantity and are anisotropic so that we need to implement solid-state measurements using a single crystal to evaluate those parameters with perfect precision. In this section we summarize the derivation procedures for those three ESR parameters from a single crystal study. For this purpose we set up the crystal and principal (axis) systems and perform data acquisition around the crystal axis,

3.3 Anisotropic Parameters in Crystal

which is transformed to the principal axis system by a suitable unitary transformation. This procedure is called the "principal axis transformation" or "diagonalization". Then we obtain the principal values of those parameters and the principal axis directions with respect to the crystal system, which are characteristics of the paramagnetic centers and their environments.

The spin Hamiltonian, $\mathcal{H} = \mu_B \mathbf{S} \cdot \mathbf{g} \cdot \mathbf{H}$, is as usual. The external magnetic field is taken as $\mathbf{H} = H(l, m, n)$ with respect to the crystal axes (laboratory system). For $S = 1/2$ the matrix elements are calculated and the energies, E_+ and E_-, are evaluated. From these energy differences we can obtain the effective g-value, g_{eff}, as follows:

$$\Delta E = E_+ - E_- = g_{\text{eff}} \mu_B H \tag{3.59}$$

$$g_{\text{eff}}^2 = (g_{xz}l + g_{yz}m + g_{zz}n)^2 + (g_{xx}l + g_{xy}m + g_{xz}n)^2$$
$$+ (g_{xy}l + g_{yy}m + g_{yz}n)^2 \tag{3.60}$$

This is a general formula for the arbitrary magnetic field orientation, $\mathbf{H} = H(l, m, n)$. Now we carry out ESR observations on the xy-, yz-, and zx-planes. For instance, $\mathbf{H} = H(\cos\theta, \sin\theta, 0)$ holds on the xy-plane, leading to

$$g_{\text{eff}}^2 = \cos^2\theta (g_{xx}^2 + g_{xy}^2 + g_{xz}^2) + 2\cos\theta\sin\theta (g_{xx}g_{xy} + g_{xy}g_{yy} + g_{xz}g_{yz})$$
$$+ \sin^2\theta (g_{xy}^2 + g_{yy}^2 + g_{yz}^2) \tag{3.61}$$

From this angular dependence around the z-axis, the following parameters are fitted to the experimental variation:

$$G_{xx} = g_{xx}^2 + g_{xy}^2 + g_{xz}^2, \quad G_{xy} = g_{xx}g_{xy} + g_{xy}g_{yy} + g_{xz}g_{yz}, \quad G_{zz} = g_{xy}^2 + g_{yy}^2 + g_{yz}^2 \tag{3.62}$$

Together with the additional rotations around the remaining two axes all the matrix elements of the **G**-tensor may be determined. It is easily assured that $G_{ij} = G_{ji}$ and $G_{ij} = (\mathbf{g}^2)_{ij}$. For the next step, the **G** matrix is diagonalized through the principal axis transformation. Then we have diagonalized $(\mathbf{g}^2)_{ii}$, which gives the principal g-values, g_x, g_y, and g_z in the principal axis:

$$\mathbf{g} = \begin{bmatrix} g_x & 0 & 0 \\ 0 & g_y & 0 \\ 0 & 0 & g_z \end{bmatrix} \tag{3.63}$$

The principal-axis representation is made by the following unitary matrix, **L**.

$$\mathbf{L} = \begin{bmatrix} l_{x1} & l_{x2} & l_{x3} \\ l_{y1} & l_{y2} & l_{y3} \\ l_{z1} & l_{z2} & l_{z3} \end{bmatrix} \tag{3.64}$$

Table 3.1 g- and A_N-tensors of nitroxides.

Radicals	g_x	g_y	g_z	g_{iso}	g_{sol}
DTBNO	2.00872	2.00616	2.00270	2.00568	2.00606
DPNO	2.0092	2.0056	2.0022	2.0057	2.0055
DANO	2.0089	2.0059	2.0022	2.0057	2.0056
H$_2$NOa	2.0091	2.0062	2.0023	2.0059	–
DPNOa	2.0086	2.0046	2.0023	2.0051	–

Radicals	A_x/mT	A_y/mT	A_z/mT	A_{iso}/mT	A_{sol}/mT
DTBNO	0.759	0.595	3.178	1.511	1.51
DPNO	0.19	0.36	2.38	0.977	0.977
DANO	0.94	0.44	2.46	1.28	1.007

a) Theoretical calculation; Kikuchi, O. (1969) *Bull. Chem. Soc. Jpn.*, 42, 1187–1191.

In this expression the direction cosine of the g_x-axis is $l = (l_{x1}, l_{x2}, l_{x3})$ and the following relations hold.

$$\mathbf{LGL^t} = (\mathbf{G})_{\text{diagonal}}, \quad \mathbf{Lg^2L^t} = (\mathbf{g^2})_{\text{diagonal}} \tag{3.65}$$

Here, $\mathbf{L^t}$ is a transposed matrix of $\mathbf{L}$. From this diagonalization all principal-axis directions may be determined. As an example of the analysis of g-anisotropy of nitroxides, the data obtained from the experiments are tabulated in Table 3.1. Discussions will be made in Section 3.3.2, together with the anisotropic nitrogen hfcc, A_N. It is remarked that the isotropic g-value observed in solution ESR (g_{sol}) is equal to g_{iso} given by

$$g_{iso} = (g_x + g_y + g_z)/3 \tag{3.66}$$

3.3.2
A-Anisotropy

The spin Hamiltonian is generally given by

$$\mathcal{H} = \mu_B \mathbf{S} \cdot \mathbf{g} \cdot \mathbf{H} + \mathbf{S} \cdot \mathbf{A} \cdot \mathbf{I} \tag{3.67}$$

However, it is quite difficult and complicated to treat this general case because the principal-axis systems do not necessarily coincide with each other to derive general formulae in an analytical expression. Thus, here the g-anisotropy is small enough to treat it isotropically. Besides, if we assume that the Zeeman term contains only S_z components, then we have

$$\mathcal{H} = g\mu_B H S_z + A_x S_z I_x + A_y S_z I_y + A_z S_z I_z \tag{3.68}$$

$$A_x = lA_{xx} + mA_{xy} + nA_{xz}, \quad A_y = lA_{xy} + mA_{yy} + nA_{yz}, \quad A_z = lA_{xz} + mA_{yz} + nA_{zz}$$

Here again $H = H(l, m, n)$. Applying the basis set of $|\alpha\alpha_n\rangle$, $|\alpha\beta_n\rangle$, $|\beta\alpha_n\rangle$, $|\beta\beta_n\rangle$ for $S = 1/2$ and $I = 1/2$, the secular determinant of the above Hamiltonian gives four energy levels, from which the effective hyperfine splitting, A_{eff}, is given by

$$A_{\text{eff}}^2 = (lA_{xx} + mA_{xy} + nA_{xz})^2 + (lA_{xy} + mA_{yy} + nA_{yz})^2 + (lA_{xz} + mA_{yz} + nA_{zz})^2 \quad (3.69)$$

This relation corresponds to the g_{eff} in the g-tensor analysis (3.60). The angular dependence on the xy-plane is expressed in a same manner as g_{eff}.

$$\begin{aligned}A_{\text{eff}}^2 &= \cos^2\theta(A_{xx}^2 + A_{xy}^2 + A_{xz}^2) + 2\cos\theta\sin\theta(A_{xx}A_{xy} + A_{xy}A_{yy} + A_{xz}A_{yz}) \\&\quad + \sin^2\theta(A_{xy}^2 + A_{yy}^2 + A_{yz}^2) \\&= T_{xx}\cos^2\theta + 2T_{xy}\cos\theta\sin\theta + T_{yy}\sin^2\theta \end{aligned} \quad (3.70)$$

Thus, we obtain the **T**-tensor from the fitting of the experimental observation with this relation. Again, $T_{ij} = T_{ji}$ and $T_{ij} = (\mathbf{A}^2)_{ij}$ are valid. The remaining procedures are the same as those of the **g** diagonalization, resulting in the diagonalized $(\mathbf{A}^2)_{ii}$. Now we obtain the **A**-matrix given by (3.38). The principal A-values are divided into the isotropic term, A_0, and dipolar term, A_{dip}:

$$A_{\text{dip},i} = A_i - A_0 \quad (i = x, y, z) \quad (3.71)$$

The isotropic A-value observed in the solution ESR (A_{sol}) is equal to A_{iso} given by (3.39).

Experimental data acquisitions of the g- and A-anisotropies can be performed using single crystals, in which the paramagnetic species concerned must be dispersed in an appropriate diamagnetic matrix. This method is intended to reduce the spin–spin interactions and to deduce the isolated spin properties on the g- and A-anisotropies, which would be otherwise impossible to obtain. The first application to organic radicals was made by McConnell's group with the objective of initiating spin-labeling studies with use of nitroxides [25, 26]. DTBNO (7 in Figure 2.1) was dispersed in a diamagnetic matrix of tetramethylcyclobutanedione, of which the ESR spectra are shown with respect to the magnetic field orientations in the principal axis systems (Figure 3.15). The data are summarized, together with other nitroxides, in Table 3.1. As the insert in Figure 3.15 indicates, the principal axes coincide with the N–O bonding axis (x-axis) or $2p_z$ orbital direction (z-axis). Judging from the comparison of the resonance positions on the basis of the dotted position the g_z is lowest and nearest to the free electron, whereas g_x along the bond axis is the highest of all. In contrast, the hfcc of the nitrogen is the largest for H//z and the smallest for H//y. These observations suggest the validity of the theory as follows. The g-anisotropy is calculated based on the matrix elements, Λ_{ii} (i, = x, y, z), of the operator for the orbital angular momenta, ($\mathbf{L}_x$, $\mathbf{L}_y$, $\mathbf{L}_z$), between the ground and excited molecular orbitals, as was previously derived

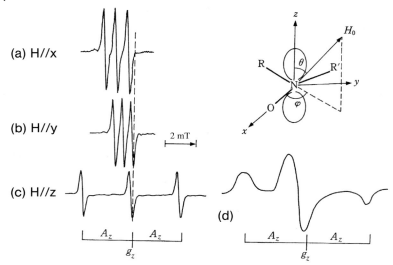

Figure 3.15 ESR spectra of DTBNO in a diamagnetic matrix depending on (a)–(c) the magnetic field directions and (d) its powder pattern. The insert shows the principal axis system in the N–O bond.

in Section 1.3.6. L_z is non-mixing, leading to the free electron g-value and L_x, rather than L_y, which induces the mixing with the lower excited state. As for nitrogen hyperfine anisotropy, the isotropic term, A_0, contributes similarly to A_x, A_y, and A_z. The dipolar term, however, has something to do with the direction of the dipolar interaction between the electron and the nuclear spins, in which the electron spreads on the $2p_z$ orbital. Because of the electron distribution along the z-direction due to the $2p_z$ orbital the dipolar interaction along the z-direction becomes the largest, giving the largest hfcc. The angular dependence of the dipolar term indicates half of the z-contribution when $H//x$ and $H//y$. The details are clearly comprehensible from the separation procedure of the anisotropies for DTBNO.

$$\begin{bmatrix} 1.511 & 0 & 0 \\ 0 & 1.511 & 0 \\ 0 & 0 & 1.511 \end{bmatrix} + \begin{bmatrix} -0.752 & 0 & 0 \\ 0 & -0.916 & 0 \\ 0 & 0 & 1.667 \end{bmatrix}$$

Two additional remarks are useful. First, from solution ESR we obtain g_{sol} and A_{sol}, listed in Table 3.1, which are almost completely identical to the averaged g_{iso} and A_{iso} calculated from (3.66) and (3.39), respectively. Second, in Figure 3.15, the ESR pattern for the powder samples is added, and the most-separated two peaks are evidently the result of for H parallel to the principal z-axis. Thus, even from the powder pattern we can deduce, at least, g_z and A_z, based on the inserted indication.

3.3.3
D-Anisotropy

We have to solve the following Hamiltonian.

$$\mathcal{H} = \mu_B S \cdot g \cdot H + S \cdot D \cdot S \tag{3.72}$$

But we treat this problem using the same approximation as the previous hyperfine case, and only the results necessary for the experimental procedure for the D-tensor analysis are described. The observed fine-structure splitting, d, is given using the D-tensor components in the crystal-axis system.

$$d = (3/g\mu_B)(D_{xx} \sin^2\theta + 2D_{xz} \sin\theta\cos\theta + D_{zz} \cos^2\theta) \tag{3.73}$$

The six independent matrix elements (because of $D_{ij} = D_{ji}$) are thus obtained and diagonalized, giving D_x, D_y, and D_z.

$$\mathbf{D} = \begin{bmatrix} D_x & 0 & 0 \\ 0 & D_y & 0 \\ 0 & 0 & D_z \end{bmatrix} \tag{3.74}$$

The diagonal sum vanishes, $D_x + D_y + D_z = 0$. Therefore, we have two forms of spin Hamiltonian in the principal axis coordinates given by (1.77) and (1.79).

We deal with an example of the analysis of D-anisotropy of DANO (**9** in Figure 2.1) in a diamagnetic matrix of di-p-anisyl benzophenone [27]. In addition to the g- and A-anisotropies of the isolated species, the ESR spectra comprises dipolar splitting owing to a radical-pair in the crystal. The D-values obtained were $D_x = 19.11$, $D_y = 17.91$, and $D_z = -37.32$ mT, and therefore, $D = 18.6$ and $E = 0.2$ mT. From the point-dipole approximation, (1.82), the radical distance is calculated to be 0.54 nm, which, together with the principal axis directions, agrees well with crystallographic data of the host crystal.

3.3.4
Anisotropic Parameters from ESR Powder Pattern

Through elaborate work using a single crystal sample mentioned above, not only the principal g-values, hyperfine coupling constants, and zero-field splitting parameters but also the principal-axis orientation in the crystal system are completely evaluated and used for the analyses and interpretations of material properties. A simple way of evaluating the magnitude of these principal values may be performed using ESR measurements of polycrystalline powder samples or frozen-solution samples. We call this ESR spectrim a "powder pattern", which characterizes several spectral singularities indicating the resonance points corresponding to the principal values.

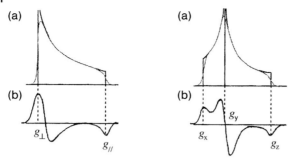

Figure 3.16 ESR powder patterns with g-anisotropy, axial ($g_{\parallel}$ and $g_{\perp}$, left figure) and orthorhombic (g_x, g_y, and g_z, right figure). Figure (a) shows theoretical line-distributions due to g-anisotropy and figure (b) comprises their derivative curves indicating the principal values with dotted lines.

First, let us examine the powder patterns of the g-anisotropies classified into (1) three-axis anisotropy (all g-values are different) and (2) axial anisotropy (two of the g-values are the same, thus we have $g_{\parallel}$ and $g_{\perp}$). Both the powder patterns are depicted theoretically in Figure 3.16 [28]. The integrated line-shapes imply the accumulation of each angular dependent line considering the transition probability and are smoothed by a certain line-width. From some of the featured points we obtain the principal g-values. It is noted that no hyperfine contribution to the line-shape is taken into account in Figure 3.16.

Hyperfine splitting anisotropy together with g-value anisotropy may be the most general situations. Both influences are, in principle, additive in the powder pattern, and, as a result, it becomes rather complicated, but we can draw some of the anisotropic values from the powder-pattern spectrum. Nitroxide radicals showed typical powder patterns in combination with g- and nitrogen hyperfine anisotropies (Figure 3.15). A slightly similar powder ESR pattern is depicted theoretically in Figure 3.17, where we assume the axial g-anisotropy ($g_{\parallel}$ and $g_{\perp}$) and the nuclear hyperfine $I = 1$ (nitrogen) anisotropy ($A_{\parallel}$ and $A_{\perp}$), and all values are taken to be nearly the same as those of nitroxide radicals (see Table 3.1), and the derivative pattern comprises several peaks. The important data deduced from this spectrum are $g_{\parallel}$ and $A_{\parallel}$, which are obtained from the magnetic field position of the center of

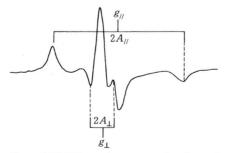

Figure 3.17 ESR powder pattern of axial g- and A-anisotropies of nitroxide radicals.

the dotted lines and half the separation between the dotted lines, respectively. These two parameters provide very useful information for spin-proving and spin-labeling studies, and they are used as a measure for dynamic phenomena.

An important remark is added here, regarding what are called "extra-lines", which are observed exceptionally and originate from a specific angular dependence possessing an angular turning point except the principal-axis directions.

Regarding the zero-field splitting parameters, the powder pattern is also advantageous as a simple manipulation without using a single crystal. Three separations of the fine structure are calculated to be $2D$ and $D \pm 3E$ when the magnetic field is applied parallel to the three principal axes. These specific points are discernible on the powder pattern of the ESR spectrum (Figure 3.18). Figure 3.18a is a summed result of each line component for all spherical orientations of the magnetic field directions, and its derivative ESR pattern is shown in Figure 3.18b for a D parameter only ($E = 0$). In the case of the orthorhombicity the spectrum contains several characteristic features, from which the parameters, D and E, can be estimated,

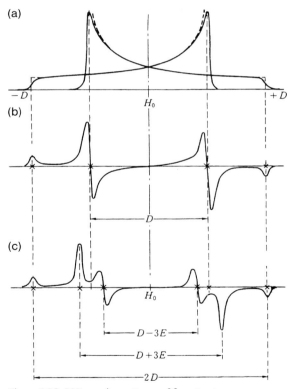

Figure 3.18 ESR powder patterns of fine structure:
(a) theoretical line distribution due to fine structure splitting;
(b) its derivative curve with axial case; (c) a derivative curve with orthorhombic case.

based on the indicated separations in Figure 3.18c. Despite a lack of detailed knowledge of the principal axis directions, this is sometimes sufficient to derive information on the evidence of $S = 1$, the distance value, and the structure variations influenced by internal and external factors. This method takes advantage of the verification of an S–T model (see Section 1.3.8), and because the intensity variation of the spectrum corresponds to the magnetic susceptibility of the electron pair system of the S–T model, the determination concerning which is a ground state of the system, the triplet or singlet state, and its energy gap may be assured.

As an example of high spin molecules with $S > 1$, tris(diphenylamino)benzene cation radical with $S = 3/2$ ground state is referred, where the fine structure consists of $4D$ and $2D$ paired splitting on both sides of the center absorption due to $m_s = 1/2 \Leftrightarrow -1/2$ [29].

3.4
Pulsed ESR

3.4.1
Fundamental Concept of FT-ESR

So far we have studied the basic principles of CW-ESR, in which the linearly polarized microwave component, H_1, is irradiated continuously through the resonance. In this section we restrict ourselves to pulsed ESR. Regarding the terminology of pulse, the microwave irradiation as a 90° or 180° pulse has already been explained previously in Section 3.2.2. The result of applying a short intense microwave pulse is to rotate the magnetization, M_z, through the flip angle, $\alpha = 90°$ or $180°$ using the specific pulse width given by the two relations (3.11).

In the simplest type of pulsed ESR a single 90° pulse transforms the original longitudinal magnetization, M_z, to the transverse direction of the y-axis of the rotating frame, then keeping the magnetic moments rotating on the xy-plane if there are no relaxation mechanisms. However, the detection systems aligned in the xy-plane pick up the oscillating signal, which decays owing to the spin–spin relaxation or inhomogeneous line-broadening. This is called free induction decay (FID), giving rise to the FT-ESR spectrum by performing the Fourier transformation (FT), into a frequency domain, which is identical to the CW-ESR spectrum. In principle, therefore, the FT-ESR method makes it possible to record the entire spectrum within a few microseconds, whereas in the CW method the spectrum has to be recorded by a slow passage through the resonances. In Figures 3.19 and 3.20, FIDs of Fremy's salt in water $(NO(SO_3^-)_2)$ and its FT-ESR [30] and the time-evolution of the FT-ESR spectra for a short-lived paramagnetic radical [31] are depicted. As the latter example shows, the FT-ESR features, in particular, its high time-resolution, so that it is very useful for investigations of short-lived paramagnetic species.

Concerning the detection techniques, some comments on the quadrature detection described in Figures 3.19 and 3.20 and on the dead time of the pulsed ESR

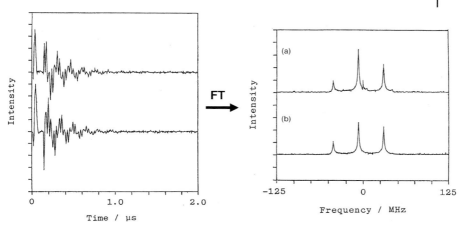

Figure 3.19 Quadrature-detected FID signals obtained from Fremy's salt at room temperature (left) and FT spectra without (a) and with (b) phase alternation scheme (right).

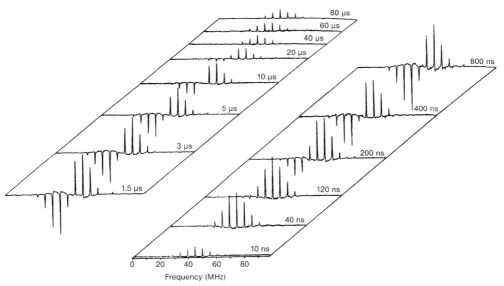

Figure 3.20 Quadrature-detected FT-RSR spectra of a photogenerated duroquinone anion radical in ethanol at 245 K. The spectra (absorption up and emission down) vary on time delay after the laser irradiation.

are appropriate. A short microwave pulse and rapid data acquisition may make difficult to excite all spins in the CW-ESR spectrum, consequently resulting in the distortion of the intensity and phase. The detection of the x- and y-components of the magnetization will improve the problem, which is a quadrature detection. An intrinsic problem in the pulsed spectrometer may be a dead time, during which the data acquisition is prevented from observing the FID. The reason is that the pulsed microwave induced in the resonator remains for a while, that is, the dead time, which ranges up to 50~100 ns after the 90° pulse. If the time becomes longer, it is impossible to detect a broad resonance line. This fact is, in some senses, analogous to the fact that the time profile of the FID is concerned with the relaxation time, T_2, and therefore, a short relaxation time (broad line-width) is equivalent to a rapid disappearance of the FID. Thus, applications to the materials having intense magnetic interactions or containing transition metal ions seem unlikely except at very low temperatures.

3.4.2
Electron Spin Echo (ESE)

3.4.2.1 Two-Pulse Method

The first spin-echo detection was made by Hahn for the nuclear spin case [6]. The rapid decay of the FID caused by inhomogeneous broadening is a basis of the spin-echo experiment in a magnetic ensemble of each spin rotating with different frequencies. In the rotating frame the time-evolution of each spin may be viewed after a 90° pulse, as is shown in Figure 3.21. Owing to different local fields acting on each spin, the rotation phase gradually spreads out, to some extent, after a time of τ. Subsequently, a second 180° pulse along the x'-axis (H_1 direction in the rotating frame) is applied in order to turn all the situations into $-y'$ direction. Since the directions of the individual rotations around the z-axis remains unaltered, as if time were reversed, the spread-out spin ensemble starts to focus again along the $-y'$ direction; the refocussing after the short time τ is an appearance of spin echo. The sequence of events and pulses are shown in Figures 3.21 and 3.22a. This method is called the "two-pulse method" or "Hahn's echo method". The longer the waiting time τ the smaller the echo. The decay of the echo intensity is fitted to an exponential function of τ as

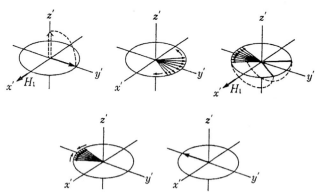

Figure 3.21 Principle image of spin echo in the rotating frame. The first figure indicates π/2 pulse and the third π pulse. The final figure is a shot at the echo peak.

$$I(2\tau) \propto \exp(-2\tau/T_M) \qquad (3.75)$$

Here, the time constant of the decay, T_M, is called the "phase memory time". The evaluation technique for this relaxation-related parameter is, in principle, different from that of time-profile of FID or the line-width of CW-ESR. However, the phase memory time may be used for characterizing materials of the spin ensemble similar to the spin–spin relaxation time, T_2.

3.4.2.2 Three-Pulse Method

The two-pulse method utilizes the pulse sequence 90°–τ–180°–τ(echo). The three-pulse method comprises three 90° pulses of (Figure 3.22b), which means the dividing of the 180° pulse in the two-pulse method into half, that is, two 90° pulses with a time interval of T. This three-pulse sequence also yields an echo called the "stimulated echo" with some influences added in the echo intensity. Although the stimulated echo may be formed by rather complicated mechanisms, the important situation takes place after the second 90° pulse, which causes inversion of the magnetization, M_z. During the evolving time, T, after the second pulse, spin–lattice relaxation may be influential to turn back the magnetization vector to its equilibrium direction. Thus, a third 90° pulse applied after a time $T \leq T_1$ would generate the FID of this reduced magnetization, M_z and at a time τ after the third 90° pulse would be detected as a stimulated echo. The three-pulse sequence is written as 90°–τ–90°–T–90°–τ(echo). The decay of the echo intensity is also fitted to an exponential function of T as

$$I(2\tau + T) \propto \exp(-T/T_1) \qquad (3.76)$$

Here the time constant of the decay, T_1, is the same as the spin–lattice relaxation time.

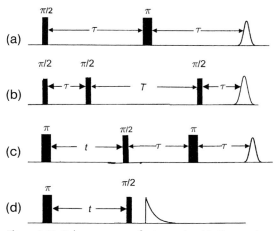

Figure 3.22 Pulse sequences for two-pulse (a), three-pulse (b), and inversion recovery (c, d) methods.

3.4.2.3 Inversion Recovery Method

The inversion recovery method for determining the relaxation time, T_1, utilizes echo detection after the different three-pulse sequence, 180° –t–90° –τ–180° τ(echo) (Figure 3.22c). The first 180° pulse produces the opposite magnetization, M_z, along the –z-axis, then one waits for a time evolution during a certain time, t, which is followed by the detection pulse sequence (the two-pulse method). The time profile of the detected echo intensity indicates the magnetization recovery characterized by the spin–lattice relaxation time, T_1.

Research examples using relaxation times are described here. Two- and three-pulse methods were applied to TTBP (**14** in Figure 2.1) dispersed in a phenol matrix [32]. The temperature dependences of T_1 and T_2 (Figure 3.23) imply molecular motional effect to the spin relaxation, giving a rotational correlation time and its activation energy, like a BPP theory [33]. For studies of conductive polypyrrole, an FID detection and its inversion recovery method (Figure 3.22d) were utilized because of the lack of the echoes. Both relaxation times exhibited characteristic temperature dependences (Figure 3.24), suggesting dopant and isotope effects. It was found that, by applying the derived relation for $1/T_1$ and $1/T_2$, (3.55), the correlation time τ_C and the fluctuating local field can be estimated, concluding that the relaxations of the electrons in the conductive materials are governed by the fluctuation of the hyperfine field and the metallic Elliot mechanism. The details of the results can be found in the literature [34–37].

3.4.2.4 Echo-Detected ESR (ED-ESR)

As long as any kind of echo detection is available, echo-detected ESR is no problem. Then echo intensities as a magnetic field varies draw an integrated ESR pattern similar to the CW-ESR experiments. In this context ED-ESR is an alternative to CW-ESR. However, it is most advantageous for very short-lived free radicals

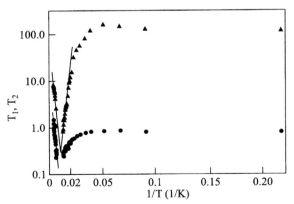

Figure 3.23 Temperature-dependent T_1 and T_2 of TTBP dispersed in a diamagnetic matrix. T_1 (▲) and T_2 (●) are plotted in μs unit and the solid lines are the fitting lines for determining motional activation energies.

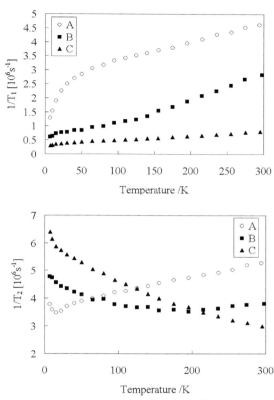

Figure 3.24 Temperature dependences of T_1^{-1} and T_2^{-1} of conductive polypyrrole. The conductivities at room temperature for samples A, B, and C are 5, 4×10^{-1}, 5×10^{-3}, respectively, in which the doping level (dopant ClO_4^-) are different, (A) 34%, (B) 15%, and (C) 9.7%.

because of the very short time-resolution of ED-ESR, comparable to the time interval between the pulses. Another interesting point about ED-ESR is that the spectrum is not necessarily the same as for CW-ESR because the echo-intensity is highly influenced by the relaxation time. Whether a short or long evolution time in the ED-ESR is used causes a change in the spectrum, or one can select a spectrum depending on the relaxation time from the mixed species present in the CW-ESR spectra. This situation is very effective for transition metal ion samples, which show very broad ESR lines and are found to possess a wide range of relaxation times.

3.4.2.5 Nutation Spectroscopy

For FID or ESE experiments one has to adjust a pulse-width to 90° or 180° as much as possible. When we take the pulse-width as a variable, the signal intensities are oscillating owing to varying nutation angles of the magnetization. The nutation spectrum can be obtained from FT of this modulation, giving rise to the nutation frequency, ω_{NT}, which is related to the spin quantum numbers, S and m_S.

$$\omega_{NT} = \sqrt{S(S+1) - m_S m_S'}\, \omega_1, \quad \omega_1 = g\mu_B H_1/\hbar \quad (3.77)$$

This nutation spectroscopy is useful when we want to discriminate mixed spin states in materials.

3.4.3
ESEEM

In a two-pulse or three-pulse experiment, in which τ or T is varied, the hyperfine interaction causes a modulation of the echo intensity. This is a nuclear modulation effect and called the electron spin echo envelope modulation (ESEEM). This may be understandable when we examine the hyperfine local field from the nucleus acting on the electron spins during the evolution and detection times. After the first pulse, the direction of the electron spin suddenly changes, and consequently the nuclear spins near the electron spin start to precess around a newly generated effective field, adding the modulation upon the echo intensity depending on the two parameters, τ and T in the pulse sequences, $V(\tau, T)$. $V(\tau, 0)$ is ESEEM for the two-pulse method.

The ESEEM obtained in the time domain is transformed into the frequency domain by FT, then we have an ESEEM spectrum, which is said to correspond to the ENDOR spectrum and is convenient for the analysis of the multiple nuclear contributions. The ESEEM spectrum rather presents supplementary data for the CW-ESR because the ESEEM effect is mainly observed for weakly coupled hyperfine interaction between the unpaired electron and the surrounding nuclei. It means the very small hfcc is hidden in the line-width. Furthermore, from the ESEEM spectrum one can deduce the quadrupolar interaction for the nuclei of $I \geq 1$, such as ^{2}D and ^{14}N. In practice, ESEEM is a powerful means of determining the nuclear coordination of ^{1}H, ^{2}D, ^{14}N, ^{19}F, ^{23}Al, and ^{31}P, especially in researches on trapped sites of radicals or active sites of metalloenzymes. Readers are advised to consult a very instructive book for ESEEM [38, 39].

3.5
Double Resonance

3.5.1
ENDOR

The first ENDOR experiment dates back to 1956, which was applied in order to resolve the complicated or smeared hyperfine lines in solids [40]. Technically ENDOR belongs to a double resonance using two electromagnetic irradiation fields for simultaneous excitation. As the abbreviation ENDOR (electron-nuclear double resonance) indicates, it is a combination of ESR and NMR. In addition to the usage of microwave for ESR and radiofrequency wave for NMR transitions, the powers of these irradiations are sufficiently strong to disturb the populations in the energy levels far from those of the thermal equilibrium. Therefore, of the two resonances, one utilizes ESR for observing or monitoring transition and NMR for simultaneously exciting transition. The latter is called the "pumping system" or "pumped transition". In general, a pumped transition of high-power may cause, to some extent, a population change in the monitoring energy levels because some cross-relaxations between the levels work effectively, regardless of whether or not the pumped transition has a level in common with the observing levels. The monitoring and pumping transitions are summarized for the four sublevels of $S = 1/2$ and $I = 1/2$ in Figure 3.25a, where the observing ν_{ESR} is influenced by the two concurrent ν_{NMR} transitions.

In the ENDOR spectrum, the pumping NMR frequency is swept, and simultaneously the ESR intensity change of the monitoring transition is recorded. Because the resonance frequencies of ESR and NMR at the magnetic field, H_0, are given by

$$\nu_{ESR,mI} = |g\mu_B H_0/h + Am_I| = |\nu_e + Am_I|$$
$$\nu_{NMR,mS} = |g_n\mu_n H_0/h - Am_S| = |\nu_n - Am_S| \tag{3.78}$$

the ENDOR absorptions are observed at two positions of $\nu_{NMR,mS}$. In these relations, the hyperfine coupling constant, A, is expressed in frequency units. The paired absorption pattern of the ENDOR spectrum varies for $\nu_n > A/2$ or $\nu_n < A/2$ (Figure

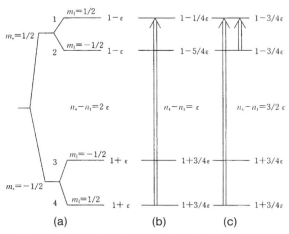

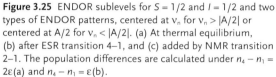

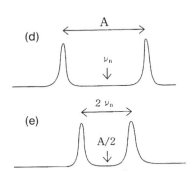

Figure 3.25 ENDOR sublevels for $S = 1/2$ and $I = 1/2$ and two types of ENDOR patterns, centered at ν_n for $\nu_n > |A/2|$ or centered at $A/2$ for $\nu_n < |A/2|$. (a) At thermal equilibrium, (b) after ESR transition 4–1, and (c) added by NMR transition 2–1. The population differences are calculated under $n_4 - n_1 = 2\varepsilon$ (a) and $n_4 - n_1 = \varepsilon$ (b).

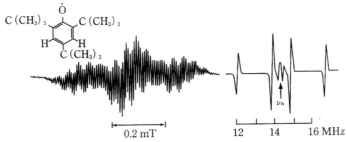

Figure 3.26 ESR and ENDOR spectra of TTBP. In the ENDOR spectrum the splittings are caused by m-protons (A_H = 0.1736 mT), p-t-butyl protons (A_H = 0.0365 mT), and o-t-butyl protons (A_H = 0.0072 mT) [39].

3.25d, e). For instance, ν_n of ^{1}H at the ESR magnetic field for an X-band region (9.4 GHz) amounts to ca.14 MHz, and usually corresponds to the case of $\nu_n > A/2$ (consider that 2.8 MHz is equivalent to 0.1 mT), so that ENDOR spectra is centered at ca.14 MHz and the paired separation equals the hyperfine coupling constant, A, in the frequency unit. A typical example is given in Figure 3.26, in which the ENDOR spectrum of TTBP (**14** in Figure 2.1) is depicted [41], together with its ESR spectrum. The ENDOR spectrum consists of only three paired lines (six lines), corresponding to the number of the groups of the hyperfine splittings with different hfccs (the m-protons, the p-tert-butyl, and the o-tert-butyls have larger hfccs in this order). On the other hand, the ESR spectrum consists of a number of lines, theoretically 570 lines in this phenoxyl radical. As this result indicates, one of the advantages of ENDOR is resolution enhancement.

3.5.2
TRIPLE

TRIPLE resonance consists of literally "triple" resonances, but is based on ENDOR spectroscopy. This means that the triple resonance methods use an additional radiofrequency irradiation to those used in ENDOR. One expects the monitoring ESR to vary due to the concomitant NMR transitions. There are some triple modes, depending on how the two radiofrequencies for the NMR transitions are applied. The main methods are called General TRIPLE (GTR) and Special TRIPLE [42]. GTR, which is exclusively described in this section, is able to deduce the sign of the hfcc, which is unknown from the CW-ESR analysis. The experimental example is taken from DPNO (**8** in Figure 2.1), of which ENDOR and GTR spectra are shown in Figure 3.27 [43, 44]. As its solution ESR indicates, there are two kinds of equivalent protons, one ^{1}H group consisting of the o- and p-positions (A_{H1}) and the other one of the m-positions (A_{H2}). From the paired lines centered at the free proton frequency, ν_H, A_{H1} = 0.1913 mT and A_{H2} = 0.082 mT were obtained. In TRIPLE resonance, first a 1.881 MHz line is pumped, then we have the spectrum (b), resulting in the intensity variations for each line compared to the above

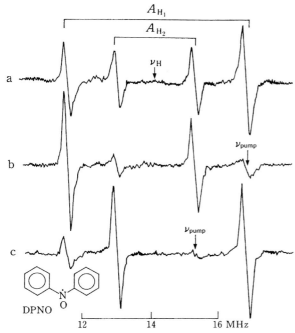

Figure 3.27 ENDOR (a) and GTR (b and c) spectra of DPNO. A_{H1} pair is attributed to the o- and p-protons and A_{H2} pair to the m-protons. The intensity ratios in each pair are opposite in (b) and (c) spectra, implying an opposite sign of the A_{H1} and A_{H2}.

ENDOR spectrum. One of the lines at low-frequency pairing with the pumped line increased a lot in intensity. On the other hand, the inner paired lines were also influenced, but the intensity relations of increase or decrease is opposite. In spectrum (c) the pumping position was changed to one of the inner pair on the high-frequency side (15.348 MHz), and then all the intensity variations mentioned above are completely reversed. These facts are explained on the assumption of energy sublevels with the opposite sign to the hyperfine coupling constants, A_H, among the protons of the o- and p-positions (A_{H1}) and those of the m-positions (A_{H2}). This conclusion may be evinced from the theoretical MO calculations, implying positive spin density on the o- and p-positions and negative spin density on the m-positions. Thus, GTR experiments feature the relative sign determination of the delocalized spin density.

3.5.3 ELDOR

ELDOR (electron-electron double resonance) indicates double resonances of ESR and ESR using two kinds of microwave frequencies. The powers of these irradiations also have to be sufficiently strong to disturb the populations in the spin levels far from those of the thermal equilibrium. The ELDOR experiment needs two kinds of microwave frequencies, one for pumping and another for monitoring. In the example of the four sublevels in Figure 3.28 the transition of the monitoring ESR between the levels 2 and 3 may be influenced by the microwave irradiation of the pumping ESR between the levels 1 and 4, both the microwave frequencies, v_o and v_p, being given by

$$v_o = g\mu_B H_0/h - A/2, \quad v_p = g\mu_B H_0/h + A/2 \tag{3.79}$$

Therefore, we obtain the hyperfine coupling constant from $v_p - v_o = A$. An ELDOR experiment in solution was designed and implemented for a DPPH solution by Hyde [45]. DPPH comprises two neighboring nitrogens (4 in Figure 2.1) on which the majority of the unpaired electrons are delocalized. The CW-ESR spectrum of DPPH, however, never splits the hyperfine coupling constants of the

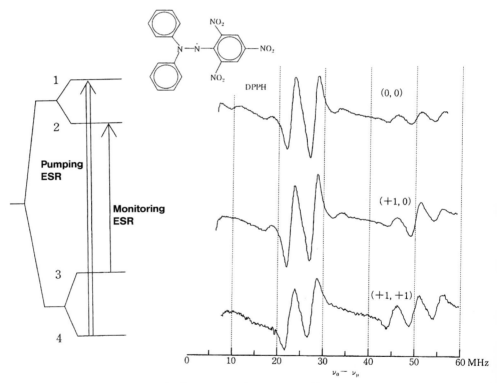

Figure 3.28 ELDOR sublevels of pumping and monitoring ESR ($S = 1/2$ and $I = 1/2$) and its application to DPPH, two-nitrogen ($I = 1$) case. The nitrogen-α (left) and nitrogen-β (right) of DPPH are well resolved. Monitoring nitrogen hf lines are discriminated by the nitrogen hf components, (0, 0), (+1, 0), and (+1, +1).

two nitrogens. The spectral sweep was carried out through the variation of $v_o - v_p$, as is shown in Figure 3.28, in which the two lines ranging from 20 to 30 MHz can be assigned to the two nitrogens, N_α and N_β, in the inserted molecular structure:

$$A_{N_\alpha} = 0.7933 \pm 0.007 \,\mathrm{mT}, \qquad A_{N_\beta} = 0.9739 \pm 0.014 \,\mathrm{mT} \qquad (3.80)$$

In addition to the clear separation of the hyperfine splitting, a higher accuracy of more than one order of magnitude convinces us of the advantage of the double resonance method.

3.5.4
Pulsed Methods for Double Resonance

As for the double resonance using the pulsed ESR techniques, two irradiation fields in the ENDOR and ELDOR are applied as pulses, detecting the response as electron-spin-echo (ESE). The pulsed schemes contrive to solve complicated relaxation problems, including electron and nuclear spins, which make it frequently impossible to observe the response from the pumping transition in the CW-ESR. Regarding pulsed ENDOR, we have two methods, the Davis and Mims methods, based on the two- and three-pulse methods, respectively (Figure 3.29a, b). The Mims method utilizes a three-pulse sequence for the microwave and the radiofrequency NMR pulse applied over a rather long time between the second and third 90° pulses. Then we observe a reduced echo intensity when the NMR transition is induced. Accordingly, the frequency of the radio frequency pulse is swept as a function of the echo intensity.

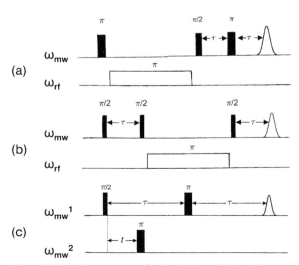

Figure 3.29 Pulse sequences for Davis (a) and Mims (b) ENDOR and pulsed ELDOR (PELDOR or DEER) methods.

Three-pulse echo-detected ELDOR (pulsed ELDOR) is known as "DEER" or "PELDOR" [38]. The experiment consists of a two-pulse echo sub-sequence with a mixed interpulse delay τ at the observer frequency, ω_{mw}^1, and a pump pulse of flip angle, π, at frequency, ω_{mw}^2, with a variable delay time, t, with respect to the first pulse of the observer sequence (Figure 3.29c). By this pulse sequence the dipolar interaction of two spins may be observed as a modulation in the time domain and its FT yields a so-called Pake doublet that contains the distance information between the spins.

3.6
ESR of Magnetic Materials

3.6.1
Low-Dimensional Magnetic Materials

ESR characteristics in magnetic materials mentioned in Chapters 1 and 2 are summarized. Different from the phenomenological treatment such as the Bloch equation (3.12), the theory of the resonance and the relaxation on the basis of quantum statistics has been completed by Kubo and Tomita [46]. This is a kind of linear response theory in the resonance phenomena, in which the Kubo–Tomita theory defines the autocorrelation function with respect to the $M_x(t)$.

$$G(t) = <M_x(t)\, M_x(0)>/<M_x(0)^2> \quad (3.81)$$

This autocorrelation function is normalized at $t = 0$. Then assuming a local fluctuating field on each spin, the resonance frequency, ω, also fluctuates and its spin correlation function is written as

$$\Psi(\tau) = <\omega(\tau)\, \omega(0)>/<\omega(0)^2> \quad (3.82)$$

According to Kubo–Tomita theory, the FT of the resonance line $I(\omega)$ is given by $\phi(t)$ as

$$\phi(t) = \exp(-<\omega(0)^2> \int (t-\tau)\Psi(\tau)d\tau \quad (3.83)$$

When $\Psi(\tau)$ is constant, then we have a Gaussian function, $\phi(t)$, resulting in a Gaussian $I(\omega)$. On the other hand, in the effective exchange-interaction system, $\Psi(\tau)$ shows a rapid decay, giving rise to a Lorentzian $I(\omega)$. What happens in the low-dimensional system? We assume a spin diffusion process reflecting the dimensionality; that is,

$$\phi(t) \sim t^{-d/2} \quad (d = 1, 2) \quad (3.84)$$

The suggested line shape comes between the Lorentzian and the Gaussian, as was first shown for $(CH_3)_4NMnCl_3$ [47]. The spin diffusion model was found to be in good agreement and characterizes the low dimensionality of the magnetic materials. Besides, the angular variation of the line-width is shown to be different from the usual Gauss approximation, $(1 + \cos^2\theta)$, but, rather, is expressed by

$$\Delta H = a + b|3\cos^2\theta - 1| \tag{3.85}$$

Here the angle, θ, is measured from the direction of the linear chain.

The deviation of g-value depending on the direction of the applied magnetic field specifies the direction of the linear chain, which is a critical phenomenon owing to the highly developed spin correlation upon approaching the phase transition. Taking into account the exchange interaction within the linear chain in the general equation of motion, the shift of the resonance position may be estimated to be

$$h\nu = g\mu_B H(1+2\delta), \quad h\nu = g\mu_B H(1-\delta) \tag{3.86}$$

where $\delta = (3g^2\mu_B^2/Hr^3) < S_{zi}S_{zi+1} - S_{xi}S_{xi+1} > / <S_z^2>$ with the distance, r, between the spins in the linear chain. This model was demonstrated for one-dimensional $CsMnCl_3 \cdot 2H_2O$ [48, 49]. These ideas are also applicable to a two-dimensional plane network, such as DANO [50] described in Section 2.4.3 of Chapter 2.

3.6.2
Ferromagnetic Resonance (FMR)

The magnetization of a ferromagnet, $\mathbf{M}$, follows $d\mathbf{M}/dt = \gamma \mathbf{H} \times \mathbf{M}$, where $\mathbf{H}$ represents, in addition to the external magnetic and microwave oscillating fields, the magnetic anisotropic, and demagnetizing fields. The exchange field proportional to the magnetization dose not need to be considered because it would vanish in the right side of the above equation, $\mathbf{H} \times \mathbf{M}$. Demagnetization is a specific concept inherent in the ferromagnet, and it is induced by spontaneous magnetization and is dependent upon the shape of the sample. Neglecting the anisotropic field, $\mathbf{H}$ is then given by

$$\mathbf{H} = (2H_1 e^{i\omega t} - N_x M_x, -N_y M_y, H_0 - N_z M_z) \tag{3.87}$$

Here N_x, N_y, and N_z are the demagnetization coefficients for an ellipsoid sample and $N_x + N_y + N_z = 1$ holds. Neglecting $dM_z/dt = 0$ because it contains only the second order of the small quantities of M_x and M_y, and assuming M_x and M_y have a time-dependence of $e^{i\omega t}$, we attain an approximate solution:

$$\omega = \gamma\sqrt{\{H_0 + (N_y - N_z)M_z\}\{H_0 + (N_x - N_z)M_z\}} \tag{3.88}$$

This resonance equation for a ferromagnet is called Kittel's equation [7]. For a spherical sample satisfying $N_x = N_y = N_z = 1/3$ it leads to

$$\omega = \gamma H_0 \tag{3.89}$$

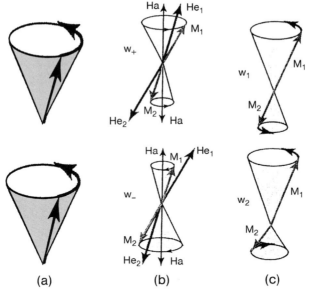

Figure 3.30 Precession modes of FMR (a), AFMR (b), and ferrimagnetic resonance (c). Mode (a) is all-in-phase homogeneous precession and ω_1 in (c) is analogous to a ferromagnetic mode.

This equation is identical to the resonance equation of paramagnets. From a standpoint of Larmor precession, all spins precess in phase, so that this solution is said to be a homogeneous mode (Figure 3.30a). More strict explanations need spin-wave concepts, but this Kittel mode corresponds to the wave vector $k = 0$. For ferromagnets we have another interesting mode called the Walker mode, which in general provides additional resonance lines. The detail is not explanied here.

3.6.3
Antiferromagnetic Resonance (AFMR)

We discuss the magnetic resonance of antiferromagnets, that is, two sublattice systems, which is classified into two types, antiferromagnetic resonance (AFMR) and ferrimagnetic resonance. Usual antiferromagnets are composed of sublattices with identical g-values (gyromagnetic ratios) and sublattice magnetizations. On the other hand, ferrimagnetism is characterized by different sublattices with different g-values and sublattice magnetizations. Ferrimagnets exhibiting a ferrimagnetic compensation temperature will be approximated antiferromagnets over a limited range of temperature, which will be briefly mentioned in the next section.

Coupled equations of motion are set up by applying Kittel's equation of motion for a ferromagnet to each of the sublattice magnetizations [8, 51].

$$d\mathbf{M}_i/dt = \gamma \mathbf{H}_i \times \mathbf{M}_i \quad (i=1,2) \tag{3.90}$$

Here $M_1 = -M_2$ and H_i is given below.

$$H_i = H_0 + H_{ai} + H_{ei} \quad (i = 1, 2) \tag{3.91}$$

The effective field includes the anisotropic and exchange fields, H_a and H_e, respectively. The directions of vectors of H_{a1} and H_{a2} are opposite, that is, $|H_{a1}| = |H_{a2}| = H_a$. As for H_{ei} we have

$$H_{e1} = -\Gamma M_2, \quad H_{e2} = -\Gamma M_1, \quad |H_{e1}| = |H_{e2}| = H_e \tag{3.92}$$

Here Γ is the molecular field coefficient introduced in (1.58) and is given by the simple approximation using the number, z, of the nearest spins as $\Gamma = 2z|J|S/g\mu_B$. Analogously to the FMR case, the AFMR equations for a uniaxial anisotropy can be derived. $H_e \gg H_a$ is a general case so that they are approximated to be

$$\omega_\pm = \gamma\sqrt{2H_a H_e} \pm H_0 \tag{3.93}$$

We have two modes with different precession directions as shown in Figure 3.30b. For an orthorhombic anisotropy with an easy z-axis several branches come out, which can be characterized by a ω–H diagram [8]. The first AFMR investigation for organic antiferromagnets has been carried out for p-Cl-BDPA (see **3** in Figure 2.1 and Table 2.1), of which an ω–H diagram is shown in Figure 3.31 [52]. The organic antiferromagnet features $H_a = 1.7\,\text{mT}$, $H_e = 6.5\,\text{T}$, concluding the spin-flop field of $H_{cr} = 150\,\text{mT}$. As is demonstrated here, magnetic resonance supplies very useful information which it is otherwise impossible to evaluate.

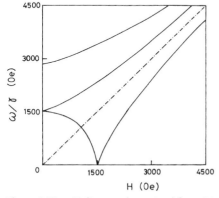

Figure 3.31 ω-H diagram determined from AFMR data of the antiferromagnet, p-Cl-BDPA. This organic radical has $T_N = 3.25\,\text{K}$ and shows an orthorhombic anisotropy, giving $H_{cr} = 1500\,\text{Oe}$.

3.6.4
Ferrimagnetic Resonance

Ferrimagnetic materials such as ferrites or garnets are composed of two sublattices with different spins or species from each other, and therefore, the equations of motion of the two sublattices must be modified upon the AFMR paired equations (3.90). We have to use different gyromagnetic ratios, γ_1 and γ_2, and different magnetizations, M_1 and M_2. Besides, ferrimagnetic materials, in spite of the spin antiparallel orientation, in principle belong to ferromagnets because of the presence of spontaneous magnetization. There appear also to be two characteristic modes (Figure 3.30c); one is the same as Kittel's mode in FMR and the other is a unique mode in ferrimagnetic resonance. The ω–H relations can be explored in a text [53].

References

1 Zavoisky, E. (1945) *Journal of Physics USSR*, **9**, 245.
2 Zavoisky, E. (1946) *Journal of Physics USSR*, **10**, 197–8.
3 Zavoisky, E. (1947) *Journal of Physics USSR*, **11**, 184–8.
4 Purcell, E.M., Torrey, H.C. and Pound, R.V. (1946) *Physical Review*, **69**, 37–8.
5 Bloch, F., Hansen, W.W. and Packard, M. (1946) *Physical Review*, **70**, 474–85.
6 Hahn, E. (1950) *Physical Review*, **80**, 580–94.
7 Kittel, C. (1951) *Physical Review*, **82**, 565.
8 Nagamiya, T., Yoshida, K. and Kubo, R. (1955) *Advances in Physiology Education*, **4**, 1–112.
9 Bloch, F. (1946) *Physical Review*, **70**, 460–74.
10 Atherton, N.M. (1973) *Electron Spin Resonance*, John Wiley & Sons, Inc., p. 18.
11 Gutowsky, H.S., McCall, D.W. and Slichter, C.P. (1953) *The Journal of Chemical Physics*, **21**, 279–92.
12 McConnell, H.M. (1958) *The Journal of Chemical Physics*, **28**, 430–1.
13 Atherton, N.M. (1973) *Electron Spin Resonance*, John Wiley & Sons, Inc., 309–13.
14 Nakashima, K. and Yamauchi, J. (2005) *Journal of the American Chemical Society*, **127**, 1606–7.
15 McConnell, H.M. (1956) *The Journal of Chemical Physics*, **24**, 764–6.
16 Slichter, C.P. (1990) *Principles of Magnetic Resonance*, Springer-Verlag, pp. 206–15.
17 Rogers, R.N. and Pake, G.E. (1960) *The Journal of Chemical Physics*, **33**, 1107–11.
18 Kivelson, D. (1960) *The Journal of Chemical Physics*, **33**, 1094–106.
19 Wilson, R. and Kivelson, D. (1966) *The Journal of Chemical Physics*, **44**, 154–68.
20 Freed, J.H. (1989) Spin labeling theory and applications, in *Biological Magnetic Resonance*, Vol. 8, (ed L.J. Berliner), Plenum Press, pp. 1–76.
21 Weissmann, S.I. and Ward, R.L. (1957) *Journal of the American Chemical Society*, **79**, 2086–90.
22 Freed, J.H. and Fraenkel, G.K. (1962) *The Journal of Chemical Physics*, **37**, 1156–7.
23 Hirota, N. (1967) *Journal of Physical Chemistry*, **71**, 127–39.
24 Rutter, A.W. and Warhurst, E. (1968) *Transactions of the Faraday Society*, **64**, 2338–41.
25 Griffith, O.H., Cornell, D.W. and McConnell, H.M. (1965) *The Journal of Chemical Physics*, **43**, 2909–10.
26 Libertini, L.J. and Griffith, O.H. (1970) *The Journal of Chemical Physics*, **53**, 1359–67.
27 Takizawa, O., Yamauchi, J., Nishiguchi, H.O. and Deguchi, Y. (1973) *Bulletin of the Chemical Society of Japan*, **46**, 1991–5.

28 Kneubühl, F.K. (1960) *The Journal of Chemical Physics*, **33**, 1074–8.
29 Yoshizawa, K., Chano, A., Ito, A. et al (1992) *Journal of the American Chemical Society*, **114**, 5994–8.
30 Mizuta, Y., Kohno, M. and Fujii, K. (1993) *Japanese Journal of Applied Physics*, **32**, 1262–7.
31 Prisner, T., Dobbert, O., Dinse, K.P. and van Willigen, H. (1988) *Journal of the American Chemical Society*, **110**, 1622–3.
32 Yamauchi, J., Yamaji, T. and Katayama, A. (2003) *Applied Magnetic Resonance*, **25**, 209–16.
33 Bloembergen, N., Purcell, E.M. and Pound, R.V. (1948) *Physical Review*, **73**, 679.
34 Kanemoto, K., Yamauchi, J., and Adachi, A. (1998) *Solid State Communications*, **107**, 203–7.
35 Kanemoto, K. and Yamauchi, J. (2000) *Physical Review B, Condensed Matter*, **61**, 1075–82.
36 Kanemoto, K., and Yamauchi, J. (2000) *Synthetic Metals*, **110**, 65–70, **114**, 79–84.
37 Kanemoto, K. and Yamauchi, J. (2001) *Journal of Physical Chemistry. B*, **105**, 2117–21.
38 Schweiger, A. and Jeschke, G. (2001) *Principles of Pulse Electron Paramagnetic Resonance*, Oxford University Press.
39 Dikanov, S.A. and Tsvetkov, Y. D. (1992) *Electron Spin Echo Envelope Modulation (ESEEM) Spectroscopy*, CRC Press Inc.
40 Feher, G. (1957) *Physical Review*, **105**, 1122–3.
41 Yamauchi, J., Katayama, A., Tamada, M. and Tanaka, S. (2000) *Applied Magnetic Resonance*, **18**, 249–54.
42 Möbius, K. and Biehl, R. (1979) Electron-nuclear-nuclear TRIPLE resonance of radicals in solutions, in *Multiple Electron Resonance Spectroscopy*, (eds M.M. Dorio and J.H. Freed), Plenum Press, pp. 475–507.
43 Deguchi, Y., Okada, K., Yamauchi, J. and Fujii, K. (1983) *Chemistry Letters*, 1611–14.
44 Yamauchi, J., Fujita, H. and Deguchi, Y. (1991) *Bulletin of the Chemical Society of Japan*, **64**, 3620–6.
45 Hyde, J.S., Sneed, Jr. R. C. and Rist, G.H. (1969) *The Journal of Chemical Physics*, **51**, 1404–16.
46 Kubo, R., Tomita, K. (1954) *Journal of the Physical Society of Japan*, **9**, 888–919.
47 Dietz, R.E., Merritt, F.R., Dingle, R., Hone, D., Silbernagel, B.G. and Richards, P.M. (1971) *Physical Review Letters*, **26**, 1186–8.
48 Nagata, K. and Tazuke, Y. (1972) *Journal of the Physical Society of Japan*, **32**, 337–45.
49 Nagata, K. (1976) *Journal of the Physical Society of Japan*, **40**, 1209–10.
50 Takzawa, O. (1976) *Bulletin of the Chemical Society of Japan*, **49**, 583–8.
51 Nagamiya, T. (1954) *Progress of Theoretical Physics*, **11**, 309–27.
52 Yamauchi, J. (1977) *The Journal of Chemical Physics*, **67**, 2850–5.
53 Foner, S. (1963) Antiferromagnetic and ferrimagnetic resonance, in *Magnetism I*, (eds G.T. Rado and H. Suhl), Academic Press, pp. 383–447.

4
Recent Advances in ESR Techniques and Methods Employed in Nitroxide Experiments
Alex I. Smirnov

4.1
Introduction

The main purpose of this chapter is to review recent developments in ESR techniques and methods in relationship to experiments with nitroxide free radicals. During the current decade the field of nitroxides blossomed even more from a renewed growth in applications of these compounds in diverse areas such as magnetic materials, polymers, medicine, and biophysics. These and other emerging fields are reviewed in the following chapters. At the same time, advances in digital electronics, microwave and mm-wave devices that are driven by demands of the telecommunications industry, as well as recent innovations in cryo-cooling and superconducting magnet technologies, created unprecedented opportunities for advancing ESR instrumentation beyond the recent limits. These technical developments allow for many of the advanced magnetic resonance methods reviewed in the preceding chapter to be widely explored by the nitroxide research community. For example, pulsed methods are now broadly employed by biophysicists for determination of structural constraints in proteins and nucleic acids that are specifically labeled with nitroxide pairs [1–5]. Developments in high frequency or high magnetic field ESR instrumentation enabled much more detailed studies of structure and dynamics of complex spin-labeled macromolecules than has been possible with conventional X-band (9–10 GHz) ESR [1, 6–9]. Furthermore, it has been shown that more sophisticated pulse techniques, such as two-dimensional and multiquantum coherence experiments initially discovered in the field of NMR [10], could be adapted to ESR experiments [11]. All these experiments were proved to be extremely valuable for biophysicists. At the same time, site-directed nitroxide spin-labeling (SDSL) methods that were first developed to make diamagnetic proteins amenable for studying by ESR have now found new uses in solving protein structures by solution NMR (e.g. [12]). Another example is a recent development of a method for macroscopic alignment of lipid bilayers and membrane proteins by nanoporous anodic aluminum oxide substrates. After initial demonstration of this method by spin-labeling ESR [13] this technique is now employed for structure–

Nitroxides: Applications in Chemistry, Biomedicine, and Materials Science
Gertz I. Likhtenshtein, Jun Yamauchi, Shin'ichi Nakatsuji, Alex I. Smirnov, and Rui Tamura
Copyright © 2008 WILEY-VCH Verlag GmbH & Co. KGaA, Weinheim
ISBN: 978-3-527-31889-6

function studies of membrane proteins by solid-state NMR [14–16]. Clearly, many other useful experimental techniques are expected to come from synergetic developments in magnetic resonance instrumentation, experimental methods, and synthesis of novel nitroxides with chemical and magnetic properties that would tailor specific applications. This chapter provides an overview of recent synergetic developments in these fields and also paints an outlook for the future.

4.2
Macromolecular Distance Constraints from Spin-Labeling Magnetic Resonance Experiments

Structural characterization of biological macromolecules, including proteins, DNAs, and RNAs, is of undisputed significance for many areas of biophysical and biomedical sciences. While many structural problems related to biomolecules could be successfully approached by crystallography, the inherent flexibility of protein side chains and nucleic acids presents a significant challenge for this method. Generally, the packing forces in protein crystals are considered to be comparable in magnitude with the electrostatic and van der Waals interactions that shape the protein structure and determine interdomain interactions. Moreover, crystallization conditions, such as ionic strength, pH, and co-factors present, usually differ from those observed in cells and, therefore, crystallized proteins could be locked in only few of many possible conformational states. Finally, researchers face significant challenges in preparing suitable crystals of protein complexes and membrane proteins that are especially difficult to crystallize. This is the main reason why the current rate of membrane protein structures solved still lags significantly behind that of water-soluble proteins [17].

To summarize, the growing interest in structure–function studies of biomolecules fuels further developments of spectroscopic methods capable of providing molecular distance constraints without sample crystallization and, ideally, under conditions that are as close as possible to those found in cells. While many of these challenges can be successfully met with high-resolution NMR [18, 19], the method also has some limitations. Typically, the NMR structures are based on measurements of numerous short-range (<5–6 Å) constraints that are followed by computational structure optimization. Most of these constraints are obtained from ^{1}H–^{1}H NOEs (nuclear Overhauser effects) that report on through-space dipolar interactions. Unfortunately, a small number of proton contacts between different protein domains and partial averaging of dipolar interactions by dynamic effects make determination of interdomain orientation solely from NOEs a rather difficult task. Thus, researchers have focused on developing alternative approaches to derive interdomain distances and orientations. Some of these data could be obtained from residual dipolar coupling (RDC) NMR experiments [20, 21]. Others have exploited advantages offered by electronic spins that impose dipolar effects of much larger magnitudes than those of nuclear spins. Such experiments are based on selectively labeling a specific site of a biomolecule with a nitroxide (or employing an endogenous paramagnetic metal binding site) and using another spin to observe long-range interactions by ESR or NMR.

Recently, distance measurements in biological systems by ESR have been extensively reviewed [22]. Thus, we will focus on the key papers and the very recent developments in this area of nitroxide research.

4.2.1
Continuous Wave ESR of Nitroxide–Nitroxide Pairs

Continuous wave (CW) spectroscopy remains to be the most widely available ESR method for biomedical and biophysical studies. CW ESR is easy to use and it offers the best concentration sensitivity for diluted spin-labeled biological samples. Thus, many of the distance measurements between spin-labeled pairs are still carried out by CW ESR.

Generally, for two nitroxides attached to the same macromolecule, both spin exchange and dipolar interactions should be considered. However, through-space exchange integral is known to decay exponentially with the spin separation, r, [23] and, therefore, is expected to contribute negligibly for $r > 14$ Å [24]. Thus, for determination of long-range distance constraints we should rely on analysis of dipolar interactions. Such an analysis is typically carried out for two limiting cases: the rigid and the motion-narrowing limits. The fast-motion regimen for a spin system is defined as a stochastic process for which the anisotropic part of the time-dependent spin Hamiltonian, $H_1(t)$, is effectively averaged out. For a stochastic process with a correlation time τ_c the fast-motion condition can be written as:

$$|h^{-1}H_1(t)|\tau_c \leq 1 \tag{4.1}$$

Spin–spin dipolar interaction depends upon the interspin distance and the orientation of the interspin vector with respect to the external magnetic field. For many practical applications the interspin distance could be assumed to be time-independent and then the only stochastic mechanism that could average the dipolar field would be reorientation of the interspin vector in magnetic field arising from rotational tumbling of a spin-labeled macromolecule in solution. For such a process the fast motion condition (4.1) is:

$$\tau_c \leq \frac{2\hbar r^3}{3g^2\beta^2} \tag{4.2}$$

where g is the electronic g-factor and β is the Bohr magnetron. If rotational motion of the macromolecule bearing the spin pair is approximated as an isotropic and Redfield conditions are satisfied, the dipolar interaction would give rise to an additional spin relaxation characterized by the following T_{2dd} [25]:

$$\frac{1}{T_{2dd}} = \gamma^4 \hbar^2 S(S+1) \frac{\tau_c}{r^6}\left[3 + \frac{5}{1+\omega_S^2\tau_c^2} + \frac{2}{1+4\omega_S^2\tau_c^2}\right] \tag{4.3}$$

where τ_c is the correlation time, γ is magnetogyric ratio, and ω_s is the resonance frequency.

Under this fast-motion (or dynamic) model the dipolar effect would result in additional uniform Lorentzian broadening of CW ESR spectra. Such a broadening could be used to quantify the distance between two nitroxide spin labels attached to a protein [26]. This method has been described and demonstrated by Mchaourab et al. for a series of double cysteine mutants of a small globular protein T4 lysozyme [26]. The tumbling of this protein was assumed to be isotropic with the correlation time $\tau_c \approx 6$ ns. The dipolar effects were estimated by comparing CW ESR line widths of single and double-labeled mutants and relating the observed broadening $\Delta H_{dd} = 2\hbar(g\beta T_{2dd})^{-1}$ with the interspin distance r using (4.3) [26]. It has been shown that the distances calculated this way agreed well with the crystal structure data. Thus, this method is applicable to proteins in solutions and could be very useful in studying conformational states. It should be noted here that dynamic dipolar broadening effect falls off as $\Delta H_{dd} \sim r^{-6}\tau_c$ and therefore uncertainties in either τ_c or ΔH_{dd} would have only minor effects on the interspin distance r determined through such measurements. At the same time the broadening is expected to diminish very rapidly with distance, thus limiting the range to maximum of about 20 Å [26].

The range of measurable distances could be further extended under the slow motion conditions for the dipolar interaction (i.e. when inequality (4.2) is not satisfied). The largest effects are expected for the spectra in the rigid limit as initially considered by Pake for NMR [27]. In application to ESR, the earlier work has been reviewed by Likhtenstein [28]. The recent advances in this area involve computer simulations of dipolar broadening effects on nitroxide ESR line shapes in order to improve accuracy in distance determination.

One of the simplest approaches for analyzing dipolar effects in nitroxide pairs is to neglect the anisotropy of dipolar interactions in static spin pairs and to apply the Pake model originally developed for an isotropic (Gaussian) line. Then the spectrum $I(B)$ in the presence of dipolar interaction can be approximated as a convolution integral of a spectrum recorded in the absence of dipolar interaction $f(B)$ and the dipolar broadening function $G(B)$:

$$I(B) = \int f(B')G(B-B')dB' \qquad (4.4)$$

Such an approach has been initially described by Steinhoff et al. for simulating dipolar broadening effects on ESR spectra of nitroxides attached selectively to the hystedine 15 and to the lysines 13, 96, 97 of the same lysozyme molecule [29]. These authors also found that the best agreement between experimental ESR spectra broadened by the dipolar interaction and the convolution model is achieved when a Gaussian distribution of distances is assumed. Further justification of this approach to proteins has been provided using insulin as a model and by comparing interspin distances between nitroxides from ESR and the available x-ray insulin structure [30].

Independently, Rabenstein and Shin developed a Fourier deconvolution method to derive distances between two nitroxide labels attached to a macromolecule [31]. These authors suggested that an ESR spectrum from a nitroxide pair in a typical protein experiment could be approximated as a convolution integral (i.e. 4.4) of the spectrum of two non-interacting labels with the Pake dipolar pattern that is averaged out owing to a distribution of interspin distances and mutual spin orientations [31]. The method also allows for correcting for impurities of single-labeled proteins in the sample [31].

More recently, Altenbach et al. described a user-interactive convolution–deconvolution procedure to carry out empirical evaluation of the inter-residue distances in spin-labeled proteins under conditions of static dipolar interaction [32], that is, when inequality (4.2) is not satisfied [32]. Similar to the method of ref. [31] the Fourier deconvolution was carried out first, but then the recovered broadening function was interpreted as a normalized sum of Pake patterns. This interpretation was carried out with participation of a program user and the results were verified by direct convolution with the experimental spectrum [31].

It is worthwhile to note here that, although the modeling of the static dipolar interaction in the nitroxide pairs by convolution Equation 4.4 involving an ESR spectrum from non-interacting spins and a weighted/averaged Pake's dipolar profile gave good results for liquid solutions [32] and frozen samples of small peptides [31] and proteins [30], it is unlikely to be applicable for spin labels that adopt unique mutual orientations, exhibit narrow distributions of distances and orientations, or both [33]. In the latter case explicit spectral simulations such as developed by Hustedt et al. should be applied instead [34]. More recent work shows that distribution of nitroxide orientations according to a "tether-in-a-cone" model could also be included in the simulation [35]. Figure 4.1 illustrates an application of the "tether-in-a-cone" model to analysis of experimental ESR spectra of double-labeled T4 lysozyme mutants.

As has been noted in [36], one reason why the convolution methods for modeling the static dipolar interaction are so successful is that the nitroxide rigid limit ESR spectra at X-band (9.5 GHz) are dominated by the central feature, which is a superposition of A_{xx}, A_{yy}, g_{xx}, g_{yy}, and g_{zz} components. In these spectra only the A_{zz} components are clearly resolved, but the peak-to-peak amplitude of those is smaller than that of the central feature. Thus, the results of the convolution-based fitting or Fourier deconvolution will be dominated by the central component. The latter is approximately "isotropic" because it mainly contains the contributions from all the nitroxide orientations with respect to the magnetic field.

4.2.2
Continuous Wave ESR of Nitroxide–Metal Ion Pairs

When discussing dipolar interaction in nitroxide pairs in the preceding section we have considered rotational diffusion of the interspin vector as the main source of stochastic modulation of the dipolar interaction. The reason for this is that in

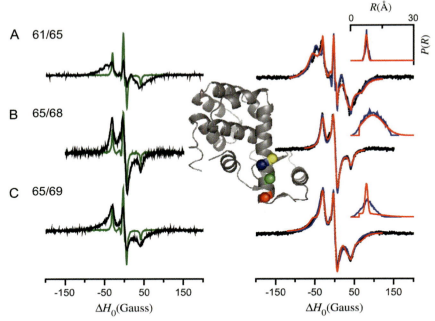

Figure 4.1 EPR spectra of spin-labeled T4 lysozyme mutants. Shown on the left are the sum of singles spectra (T4L 61 + T4L 65, T4L 65 + T4L 68, T4L 65 + T4L 69; green lines) with the spectra of the double cysteine mutants (T4L 61/65, T4L 65/68, T4L 65/69; black lines) overlaid on a fivefold expanded scale. The spectra of the double cysteine mutants are also shown on the right (black circles) with fits obtained to the tether-in-a-cone model (blue lines) and obtained by the convolution method (red lines) overlaid. The distance distributions obtained using both methods are plotted in the insets at the top right. All T4L spectra were measured at −30 °C in a desalting buffer containing 70% (w/w) glycerol. The inset at the center of the figure is a ribbon diagram representing the structure of T4L (PDB, 3LZM (40)) with the {alpha}-carbons of residues 61 (red), 65 (green), 68 (blue), and 69 (yellow) shown as colored spheres. (Reproduced with permission from [35]).

aqueous solutions rotational diffusion of proteins is known to occur on a much faster nanosecond timescale (e.g. rotational correlation time of T4 lysozyme is about 6 ns [26]) than the microsecond scale electronic relaxation time of nitroxides [37].

Under the same experimental conditions, electronic relaxation of metal ions (both T_{1e} and T_{2e}) is generally much shorter than that of organic free radicals [37]. For example, some of the longest reported spin-lattice electronic relaxation times for metal ions observed at room temperature are for Mn^{2+} and Cu^{2+} ions and range from ca. 1 ns to 8.4 ns (see Table 3 in [37] and references therein). Thus, even for these long-relaxing metal ions the dominant stochastic mechanism for modulation of dipolar field experienced by the nitroxide label in the nitroxide–metal ion pairs would be the electronic relaxation rather than the rotational diffusion of the interspin vector. For such a two-spin system, Leigh presented a theory to describe relaxation effects of a metal ion on the nitroxide spectra [38]. The earlier work involved applications of this theory to several proteins and protein cofactors containing indigenous paramagnetic metal sites as well as model spin-labeled metal ion complexes – all under the rigid lattice conditions achieved by sample freezing [39, 40].

More recently, Voss et al. described an application of the Leigh method to solutions of proteins under ambient conditions [41, 42]. The method has been verified using a series of T4 lysozyme mutants [41]. A metal binding site has been designed by introducing two histidine replacements in a T4 lysozyme helix and the nitroxide labels have been positioned at three different distances from that site. Analysis of the nitroxide ESR spectra in the presence and absence of Cu^{2+} has been carried out both for room-temperature and for frozen (−20 °C) protein solutions [41]. It has been shown that the distances determined from the frozen spectra using normalized amplitude analysis agree well with those calculated from the crystal structure when T_{1e} = 3 ns for the complexed Cu^{2+} ion was assumed. Distances measured at room temperature were in good agreement with those determined in frozen solutions, although they were systematically 10–20% shorter [41]. This discrepancy could be related to an increased flexibility of the nitroxide side chains with the temperature and also to an additional rotational modulation of the dipolar field from the protein tumbling in solution. Finally, we cannot exclude some decrease in T_{1e} of Cu^{2+} with temperature. Consequently, this method has been utilized for deriving distance constraints for membrane protein lactose permease of *Escherichia coli* [42] and metal transporter Nramp1 (natural resistance-associated macrophage protein 1) from *Mycobacterium leprae* [43]. It has been also suggested to employ the bound metal ion as a reference point for deriving the protein helix packing in membrane proteins by triangulation of distances measured to spin labels attached to single cysteine residues in each transmembrane helix [42].

So far the demonstrated distance range of CW ESR nitroxide–metal ion method was 10–23 Å [41–43] in accord with the theory predictions for Cu^{2+} [41]. A metal ion with a larger magnetic moment, a longer relaxation time, or both, is also expected to extend this range [41]. The first step in this direction has been recently reported [44]. Gd^{3+} is an s-state metal ion that has a large electron magnetic moment (S = 7/2) and a relatively long electronic relaxation time in aqueous solutions. Moreover, ESR line width of Gd^{3+} decreases with magnetic field owing to an increase in the electronic relaxation time [45, 46]. Although no direct measurements of the electronic relaxation of aqua Gd^{3+} ions and their complexes have been reported for liquid solutions, it was predicted to reach about 10–30 ns at magnetic field of 3.4 T corresponding to 94 GHz (W-band) ESR frequency [45, 47]. These electronic properties are considered to be very favorable for determination of long range distance constraints through measurements of dipolar effects on ESR spectra of nitroxide spin labels [44]. It could also be used to map potential locations of binding sites for metal ions and their complexes as has been demonstrated in [44] on example of binding a lipophilic MRI contrast agent Gd-DOTAP (1,4,7,10-tetraazacyclododecane-N-(n-pentyl)-N′,N″,N‴-triacetic acid) to human serum albumin (HSA). The fatty acid (FA) binding channel of HSA was labeled with a series of n-doxyl stearic acids and room temperature W-band ESR spectra were recorded in absence and presence of Gd-DOTAP [44]. It has been shown that the Gd-DOTAP effect on ESR spectra of n-doxyl stearic acids immobilized in the FA channel are well described as an additional homogeneous (Lorentzian) broadening (4.4) that could be analyzed in terms of Solomon–Bloembergen equation for the two unlike spins [44]. Such measurements indicated that the binding of Gd-DOTAP to HSA occurs in the vicinity of the main FA binding site [44].

4.2.3
Time Domain Magnetic Resonance of Spin Pairs

Time-domain methods provide several attractive alternatives to CW ESR spectroscopy for distance determination in nitroxide–nitroxide and nitroxide–metal ion pairs. For example, in pulse experiments the dipolar effects on the relaxation rates could be directly measured, whereas in CW ESR those changes are assessed indirectly usually through line-width or rollover power saturation curves. Moreover, in time-domain experiments manipulation of the spins could be designed in a way to suppress all but the interspin dipolar effects. This allows one to evaluate the magnitudes of dipolar interactions that are much smaller than the CW ESR linewidth. Such approaches significantly extend the range of measurable constrains and eliminate some ambiguities associated with CW ESR line-shape analysis.

4.2.3.1 Nitroxide Spin Labels in Protein Structure Determination by NMR

The dipolar field produced by an electronic spin of a nitroxide could be observed with the help of another spin belonging to a different nitroxide or a magnetic nucleus. In NMR the effect of nitroxide spin label is usually observed through changes in the relaxation rate and also from distance-dependent chemical shift [48]. The method of paramagnetic relaxation enhancement (PRE) has been known in NMR for many years to provide long-range distance constraints in proteins [49] but its application has been limited because many proteins are diamagnetic. With advent of SDSL researchers received a versatile tool to insert a relatively small nitroxide side chain into a specific site in the protein. Then the nitroxide could be used as a reference point for measuring a series of long-distance constraints to magnetic nuclei through PRE. The method has been demonstrated by determination of the global folds of two monocysteine derivatives of barnase, H102C and H102A0/Q15C. The proteins were ^{15}N-enriched and labeled with a cysteine-specific nitroxide spin-label MTSL (with 1-oxyl-2,2,5,5-tetramethyl-3-pyrroline-3-methyl) at the single (per protein) Cys residues [50]. ^{1}H longitudinal relaxation times of amides were measured for both paramagnetic (spin-labeled) and diamagnetic (i.e. reduced with ascorbic acid) protein samples to calculate the magnitude of PRE effects. The correlation time, τ_c, for the electron–amide proton vectors were calculated from the frequency dependence of PRE effects measured at 800, 600, and 500 MHz using the Solomon–Bloembergen approximation [50]. The values of τ_c ranged from 0.6 to 1.8 ns and were treated as an average correlation time of several stochastic processes, such as electronic relaxation, global protein tumbling, and chemical exchange, that are expected to contribute to modulation of the dipolar interaction. Distance constraints in the range of 8 Å to 35 Å were calculated from the magnitude of the paramagnetic contribution to the relaxation processes and the individual amide ^{1}H correlation times. These long-range constraints and known secondary structure restraints were employed to calculate the barnase global folds that were found to have backbone root–mean-square deviations <3 Å from the crystal structure. This work demonstrates the utility of a limited number of paramagnetic distance constraints for rapid determination of the overall protein topology from only PRE effects [50].

More recently the site-directed spin-labeling PRE approach has been further critically evaluated as a general method for obtaining long-range distance constraints for integral membrane proteins by solution NMR methods [12]. On example of the outer membrane protein A (OmpA) it was found that in a lipid-like environment the nitroxide spin-labels are difficult to reduce completely and this could lead to erroneous distances [12]. In order to avoid these complications a parallel labeling with a diamagnetic analog of MTSL has been described. Such an approach has been proven to yield more reliable measurements of the PRE effects. It has been shown that parallel SDSL and PRE can be employed to successfully refine OmpA structures using a sufficient number of strategically placed nitroxides. Moreover, it has even been possible to obtain structures of reasonable quality from PRE distances only, in the absence of any NOEs [12].

The nitroxide spin-labeling approach has also been shown to be useful for mapping protein binding sites (for earlier examples see refs. in [48, 49]) and for studying DNA-protein complexes where the need for long-range distance constraints is especially critical. Some very recent examples of combining NMR and spin-labeling methods for studying ligand binding have been provided by Prestegard and coworkers [51]. These authors prepared a nitroxide-labeled analog of N-acetyllactosamine and mapped the binding sites of this small oligosaccharide on the surface of galectin-3, a mammalian lectin of a 26 kD size [51]. Identification of protein residues proximate to the binding site for this oligosaccharide has been carried out from perturbation of intensities of cross-peaks in the ^{15}N heteronuclear single quantum coherence (HSQC) spectrum of the full-length galectin-3. An important contribution of this work is in describing a protocol that could be employed as a part of a drug design strategy, in which subsequent perturbation of chemical shifts of distance mapped amide cross-peaks can be used effectively to screen a library of compounds for other ligands that bind to the target protein at distances suitable for chemical linkage to the primary ligand. The main advantage of such an approach is that it bypasses the need for structure determination and resonance assignment of the target protein [51]. Another recent example was provided by Cai *et al.*, who described determination of three-dimensional structure of the Mrf2-DNA complex by combining DNA-protein docking calculations with experimental PRE constraints derived from DNA labeled with a nitroxide at a single site [52]. Such an approach could be very useful in studying other protein complexes with limited number of intermolecular contacts such as RNA-protein complexes.

It is expected that the number of NMR structural studies that make use of the PRE effects from spin-labels for deriving long-range distance constraints would continue to grow and also to expand to solid-state NMR. Very recently Jaroniec and coauthors reported on deriving distances ranging from ca. 10 Å to 20 Å from magic-angle spinning (MAS) solid-state nuclear magnetic resonance (ssNMR) spectroscopy of ^{13}C, ^{15}N-entiched B1 immunoglobulin-binding domain of protein G (GB1) [53]. While the PRE distance constraints obtained in [53] were still considered to be rather qualitative than quantitative at this stage, these restraints provide valuable information about the protein fold on length scales inaccessible to traditional ssNMR. Further improvements of this method are expected through more accurate site-resolved measurements of the nuclear relaxation rates (using 3D or 4D pulse schemes incorporating variable relaxation delays) as well as additional data on electronic correlation times and possible conformations of spin-labeled side chains.

4.2.3.2 Electronic Relaxation Enhancement in Spin Pairs

Similar to PRE NMR, relaxation enhancement effects from a spin-label and/or paramagnetic metal ion could also be detected by another electronic spin from changes in the relaxation rate. Initial work was carried by the Novosibirsk group in the late 1970s and focused on analysis of effects of dipolar interactions on the shape and time constants of the electronic spin echo [54, 55]. At first, these methods have been initially applied to a distribution of radicals [54, 55] and have then been extended to studying immersion depth of paramagnetic centers in biological systems (reviewed in this book and in ref. [56]). The relaxation enhancement method has also been employed to study distances between indigenous paramagnetic centers in metalloproteins and other spins such as nitroxides (reviewed in [57]).

With the exception of electronic spin immersion or accessibility studies of membrane protein (e.g. [58, 59]), ESR relaxation enhancement experiments are typically carried out with frozen samples at low temperatures. However, recent technological and methodological advances in Fourier transform ESR enabled routine measurement of free induction decay signals from spin-labeled macromolecules, even in aqueous solutions at ambient temperatures. Saxena and coworkers demonstrated that new commercial pulsed ESR spectrometers can be successfully employed for measuring metal ion to nitroxide distances through enhancement in the nitroxide relaxation rate at room temperatures [60]. Specifically, alanine-based (PPHGGGWPAAAAKAAAAKCAAAAKA) and proline-based (PPH-GGGWPPPPPPPPCPPK) peptides were designed to contain a Cu^{2+} binding PHGGGW sequence at one end while the single cysteine at another end was labeled with the nitroxide MTSL [60]. Pulsed inversion recovery ESR experiments have been carried out in the presence and absence of Cu^{2+} as a function of temperature and related to the thermal unfolding of the peptides' α-helix. This work demonstrated that distances up to about 25 Å are accessible by such a method [60].

It should be noted that while pulsed methods (both NMR and ESR) that are based on measurements of paramagnetic enhancement extend the range of accessible distances over that of CW ESR, they all require the baseline measurements of relaxation rates in absence of the paramagnetic relaxer(s). The same problem exists in CW ESR methods which are based on the line-shape analysis: very often a control spectrum from a single-labeled protein is required. Thus, one has to prepare and label both single- and double-cysteine mutants of the same proteins. For high-resolution NMR experiments the best solution for the reference diamagnetic spectrum, perhaps, involves a parallel labeling with a diamagnetic analog of the nitroxide MTSL [12].

In experiments with metalloproteins we could substitute a paramagnetic metal ion with a diamagnetic analog (e.g. Co^{2+} for Zn^{2+} and Mn^{2+} for Mg^{2+}). Nevertheless, these extra steps in preparation and experiments with additional samples are both time-consuming and undesirable because of concerns of changing protein stability and/or affecting conformational states. Thus, it is highly desirable to obtain the distance information from a single experiment with double-labeled proteins. A brief overview of such methods and recent developments is provided in the next section.

4.2.3.3 Pulsed Double Electron-Electron Resonance

Pulsed double-electron electron resonance (DEER) is an example of an ESR method that, after having been initially available only at a few laboratories in the world, has become, perhaps, one of the most popular tools in structural protein research by SDSL. Initial work on development of pulsed ESR methods for measuring the magnitude of static dipolar interactions has been carried out at the Institute of Chemical Kinetics and Combustion in Novosibirsk, Russia (for comprehensive reviews see, e.g. [3, 61]). This work led to a development of a so-called "2 + 1" pulse sequence that results in an electron spin echo (ESE) being modulated by the static dipolar interactions (e.g. see [61] and references therein). This three-pulse sequence was further improved by addition of another $\pi/2$ pump pulse and formation of a pulse train for improved sensitivity (Figure 4.2) [62]. The use of deuterated solvents and varying the echo time improved sensitivity of the DEER method even further [63].

Nowadays the four-pulse DEER experiments could be carried out with a commercial Bruker Biospin (Germany) E 580–400 pulse-ELDOR accessory and analyzed by computer-assisted modeling methods (e.g. see ref. [64] for practical introduction to DEER). This method is becoming very popular among biophysicists as it enables them to obtain both reliable long-range (generally, up to about 75 Å) distance constraints for complex protein systems and also to evaluate existence of multiple distance populations and distance distributions. Such data are of critical value for understanding structural organization of large protein assemblies and membrane protein complexes. Recent developments include demonstration of the DEER method for deriving accurate long-range distance constraints for DNA and RNA molecules [65, 66]. Figure 4.3 illustrates measurement of long-range distance constraints in a large DNA that was selectively labeled with

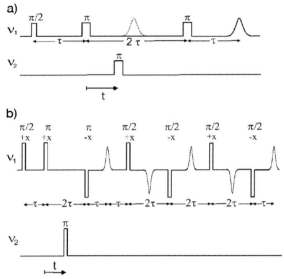

Figure 4.2 Pulse patterns for dead-time-free DEER.
(a) Four-pulse DEER. The time *t* between the second and third pulse of the sequence is incremented in steps. (b) Pulse-train DEER. (Reproduced with permission from [62]).

two nitroxides by DEER [65]. Figure 4.3b (left) shows original ESE signal that is modulated by static dipolar interaction. The DEER signal is corrected by subtracting intramolecular dipolar contribution and analyzed to reveal the distance distributions between the two nitroxide labels (Figure 4.3b).

While some distribution of distances between two spin labels arises from inherent flexibility of biopolymers, the rest of the uncertainty is simply caused by the flexibility of the nitroxide tether that could adopt several conformations with respect to, for example, the protein backbone. The latter uncertainty arises in all the NMR and ESR distance measurement methods that are based on labeling biomolecules with flexible nitroxide chains. This uncertainty could be reduced and, thus, the accuracy of distance measurements could be improved through modeling the nitroxide side chains using the "tether-in-a-cone" approach we discussed [35] or through molecular dynamics simulations of the nitroxide side chain conformations [67]. It has been shown that the latter approach improves agreement between the measured and modeled distances by a factor of 2 across the full range, having a mean error of only 3 Å, but, more importantly, the correlation increases by a factor of 4 over the shorter distance range [67]. Further refinements of molecular dynamics methods are discussed in Section 4.5.1.2 of this chapter and improvements in models of dipolar interaction are expected to improve accuracy even further.

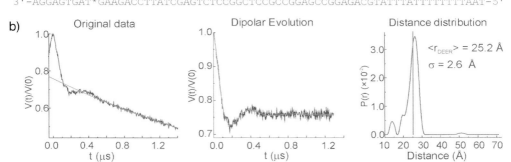

Figure 4.3 DEER measurement in a large DNA. (a) The sequence of a 68 bp duplex containing the entire DNA origin of replication in SV40, with "*" indicates the R5 labeling sites. (b) DEER data. (Reproduced with permission from [65]).

4.2.3.4 ESR Double Quantum Coherence Experiments

Another way to directly extract dipolar interaction from a spin-pair system is a double quantum coherence (DQC) experiment. DQM and multiple-quantum (MQC) experiments are well-known NMR methods [10]. However, in application to ESR these methods are rather difficult to employ because of very demanding requirements of this technique on the resonator bandwidth and the spectrometer dead time. Initial work of Saxena and Freed demonstrated that DQC could be generated for pairs of nitroxides by using "forbidden" coherence passways [68, 69]. Further developments of pulsed ESR instrumentation at the ACERT ESR Center led to much shorter dead-times, increased B_1 without compromising the resonator bandwidth, increased signal-to-noise ratio [70], and made it possible to directly produce MQC by allowed passways [11]. The main advantage of the DQC method is in producing experimental signals that are free of any undesirable contributions. These signals yield Pake doublets that can be effectively used to determine the broad range of internitroxide distances.

The DQC method has been verified for a variety of systems (e.g. [71] and refs. therein) including a series of spin-labeled T4 lysozyme double cysteine mutants [72]. The demonstrated distance range was from c. 20 Å to 50 Å [72], which is similar to that demonstrated by DEER. In order to increase the DQC distance range a variant of this method has been developed [73]. This variant employs the same six-pulse sequence, but in a format which simply refocuses the primary echo after it is passed through a double quantum filter. The method was applied to a long (26 base pairs) double-stranded A-type RNA that was spin-labeled at both ends. The distance measured by DQC was 72 ± 4 Å that agrees well with 70 ± 5 Å estimated from molecular modeling [73]. Note that the main source of uncertainty in molecular modeling was the flexibility of the nitroxide linker employed.

Further details on the DQC method and examples of applications could be found in several reviews [71, 74, 75].

4.2.4
Distance and Angular Constraints by ESR of Spin Pairs at High Magnetic Fields

So far we have focused on development of ESR methods for deriving accurate distances and distance distributions between spin-labeled side-chains of biomolecules. However, in principle, spin-labeling ESR is also capable of providing information on orientation of the nitroxide rings. In cases when the nitroxide attachment is rigid and/or nitroxide ring orientation could be related to the geometry of, for example, protein backbone; such information is unique among other spectroscopic methods and considered to be extremely valuable for understanding molecular mechanisms of protein function.

The reason for angular selectivity of nitroxide ESR spectra arises from the anisotropy of magnetic interactions. It should be noted here that at magnetic fields of conventional X-band (9–10 GHz) ESR the nitroxide spectra are mainly determined by the axial nitrogen hyperfine coupling tensor A, while contributions from the rhombic g-matrix are only moderate. With increase in the magnetic field the contribution of the Zeeman term grows proportionally and at the fields corresponding to W-band (95 GHz) the anisotropy of the g-matrix prevails over that of the hyperfine tensor A. This leads to greatly improved angular selection of the method.

Perhaps, some of the first studies of nitroxide pairs with high field (HF) ESR have been carried out by Lebedev and co-workers in Moscow [76, 77]. The Moscow researchers noted that, although the number of spectral components of the rigid-limit 140 GHz ESR spectra of a nitroxide biradical is greater than at conventional X-band, the spectrum is still easier to interpret because of the excellent angular resolution which allowed for sorting out the spectral components and the corresponding anisotropic dipolar couplings [77]. Basically, for well-resolved spectra, the initial estimates for the components of the dipolar tensor may be deduced from measurements of dipolar splittings directly from the spectrum.

The angular selection of dipolar spin-labeling HF ESR was further explored by Hustedt and coworkers in a study of ^{15}N-spin-labeled coenzyme NAD$^+$ (SL-NAD$^+$) bound to a microcrytalline, tetrameric glyceraldehyde-3-phosphate dehydrogenase (GAPDH) [34]. For such a complex the spin-label tethers adopt unique conformations and remain virtually immobilized with respect to the protein. ESR spectra were recorded at 4°C at X-, Q-, and W-bands to be fit simultaneously with a least-squares software package developed at Vanderbilt University. Such rigorous spectral analysis yielded both the interspin distance and the mutual orientations of the protein domains [34].

It should be noted here that the distance range of HF CW ESR method is essentially the same as at X-band assuming that the rigid-limit nitroxide line-width is determined solely by unresolved hyperfine interactions. Similar to X-band ESR we would expect that development of pulsed methodologies at high fields should lift this limitation but, at the same time, permit determination of additional structural information – mutual orientation of two radicals. The first step in this direction was undertaken by Bennati and coworkers, who reported on high-field (180 GHz) pulsed electron–electron double resonance (ELDOR) experiments to derive the orientation of the tyrosyl radicals in ribonucleotide reductase from *Escherichia coli* [78]. Specifically, these authors observed a correlation between the orientation-dependent dipolar interaction and their resolved components of the *g*-matrices. These data allowed them to derive mutual orientations of two radicals embedded in the active homodimeric protein in solution [78]. Although tyrosyl radicals are not nitroxides, the anisotropy of *g*-matrix that determines angular selection at high magnetic fields is essentially the same as for nitroxide radicals in aqueous media. Thus, the new HF ESR pulsed ELDOR method could also be applied to spin-labeled biomolecules providing new type of structural constraint – a mutual orientation of two nitroxide rings that are coupled by the dipolar field.

4.3
Multiquantum ESR

Electronic relaxation times of nitroxide spin labels are very informative spectral parameters that report directly on immediate molecular environment, nitroxide dynamics, and spin-spin interactions. Quantitative measurements of nitroxide spin-spin interactions are the basis of several biophysical methods – most notably for determination of long-range distance constraints reviewed in the preceding section of this chapter and for assessing the immersion depth of the nitroxide (i.e. molecular accessibility experiment). Typically, electronic spin-lattice relaxation time T_1 of nitroxides is evaluated from rollover power saturation CW ESR or from pulsed saturation recovery experiments. Recently, it has been demonstrated that qualitative data on nitroxide-labeled proteins could also be obtained from multiquantum (MQ) ESR [79].

MQ ESR is based on irradiating a homogeneous ESR transition by two closely spaced microwave sources of equal intensity and a common time-base. Such a two frequency irradiation generates new microwave frequencies, termed intermodulation sidebands, when the resonance condition is satisfied. An odd number of quanta are involved in this process, with one more quantum absorbed than emitted, which leads to the nomenclature multiquantum ESR spectroscopy. A third microwave reference frequency on the same time-base is used for detection of the intermodulation sidebands [79–81].

Some of the useful features of MQ ESR are in providing ESR signal without magnetic field modulation in both pure absorption and dispersion displays [79–81]. Another advantage of the method is that its spectral displays are proportional to the relaxation parameters $T_1 T_2^2$ for three-quantum (3Q) and $T_1^2 T_2^3$ for five-quantum (5Q) spectra [81]. This exceptional sensitivity of MQ ESR displays to spectral parameters have been exploited in a recent study of spin-labeled arrestin mutant K267C using a newly constructed MQ spectrometer operating at Q-band (35 GHz)

[79]. Enhanced sensitivity of the MQ ESR displays to nitroxide T_1 and T_2 was evident as well as the presence of two states of the label with different relaxation times [79]. This work demonstrated for the first time the feasibility of utilizing MQ ESR for site-directed spin-labeling studies of biologically relevant samples. It is anticipated that future development of MQ ESR would yield quantitative data on electronic relaxation parameters that would be useful assets in spin-labeling studies of protein structure and dynamics.

4.4
Spin-Labeling ESR of Macroscopically Aligned Lipid Bilayers and Membrane Proteins

Substrate-supported phospholipid bilayers are of interest for many reasons. Firstly, such bilayers represent a convenient and versatile model of cellular membranes [82, 83]. Secondly, many inorganic substrates can be biofunctionalized by self-assembling lipid bilayers on the surfaces [83–85]. In application to spin-labeling ESR substrate-supported bilayers provide macroscopic alignment of phospholipids and embedded membrane proteins. Because nitroxide magnetic parameters are anisotropic, such a macroscopic alignment decreases the range of angles over which these parameters are averaged by static and dynamic processes resulting in an increased intensity of observed ESR lines. More importantly, such a macroscopic alignment allows for determination of orientation of nitroxides with respect to the alignment axis. This puts spin labeling ESR in a unique position among other spectroscopic methods because only a few of these techniques are capable of determination of molecular orientations in (partially) aligned samples. Currently, there is a particular interest in studying orientations of biological molecules such as peptides and proteins with respect to membrane surfaces. Additionally, there is a growing interest in inferring the orientations of proteins and peptides imprinted on the surface of hybrid nanoscale devices. Here we review the recent progress in developing lipid bilayer alignment methods and the use of those for studying lipid bilayers and membrane proteins.

4.4.1
Mechanical Alignment of Lipid Bilayers on Planar Surfaces

Since the initial reports on formation of macroscopically aligned lipid multibilayers on glass slides by spin-labeling ESR [86] and the consequent studies of McConnell and coworkers [82, 87], glass substrates became the most commonly used platform for stabilizing lipid bilayers. Typically, several hundreds or even thousands of lipid bilayers are deposited on top of each other onto a surface of a glass slide providing sufficient quantity of spin-labeled molecules for ESR studies. It should be noted here that with advances in commercial ESR instrumentation and improvements in spectrometer sensitivity, particularly, at X-band, it is becoming feasible to study even single monolayers of spin-labeled proteins absorbed on planar surfaces. For example, Risse, Hubbell and coworkers reported on ESR studies of a series of spin-labeled T4 lysozyme cysteine mutants selectively tethered via a His-tag to the chelating headgroups (NTA Ni) of planar quartz-supported lipid bilayers [88] or non-specifically absorbed in the partially unfolded state on quartz [89]. In the first case a vectorially oriented ensemble of proteins on the surface that gives rise to angular-dependent electron paramagnetic resonance spectra has been formed [88]. These studies demonstrated utility of site-directed spin labeling methods for studying secondary and tertiary structure of adsorbed proteins in monolayers and also protein orientation with respect to the surface [88].

It is worthwhile to note here that such studies are only possible with state-of-the-art spherical SHQ (super-high-Q) resonator from Bruker Biospin (Germany) and upon achieving essentially complete coverage of the surface with a monolayer of a small protein. Even at such conditions the spectra were accumulated for up to 800 min in less favorable cases [88]. Clearly, such long data acquisition times may be undesirable for high throughput measurements and/or due to problems with protein stability. Thus, researchers have focused on development of efficient methods of fabricating well-aligned multilayer structures on planar support. One of the methods to produce such samples is isopotential spin-dry ultracentrifugation (ISDU) [90]. It involves sedimentation of the membrane fragments (in the gel phase) with simultaneous evaporation of the water phase in a vacuum ultracentrifuge. One of the examples of the utility of this method has been provided by a study of a membrane ion channel gramicidin A in a lipid bilayer environment of various compositions [91]. The method has been also utilized for preparing macroscopically aligned lipid bilayers for HF ESR at 250 GHz [92]. The bilayers were composed of mixtures of DMPC and 1,2-dimyristoyl-*sn*-glycero-3-[phospho-L-serine] (DMPS) and doped with a cholestane spin probe which served as a cholesterol analog [92]. Cholestane is a so-called "γ-axis" membrane probe because the nitroxide molecular y-axis approximately coincides with the long axis of this molecule. In bilayers composed of phosphatidylcholine lipids (such as DMPC) this molecule aligns with the lipids and the nitroxide y-axis becomes approximately parallel to the bilayer director. Then the 250 GHz ESR spectrum from macroscopically aligned lipid bilayers with director vector parallel to the magnetic field should resemble that of a single crystal when the field is directed along the nitroxide y-axis. The exceptional angular resolution of 250 GHz ESR permitted detailed study of effects of bilayer composition on orientation this spin-labeled cholesterol analog. Specifically, upon increasing the fraction of DMPS in DMPC significant changes in 250 GHz ESR spectra were observed and interpreted in terms of a strong local biaxial environment. This biaxiality was initially predicted from molecular dynamics simulations but spin-labeling 250 GHz ESR provided the first experimental evidence for such cholesterol behavior [92]. It should be noted here that orientational resolution of conventional X-band ESR was insufficient for unambiguous data interpretation for such a lipid system.

Further advances in preparing mechanically-aligned lipid bilayers have been reported by the researchers of the ACERT ESR Center at Cornell [93, 94]. Specifically, the Cornell group has developed a technique with which one can prepare individual samples that are oriented at a range of fixed angles with respect to the applied magnetic field, yet can be positioned and aligned for minimum dielectric losses in standard ESR resonators. The basis of this technique lays in preparing a standard oriented lipid bilayer using conventional isopotential spin-dry ultracentrifugation, freezing the sample and then taking thin slices at a specified angle from the sample normal. The slices are taken with a scalpel under the control of a fine mechanical mechanism. This method expands the angular range at which ESR spectra of spin-labeled samples could be collected using optimal (in term of sensitivity) geometry of the overall planar sample/sample holder in the ESR resonator [93, 94].

4.4.2
Alignment of Discoidal Bilayered Micelles by Magnetic Forces

Solid-state NMR (ssNMR) is among the other methods that benefit from macroscopical alignment of lipid bialyers and membrane protein samples. For ssNMR the gain in spectral resolution that could be achieved by employing the samples that macroscopically aligned with respect to the magnetic field, B_0, is even greater than in spin-labeling ESR. This justifies the efforts in preparing samples with an exceptional degree of alignment. Currently, NMR alignment methods are based on sandwiching several hundreds of lipid bilayers between ultra thin glass plates [95] or by using magnetic interactions to align bilayer discs composed of a mixture of long- and short-chain phospholipids (bicelles) in the external magnetic field [96, 97]. Typically, these discoids – also called bicelles – are made from circa 3:1 mixture of a long-chain 1,2-dimyristoyl-*sn*-glycero-3-phosphocholine (DMPC) and a short-chain 1,2-dihexanoyl-*sn*-glycero-3-phosphocholine (DHPC).

Magnetic alignment forces are proportional to B_0^2 and, therefore, bicelles would align only at magnetic fields that exceed the threshold. Typically, bicelles undergo spontaneously macroscopic ordering at magnetic fields above 2.5 T and temperatures of 37–40 °C [98]. These conditions are suitable for W-band ESR as the resonant field for nitroxide at this frequency is 3.4–3.5 T; that is, sufficient for bicelles' alignment. Such an ESR experiment has been first demonstrated by Mangels, Harper, Smirnov, Howard, and Lorigan in a W-band study of order parameter of a cholestane spin probe in mixed DMPC: bicelles [99]. These authors determined that the order parameter for cholestane in DMPC/DHPC bicelles containing 10% of cholesterol is S = 0.64 ± 0.04 which is very similar to that determined for mechanically aligned DMPC bilayers. W-band EPR spectra of 5-doxyl steric acid in DMPC/DHPC bicelles at two orientations also have been reported [100].

In ongoing efforts Lorigan and coworkers extended the method of magnetic field alignment to X-band ESR by utilizing lanthanides to increase magnetic anisotropy of bicelles and, therefore, the magnitude of magnetic alignment forces [101, 102]. In order to initiate the alignment the magnetic field was ramped up to 0.72 T and then lowered down to circa 0.34 T in order to observe nitroxide ESR signals at X-band [101, 102]. The Lorigan group has also carried out a series of studies of nitroxide-labeled fatty acids and cholesterol analogues in bicelles and searched for optimized conditions of bicelle alignment at X- and Q-bands [103–107]. More recently, this group reported on measuring the tilt of the pore-lining transmembrane domain (M2δ) of the nicotinic acetylcholine receptor (AChR) that contained a rigid unnatural amino acid 2,2,6,6-tetramethylpiperidine-1-oxyl-4-amino-4-carboxylic acid (TOAC) incorporated into the peptide backbone [108]. The helical tilt was directly calculated from the orientation-dependent hyperfine splitting of the TOAC ESR spectra. This method could be very advantageous in studies of membrane peptides because ESR is circa 1000 fold more sensitive than ^{15}N ssNMR. Thus, it is feasible to determine the helical tilt of an integral membrane peptide by EPR of just 0.1 mg of spin-labeled sample. The helical tilt can be determined more accurately by placing TOAC at several backbone positions [108].

4.4.3
Nanopore-Confined Cylindrical Bilayers

Recently, another type of substrate-supported lipid bilayers that provides an attractive way for aligning membrane protein samples for magnetic resonance experiments has been introduced [109]. In brief, it has been demonstrated that lipids assemble into nanotubular bilayers when placed inside nanoporous anodic aluminum oxide membranes (AAO). These structures – which are called lipid nanotube arrays – have a high density of the nanoporous channels, thus, providing at least a 600 fold gain in the bilayer surface area for a similar sized planar substrate chip. It has also been shown that these new substrate-supported bilayers retain many biophysical properties of unsupported bilayers [110, 111] and they are suitable for aligning membrane proteins for high resolution 2D ssNMR studies [112]. Recent ssNMR [112–116] and DSC [110, 111] studies provided detailed characterizations of lipid properties in such nanotubular bilayers. Figure 4.4 shows a cartoon of a

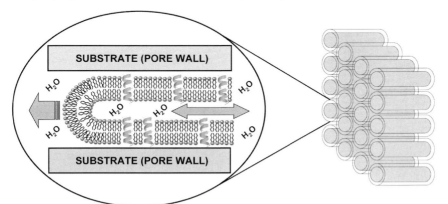

Figure 4.4 A cartoon of a single lipid bilayer aligned inside a nanopore as a multilamellar aqueous dispersion of phospholipids is drawn inside the pores by capillary action. Arrayed structure of nanoporous channels of Anodic Aluminum Oxide membranes is illustrated on the right. (Reproduced with permission from [62]).

single lipid bilayer aligned inside a nanopore. For nanotubular bilayers formed from synthetic DMPC lipids both leaflets were found to be fully accessible to water-soluble molecules [116]. Moreover, high hydration levels of these structures as well as pH and desirable ion and/or drug concentrations can be easily maintained allowing for structure-function studies of membrane proteins by ssNMR spectroscopy. The first demonstration of such experiments at high magnetic field (19.6 T) using ^{17}O NMR anisotropic chemical shift effects of ion binding to the gramicidin A channel has been reported [116].

Initial demonstration of self-assembly of phospholipids into cylindrical structures when confined to nanoporous AAO channels was obtained with spin-labeling ESR at 95 GHz [109]. Because the local order parameter S of the lipid fatty acid chains is known to decrease progressively towards the bilayer center, 1-palmitoyl-2-stearoyl-(5-doxyl)-sn-glycero-3-phosphocholine (5PC) lipids containing nitroxide moiety at position 5 were chosen for the experiments. Dynamic lipid disorder and partial averaging of spectral anisotropies were further reduced by taking ESR spectra at a low temperature (150 K) at which dynamics of the phospholipids is approaching the rigid limit. The largest changes in 5PC spectra upon reorientation of the AAO substrate in the magnetic field were observed at 95 GHz (W-band) because of an enhanced angular resolution of high-field ESR over conventional X-band. The relative intensities of characteristic peaks of these spectra, Figure 4.5, are clearly different for two orientations of the substrate in the magnetic field. Particularly noticeable are the changes in the g_z-region (i.e. high-field component which spreads from ca. 3.3850 to 3.3940 T): when the surface of the AAO substrate is perpendicular to the magnetic field (bottom spectrum), the z-component almost

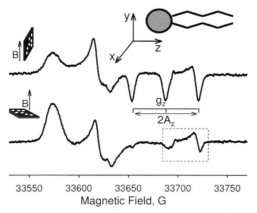

Figure 4.5 Experimental rigid limit (T = 150 K) high resolution 94.4 GHz (W-band) EPR spectra of AAO substrate with deposited DMPC: 5PC (100:1 molar ratio) at two orientations of the substrate surface in the magnetic field. A cartoon on the top shows orientations of the magnetic axes with respect to the phospholipid. Note that the bottom EPR spectrum has a low intensity in the g_z-region (the feature in the dashed box is due to a paramagnetic AAO impurity) indicating that at this substrate orientation the lipid chains are perpendicular to the magnetic field. (Reproduced with permission from [100]).

completely disappears (the signals inside the dashed box are mainly due to paramagnetic impurities in the AAO substrate). This means that at this substrate orientation only a very small fraction of molecules has the z-axis of the N–O frame aligned with the external magnetic field. Thus, it must be concluded that a majority of the phospholipids inside the nanopores (since the surface phospholipids were mechanically removed during sample preparation) are positioned with their long axis perpendicular to the magnetic field and therefore perpendicular to the direction of pores. Detailed ssNMR studies of nanotubular bilayers [114] confirmed these initial spin-labeling ESR findings.

It appears that lipid nanotube arrays have several important advantages over conventional mechanically aligned planar bilayers. Among those advantages are very high surface area, long-term stability of aligned lipid assemblies under exceptionally wide range of temperatures, pH, and salt concentration, high hydration level of lipid bilayers, and protection from surface contaminations. Another feature is the excellent accessibility of the bilayer surface for exposure to solute molecules. This makes it possible to expose membrane protein samples to different solution media and to repeat this procedure multiple times with the sample. In addition, fully hydrated membrane proteins could be exposed, for example, to buffers at various pH, ion and drug concentrations facilitating a wider range of structural and functional studies in the native-like environment by means of magnetic resonance (as demonstrated in ref. [116]) and spin-labeling ESR. The lipid nanotube arrays could also be used to derive the helical tilt of transmembrane peptides containing rigid unnatural amino acid TOAC [117].

4.5
Spin-labeling ESR at High Magnetic Fields

While applications of nitroxides in chemical, biomedical, and materials science fields continue to grow at a rather rapid pace, the majority of ESR experiments with nitroxides are still carried out at X-band (9–10 GHz). However, high-field or high-frequency spin-labeling ESR utilizing microwaves above ca. 90 GHz are uniquely positioned to offer new quantitative and qualitative information on spin-labeled species. With an increase in magnetic field the ESR spectra from nitroxides undergo significant changes. The main transformations in shapes of the spectra occur at the resonant magnetic fields of ca. 1.1 T. Above those fields the magnitude of the Zeeman anisotropy in the spin Hamiltonian prevails over the anisotropy of the nitrogen hyperfine coupling interaction. With a further increase in the magnetic field or frequency the g-factor resolution continues to increase. Thus, while at the magnetic fields of conventional X-band ESR (0.3 T, 9 GHz) the nitroxide spectrum is determined by an axial hyperfine term and an averaging from molecular motion, above 3.4 T (95 GHz, W-band) the rhombic Zeeman term becomes dominant providing new information on protein structure and dynamics that is inaccessible by traditional ESR spectroscopy.

The technical problems of developing HF ESR were first addressed successfully by Professor Yakov S. Lebedev and coworkers from the Institute of Chemical

Physics (Moscow, Russia). Using a 140 GHz (5 Tesla) ESR spectrometer, they were clearly the first who explored the enhanced sensitivity of spin-label high-field ESR spectra to molecular motion, local polarity, and pioneered many other useful applications (see ref. [118] for a review). Nowadays, HF ESR is undergoing rapid development powered by the efforts of many research groups to advance the field. At the same time, advances in digital electronics, microwave, and mm-wave devices that are driven by demands of telecommunication industry, provide researchers in the HF ESR field with new high-performance components, including powerful and stable mm-wave sources, switches, low-noise amplifiers, and various low-noise passive components. At the same time recent innovations in cryocooling and superconducting magnet technologies produced the first cryogen-free magnets that require no refilling service with liquid cryogens and could provide accurate scanning of magnetic field of up to, ca. 12.5 T [119]. These recent development created unprecedented opportunities for advancing HF ESR instrumentation beyond recent limits and making HF ESR instrumentation more widely available.

In view of this author, high-field ESR in general, and its application to proteins and other complex biomolecules and assemblies in particular, have tremendous potential and we confidently expect to see ever-increasing activity in the future. A comprehensive volume describing various activities in HF ESR, including nitroxide studies has been published [120]. An introduction to spin-labeling HF ESR in membrane and protein biophysics is also available [100]. Here, we provide a brief overview of recent development of HF ESR in application to nitroxides that are of interest to the author of this chapter.

4.5.1
Spin-Labeling HF and Multifrequency ESR in Studying Molecular Dynamics

One of the principal features of spin-labeling ESR methods and, in particular, HF ESR, is its ability to provide detailed information on complex rotational motion of spin-labeled molecules. Let us consider a molecule bearing an electronic spin and undergoing a random motion. One of the results of such a motion is a stochastic reorientation of a molecule with respect to the external magnetic field. Thus, for such a molecule the polar angles $\{\theta, \phi\}$ between the molecular frame and the external (laboratory) magnetic field are not constants but time-dependent functions: $\theta = \theta(t)$ and $\phi = \phi(t)$. Analogously, the spin Hamiltonian for such a system becomes time-dependent because magnetic parameters are replaced by time-dependent matrices: $g = g(t)$ and $A = A(t)$. It is convenient to break such a time-dependent spin Hamiltonian into an isotropic H_0 and an anisotropic $H_1(t)$ part:

$$H = H_0 + H_1(t) \tag{4.5}$$

The anisotropic part is given by:

$$H_1(t) = \mu_B \vec{B} \cdot [g(t) - g_{iso}] \cdot S + hS \cdot [A(t) - A_{iso}] \cdot I \tag{4.6}$$

where $\hat{S}$ and $\hat{I}$ are electronic and nuclear spin operators, respectively, and the isotropic parameters are defined as:

$$g_{iso} = \frac{1}{3}Tr\{g\} = \frac{1}{3}(g_{xx} + g_{yy} + g_{zz}) \quad (4.7)$$

$$A_{iso} = \frac{1}{3}Tr\{A\} = \frac{1}{3}(A_{xx} + A_{yy} + A_{zz}) \quad (4.8)$$

whereas the anisotropic time-dependent matrices $g(t)$ and $A(t)$ are traceless.

In liquid solutions spin-labeled fragments of biomolecules undergo a complex reorientational motion. While the isotropic part of the spin Hamiltonian is invariant to such fluctuations, the anisotropic part $H_1(t)$ fluctuates in time. Such fluctuations modulate the energy levels and therefore the transition frequencies. Thus, due to stochastic rotational motion the ESR transitions for the same spin should occur at different field or frequency, resulting in broadening of the spectral lines.

Note that the first term in the time-dependent Hamiltonian $H_1(t)$ given by 4.6 is proportional to the magnetic field B while the second hyperfine term is field-independent. The first consequence of this observation that the fast motion condition given by inequality (4.1) would break down at shorter correlation times as an increase in magnetic field would increase the magnitude of $H_1(t)$. Therefore for almost all spin-labeled protein and membrane systems with, perhaps, exception of small spin-labeled peptides, we would expect HF ESR spectra to fall into slow or intermediate motion regimens. The increase in $H_1(t)$ would also modulate the electronic spin transition frequencies to a larger degree increasing the spectral width and also increasing the sensitivity of ESR spectra to molecular motion. Another observation is that because of different symmetry of contributing matrices $g(t)$ and $A(t)$ the HF ESR spectra would have different sensitivity to anisotropic rotational fluctuations than conventional X-band ESR and superior angular resolution because of the larger spread of anisotropic spectra in magnetic field or frequency.

4.5.1.1 Stochastic Liouville Theory of Slow Motion ESR Spectra Simulations

Detailed understanding of ESR line shapes of nitroxides over the entire range of rotational correlation times could be obtained from a slow motion theory that is based on solving the stochastic Liouville equation (SLE). This approach that has been the research focus of the group of Prof. Freed (Cornell) was proven to be an important means of analyzing rather complex spectra that can be observed both in biological and model systems (for a review see [121]).

In brief, the slow motion ESR line-shape theory employs the density matrix formalism to describe the coupled spin and spatial degrees of freedom. The equation of motion for the density matrix ρ is governed by the Liouville–von Neumann equation:

$$\frac{d}{dt}\rho = \frac{1}{i\hbar}[H_0 + H_1(t), \rho] \qquad (4.9)$$

While for short correlation times τ_c (fast-motion regimen) the effects of $H_1(t)$ may be treated using perturbation theory (motional narrowing theory), more general methods of analysis should be used for longer τ_c. By performing an ensemble average over the density matrix, it is possible to show that the resulting equation of motion for the density matrix is:

$$\frac{d}{dt}\langle\rho\rangle = \frac{1}{i\hbar}[H,\langle\rho\rangle] + \Gamma\langle\rho\rangle \qquad (4.10)$$

where Γ is a stochastic operator describing stochastic processes giving rise to electronic relaxation in the system and H is the full quantum mechanical Hamiltonian operator of the system. Then the main strategy for determining the slow-motion ESR line shape involves (1) choosing the proper model for the stochastic process and to derive the analytical form of the operator Γ and (2) solving Equation 4.10 digitally in order to derive the ESR line-shape.

Several simple forms of the operator Γ, such as, for example, those corresponding to an isotropic Brownian diffusion, have been discussed [7]. However, local dynamics of spin-labeled protein chains and/or membrane proteins could be rather complicated. For example, for spin-labeled proteins the rotational modes of the tether that links the nitroxide to the protein can be also affected by slowly fluctuating protein and/or membrane environment as well as the overall tumbling modes of the protein. These complex reorientational dynamics of the nitroxide can be described by a slowly relaxing local structure (SRLS) model [122]. In the SRLS model, the spin probe is assumed to be reorienting in a local environment which itself relaxes on a longer timescale. In applications to macromolecular systems, the faster motion describes the internal dynamics, while the slower motions account for the global rotation of the macromolecule. This model has been successfully applied in studies of spin-labeled proteins and other model systems (reviewed in ref. [7]).

The studies of the Cornell group have shown that one of the principal advantages of multifrequency ESR approach is in its ability to "tune" the spectral response of the ESR line-shapes to a specific timescale of spin-label motion by choosing the appropriate magnetic field or resonant frequency for the experiment. For example, X-band ESR spectra of spin-labeled proteins are generally sensitive to the overall protein tumbling modes, whereas these modes are often "frozen out" in high-field ESR spectra because of the larger magnitude of anisotropic Hamiltonian that is subject to averaging. For the same reasons, the diffusion modes corresponding to rapid internal motions may be studied in great detail by analyzing the high-field ESR line-shapes. To summarize, the fast-tumbling modes are much easier to analyze from high-field ESR spectra while the slower modes have the greatest effect on the line-shape at lower frequencies. This is a manifestation of the so-called "snapshot" effect [1] that is illustrated in Figure 4.6 by simulations of ESR

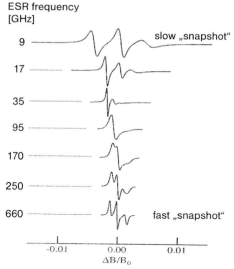

Figure 4.6 Simulations of EPR spectra of a typical nitroxide tumbling isotropically with the rotation diffusion rate $R = 10^8 \, s^{-1}$ as a function of the resonant frequency of the EPR experiment. The large variations that the spectrum undergoes as the frequency is changed are very sensitive to the character of the motion. While for the given rate of rotational diffusion the ESR spectrum at conventional 9.0 GHz (top) is in motionally narrowed regime, the spectra at high frequencies (bottom) display very slow motion, almost at the rigid limit. (Reproduced with permission from [1]).

spectra of a typical nitroxide tumbling isotropically with the rotation diffusion rate $R = 10^8 \, s^{-1}$ as a function of the resonant frequency of the ESR experiment.

4.5.1.2 Molecular Dynamics Simulation Methods

Detailed understanding of dynamics of protein side chain containing nitroxide label is important for many reasons from improving quality of long-range distance constraints to extracting information on protein backbone dynamics and conformations. While such data could be obtained from rigorous simulations of ESR spectra preferably at multiple magnetic fields or frequencies, such an analysis requires the analytical form of operator Γ for realistic models of the nitroxide tether tumbling. Such models could be investigated computationally using molecular dynamics (MD) methods.

Initial work on MD in application to ESR of nitroxides has been carried out using a Brownian dynamics (BD) model to generate the time evolution of the magnetization operator from a trajectory of molecular orientations [123–125]. Recently, a significant progress has been achieved in employing more realistic MD models in spectral simulations of slow-motional ESR line-shapes. Specifically, Budil *et al.* utilized MD trajectories to derive the nitroxide diffusion parameters such as rotational diffusion tensor, diffusion tilt angles, and expansion coefficients of the orienting potential, which were then used as direct inputs to the SLE line-shape program [126]. The method also provides a basis for separating effects of local spin-label motion and the overall motion of a labeled molecule or a domain by multifrequency ESR: once the local motion has been characterized by this approach using HF ESR data, the label diffusion parameters may be used in conjunction with line-shape analysis at lower ESR frequencies to characterize the global motions [126].

An introduction to MD modeling of nitroxide spin-labels could be found in ref. [127]. The efforts in improving MD simulations of spin-labels are expected to continue (see, for example, ref. [128] for development of simulated scaling method) and are expected to result in better understanding complex dynamics of protein side chains.

4.5.2
High Field ESR in Studying Nitroxide Microenvironment

Sensitivity of nitroxide magnetic parameters to intermolecular interactions and, in particular, to hydrogen bonding and local solvent polarity has been well documented in the literature [129–131]. In general, when a spin-labeled molecule is transferred from an aqueous (polar) to a hydrocarbon (non-polar) environment, the isotropic nitrogen hyperfine constant A_{iso} decreases by up to 0.1–0.2 mT while the isotropic g-factor, g_{iso}, increases only slightly (by ≈0.0004). At X-band this change in g_{iso} corresponds to ≈0.68 G shift of the ESR line, which is smaller than the typical nitroxide line-width (ca. 1 G). The shift in the resonance line positions attributed to the g-factor differences grows proportionally with increasing the magnetic field or resonant frequency of the experiment. For example, at W-band (95 GHz, 3.4 T) the shifts of nitroxides ESR specra due to g-factor are typically about threefold greater than those due to A_{iso}. This difference is increasing even further with increase in magnetic field. The sensitivity on nitroxide ESR spectra to hydrophobicity of the immediate molecular environment could be utilized to map local polarity and proticity in lipid bilayers and over protein surfaces. Another more direct approach to probing local electrostatic environment of biomolecules is based site-specific incorporation of a nitroxide that may be reversibly protonated and which protonation state could be directly assessed from ESR spectra.

4.5.2.1 Probing Local Polarity and Proticity of Membrane and Proteins
Both the local electric field and the formation of hydrogen bonds between the oxygen atom of nitroxide and solvent molecules contribute to effects observed for CW ESR spectra of nitroxides. These interactions are recorded as changes in the nitrogen hyperfine coupling A tensorial and g-matrix components [129–134]. For nitroxide ESR spectra in fast-motional regime, these solvent effects are assessed from the isotropic magnetic parameters A_{iso} and g_{iso} [129, 130]. This approach is applicable to small nitroxide probes partitioning between phases of different polarity and is widely employed for examining phase behavior of phospholipid bilayers by X- (9.5 GHz) and W-band (95 GHz) ESR [132, 135–137]. Even if the fast motion condition is not satisfied for nitroxide ESR spectra, the isotropic magnetic param-

eters can still be evaluated from (partially) averaged components of magnetic tensors [138]. If both A_{iso} and g_{iso} are determined in the course of such measurements, an empirical linear correlation plot for g_{iso} *versus* A_{iso} measured for a specific nitroxide in a set of solvents can be used to deduce polarity of the local environment surrounding the probe. Alternatively, solvent effects can be evaluated from anisotropic magnetic parameters measured from rigid limit nitroxide ESR spectra. The component of the electronic **g**-matrix directed along the NO bond, g_x, and the component of nitrogen hyperfine coupling **A** tensor directed along the nitroxide $2p_z$-orbital, A_z, are the most sensitive indicators of local electric fields and hydrogen-bond formation to the nitroxide moiety [131, 139, 140]. A linear correlation between those two nitroxide magnetic parameters has been reported for a wide range of solvents of various polarities [131, 140].

The sensitivity of nitroxide magnetic parameters to solvent effects has been previously used to establish and characterize polarity gradients across phospholipid bilayers. Those studies employed a series of stearic acids or, alternatively, phospholipids labeled with a nitroxide at a specific carbon of the acyl chain [130, 138, 141]. Linear correlation between nitroxide anisotropic magnetic parameters measured from rigid limit 95-GHz ESR spectra was used to probe conformational changes in the bacteriorhodopsin channel [133, 134]. A study of local polarity of spin-labeled side chains in azurin using high-resolution ESR spectroscopy at 275 GHz was also recently reported [142].

Several theoretical studies have been devoted to understanding the basis of solvent effects on nitroxide magnetic parameters. Specifically, shifts between the isotropic nitrogen hyperfine coupling constant, A_{iso}, and the local charge on the nitroxide have been calibrated [143]. These effects were analyzed within a standard Hückel theoretical framework coupled with continuum model approximations of aqueous solvents. More detailed studies of electric field effects on the electronic **g**-matrix indicated that anisotropic magnetic parameters are exceptionally sensitive to both the magnitude and to the orientation of the electric field [144]. Results of *ab initio* calculations, utilizing an intermediate level Rayleigh–Schrödinger perturbation theory, predicted the electrical field effect on g-matrices of nitroxide radicals that agreed well with experimental results [145].

Effects of hydrogen bonding between the nitroxide moiety and the solvent were also evaluated from quantum mechanical calculations. Results obtained from restricted open-shell Hartree–Fock (ROHF) methods with atomic mean-field approximations demonstrate that hydrogen bonding reduces the g_x component of the nitroxide spectrum [146]. Semiempirical molecular orbital methods were also used to calculate the shift in gx resulting from formation of a single hydrogen bond with the water molecule. The shift is predicted to be $\Delta g_x \approx 4 \times 10^{-4}$ for a single hydrogen bond formed [139].

Recently, Smirnova and coworkers reported on HF ESR experiments to probe local polarity and proticity of the interior of the phospholipid-binding pocket of the major yeast phosphatidylinositol/phosphatidylcholine (PtdIns/PtdCho) transfer protein Sec14p [147]. Sec14p is the prototypical member of the large eukaryotic protein Sec14 superfamily and is operationally defined by its ability to efficiently promote the energy independent transfer of either PtdIns or PtdCho between membrane bilayers in vitro [148, 149]. A structurally similar compound, 5-doxyl stearic acid dissolved in a series of solvents, was used for experimental calibration. The experiments yielded two-component rigid limit 130- and 220-GHz ESR spectra with excellent resolution in the g_x region (Figure 4.7a). Those components were assigned to hydrogen-bonded and non-hydrogen-bonded nitroxide species. Partially resolved 130-GHz ESR spectra from n-doxyl-PtdCho bound to Sec14p (Figure 4.7b) were analyzed using this two component model and allowed quantification of two parameters [147]. First, the fraction of hydrogen-bonded nitroxide species for each n-doxyl-PtdCho was calculated. Second, the proticity profile along the phospholipid-binding cavity of Sec14p was characterized. The data suggest the polarity gradient inside the Sec14p cavity is a significant contributor to the driving molecular forces for extracting a phospholipid from the bilayer. Finally, the enhanced g-factor resolution of ESR at 130 and 220 GHz provides researchers with a spectroscopic tool to deconvolute two major contributions to the x-component of the nitroxide g-matrix: hydrogen-bond formation and local electrostatic effects [147].

Further insight into the fundamental nature of the hydrogen bonds between the nitroxides and the likely donors has been obtained from HF (D-band, 130 GHz) pulsed ENDOR studies of a lipid bilayer spin probe 5-doxyl stearic acid (5DSA)

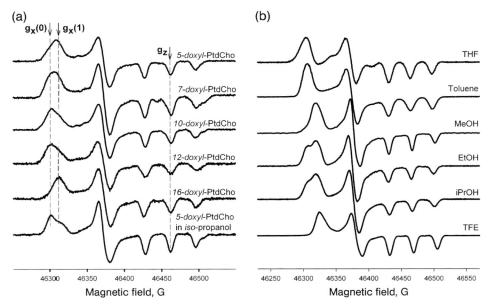

Figure 4.7 Rigid limit 130-GHz echo-detected EPR spectra (T = 25 K) from: (A, left panel): n-doxyl-stearic acid bound to PtdCho-Sec14, (B, right panel): 5-doxyl-stearic acid dissolved in a series of protic and aprotic solvents of various polarities. (Reproduced with permission from [147]).

[150]. A series of polar hydrogen-bond donor alcohols (such as deuterated 2-propanol-d_1, butanol-d_1, ethanol-d_1, and methanol-d_1) and nonpolar deuterated toluene-d_8 were used as solvents. In order to directly probe the structure of 5DSA hydrogen bonds to the likely donors, pulsed Mims-type ^{2}H-ENDOR spectra were recorded from this protonated spin label at three magnetic positions corresponding to principal axis orientations of the g-matrix, g_{xx}, g_{yy}, and g_{zz}. Selected experimental HF ENDOR spectra are shown in Figure 4.8. All experimental ^{2}H-ENDOR spectra recorded in four alcohol solvents were found to be similar: a central peak observed at the ^{2}H-Larmor frequency was superimposed with a doublet of symmetric lines (Figure 4.8b). The relative intensity of the central peak with respect to the doublet varied with the solvent, indicating that the former originated from the "matrix" deuterons. Intensity of the central line was found to correlate with the number of deuterons in the remote coordination spheres of the nitroxide moiety. Furthermore, for nonpolar solvent toluene-d_8, only the central matrix ENDOR line was detected, thus, reassuring the initial assignment. Therefore, it was concluded that

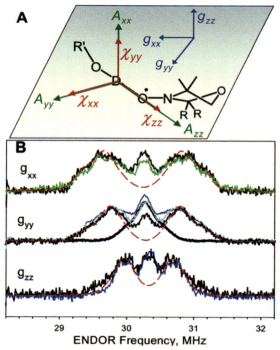

Figure 4.8 (a) Orientation of the 5DSA g-matrix and the bridging ^{2}H hyperfine (A_{xx}, A_{yy}, A_{zz}) and nuclear quadrupole (χ_{xx}, χ_{yy}, χ_{zz}) tensors in the molecular frame of 5DSA coordinated with an alcohol molecule D-O-R'. Note that vectors g_{xx}, g_{yy}, A_{yy}, and A_{zz} lie in the same plane, and that the tensors A and χ are collinear. (b) Superimposed Mims-type HF ENDOR spectra of 5DSA in alcohols (black, 2-propanol-d_1; green, butanol-d_1; cyan, methanol-d_1; blue, ethanol-d_1) and toluene-d_8 (bold line for g_{yy} orientation). Dashed red lines are least-squares simulations. (Reproduced with permission from [150]).

the doublet ENDOR line should arise from the splitting on the ^{2}H of the hydrogen bond. This splitting was found to be different for the x-, y-, and z-principal axis orientations (Figure 4.8b). For all four alcohols, the magnitudes of the splittings as well as line shapes were nearly identical. This indicated that both the strength of the hydrogen bond (i.e. spin to ^{2}H distance) and the bond geometry remain essentially the same for all four alcohol solvents studied. Direct calculations of the hyperfine tensor yielded geometrical parameters of the hydrogen bond formed [150]. Overall, this work demonstrated that HF ESR spectra are exclusively sensitive to formation of hydrogen bonds and could be used to probe the hydrogen-bond network in complex biomolecular assemblies and lipid bilayers by site-directed spin-labeling.

4.5.2.2 Site-Directed pH-Sensitive Spin-labeling: Differentiating Local pK and Polarity Effects by High-Field ESR

As we already discussed in the preceding section, both local electric fields and hydrogen bonding affect magnetic parameters of nitroxide ESR spectra and could be measured with much higher precision by HF ESR. However, nitroxides that are traditionally employed in spin-labeling studies, such as, for example, MTSL, cannot be used for probing another important parameter that affects intermolecular interactions – local pH – simply because the pKa's of these nitroxides lay outside the useful range. Currently, there is a considerable interest in assessing local proton concentration (reported as pH) because this parameter plays a very important role in many biological processes.

Recently, Smirnov and coworkers described a general approach to mapping local pKa values of peptides with high field ESR using MTS (methanothiosulfonate) derivatives of imidazolidine nitroxides [151]. The tertiary amine nitrogen N3 of this label readily participates in proton exchange reactions, which are monitored through changes in ESR spectra of nitroxide moiety (Figure 4.9).

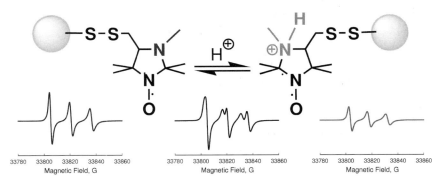

Figure 4.9 An imidazolidine nitroxide (IMTSL) covalently attached to a peptide through a disulfide bridge shows pH-dependent 95 GHz ESR spectra. See text and [151] for more details. (Reproduced with permission from [151]).

It is worthwhile to note here that reversible pH-effects on ESR spectra of stable nitroxide radicals have been known for more than 20 years (reviewed in ref. [152]), the covalent attachments of pH-sensitive nitroxides to biomacromolecules employed in those studies were not fully specific. To achieve the high specificity of covalent attachment of a pH-sensitive imidazolidine nitroxide to thiols, the Novosibirsk group of nitroxide chemists synthesized the MTS derivative, IMTSL, methanethiosulfonic acid S-(1-oxyl-2,2,3,5,5-pentamethyl-imidazolidin-4-ylmethyl) ester. This is the same attachment group as in the spin-label MTSL that is typically employed in SDSL ESR studies. Moreover, the nitroxide side chain formed upon attachment of IMTSL to a cysteine is very similar to that of MTSL and therefore exhibits very similar rotational dynamics as verified by ESR studies.

To calibrate g_{iso} and A_{iso} of IMTSL for pH-dependence a series of room temperature W-band ESR spectra of aqueous solutions of the free label and the label attached to a cysteine (IMTSL-cys) and glutathione (IMTSL-glu) were measured as a function of pH. At intermediate pH the W-band ESR spectra of these compounds consisted of two partially resolved nitroxide components (such the bottom middle spectrum shown in Figure 4.9) characteristic of slow exchange conditions. The A_{iso} titration data and corresponding least-squares Henderson–Hasselbalch titration curves are shown in Figure 4.10.

The ESR titration curves in Figure 4.10 demonstrate that for all compounds studied A_{iso} undergoes large (>0.13 mT) changes upon protonation of the tertiary amine nitrogen N3. Also, the pK_a's of IMTSL-cys and IMTSL-glu were respectively shifted to more basic pK_a = 3.21 ± 0.04 and 3.15 ± 0.03 from that of the free label (1.58 ± 0.03 units) while A_{iso}(base) and A_{iso} (acid) were not affected. It is likely that after the labeling the electronegativity of the side chain is reduced. Since the side chain is closer to N3 than to the nitroxide moiety, this decrease in the electronegativity shifts the N3 pK_a to more basic values without any measurable effects on the unpaired electron spin density. Thus, for these small spin-labeled peptides that

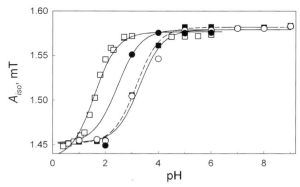

Figure 4.10 Weighted nitrogen hyperfine coupling constant A_{iso} for two-component W-band (95 GHz) EPR spectra as a function of pH and corresponding Henderson–Hasselbalch titration curves (shown as lines): open squares – IMTSL; filled squares – IMTSL-glutathione; filled circles – IMTSL-P11 (all – solid lines); open circles – IMTSL-cysteine (dashed line). (Reproduced with permission from [151]).

are not expected to acquire any globular structure it was observed that the presence of other ionizable groups in the side chain have no effects on the IMTSL ESR spectra, which showed only one transition caused by proton exchange reaction at the N3 position.

To illustrate utility of IMTSL and HF ESR in site-directed pK_a measurements, these authors have selectively labeled the thiol group of a synthetic P11 peptide fragment of the laminin B1 chain (Cys-Asp-Pro-Gly-Tyr-Ile-Gly-Ser-Arg). W-band titration experiments with IMTSL-P11 have been carried out and analyzed for changes in pK_a and polarity effects using a correlation A_{iso} vs. g_{iso} plot [151]. Such an analysis demonstrated that for IMTSL-P11 $pK_a = 2.5 \pm 0.1$ is less than that of either IMTSL-glu or IMTSL-cys ($pK_a \approx 3.18$) and is not associated with local polarity changes but rather with local partial charges of the peptide backbone [151].

Overall, this work demonstrated an extension of SDSL ESR to protein site-specific pH measurements. It is anticipated that the next applications of this method would be in studies of ion channels. The method will be also very useful for studying pH-triggered conformations in biomolecules and for better understanding of the role of hydrophobic and electrostatic interactions in membrane protein structure and function.

4.6
Perspectives

This chapter demonstrated that instrumentation and methodology development in ESR of nitroxides go hand-to-hand with applications: very often advances in ESR instrumentation would spur new research areas while the needs of specific biomedical and other studies would also stimulate spectroscopists and spectrometer manufacturers.

Perhaps, one of the most notable examples of such synergetic developments between technology and applications is given by experiments on distance measurements by ESR of nitroxides. One of the first accounts of employing nitroxides and ESR for distance determinations in biomacromolecules has been provided by Kokorin *et al.* in 1972 [153]. In the following years Kokorin and coauthors investigated mechanisms of spin-spin interactions in nitroxides pairs and developed methods of analyzing powder-pattern continuous wave ESR spectra of nitroxide biradicals [154–157]. Their work provided seminal ideas that have been employed much later in developing magnetic resonance methods for distance constraint determination from spin-labeling nitroxide experiments. Over the past decade the growing interest in structure-function of biomolecules fueled further development of ESR distance methods and related ESR instrumentation. As reviewed in Chapter significant extension of the ESR distance range have been achieved through development of pulse ESR methods such as DEER and DQC. The popularity of DEER experiment alone justified development of commercial spectrometers that are now routinely capable of such experiments.

The second, perhaps, less notable example is provided by nitroxide biradicals. From prospective of using biradicals as spin probes, all the magnetic interactions present in a nitroxide monoradical would also occur in the corresponding biradical and, therefore, the biradicals are expected to report on microenvironment in a very similar way as the monoradical does [158]. Moreover, additional magnetic interaction that couples two nitroxide spins is also expected to be influenced by the surrounding, thus, providing another faucet for examining the latter [ibid.]. While the basic theory underlying the use of nitroxide biradicals as spin-probes has been known for many years [153–158], only recently some new potentials of using both flexible and rigid bilaradicals as spin-probes have been recognized [159, 160]. Even more exciting are the prospects of employing nitroxide biradicals as polarizing agents for dynamic nuclear polarization experiments in aqueous media [161, 162]. Specifically, recently synthesized biradical TOTAPOL (1-(TEMPO-4-oxy)-3-(TEMPO-4-amino)propan-2-ol) was shown to increase the average electron-electron dipole coupling constant to ~30 MHz when compared to ~0.5 MHz for the typical 40 mM solution of monoradical TEMPO used in previous DNP experiments [162]. At 140 GHz EPR frequency this yielded a maximum enhancement of ~290 at a reduced concentration of 6 mM of the biradical polarizing agent [ibid.]. It is expected that a rational design of biradical polarizing agents would improve the DNP further to the theoretical limit.

Another important observation to make is a recent resurgence of interest in synthesis of novel nitroxide with modified or, in some cases, rather unusual properties. Such synthesis often inspires novel applications. For example, for many years MTSL has always been the spin-label of choice for SDSL experiments. While other methanethiosulfonate-derivatized nitroxide side-chains have been discussed in the literature (e.g., see [163]), only now several of such cysteine spin-labels became commercially available. One of such labels is 4-bromo-MTSL (4-bromo-(1-oxyl-2,2,5,5-tetramethyl-Δ3-pyrroline-3-methyl) methanethiosulfonate. Peptide side-chains modified with such a label have shown a somewhat restricted rotational dynamics as compared with traditional MTSL [164]. Such label properties could be advantageous in studies of membrane insertion and bilayer perturbation by antimicrobial peptides [ibid.]. A very different type of cysteine spin-label—a pH-sensitive IMTSL—has been introduced recently [151] and discussed in this chapter. Additional progress in synthesis of pH-sensitive nitroxides is expected to extend pK_a of such label further into physiological pH range [165].

Development of new materials that are based on nitroxides could be assisted by biological self-assembly and/or result in unusual applications. For example, recent studies have shown that spin labeled 2′-deoxyuridine, in which a significant fraction of the spin density is delocalized from a nitroxide radical to the DNA base residue, would pack into one-dimensional chains with predominant intra-chain ferromagnetic coupling and weaker inter-chain antiferromagnetic coupling [166]. Recently, a very promising progress in fabrication of organic field-effect transistors (OFETs) from nitroxides has been shown by Liu and coworkers [167]. These authors have shown that OFETs fabricated with vapor-deposited films of 1-imino nitroxide pyrene demonstrated excellent p-type FET characteristics, low operating voltage due to the low threshold voltage (about -0.6 V) and inverse sub-threshold slope (about 540 mV decade^{-1}) [ibid.]. It is expected that such devices would serve as a low-cost alternative to conventional silicon transistors for electronic applications. As the field of nitroxide applications in materials science and biomedicine continues to grow, an additional attention should be paid to preparing enantiopure nitroxides. In some cases, this could be achieved through a crystallization and analysis of the crystal by X-ray or EPR as has been demonstrated very recently [168].

All these recent developments demonstrate that both nitroxide methodology and related methods and instrumentation continue to evolve synergetically and a rather rapid pace. It is our hope that even more progress would follow in the coming years and that this volume would inspire at least some of new discoveries in the nitroxide field!

Acknowledgements

The author would like to thank DOE Contract DE-FG02-02ER15354, NIH 1R01GM072897, and NSF ECS 0420775 for providing partial support to this work. The author would also like to express his gratitude to Prof. Smirnova and all the members of the interdisciplinary ESR group at NCSU. Special thanks to Dr. Maxim Voinov and Mr. T. Gray Chadwick (NCSU) for the expert help in preparing this chapter.

References

1 Borbat, P.P., Costa-Filho, A.J., Earle, K.A., Moscicki, J.K. and Freed, J.H. (2001) *Science*, **291**, 266–9.
2 Park, S., Borbat, P.P., Gonzalez-Bonet, G., Bhatnagar, J., Pollard, A.M., Freed, J.H., Bilwes, A.M. and Crane, B.R. (2006) *Nature Structural and Molecular Biology*, **13**, 400–7.
3 Tsvetkov, Yu.D. (2004) *Biological Magnetic Resonance*, Vol. 21 (eds L.J. Berliner and C.J. Bender), Kluwer, New York, pp. 385–433.
4 Cai, Q., Kusnetzow, A.K., Hubbell, W.L., Haworth, I.S., Gacho, G.P., Van Eps, N., Hideg, K., Chambers, E.J. and Qin, P.Z. (2006) *Nucleic Acids Research*, **34**, 4722–30.
5 Fanucci, G.E. and Cafiso, D.S. (2006) *Current Opinion in Structural Biology*, **16**, 644–53.
6 Liang, Z., Lou, Y., Freed, J.H., Columbus, L. and Hubbell, W.L. (2004) *Journal of Physical Chemistry*, **108**, 17649–59.
7 Earle, K.A. and Smirnov, A.I. (2004) *Biological Magnetic Resonance*, Vol. 22 (eds L.J. Berliner and O.G. Grinberg), Kluwer, New York, pp. 469–517.
8 Earle, K.A., Hofbauer, W., Dzikowski, B., Moscicki, J.K. and Freed, J.H. (2005) *Magnetic Resonance in Chemistry*, **43**, S256–66.
9 Smirnova, T.I. and Smirnov, A.I. (2007) *Biological Magnetic Resonance*, Vol. 27 (eds M.A. Hemminga and L.J. Berliner), Springer, New York, pp. 165–253.
10 Ernst, R.R., Bodenhausen, G. and Wokaun, A. (1987) *Principles of Nuclear Magnetic Resonance in One and Two Dimensions*, Clarendon, Oxford.
11 Borbat, P.P. and Freed, J.H. (1999) *Chemical Physics Letterss*, **313**, 145–54.
12 Liang, B., Bushweller, J.H. and Tamm, L.K. (2006) *Journal of the American Chemical Society*, **128**, 4389–97.
13 Smirnov, A.I. and Poluektov, O.G. (2003) *Journal of the American Chemical Society*, **125**, 8434–5.
14 Chekmenev, E.Y., Hu, J., Gor'kov, P.L., Brey, W.W., Cross, T.A., Ruuge, A. and Smirnov, A.I. (2005) *Journal of Magnetic Resonance*, **173**, 322–7.

15 Chekmenev, E.Y., Gor'kov, P.L., Cross, T.A., Alaouie, A.M. and Smirnov, A.I. (2006) *Biophysical Journal*, **91**, 3076–84.
16 Lorigan, G.A., Dave, P.C., Tiburu, E.K., Damodaran, K., Abu-Baker, S., Karp, E.S., Gibbons, W.J. and Minto, R.E. (2004) *Journal of the American Chemical Society*, **126**, 9504–5.
17 White, S.H. (2004) *Protein Science: A Publication of the Protein Society*, **13**, 1948–9.
18 Wuthrich, K.J. (1990) *Biological Chemistry*, **265**, 22059–62.
19 Cavanagh, J., Fairbrother, W.J., Palmer, A.G., III and Skelton, N.J. (2007) *Protein NMR Spectroscopy, Principles and Practice*, 2nd edn, Vol. 885, Elsevier, New York.
20 Tolman, J.R., Flanagan, J.M., Kennedy, M.A. and Prestegard, J.H. (1995) *Proceedings of the National Academy of Sciences of the United States of America*, **92**, 9279–83.
21 Tjandra, N. and Bax, A. (1997) *Science*, **278**, 1111–14.
22 Distance measurements in biological systems by EPR, in *Biological Magnetic Resonance*, Vol. 19 (eds L.J. Berliner, S.S. Eaton and G.R. Eaton) (2000) Kluwer, New York.
23 Closs, G.L., Forbes, M.D.E. and Piotrowiak, P. (1992) *Journal of the American Chemical Society*, **114**, 3285–94.
24 Fiori, W.R., Miick, S.M. and Millhauser, G.L. (1993) *Biochemistry*, **32**, 11957–62.
25 Abragam, A. (1961) *The Principles of Nuclear Magnetism*, Chapter VIII, Oxford, London, pp. 289–304.
26 Mchaourab, H.S., Oh, K.J., Fang, C.J. and Hubbell, W.L. (1997) *Biochemistry*, **36**, 307–16.
27 Pake, G.E. (1948) *The Journal of Chemical Physics*, **16**, 327–36.
28 Likhtenstein, G.I. (1976) *Spin Labeling Methods in Molecular Biology*, Chapter 3, John Wiley & Sons, New York.
29 Steinhoff, H.-J., Dombrowsky, O., Karim, C. and Schneiderhahn, C. (1991) *European Biophysics Journal: EBJ*, **20**, 293–303.
30 Steinhoff, H.-J., Radzwill, N., Thevis, W., Lenz, V., Brandenburg, D., Antson, A., Dodson, G. and Wollmer, A. (1997) *Biophysical Journal*, **73**, 3287–98.
31 Rabenstein, M.D. and Shin, Y. (1995) *Proceedings of the National Academy of Sciences of the United States of America*, **92**, 8239–43.
32 Altenbach, C., Oh, K.-J., Trabanino, R.J., Hideg, K. and Hubbell, W.L. (2001) *Biochemistry*, **40**, 15471–82.
33 Hustedt, E.J. and Beth, A.H. (1999) *Annual Review of Biophysics and Biomolecular Structure*, **28**, 129–53.
34 Hustedt, E.J., Smirnov, A.I., Laub, C.F., Cobb, C.E. and Beth, A.H. (1997) *Biophysical Journal*, **72**, 1861–77.
35 Hustedt, E.J., Stein, R.A., Sethaphong, L., Brandon, S., Zhou, Z. and Desensi, S.C. (2006) *Biophysical Journal*, **90**, 340–56.
36 Smirnov, A.I. and Smirnova, T.I. (2004) *Biological Magnetic Resonance*, Vol. 21 (eds L.J. Berliner and C. Bender), Kluwer, New York, pp. 277–348.
37 Eaton, S.S., Eaton, G.R. and Eaton, G.R. (2000) *Biological Magnetic Resonance*, Vol. 19 (eds S.S. Eaton and L.J. Berliner), Kluwer, New York, pp. 29–154.
38 Leigh, J.S. (1970) *The Journal of Chemical Physics*, **52**, 2608–12.
39 Hyde, J.S., Swartz, H.M. and Antholine, W.E. (1979) *Spin-Probe-Spin Label Method* (ed. L.J. Berliner), Academic Press, New York, pp. 71–113.
40 Eaton, S.S. and Eaton, G.R. (1988) *Coordination Chemistry Reviews*, **83**, 29–72.
41 Voss, J., Salwinski, L., Kaback, H.R. and Hubbell, W.L. (1995) *Proceedings of the National Academy of Sciences of the United States of America* *Proceedings of the National Academy of Sciences of the United States of America*, **92**, 12295–9.
42 Voss, J., Hubbell, W.L. and Kaback, H.R. (1995) *Proceedings of the National Academy of Sciences of the United States of America*, **92**, 12300–3.
43 Reeve, I., Hummel, D., Nelson, N. and Voss, J. (2002) *Proceedings of the National Academy of Sciences of the United States of America*, **99**, 8608–13.
44 Smirnova, T.I. (2007) *Applied Magnetic Resonance*, **31**, 431–46.
45 Powell, D.H., Merbach, A.E., Gonzalez, G., Brucher, E., Micskei, K., Ottaviani, M.F., Kohler, K., von Zelewsky, A.,

Grinberg, O.Ya. and Lebedev, Ya.S. (1993) *Helvetica Chimica Acta*, **76**, 2129–46.

46 Smirnova, T.I., Smirnov, A.I., Belford, R.L. and Clarkson, R.B. (1998) *Journal of the American Chemical Society*, **120**, 5060–72.

47 Powell, D.H., Dhubhghaill, O.M., Pubanz, D., Helm, L., Lebedev, Y.S., Schlaepfer, W. and Merbach, A.E. (1996) *Journal of the American Chemical Society*, **118**, 9333–46.

48 Dwek, R.A. (1973) *Nuclear Magnetic Resonance (N.M.R.) in Biochemistry. Applications to Enzyme Systems*, Chapter 12, Claderon Press, Oxford, London, pp. 285–327.

49 Kosen, P.A. (1989) *Methods in Enzymology*, **177**, 86–121.

50 Gaponenko, V., Howarth, J.W., Columbus, L., Gasmi-Seabrook, G., Yuan, J., Hubbell, W.L. and Rosevear, P.R. (2000) *Protein Science: A Publication of the Protein Society*, **9**, 302–9.

51 Jain, N.U., Venot, A., Umemoto, K., Leffler, H. and Prestegard, J.H. (2001) *Protein Science: A Publication of the Protein Society*, **10** (2), 393–2400.

52 Cai, S., Zhu, L., Zhang, Z. and Chen, Y. (2007) *Biochemistry*, **46**, 4943–50.

53 Nadaud, P.S., Helmus, J.J., Höfer, N. and Jaroniec, C.P. (2007) *Journal of the American Chemical Society*, **129**, 7502–3.

54 Dzuba, S.A., Raitsimring, A.M. and Tsvetkov, Yu.D. (1979) *The Journal of Chemical Physics*, **44**, 357–65.

55 Dzuba, S.A., Raitsimring, A.M. and Tsvetkov, Yu.D. (1980) *Journal of Magnetic Resonance*, **40**, 83–9.

56 Likhtenstein, G.I. and Eaton, G.R. (2000) *Biological Magnetic Resonance*, Vol. 19 (eds G.R. Eaton, S.S. Eaton and L.J. Berliner), Kluwer, New York, pp. 309–46.

57 Eaton, S.S. and Eaton, G.R. (2000) *Biological Magnetic Resonance*, Vol. 19 (eds G.R. Eaton, S.S. Eaton and L.J. Berliner), Kluwer, New York, pp. 347–81.

58 Subczynski, W.K., Hyde, J.S. and Kusumi, A. (1989) *Proceedings of the National Academy of Sciences of the United States of America*, **86**, 4474–8.

59 Nielsen, R.D., Che, K., Gelb, M.H. and Robinson, B.H. (2005) *Journal of the American Chemical Society*, **127**, 6430–42.

60 Jun, S., Becker, J.S., Yonkunas, M., Coalson, R. and Saxena, S. (2006) *Biochemistry*, **45**, 11666–73.

61 Raitsimring, A. (2000) *Biological Magnetic Resonance*, Vol. 19 (eds G.R. Eaton, S.S. Eaton and L.J. Berliner), Kluwer, New York, pp. 461–92.

62 Pannier, M., Veit, S., Godt, A., Jeschke, G. and Spiess, H.W. (2000) *Journal of Magnetic Resonance*, **142**, 331–40.

63 Jeschke, G., Bender, A., Paulsen, H., Zimmermann, H. and Godt, A. (2004) *Journal of Magnetic Resonance*, **169**, 1–12.

64 Fajer, P.G., Brown, L. and Song, L. (2007) *Biological Magnetic Resonance*, Vol. 27 (eds M.A. Hemminga and L.J. Berliner), Springer, New York, pp. 95–128.

65 Cai1, Q., Kusnetzow, A.K., Hubbell, W.L., Haworth, I.S., Paola, G., Gacho, C., Van Eps, N., Hideg, K., Chambers, E.J. and Qin, P.Z. (2006) *Nucleic Acids Research*, **34**, 4722–30.

66 Piton, N., Schiemann, O., Mu, Y., Stock, G., Prisner, T. and Engels, J.W. (2005) *Nucleosides, Nucleotides and Nucleic Acids*, **24**, 771–5.

67 Sale, K., Song, L., Liu, Y.S., Perozo, E. and Fajer, P. (2005) *Journal of the American Chemical Society*, **127**, 9334–5.

68 Saxena, S. and Freed, J.H. (1996) *Chemical Physics Letters*, **251**, 102–10.

69 Saxena, S. and Freed, J.H. (1997) *The Journal of Chemical Physics*, 1317–40.

70 Borbat, P.P., Crepeau, R.H. and Freed, J.H. (1997) *Journal of Magnetic Resonance*, **127**, 155–67.

71 Borbat, P.P. and Freed, J.H. (2000) *Biological Magnetic Resonance*, Vol. 19 (eds G.R. Eaton, S.S. Eaton, and L.J. Berliner), Kluwer, New York, 383–459.

72 Borbat, P.P., Mchaourab, H.S. and Freed, J.H. (2002) *Journal of the American Chemical Society*, **124**, 5304–14.

73 Borbat, P.P., Davis, J.H., Butcher, S.E. and Freed, J.H. (2004) *Journal of the American Chemical Society*, **126**, 7746–7.

74 Bhatnagar, J., Freed, J.H., Crane, B.R. and Simon, M., (2007) *Methods in*

Enzymology, 423, Chapter 4 (eds B.R. Crane and A.B. Crane), Academic Press, pp. 117–33.
75 Borbat, P.P., Freed, J.H. and Simon, M., (2007) *Methods in Enzymology*, 423, Chapter 3 (eds B.R. Crane and A.B. Crane), Academic Press, pp. 52–116.
76 Lebedev, Ya.S. (1990) *Modern Pulsed and Continuous-Wave Electron Spin Resonance* (eds L. Kevan and M.K. Bowman), John Wiley & Sons, New York, pp. 365–404.
77 Ondar, M.A., Dubinskii, A.A., Grinberg, O.Ya., Grigorev, I.A., Volodarskii, L.B. and Lebedev Ya.S. (1981) *Zh Strukt Khim in Russian*, **22**, 59–66.
78 Denysenkov, V.P., Prisner, T.F., Stubbe, J. and Bennati, M. (2006) *Proceedings of the National Academy of Sciences of the United States of America*, **103**, 13386–90.
79 Klug, C.S., Camenisch, T.G., Hubbell, W.L. and Hyde, J.S. (2005) *Biophysical Journal*, **88**, 3641–7.
80 Sczaniecki, P.B., Hyde, J.S. and Froncisz, W. (1991) *The Journal of Chemical Physics*, **94**, 5907–16.
81 Mchaourab, H.S. and Hyde, J.S. (1993) *The Journal of Chemical Physics*, **98**, 1786–96.
82 Brian, A.A. and McConnell, H.M. (1984) *Proceedings of the National Academy of Sciences of the United States of America*, **81**, 6159–63.
83 Sackmann, E. (1996) *Science*, **271**, 43–8.
84 Boxer, S.G. and Cremer, P.S. (1999) *Journal of Physical Chemistry B*, **103**, 2554–9.
85 Keller, C.A., Glasmästar, K., Zhdanov, V.P. and Kasemo, B. (2000) *Physical Review Letters*, **84**, 5443–6.
86 Schreier-Muccillo, S., Marsh, D., Dugas, H., Schneider, H. and Smith, I.C.P. (1973) *Chemistry and Physics of Lipids*, **10**, 11–27.
87 McConnell, H.M., Tamm, L.K. and Weis, R.M. (1984) *Proceedings of the National Academy of Sciences of the United States of America*, **81**, 3249–53.
88 Jacobsen, K., Oga, S., Hubbell, W.L. and Risse, T. (2005) *Biophysical Journal*, **88**, 4351–65.
89 Jacobsen, K., Hubbell, W.L., Ernst, O.P. and Risse, T. (2006) *Angewandte Chemie (International Ed. in English)*, **45**, 3874–7.
90 Ge, M., Budil, D.E. and Freed, J.H. (1994) *Biophysical Journal*, **67**, 2326–44.
91 Dzikovski, B.G., Borbat, P.P. and Freed, J.H. (2004) *Biophysical Journal*, **87**, 3504–17.
92 Barnes, J.P. and Freed, J.H. (1998) *Biophysical Journal*, **75**, 2532–46.
93 Earle, K.A., Dzikovski, B., Hofbauer, W., Moscicki, J.K. and Freed, J.H. (2005) *Magnetic Resonance in Chemistry*, **43**, S256-S266.
94 Dzikowski, B.G., Earle, K.A., Pachtchenko, S. and Freed, J.H. (2006) *Journal of Magnetic Resonance*, **179**, 273–9.
95 Moll, F. and Cross, T.A. (1990) *Biophysical Journal*, **57**, 351–62.
96 Sanders, C.R., Hare, B.J., Howard, K.P. and Prestegard, J.H. (1994) *Progress in Nuclear Magnetic Resonance Spectroscopy*, **26**, 421–44.
97 Sanders, C.R., and Landis, G.C. (1995) *Biochemistry*, **34**, 4030–40.
98 Katsaras, J., Donaberger, R.L., Swainson, I.P., Tennant, D.C., Tun, Z., Vold, R.R. and Prosser, R.S. (1997) *Physical Review Letters*, **78**, 899–902.
99 Mangels, M.L., Harper, A.C., Smirnov, A.I., Howard, K.P. and Lorigan, G.A. (2001) *Journal of Magnetic Resonance*, **151**, 253–9.
100 Smirnova, T.I. and Smirnov, A.I. (2007) *Biological Magnetic Resonance*, Vol. 27 (eds M.A. Hemminga and L.J. Berliner), Springer, New York, pp. 165–252.
101 Garber, S.M., Lorigan, G.A. and Howard, K.P. (1999) *Journal of the American Chemical Society*, **121**, 3240–1.
102 Mangels, M.L., Harper, A.C., Howard, K.P. and Lorigan, G.A. (2000) *Journal of the American Chemical Society*, **122**, 7052–8.
103 Caporini, M.A., Padmanabhan, A., Cardon, T.B. and Lorigan, G.A. (2003) *Biochimica et Biophysica Acta*, **1612**, 52–8.
104 Cardon, T.B., Tiburu, E.K. and Lorigan, G.A. (2003) *Journal of Magnetic Resonance*, **161**, 77–90.
105 Lu, J.X., Caporini, M.A. and Lorigan, G.A. (2004) *Journal of Magnetic Resonance*, **168**, 18–30.

106 Nusair, N.A., Tiburu, E.K., Dave, P.C. and Lorigan, G.A. (2004) *Journal of Magnetic Resonance*, **168**, 228–37.

107 Inbaraj, J.J., Nusair, N.A. and Lorigan, G.A. (2004) *Journal of Magnetic Resonance*, **171**, 71–9.

108 Inbaraj, J.J., Cardon, T.B., Laryukhin, M., Grosser, S.M. and Lorigan, G.A. (2006) *Journal of the American Chemical Society*, **128**, 9549–54.

109 Smirnov, A.I. and Poluektov, O.G. (2003) *Journal of the American Chemical Society*, **125**, 8434–5.

110 Alaouie, A.M. and Smirnov, A.I. (2005) *Biophysical Journal*, **88**, L11-3.

111 Alaouie, A.M. and Smirnov, A.I. (2006) *Langmuir: The ACS Journal of Surfaces and Colloids*, **22**, 5563–5.

112 Chekmenev, E.Y., Hu, J., Gor'kov, P.L., Brey, W.W., Cross, T.A., Ruuge, A. and Smirnov, A.I. (2005) *Journal of Magnetic Resonance*, **173**, 322–7.

113 Lorigan, G.A., Dave, P.C., Tiburu, E.K., Damodaran, K., Abu-Baker, S., Karp, E.S., Gibbons, W.J. and Minto, R.E. (2004) *Journal of the American Chemical Society*, **126**, 9504–5.

114 Gaede, H.C., Luckett, K.M., Polozov, I.V. and Gawrisch, K. (2004) *Langmuir: The ACS Journal of Surfaces and Colloids*, **20**, 7711–19.

115 Wattraint, O., Warschawski, D.E. and Sarazin, C. (2005) *Langmuir: The ACS Journal of Surfaces and Colloids*, **21**, 3226–8.

116 Chekmenev, E.Y., Gor'kov, P.L., Cross, T.A., Alaouie, A.M. and Smirnov, A.I. (2006) *Biophysical Journal*, **91**, 3076–84.

117 Karp, E.S., Inbaraj, J.J., Laryukhin, M. and Lorigan, G.A. (2006) *Journal of the American Chemical Society*, **128**, 12070–1.

118 Grinberg, O.Ya. and Dubinskii, A.A. (2004) *Biological Magnetic Resonance*, Vol. 22, Chapter 1 (eds O. Grinberg and L.J. Berliner), Kluwer. New York, pp. 1–18.

119 Smirnov, A.I., Smirnova, T.I., MacArthur, R.L., Good, J.A. and Hall, R. (2006) *The Review of Scientific Instruments*, **77**, 035108.

120 Very high frequency (VHF) ESR/EPR, in *Biological Magnetic Resonance*, Vol. 22 (eds O. Grinberg and L.J. Berliner) (2004) Kluwer, New York, p. 569.

121 Schneider, D.J. and Freed, J.H. (1989) *Biological Magnetic Resonance*, Vol. 8 (eds L.J. Berliner and J. Reuben), Plenum, New York, pp. 1–75.

122 Polimeno, A. and Freed, J.H. (1995) *The Journal of Chemical Physics* **99**, 10995–1006.

123 Robinson, B.H., Slutsky, L.J. and Auteri, F.P. (1992) *The Journal of Chemical Physics*, **96**, 2609–16.

124 Steinhoff, H.J. and Hubbell, W.L. (1996) *Biophysical Journal*, **71**, 2201–12.

125 Steinhoff, H.-J., Müller, M., Beier, C. and Pfeiffer, M. (2000) *Journal of Molecular Liquids*, **84**, 17–27.

126 Budil, D.E., Sale, K.L., Khairy, K.A. and Fajer, P.G. (2006) *Journal of Physical Chemistry. A*, **110**, 3703–13.

127 Fajer, M.I., Sale, K.L. and Fajer, P.G. (2007) *Biological Magnetic Resonance*, Vol. 27, Appendix 1 (eds M.A. Hemminga and L.J. Berliner), Springer, New York, pp. 253–9.

128 Li, H., Fajer, M. and Yanga, W. (2007) *The Journal of Chemical Physics*, **126**, 024106.

129 Kawamura, T., Matsunami, S. and Yonezawa, T. (1967) *Bulletin of the Chemical Society of Japan*, **40**, 1111–15.

130 Griffith, O.H., Dehlinger, P.J. and Van, S.P. (1974) *The Journal of Membrane Biology*, **15**, 159–92.

131 Ondar, M.A., Grinberg, O.Ya., Dubinskii, A.A. and Lebedev, Ya.S. (1985) *Sov. J. Chem. Phys.*, **3**, 781–92.

132 Smirnov, A.I. and Smirnova, T.I. (2001) *Applied Magnetic Resonance*, **21**, 453–67.

133 Wegener, C., Savitsky, A., Pfeiffer, M., Möbius, K. and Steinhoff, H.-J. (2001) *Applied Magnetic Resonance*, **21**, 441–52.

134 Steinhoff, H.-J., Savitsky, A., Wegener, C., Pfeiffer, M., Plato, M. and Möbius, K. (2000) *Biochimica et Biophysica Acta*, **1457**, 253–62.

135 Shimshick, E.J. and McConnell, H.M. (1973) *Biochemistry*, **12**, 2351–60.

136 Polnaszek, C.F., Schreier, S., Butler, K.W. and Smith, I.C. (1978) *Journal of the American Chemical Society*, **100**, 8223–31.

137 Smirnov, A.I., Smirnova, T.I. and Morse, P.D., II (1995) *Biophysical Journal*, **68**, 2350–60.

138 Kurad, D., Jeschke, G. and Marsh, D. (2003) *Biophysical Journal*, **85**, 1025–33.
139 Plato, M., Steinhoff, H.-J., Wegener, C., Törring, J.T., Savitsky, A. and Möbius, K. (2002) *Molecular Physics*, **100**, 3711–21.
140 Owenius, R., Engstrom, M., Lindgren, M. and Huber, M. (2001) *Journal of Physical Chemistry*, **105**, 10967–77.
141 Earl, K.A., Moscicki, J.K., Ge, M., Budil, D.E. and Freed, J.H. (1994) *Biophysical Journal*, **66**, 1213–21.
142 Finiguerra, M.G., Blok, H., Ubbink, M. and Huber, M. (2006) *Journal of Magnetic Resonance*, **180**, 197–202.
143 Schwartz, R.N., Peric, M., Smith, S.A. and Bales, B.L. (1997) *Journal of Physical Chemistry. B*, **101**, 8735–9.
144 Gulla, A.F. and Budil, D.E. (2001) *Journal of Physical Chemistry. B*, **105**, 8056–63.
145 Ding, Z., Gulla, A.F. and Budil, D.E. (2001) *The Journal of Chemical Physics*, **115**, 10685–93.
146 Engstrom, M., Owenius, R. and Vahtras, O. (2001) *Chemical Physics Letters*, **338**, 407–13.
147 Smirnova, T.I., Chadwick, T.G., Voinov, M.A., Poluektov, O., van Tol, J., Ozarowski, A., Schaaf, G., Ryan, M.M. and Bankaitis, V.A. (2007) *Biophysical Journal*, **92**, 3686–95.
148 Cleves, A.E., McGee, T.P. and Bankaitis, V.A. (1991) *Trends in Cell Biology*, **1**, 30–4.
149 Phillips, S.E., Vincent, P., Rizzieri, K., Schaaf, G., Gaucher, E.A. and Bankaitis, V.A. (2006) *Critical Reviews in Biochemistry and Molecular Biology*, **41**, 21–49.
150 Smirnova, T.I., Smirnov, A.I., Pachtchenko, S. and Poluektov, O.G. (2007) *Journal of the American Chemical Society*, **129**, 3476–7.
151 Smirnov, A.I., Ruuge, A., Reznikov, V.A., Voinov, M.A. and Grigor'ev, I.A. (2004) *Journal of the American Chemical Society*, **126**, 8872–3.
152 Khramtsov, V.V. and Volodarsky, L.B. (1998) *Biological Magnetic Resonance*, Chapter 4 (ed. L.J. Berliner), Plenum, New York, pp. 109–80.
153 Kokorin, A.I., Zamarayev, K.I., Grigoryan, G.L., Ivanov, V.P. and Rozantsev, E.G. (1972) *Biofizika*, **17**, 34–41.
154 Parmon, V.N., Kokorin, A.I., Zhidomirov, G.M. and Zamarayev, K.I. (1973) *Molecular Physics*, **26**, 1565–9.
155 Parmon, V.N., Kokorin, A.I., Zhidomirov, G.M. and Zamarayev, K.I. (1975) *Molecular Physics*, **30**, 695–701.
156 Parmon, V.N., Kokorin, A.I. and Zhidomirov, G.M. (1977) *J. Magn. Reson.*, **28**, 339–49.
157 Parmon, V.N., Kokorin, A.I. and Zhidomirov, G.M. (1977) *J. Struct. Chem.*, **18**, 104–47.
158 Luckhurst, G.R., (ed. L. J. Berliner), *Spin Labeling. Theory and Applications* (1976), Academic Press, New York, Chapter 4, pp. 133–83.
159 Grampp, G., Rasmussen, K. and Kokorin, A.I. (2004) *Appl. Magn. Reson.*, **26**, 245–52.
160 Kokorin, A.I., Tran, V.A., Rasmussen, K. and Grampp, G. (2006) *Appl. Magn. Reson.*, **30**, 35–42.
161 Hu, K.-N., Yu, H.-H., Swager, T.M. and Griffin, R.G. (2004) *J. Am. Chem. Soc.*, **126**, 10844–5.
162 Song, C., Hu, K.-N., Joo, C.-G., Swager, T.M. and Griffin, R.G. (2006) *J. Am. Chem. Soc.*, **128**, 11385–90.
163 Mchaourab, H.S., Lietzow, M.A., Hideg, K. and Hubbell, W.L. (1996) *Biochemistry*, **35**, 7692–704.
164 Pistolesi, S., Pogni, R. and Feix, J.B. (2007) *Biophys. J.*, **93**, 1651–60.
165 Voinov, M.A., Polienko, J.F., Schanding, T., Bobko, A., Khramtsov, V.V., Gatilov, Y.V., Rybalova, T.V., Smirnov, A.I. and Grigor'ev, I.A. (2005) *J. Org. Chem.*, **70**, 9702–11.
166 Das, K., Pink, M., Rajca, S. and Rajca, A. (2006) *J. Am. Chem. Soc.*, **128**, 5334–5.
167 Wang, Y., Wang, H., Liu, Y., Di, C.-A., Sun, Y., Wu, W., Yu, G., Zhang, D. and Zhu, D. (2006) *J. Am. Chem. Soc.*, **128**, 13058–9.
168 Levkin, P.A., Kokorin, A.I., Schurig, V. and Kostyanovsky, R.G. (2006) *Chirality*, **18**, 232–8.

5
Preparations, Reactions, and Properties of Functional Nitroxide Radicals

Shin'ichi Nakatsuji

5.1
Short Historical Survey and General Preparative Methods of NRs

This chapter deals with the preparative chemistry of functional nitroxide radicals (FNR). The term "functional nitroxide radicals" is used here for nitroxide radicals which exhibit responses to outside effects, such as light, heat, electron, proton, biological stimuli, and so on. The preparations, reactions, and properties of FNRs will be described in the following sections.

As is well-known that the very first nitroxide radical (NR) to appear in the scientific literature is "Fremy's salt", reported by E. Fremy in 1845, although, of course, it was not regarded at that time as a radical species [1]. Fremy's salt [dipotassium nitrosodisulfonate, $NO(SO_3K)_2$ or disodium salt] is actually an inorganic radical salt, and was prepared by the reaction of potassium bisulfite with potassium nitrite followed by oxidation. It is still now widely used as a strong oxidizing agent [2], for example, in the oxidation of phenols to hydroquinones, known as a "Teuber reaction" [3].

The first organic NR was prepared and named by O. Piloty and B. G. Schwerin as "porphyrexide" **3** in 1901 (Figure 5.1) [4], only one year after the historical discovery of the triphenylmethyl radical by one of their contemporaries, M. Gomberg [5]. Although they could not infer it to be a radical species, much attention was paid to the unusual and formal four-valency of the oxygenated nitrogen atom, together with its high oxidizing ability, similar to Fremy's salt, as was noted in the same paper. The fact that porphirexide is actually a radical species was first elucidated by Holden *et al.*, by means of ESR spectroscopy, just half a century after its discovery [6].

The next remarkable advance in nitroxide chemistry was made by H. Wheland *et al.* [7] and K. Meyer *et al.* [8]. They prepared a large number of diaryl NRs like **5** or **7**, either by reacting nitrosobenzene derivatives with aryl Grignard reagents and, following oxidation of the resulting hydroxyamines, like **4** or by oxidizing diphenylamines, like **6**, as shown in Figure 5.2.

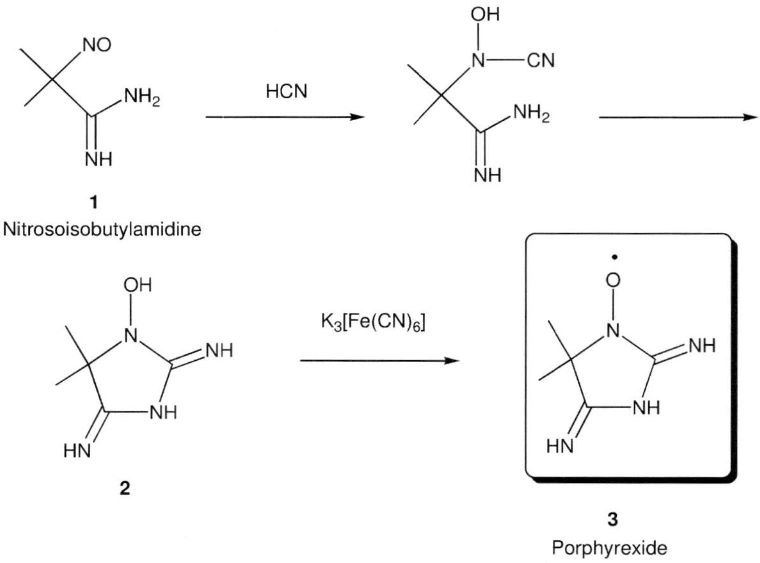

162 | 5 Preparations, Reactions, and Properties of Functional Nitroxide Radicals

Figure 5.1 Synthesis of porphyrexide.

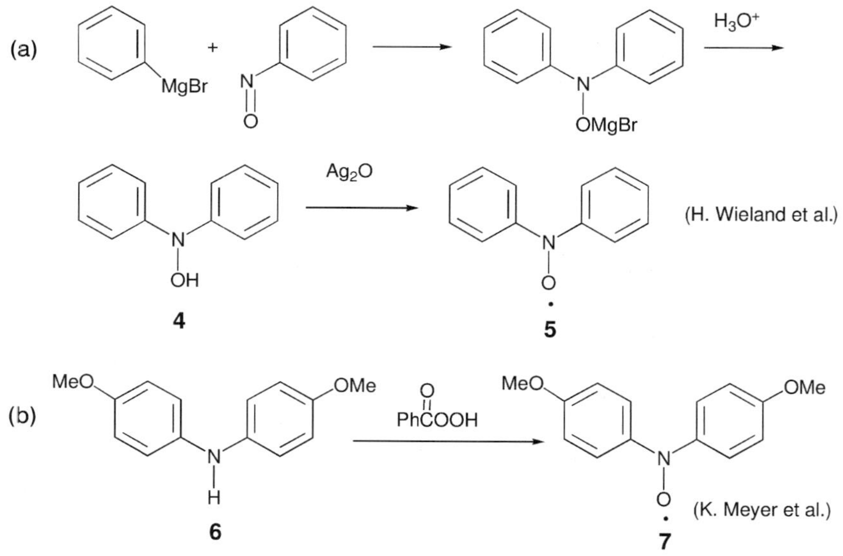

1 Nitrosoisobutylamidine

3 Porphyrexide

Figure 5.2 Preparation of diphenyl NRs.

(H. Wieland et al.)

(K. Meyer et al.)

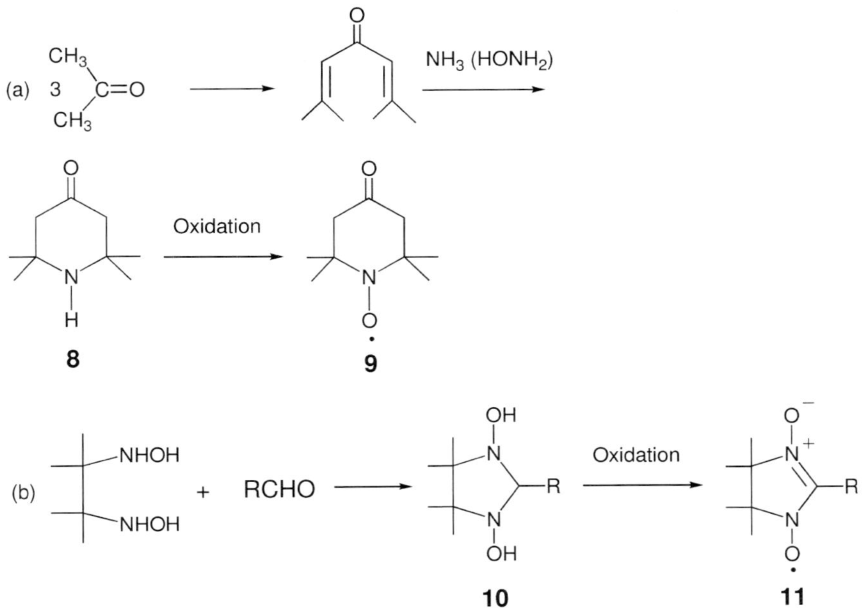

Figure 5.3 (a) Preparative scheme for 4-oxo-TEMPO; (b) general preparative scheme for NN radicals.

Several decades after Levedev et al. [9] prepared 4-oxo-TEMPO (2,2,6,6-tetramethyl-1-piperidinyloxy) **9** from acetone and ammonia [9] (Figure 5.3a), a wide variety of TEMPO derivatives may be found to be available from the radical and some of them are even available commercially. In this context, a paper by Neiman and Rozantsev [10] substantially affected the development of this area by introducing NR reactions without direct involvement of the spin center into synthetic chemistry, and since then rapid and extensive progress has been made in this field of chemistry [11]. It is to be noted that Russian schools have significantly contributed to the progress of the chemistry of NR as is impressively outlined in the books of Rozantsev, Voldarsky, and Vordarsky et al. [12]. Since NRs have been found to be unusually stable, they have further been applied to biological and related sciences, being used as spin-labels, spin-probes, and so forth [13]. Some time later, Ullman et al. prepared another important class of NRs, namely the nitronyl nitroxide (NN) **11** (NN; 2-substituted-4,4,5,5-tetramethylimidazoline-1-oxy-3-oxide) radicals, from alkyl- or arylaldehydes and 2,3-dimethyl-2,3-bis(hydroxylamino)butane (Figure 5.3b) [14]. These stable NRs have been widely used not only as spin-labels or spin-probes in biological studies but also more recently as the building blocks for organic or molecule-based magnetic materials (*vide infra*) [15].

Concerning synthetic aspects of the chemistry of NRs, there are several excellent books and reviews, which cover early as well as recent progress in this field [11, 12, 16].

5.2
Early Progress toward FNRs for Organic Magnetic Materials

In recent years, much attention has been focused on the development of organic and molecular-based magnetic materials, and NRs have played crucial roles as spin sources and building blocks [15, 17]. Among these, TEMPO radicals and NN radicals are the two representative classes of compounds which have been widely used and studied for this purpose (Chart 5.1).

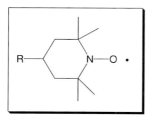
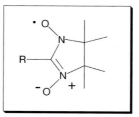

TEMPO Radicals   Nitronyl Nitroxide Radicals

Chart 5.1

Researches toward preparing organic ferromagnets have been, and still are, dreams for materials chemists, and many studies have so far been made using NRs. Soon after 4-oxo-TEMPO was reported to be very stable a lot of its derivatives were prepared and their magnetic properties were investigated. In 1972, Veyret et al. prepared and reported on the magnetic behavior of several TEMPO derivatives [18], among which the ferromagnetic interaction observed in TANOL

Figure 5.4 Structural formula of TANOL suberate **12** (upper) and polymerization of a TEMPO-substituted biradical **13** (lower).

(TEMPO) suberate **12** (Figure 5.4, upper) received much attention and was believed to be the first organic ferromagnet for some time. However, it later proved not to be a genuine ferromagnet, as the spins in ferromagnets should align in the same direction in the bulk material (*vide infra*) and this was not the case. Ovchinnikov *et al.* reported a surprising result in 1986 [19] (Figure 5.4, lower). They prepared a TEMPO-substituted biradical **13** with a butadiyne unit and polymerized it by heating or by irradiating it in the solid state to give black powders, which they claimed to be a ferromagnet: but it was later proved to be a mistake due to trace amounts of contaminating magnetic metals. However, this work evoked further progress toward the construction of organic or molecular-based magnetic materials. Before long the first molecular-based ferromagnet was reported, by Miller *et al.*, and the substance was decamethylferrocene·TCNE complex without a stable radical, which showed the T_c-value (Curie temperature) of 4.8 K [20].

5.3
Organic Ferromagnets Based on NRs

Before starting to read this section, it is recommended that readers refer to Chapters 1–3 of this book concerning theoretical considerations on the magnetism of NRs at low temperature.

In 1991, the first purely organic ferromagnet was reported by Kinoshita *et al.* [21] and the substance was a β-phase crystal of *p*-nitrophenyl NN (NPNN) **14**

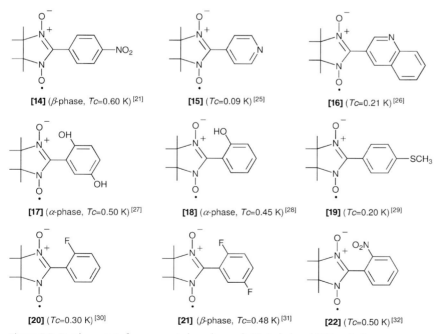

Figure 5.5 Purely organic ferromagnets based on substituted-phenyl NN radicals.

(Figure 5.5), prepared by the method of Ullman et al. [22], although the Curie temperature is extremely low (0.6 K). From crystal structure analysis, a 2D network through the weak intermolecular contacts between the O atoms in the NO groups and N atoms in the NO_2 groups is reported to be responsible for the phase giving the ferromagnetic behavior [21, 23]. The phase transition to the ferromagnetic state below Tc was rigorously evidenced by several experiments such as ac magnetic susceptibility measurement, magnetic heat capacity data, and zero-field muon spin rotation measurement [21, 24].

Since this important discovery, the search for related organic ferromagnets has developed rapidly and diversely, and over 20 kinds of organic ferromagnets have been developed mostly in the late 1990s. As for the NN derivatives, several examples of ferromagnets have so far been discovered (15–22) after p-nitrophenyl NN 14, and these are summarized in the same scheme together with their phase transition temperatures (Tc-values) [25–32]. Their Tc-values are usually very low, ranging between 0.09 K and 0.60 K, and the radical 14 has the highest Tc-value in this class of organic ferromagnets [24].

Along with the development of organic ferromagnets based on NN radicals, the search for organic ferromagnets based on TEMPO radicals has also been studied extensively. Nogami et al. found six ferromagnets 23–28 (and six metamagnets), as indicated in Figure 5.6, among 165 radicals prepared. These all are benzylideneamine derivatives (Schiff bases) and have been prepared by condensation of benzaldehyde derivatives and 4-amino-TEMPO. However, again, their Tc-values are very low ranging between 0.18 K and 0.4 K [33].

The highest Tc-value so far reported among the NR-based magnetic materials is that of 1,3,5,7-tetramethyl-2,6-diazaadamane-N,N'-dioxyl (α-phase) 30, prepared by Rassat et al. [34], and it marks 1.48 K [35]. The radical was reasonably designed

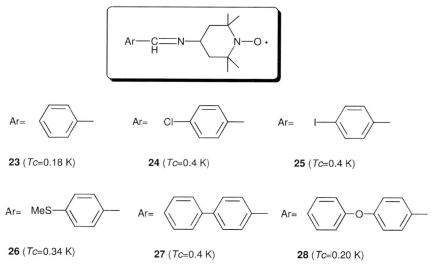

Figure 5.6 Purely organic ferromagnets based on benzylideneamino-TEMPO radicals.

to give orthogonal arrangement of the SOMO orbitals of the diradical spin centers, ensuring Hund's rule, and thus leading intramolecular ferromagnetic interaction. The elaborate preparation was carried out starting from 1,5,7-trimethylnorpseudopelletierine **29** according to Figure 5.7 and the radical **30** thus prepared in 13% yield from **29** actually showed a ferromagnetic transition at 1.48 K owing to the favorable three-dimensional network of NO chains for intermolecular interactions.

Accumulation of data on structure and property relations of organic ferromagnets has gradually revealed that designing and preparing an organic ferromagnet is quite a difficult task, since ferromagnetic order of spins is generally be necessary in a 3D manner to attain ferromagnetism; that is, it is a bulk property of materials and the realization of it depends on a very delicate balance of a crystal structure of the spin systems. Not only molecular design but also crystal engineering should be considered for designing organic ferromagnets and, in fact, only one polymor-

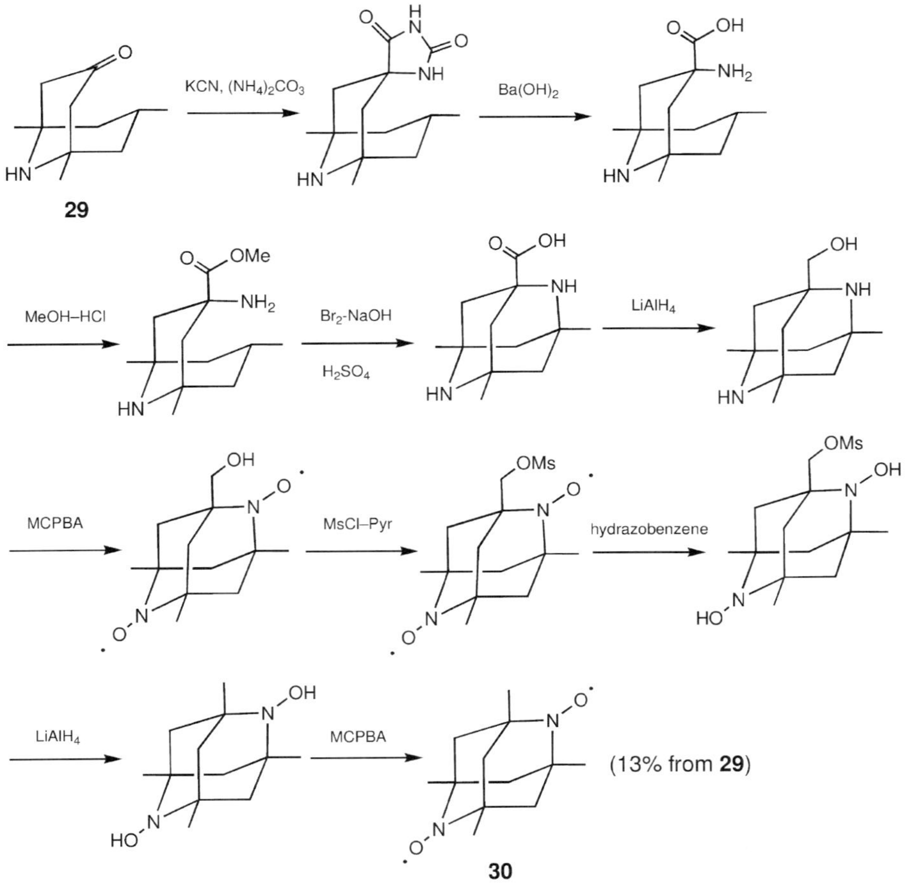

Figure 5.7 Synthesis of 1,3,5,7-tetramethyl-2,6-diazaadamane-N,N'-dioxyl (α-phase) **30**.

phic structure [36] often shows ferromagnetic order in a 3D manner as is impressively exemplified in the β-phase crystal of p-nitro-NN **14**.

5.4
Charge-Transfer Complexes/Radical Ion Salts Based on NR

Besides NR crystals, as exemplified above, ferromagnetic materials are also expected to be available from a charge-transfer complex or a radical ion salt bearing nitroxide radicals, and considerable work has been devoted to preparing such complexes and salts.

It is known that NRs react with bromine or chlorine in inert solvents, and the corresponding oxoammonium salts can be isolated in most cases [37]. The oxoammonium ions are fairly reactive and they are also used as oxidizing reagents [38], as NRs are [39]. A simple TEMPO molecule **31** or 4-hydroxy-TEMPO was found to act as a donor when reacted with strong acceptors, such as TCNQ or TCNQF$_4$, to form the corresponding oxoammonium salts **32** by electron transfer from the nitroxide moiety (Figure 5.8) [40]. On the other hand, electron transfer occurred from the amino group when 4-alkylamino-TEMPO radicals **33** were used as donors to afford the corresponding CT complexes **34** without affecting the radical moiety of the NR, even though these paramagnetic complexes showed room temperature conductivity of $<10^{-6}\,S\,cm^{-1}$. Paramagnetic complexes **35** derived from 4-alkyl-

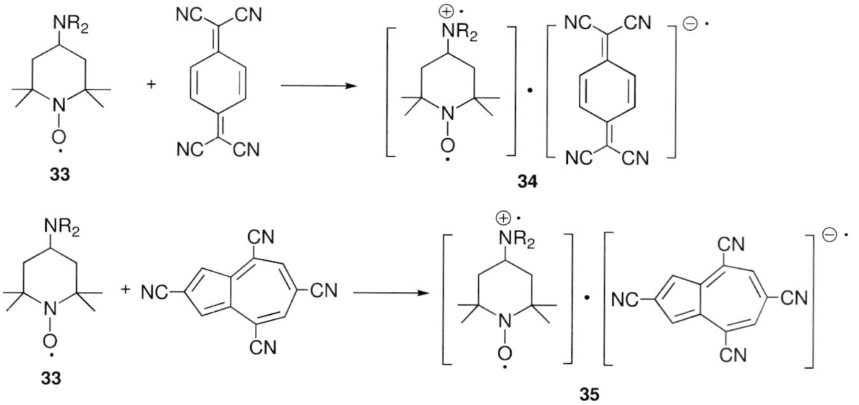

Figure 5.8 CT complex formation based on TEMPO radials.

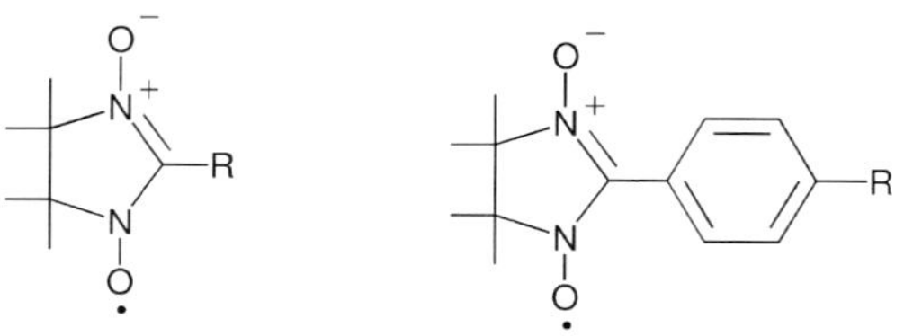

36: R=N(CH$_3$)$_2$
37: R=N(C$_2$H$_5$)$_2$
38: R=N(C$_2$H$_4$)$_2$O

39: R=N(CH$_3$)$_2$

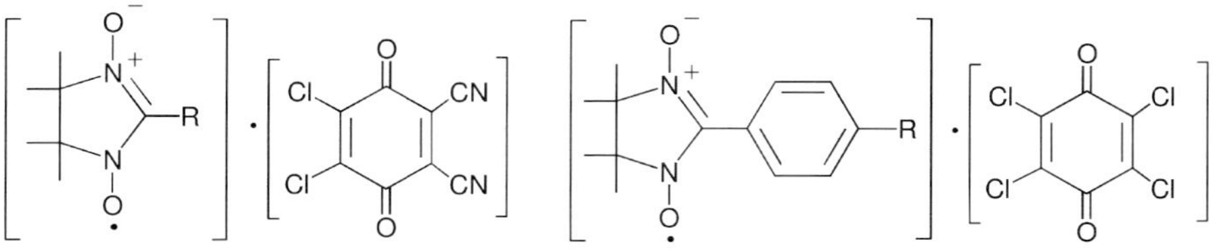

40: R=N(CH$_3$)$_2$

41: R=N(CH$_3$)$_2$

Figure 5.9 CT complex formation based on NN radicals.

amino-TEMPO radicals **33** and 2,4,6,8-tetracyanoazulene, on the other hand, proved to be semiconductive showing room-temperature conductivity $\sigma_{RT} = 1.40 \times 10^{-2} \sim 3.95 \times 10^{-5}$ S cm^{-1} with activation energy $E_a = 0.28\sim0.40$ eV [41].

Sugawara et al. investigated the possible formation of conductive CT complexes by using "spin-polarized donors" such as **36–39** (Figure 5.9) [42]. Based on the findings that one-electron oxidation of the open-shell donors **36–38** affords ground state triplet species, evidenced by ESR measurements, and that the electronic structure of the generated radical cation of **36** is well characterized by the presence of the two singly occupied orbitals, they may be regarded as a hetero-analog of trimethylenemethane (TMM) [43]. A CT complex **40** derived from **36** with DDQ (2,3-dichloro-5,6-dicyano-1,4-benzoquinone) was then prepared and the complex turned out to be an ionic radical salt, judged from the IR spectrum [44]. The complex exhibited a paramagnetic behavior with weak antiferromagnetic interaction ($\theta = -1$ K), presumably due to the large intermolecular antiferromagnetic interaction (J_2), compared with the intramolecular ferromagnetic interaction (J_1). Another CT complex **41** derived from **39** with chloranil was also prepared and shown to have a mixed-stack columnar structure. Since the donor **39** was proved to afford a ground state triplet cation diradical, the complex is considered to satisfy the requirement for the ferromagnetic CT complex [45]. The complex, however, did not exhibit any ferromagnetic interactions, because the degree of the charge transfer is not large enough to cause a sufficient contribution of the charge-transferred state into the electronic structure, owing to the large difference in redox potentials of the donor and the acceptor.

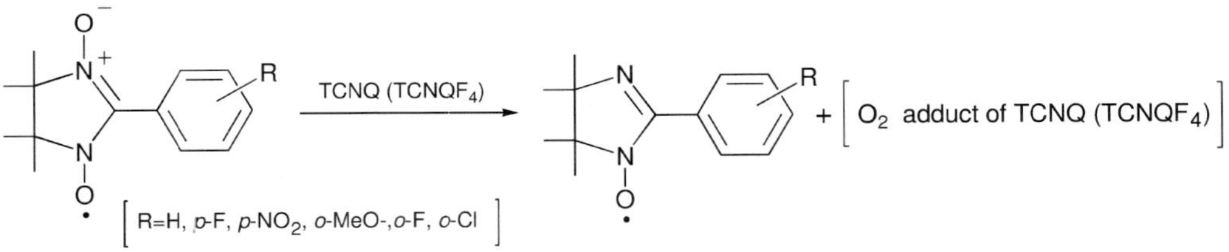

Figure 5.10 Deoxygenation reaction of substituted-phenyl NNs with TCNQ or TCNQF$_4$.

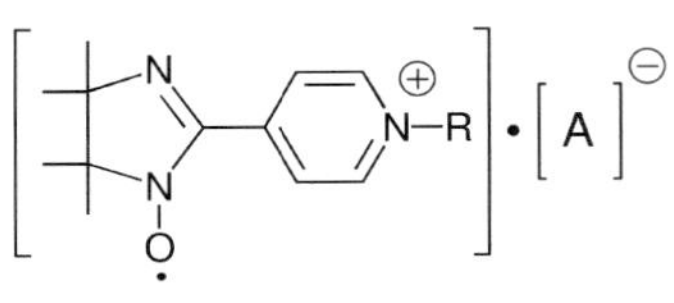

44 (R=CH$_3$, A=TCNQF$_4$, T_c=0.45 K)
45 (R=CH$_3$, A=HCBD, T_c=0.5 K)
46 (R=C$_3$H$_7$, A=TCNQF$_4$, T_c=0.6 K)

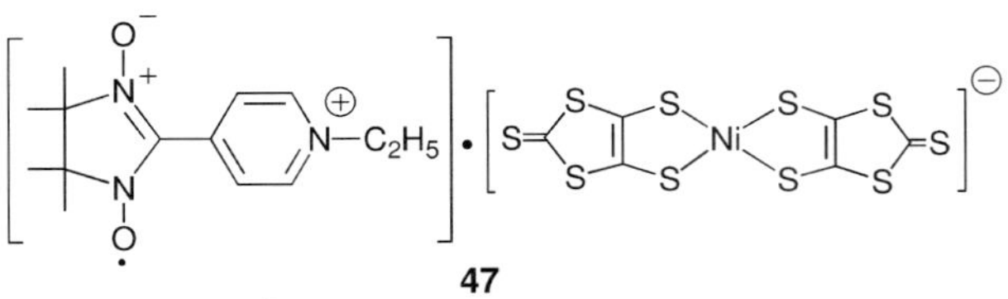

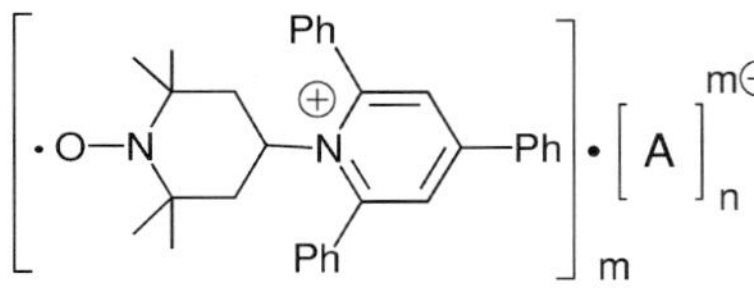

50 (A=TCNQ, m=1, n=1)
51 (A=TCNQF$_4$, m=1, n=1)
52 (A=Ni(dmit)$_2$, m=1, n=1)
53 (A=Pd(dmit)$_2$, m=2, n=1)
54 (A=Ni(dmit)$_2$, m=1, n=6)
55 (A=Pd(dmit)$_2$, n=1, n=6)

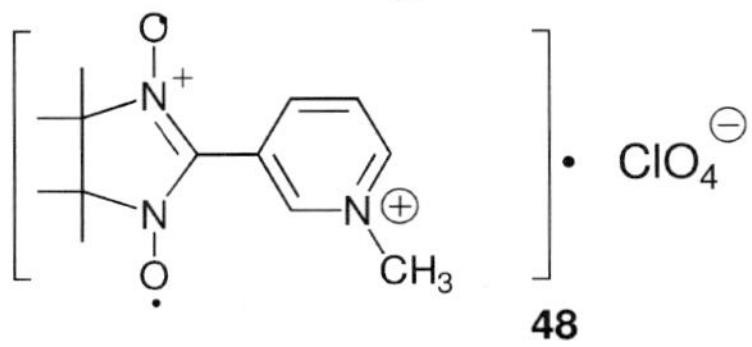

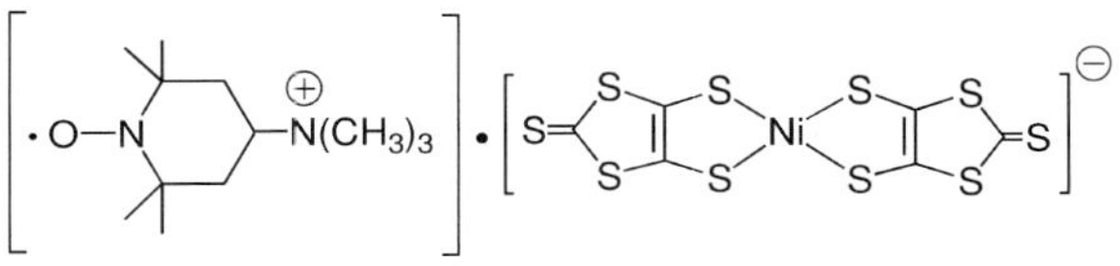

Figure 5.11 Radical cation salts with NRs.

An attempt to prepare the similar CT complexes with **41** from unsubstituted- or substituted-phenyl NNs **42** with TCNQ or TCNQF$_4$ did not give the corresponding CT complexes but instead gave imine nitroxide (IN) derivatives **43** by an anomalous deoxygenation reaction with acceptors in a selective manner (Figure 5.10) [46], though Sugano et al. could obtain TCNQF$_4$ complexes from **39** [47].

Sugimoto et al. prepared three ferromagnetic radical salts of alkylpyridiniums with IN substituents (**44–46**) and their T_c-values range from 0.45 K to 40.6 K (Figure 5.11) [48]. Some other examples of pyridinium or ammonium salts with nitroxide substituents have also been prepared mainly in order to obtain magnetic compounds with conducting property (magnetic conductors) [49–54]. Among

N-alkylaminopyridinium salts with an NN radical in p- or m-position [49], a 1:1 salt of p-N-ethylpyridinium nitronyl nitroxide and Ni(dmit)$_2$ (dmit = 1,3-dithiol-2-thione-4,5-dithiolate) **47** was elucidated by Awaga et al. as an intriguing spin ladder system [50]. Awaga et al. also found the perchlorate salt of m-N-methylpyridinium NN **48** to form a particular Kagome lattice structure [51]. Aonuma et al. prepared mono- and divalent M(dmit)$_2$ salts (M = Ni, Pd; dmit = 1,3-dithiol-2-thioxo-4.5-dithiolato) of TEMPO-substituted trimethylammonium and among them a monovalent (Me$_3$N-TEMPO)[Ni(dmit)$_2$] salt **49** exhibited a single-crystal room-temperature conductivity of $\sigma_{RT} = 4 \times 10^{-3}$ S cm^{-1} and an oxidized salt of (Me$_3$N-TEMPO)$_2$[Pd(dmit)$_2$] (=(Me$_3$N-TEMPO)[Pd(dmit)$_2$]$_4$) a higher conductivity of 1×10^{-2} S cm^{-1} [52]. Organic radical crystals (**50** and **51**) derived from 2,4,6-triphenylpyridinium carrying TEMPO radical (TPP-TEMPO) with TCNQ or TCNQF$_4$ were obtained as a couple of different crystal-forms and their magnetic properties were found to vary depending on the forms [53]. A couple of semiconducting radical salts (**54** and **55**) were prepared by oxidation of [TPP-TEMPO]$_m$[M(dmit)$_2$]$_n$ (**52**: M = Ni, m = 1, n = 1; **53**: M = Pd, m = 2, n = 1) and their room-temperature conductivities σ_{RT} were found to be relatively high, that is, $\sigma_{RT} = 1.55$ S cm^{-1} for the former ($E_a = 0.085$ eV) and $\sigma_R = 6.44$ S cm^{-1} for the latter ($E_a = 0.085$ eV), respectively [54].

In the reverse fashion of pyridiniums or ammoniums with nitroxide substituents, in which NRs are incorporated in cationic moiety to form radical salts with acceptors or anions, a series of sulfonates with nitroxide substituents (**56–72**) have been prepared, in which NRs are included in anionic moiety to afford radical salts with donors or cations (Figure 5.12).

The sodium salt of sulfonates **56** and **57** were prepared by reacting 4-amino- or 4-hydroxy-TEMPO with chlorosulfonic acid and the sulfonates were isolated as tetraphenylphosphonium salts by exchanging the counter cation. The salts were then subjected to metathesis reaction with (TTF)$_3$(BF$_4$)$_2$ to give the corresponding TTF salts of the sulfonates. Weak antiferromagnetic interactions of Curie–Weiss behavior being attributed to the TEMPO radical were observed for the former salts with conductivity of less than 10^{-6} S cm^{-1} [55].

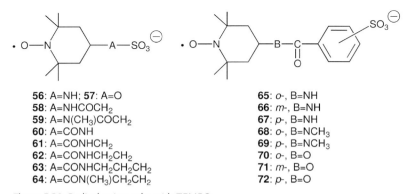

56: A=NH; **57**: A=O
58: A=NHCOCH$_2$
59: A=N(CH$_3$)COCH$_2$
60: A=CONH
61: A=CONHCH$_2$
62: A=CONHCH$_2$CH$_2$
63: A=CONHCH$_2$CH$_2$CH$_2$
64: A=CON(CH$_3$)CH$_2$CH$_2$

65: o-, B=NH
66: m-, B=NH
67: p-, B=NH
68: o-, B=NCH$_3$
69: p-, B=NCH$_3$
70: o-, B=O
71: m-, B=O
72: p-, B=O

Figure 5.12 Radical anion salts with TEMPO.

A couple of radical salts derived from the sulfonate **58**, α-(BEDT-TTF)$_3$(TEMPO-NHCOCH$_2$SO$_3$)$_2$·xH$_2$O (BEDT-TTF: bis(ethylendithio)-TTF; x = 2 or 6) showed semiconducting properties with relatively high room-temperature conductivity (σ_{RT} = 0.91 S cm^{-1} for x = 2 and 0.20 S cm^{-1} for x = 6) and small activation energy (E_a = 0.041 eV for x = 2 and E_a = 0.050 eV for x = 6) together with magnetic behavior based on 1-D Heisenberg model with weak antiferromagnetic interaction for x = 2 (J = −0.81 K) or ferromagnetic one for x = 6 (J = +0.42 K) [56]. Another BEDT-TTF-based radical salts including **59**, α-(BEDT-TTF)$_2$(TEMPO-N(CH$_3$)COCH$_2$SO$_3$)·3H$_2$O was found to be semiconductive (σ_{RT} = 0.19 S cm^{-1}, E_a = 0.65 eV) and showed semiconductor-to-semiconductor transition at 250 K, which decreased with increasing static pressure [57].

Although relatively poor conductivities were revealed in the radical salts derived from **60** to **72** and TTF, BEDT-TTF or TMTSF (tetramethyltetraselenafulvalene), the structure–property relations have been clarified [58]. In general, the intermolecular magnetic interactions were proved to be weak and the behaviors were dependent on their crystal structures; that is, packing motifs of donors/anions but unique behaviors were observed in some salts. For example, the magnetic behavior of α-(BEDT-TTF)$_5$(TEMPO-CONHCH$_2$SO$_3$)·6H$_2$O salt derived from the sulfonate **61** could be well explained by the 1 : 1 combination of a ST (singlet-triplet) model from the spins on BEDT-TTF pentamer units (J_{ST} = −41 K) and a CW (Curie–Weiss) model from those on TEMPO groups.

5.5
Donors and Acceptors Carrying NRs and the Derived CT Complexes/Radical Salts

In order to verify experimentally the theoretical prediction for constructing organoferromagnetic conductors [59], several donors and acceptors carrying NRs were prepared to give the corresponding CT complexes/radical salts as shown in Figure 5.13.

A ferrocene compound carrying a nitronyl nitroxide substituent **73** was prepared as an example and it formed a CT complex with DDQ showing antiferromagnetic coupling between the ferrocenium ion and radical spin at low temperature as evidenced by susceptibility measurements [60].

A variety of TTF derivatives with NR substituents have been prepared in recent years [61], starting from the first report of Sugano et al. on the TTF derivative **74** and its TCNQF$_4$ complex [47]. The ESR data on the radical cations of TTF derivatives **74** and **75** suggested the presence of intramolecular spin–spin exchange between the TTF radical cation and nitroxide radicals in solution, although the magnetic behavior in susceptibility measurements on the CT complexes or radical cation salts of **74** and **75** with TCNQF$_4$ or I$_2$ indicated that even the short-range spin alignment of radicals could not be achieved in them in spite of the differences of the magnetic properties depending on the kind of NR substituents [62]. A tetrakis-TEMPO-substituted TTF derivative **76** and its iodine salt were prepared but did not show eminent magnetic and conducting properties [63]. Another type of

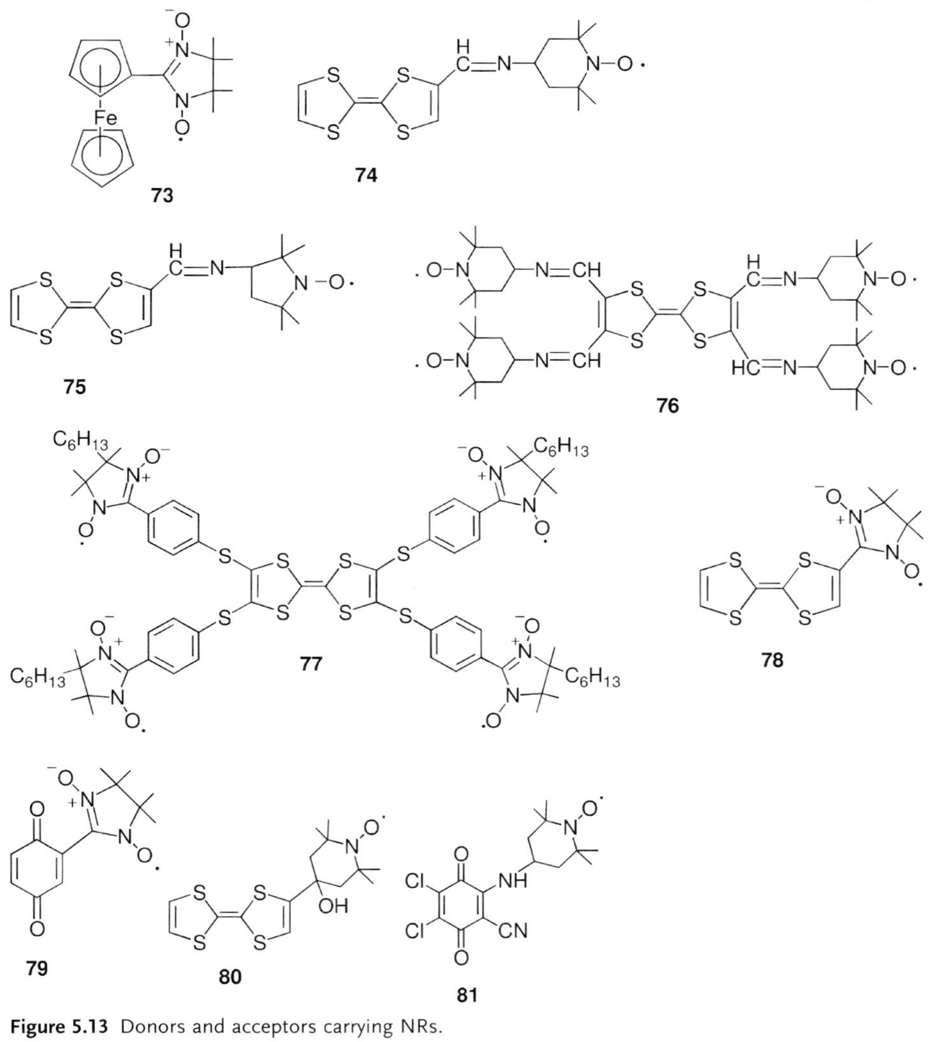

Figure 5.13 Donors and acceptors carrying NRs.

tetrakis-NN-substituted TTF **77** gave the singly oxidized species with a ground state sextet spin-multiplicity but did not give a corresponding CT complex [64].

Significant intramolecular interaction was observed for the radical salt of cross-conjugated TTF derivative **78** and the radical anion salt derived from benzoquinone derivative **79**, as estimated from their ESR measurements, in which relatively large antiferromagnetic interaction was found for the former while ferromagnetic interaction was apparent for the latter [65]. An unusual magnetic property of the DDQ complex of **78** was reported which could not be explained by simple magnetism [66].

A 4-hydroxy-TEMPO-substituted TTF derivative **80** formed CT complexes with I$_2$, DDO, TCNQF$_4$, or 4-amino-TEMPO-substituted benzoquinone derivative **81**, but the complexes showed only weak antiferromagnetic interactions in all cases with apparent decrease of magnetic susceptibilities, indicating the partial formation of singlets between the unpaired electrons despite the difference of the behaviors depending on the acceptors [67].

A TEMPO-substituted TTF derivative **82** (Figure 5.14) was prepared by Fujiwara and Kobayashi to give semiconductive radical cation salts with Au(CN)$_2$ exhibiting room temperature conductivity of about 10^{-3} S cm^{-1} with an activation energy of 0.20 eV and the magnetic behavior obeying a singlet-triplet model [68]. Another TTF derivative having PROXYL-group **83** was also prepared by the same authors to give a radical salt with iodine but the salt was found to be an insulator with a very low room temperature conductivity and showed Curie–Weiss behavior with weak antiferromagnetic interaction [69]. An extended type of **83**, a TTP (tetrathiapentalene) derivative containing a PROXYL radical **84** was prepared more recently and it gave semiconductive salts with ClO$_4$ or FeCl$_4$ showing relatively high room temperature conductivity up to 1.1×10^{-3} S cm^{-1} [70].

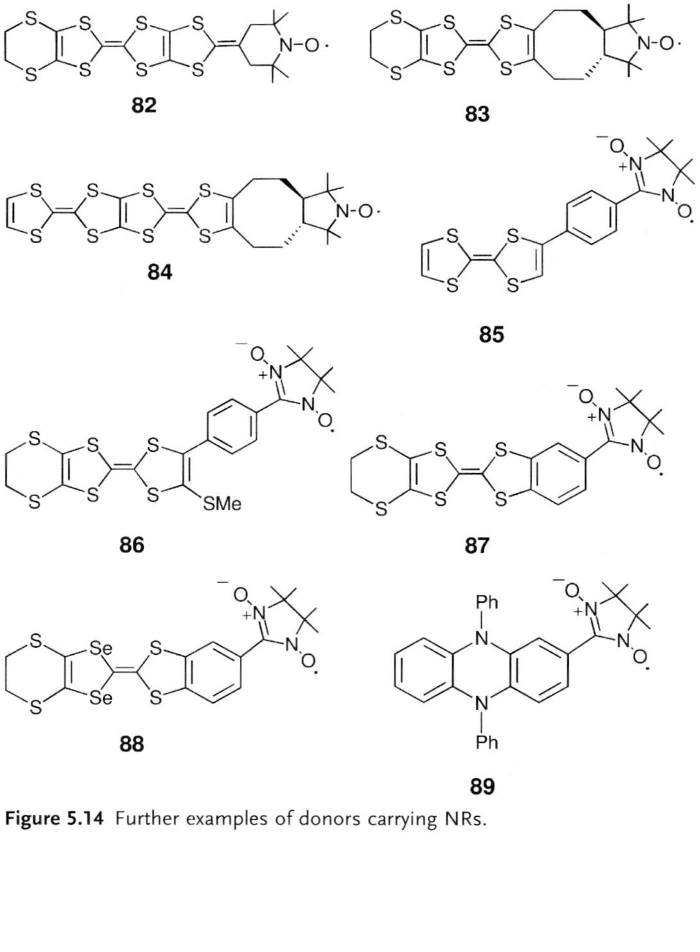

Figure 5.14 Further examples of donors carrying NRs.

In order to enhance the planarity of the radical moiety and to remove undesirable interactions observed in **78**, a couple of TTF derivatives with a NR and a p-phenylene group in between (**85, 86**) were prepared [71]. Even though they were proved to have triplet ground states upon one-electron oxidation, no crystalline radical salts were obtained. On the contrary, a benzo-annulated derivative **87** was found to give crystalline radical salt of the 2:1 donor-to-anion ratio with ClO_4 and to exhibit semi-conductivity with a room temperature conductivity of $10^{-2}\,S\,cm^{-1}$ and an activation energy of 0.16 eV [72]. Quite recently, a diselena analog of **88** was synthesized and a radical salt with ClO_4 was found to show negative magnetoresistance below 15 K, which is the first example detecting the interaction between localized spin and conducting electrons in a genuine organic system [73].

When NN-substituted-5,10-diphenyl-5,10dihydrophenazine **89** was used as a donor, it was found to give a stable radical cation with ClO_4, which showed strong intramolecular ferromagnetic coupling [74].

5.6
Suprampolecular Spin Systems Carrying NRs

Preparing supramolecular spin systems with specific function, in which NRs play indispensable rolls is also of recent interest.

A few decades ago, several crown compounds carrying NRs were prepared for use as spin-labeled ionophores. A TEMPO-substituted benzo-15-crown-5 **90** (Figure 5.15) was prepared and the potassium complex revealed a strong spin–spin interaction due to formation of the sandwiched dimer being investigated on the basis of the ESR parameters of the triplet cluster [75]. Three spin-labeled crown ethers **91–93** were then synthesized and studied by ESR [76]. The ESR studies of hyperfine splitting and Heisenberg spin exchange suggested that these spin-labeled crowns are poor complexing agents for alkali metal cations. Two stereochemically distinct chiral crown ethers with di-nitroxide substituents **94, 95** were synthesized and the presence of a spin–spin interaction for the *syn*-isomer was used as a spectral parameter to follow the binding of a potassium ion to the crown ether **94** [77]. A couple of nitroxide crown ethers **96, 97** and a nitroxide cryptand **98**, in which the NO group is thrust toward the cavity of the molecule, were prepared and their complexation properties were investigated in detail [78]. While the a_N values of these NRs were not sensitive to the presence of alkali metal cations in methanol, the nitroxide cryptand was found to form a 1:1 complex with $NaBH_4$ in $CDCl_3$.

Preparations of calix[n]arenes carrying NRs have been developed more recently. Calix[n]aryl esters **99** (n = 4) and **100** (n = 6) (Figure 5.16) carrying two TEMPO radicals on the lower rim were synthesized and metal-induced conformational changes were investigated [79]. While a Na^+-induced rotation of the carbonyl groups of the cone form in **99** was not enough to change the exchange interaction being estimated by ESR spectra, the exchange interaction in **100** was found to change in response to the metal-induced 1,2,3-alternate-to-cone conformational

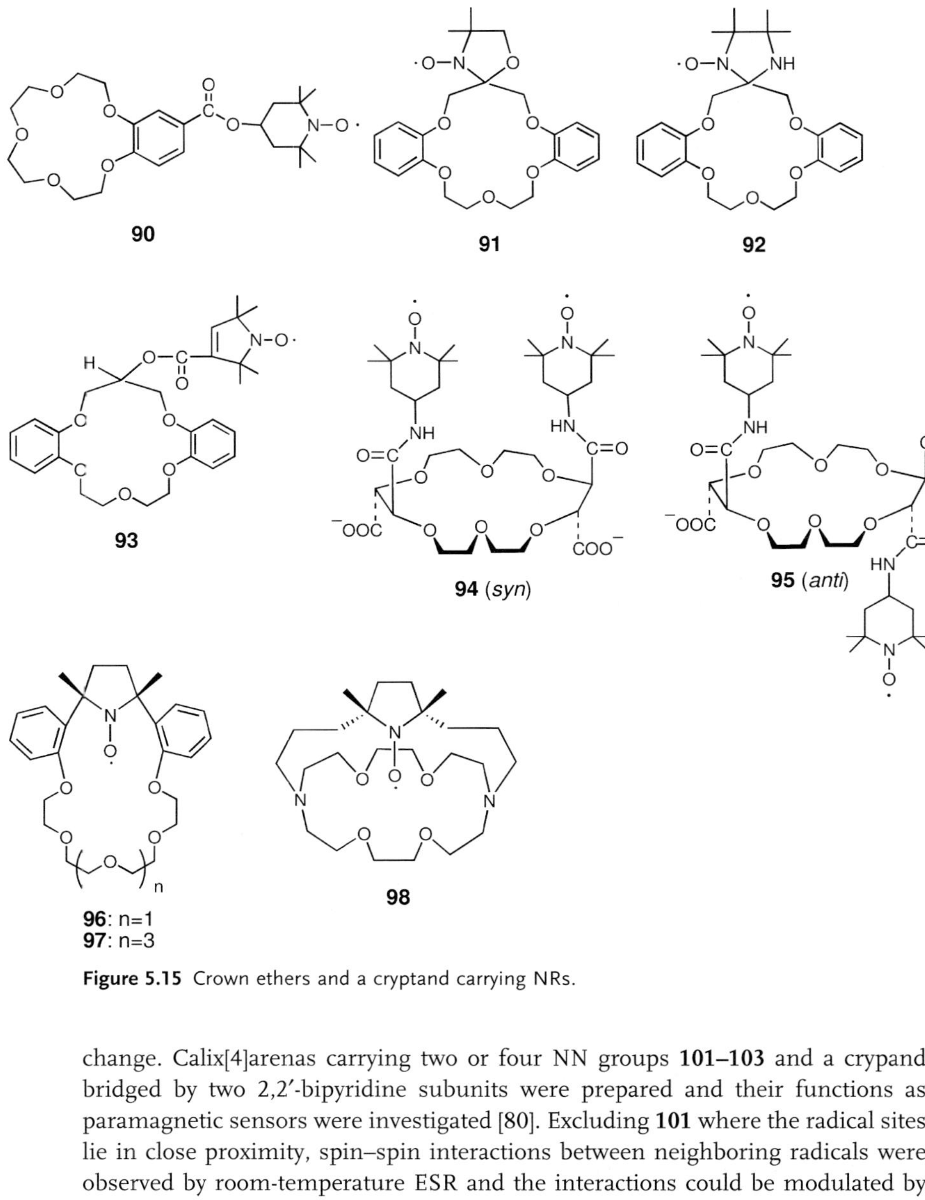

Figure 5.15 Crown ethers and a cryptand carrying NRs.

change. Calix[4]arenas carrying two or four NN groups **101–103** and a crypand bridged by two 2,2′-bipyridine subunits were prepared and their functions as paramagnetic sensors were investigated [80]. Excluding **101** where the radical sites lie in close proximity, spin–spin interactions between neighboring radicals were observed by room-temperature ESR and the interactions could be modulated by insertion of a cation like Zn^{2+} into the available coordination sites.

A stable calix[4]arene **104** carrying two *t*-butylnitroxide groups on the upper rim was prepared, which exhibited a strong intramolecular spin–spin exchange interaction and was found to undergo reversible conformational transitions upon heating [81]. Synthesis and magnetic characterization of a calix[4]arene **105** with four *t*-butylnitroxide at the upper rim was reported [82]. In solution, it has a four-

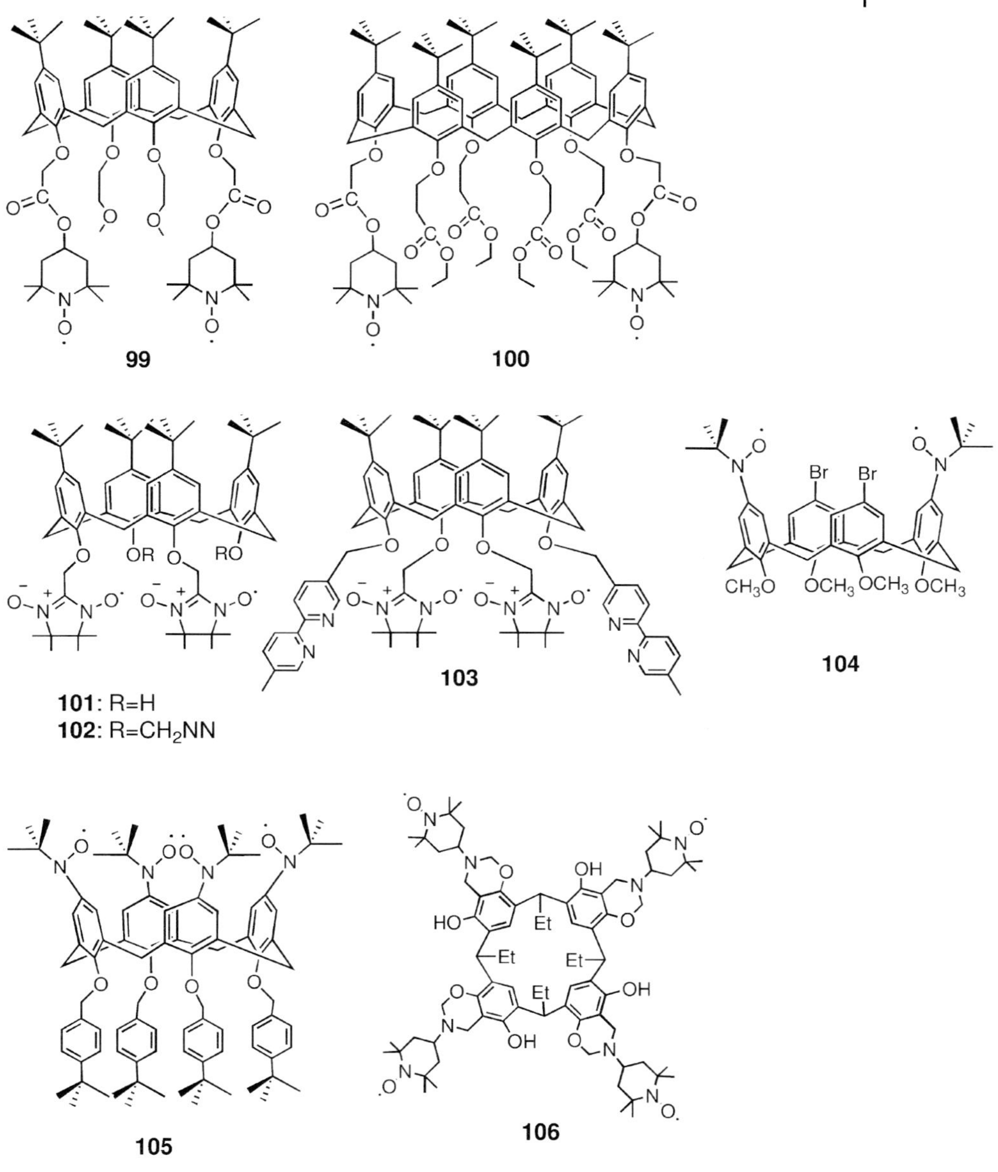

101: R=H
102: R=CH₂NN

Figure 5.16 Calixarenes carrying NRs.

fold symmetric fixed cone conformation on the NMR timescale and small exchange interactions between the radicals (30 K > $|J/k|$ >> 1.8 mK). In the solid sate, dimerization of one diagonal pair of NRs leads to a pinched cone conformation for **105** with strong intradimer antiferromagnetic coupling with $|J/k|$ = 200–300 K. A resorcinarene **106** carrying four TEMPO residues at its wider rim was prepared and it was found to recognize small molecules such as acetonitrile [83]. The effects of exchange interaction between the closely spaced radicals were used to monitor guest occupation in host–guest complexes as evidenced by ESR line-shape.

The use of nucleobases with NRs as building blocks is one of the attractive approaches for constructing supramolecular spin systems, and considerable efforts have been made in recent years along this line, even though a lot of nucleosides labeled with nitroxide radicals have so far been developed mainly to analyse DNA structures and DNA–protein interactions.

A uracil-substituted NN **107** (Figure 5.17) was prepared as an example, and was characterized chemically as well as structurally [84]. In this case, only a small amount of spin density was observed on the uradinyl ring in **107** having a high torsion angle (circa 65°) between the component heterocyclic rings. It was reported to form aggregates in solution, although its poor solubility in apolar solvents prohibited their full structural characterization. In order to improve crystal packing and to increase radical to uradinyl spin delocalization, another type of uracil derivative carrying IN **108** was designed and synthesized [85]. The radical **108** is much flatter than **107**, with a biannular dihedral angle of only 14.5° and shows antiferromagnetic spin pairing with $2J/k = -14$ K, attributable to a close contact between unpaired spin density on the imidazole-type nitrogen atoms. Hydrogen bonds aid dimer formation, even though they do not appear to play an electronic role in the magnetic behavior. Moreover, the radical was found to bind to hydrogen-bonding complement 2,6-di(propylamido)pyridine in chloroform forming a couple of supramolecular assemblies [86].

A cytosine-substituted NN radical **109** was synthesized and it was found from X-ray structure analyses and magnetic susceptibility data that the intermolecular magnetic interactions in the crystal are propagated by orbital overlaps between the phenyl NN moieties of the adjacent molecules, whereas the relative arrangement of the molecules is governed by the hydrogen bonding of the nucleobase moiety, giving a double chain structure [87]. More recently, a reverse Watson–Crick-type molecular complex of thymine and adenine bases carrying a NN radical **110**, **111** was also synthesized, which forms by intermolecular hydrogen bonds a double-chain spin system in the crystal [88]. The solid state magnetism of the complex could be well explained by the double-chain model with the periodic four-spin cluster.

Several spin-labeled nucleosides **112–114** with an N-t-butylnitroxide radical were prepared to investigate the stability and behavior of the radical on a necleobase [89]. An ESR study showed that the radicals were stable and that their hyperfine structure was dependent on the position of the NR and was also affected by electron density of pyrimidine.

A nucleoside labeled with a NN group **115** was designed and prepared together with oligonucleosides containing one or two **115** units [90]. The spin signals from

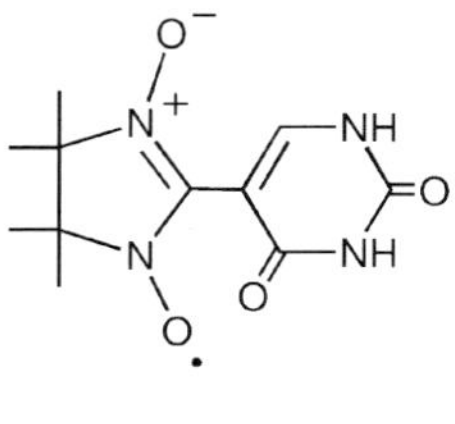

Figure 5.17 Nucleobases and nucleosides carrying NRs.

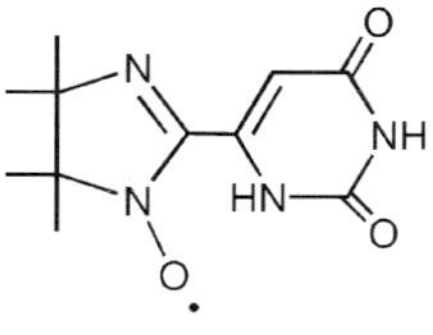

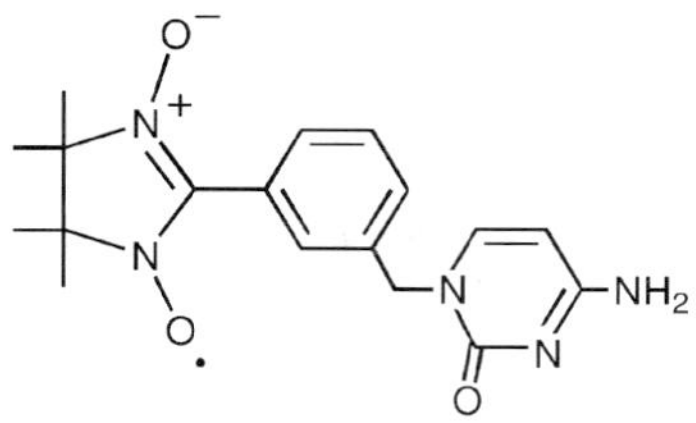

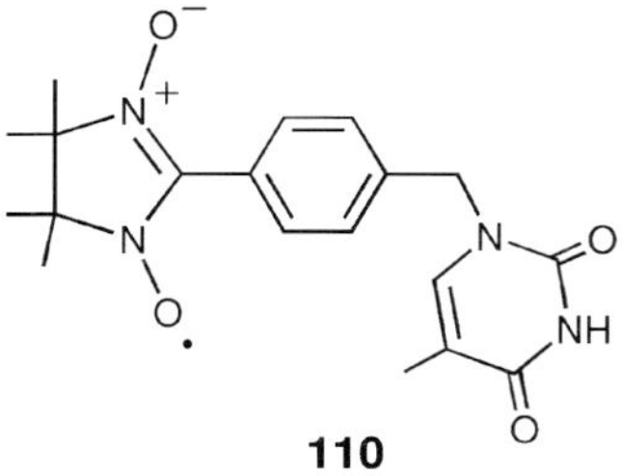

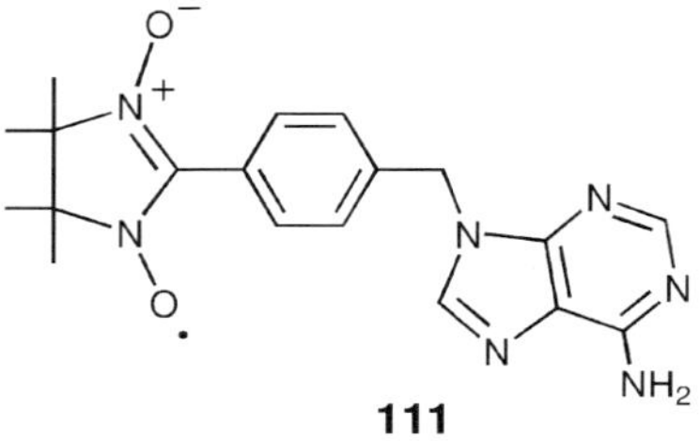

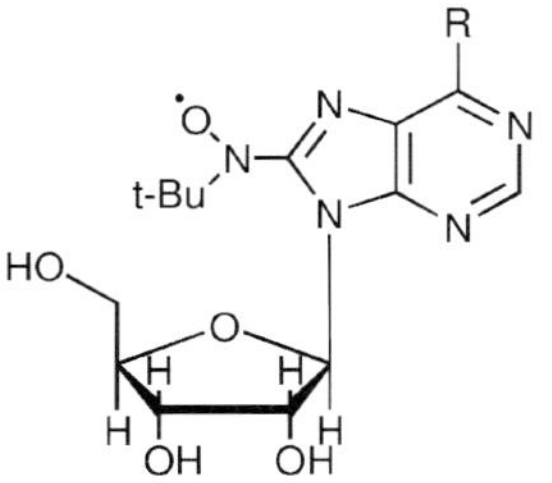

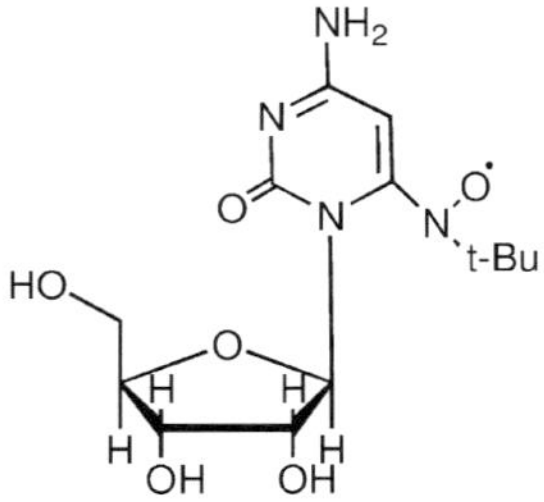

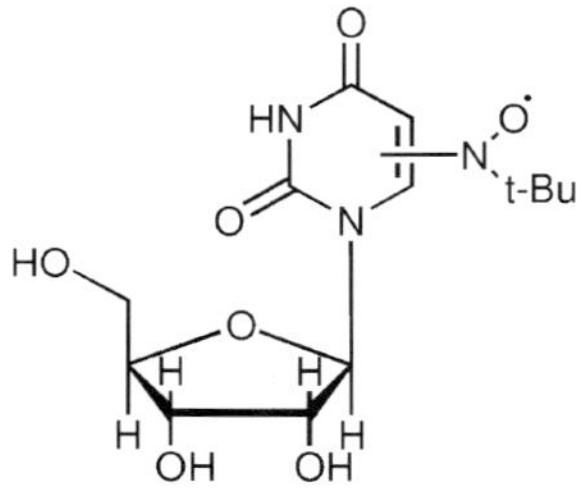

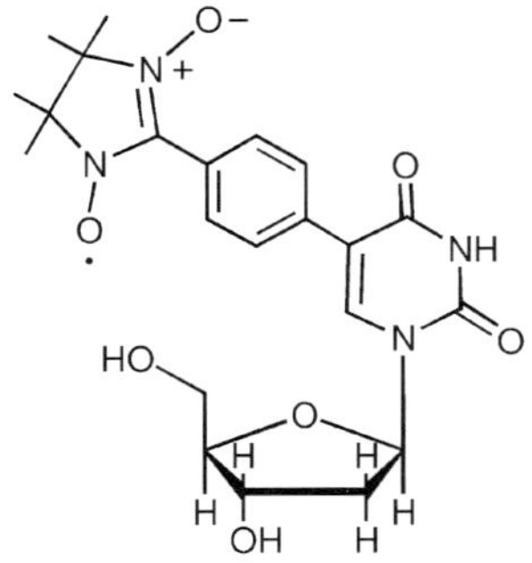

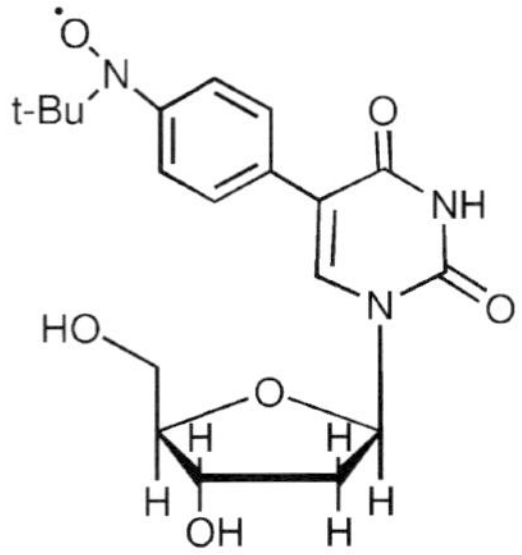

the oligonucleosides were found to vary because of the degree of hybridization of the complementary strand and the distance between the NN spins. Another spin-labeled 2′-deoxyuridine **116**, in which a significant fraction of the spin density is delocalized from a nitroxide label to the base residue, was prepared as a crystalline solid, stable at ambient conditions [91]. The crystal packing of **116**, which includes multiple hydrogen bonds, leads to one-dimensional chains of molecules with predominant intrachain ferromagnetic coupling and weaker interchain antiferromagnetic coupling.

Much attention has also been given to supramolecular wires and rods incorporating NRs in recent years, and these are often biradicaloid compounds, mainly to demonstrate intramolecular spin–spin interactions.

A couple of acyclic glycol ether ligands with two NR groups at both ends **117** (n = 3,5) (Figure 5.18) were prepared and they were found to be conformationally stabilized to allow the intramolecular spin exchange in the presence of potassium ions (when n = 3) or thiourea (when n = 5) in frozen solutions and the results could be interpreted in terms of conformational effects that mimic allosteric effects [92].

Three m-phenylene ethynylene oligomers **118** (n = 4,5,6) with two TEMPO units were synthesized to investigate the formation of the helical conformation in solution [93]. The ESR data revealed that one helical turn consists of six repeating units within the folded structure and that the spin-labels are spatially closer in the folded structure when n = 5 as compared to other oligomers of n = 4,6.

Long rod-like diradicals, such as **119** or **120** carrying TEMPO groups at both ends, were prepared and their end-to-end distances were determined by means of a new four-pulse double electron electron resonance (DEER) experiment [94]. The results revealed averaged end-to-end distance of **119** to be 2.36 nm and that of **120** 2.77 nm. For distances between paramagnetic centers larger than 3 nm, another technique, single-frequency technique for refocusing (SIFTER), provided better resolution than DEER and good agreement between distances from SIFTER measurements and force-field computations was found for shape-persistent biradicals such as **121** with distances up to 5.1 nm [95]. More recently, a new approach for the experimental characterization of the flexibility of shape-persistent molecules such as **121** has been introduced [96].

A series of organic biradical compounds **122** consisting of an aromatic core (biphenyl, naphthalene, azobenzene, or azoxybenzene) and long alkoxy groups with TEMPO or PROXYL radicals were prepared [97]. The TEMPO derivatives were found to show significant intermolecular antiferromagnetic interactions (J = −34~−45 K) being well expressed by a singlet–triplet model irrespective of the aromatic core, and the behavior is understandable by taking a hand-in-hand like assembled structure into consideration, while only weak antiferromagnetic interactions with Curie–Weiss behavior were observed in the corresponding PROXYL derivatives.

Dendrimers of variable size such as **123** and **124** (Figure 5.19) being attached by TEMPO-based stable radicals were prepared and used to control radical polymerization of styrene, vinyl acetate, and methacrylates [98]. The thermal polymerization

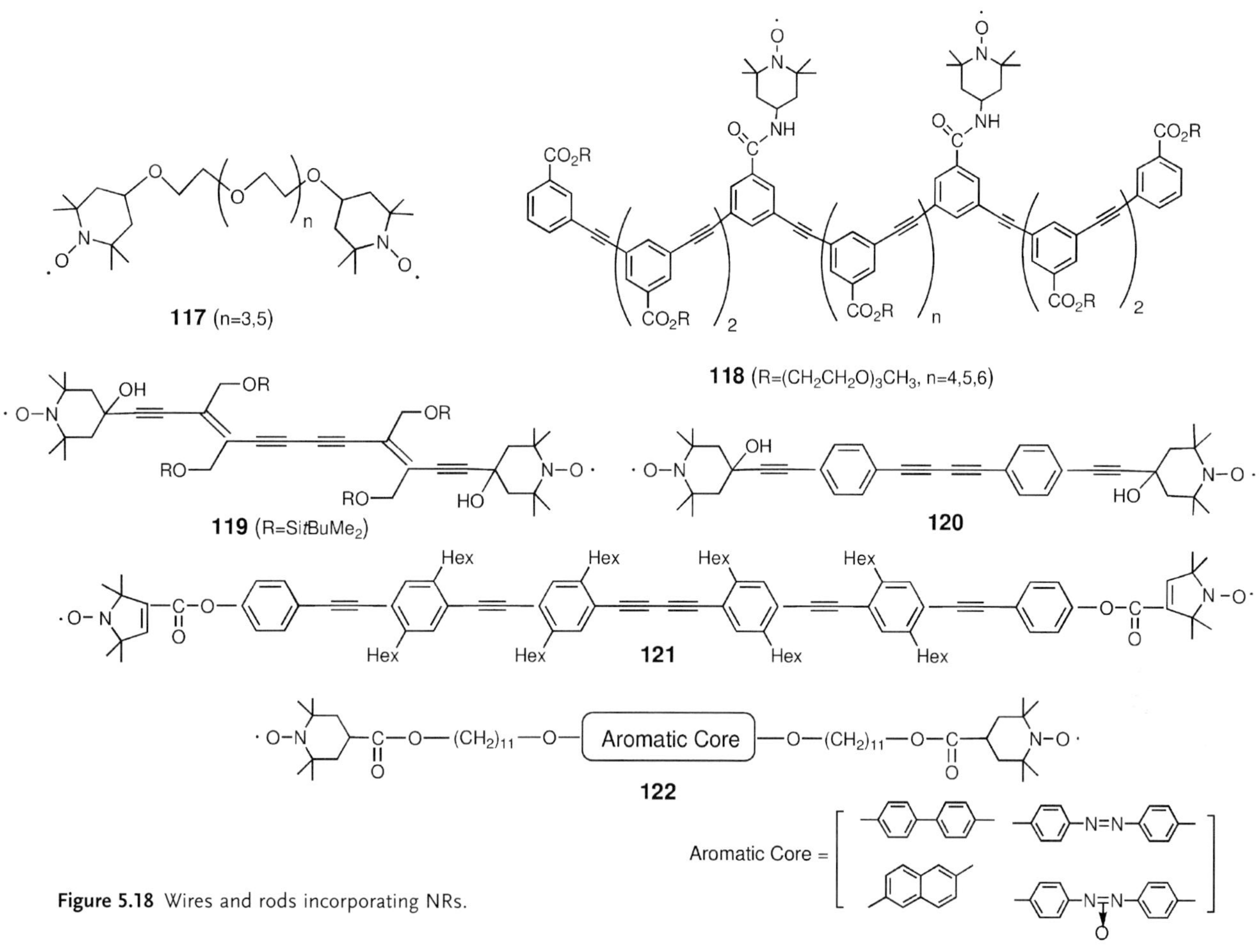

Figure 5.18 Wires and rods incorporating NRs.

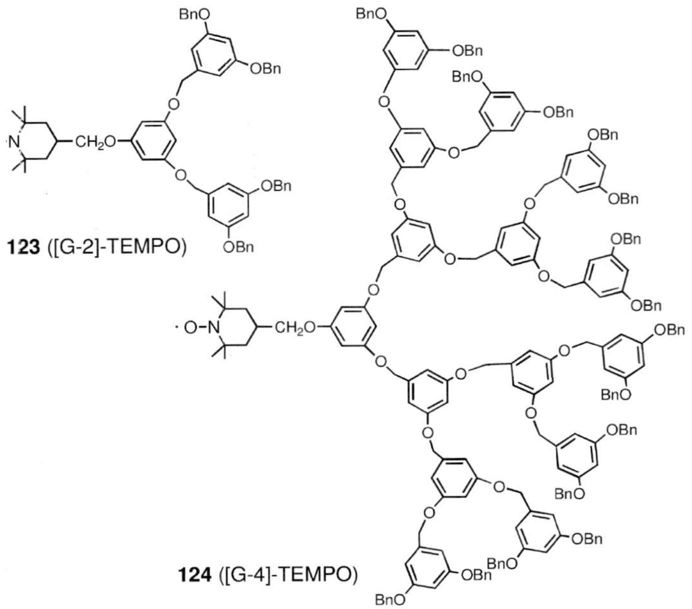

182 | *5 Preparations, Reactions, and Properties of Functional Nitroxide Radicals*

123 ([G-2]-TEMPO)

124 ([G-4]-TEMPO)

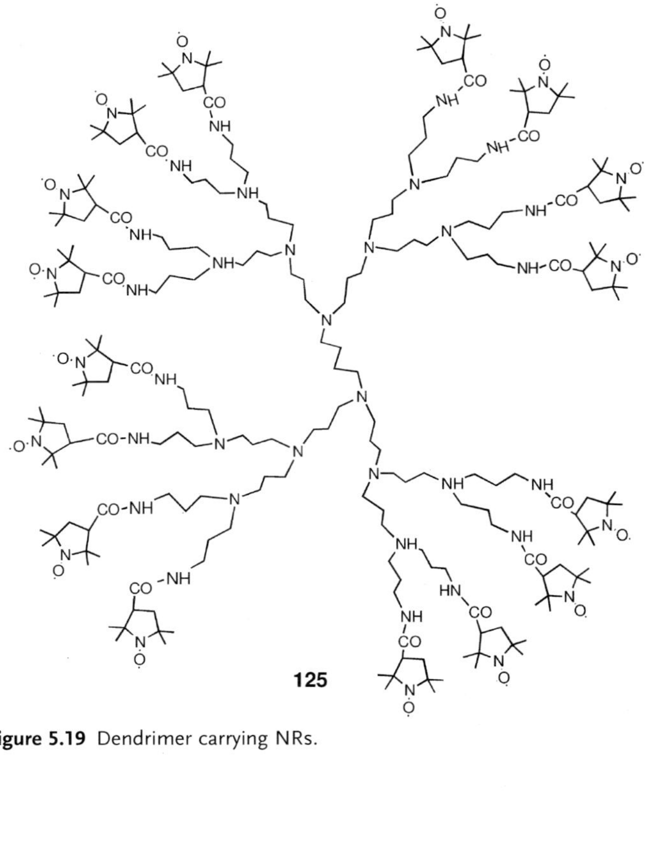

125

Figure 5.19 Dendrimer carrying NRs.

of styrene with **123** proceeded similarly to the polymerization with TEMPO alone. In the polymerization of styrene initiated by benzoylperoxide, the kinetics and the molecular weight/conversion relationships showed the same tendency as with TEMPO itself, though the polydispersity was higher than without dendrimers.

The first five generations of poly(propylene imine)dendrimers have been functionalized with pendant PROXYL radicals [99]. The NRs at the periphery of the dendrimers exhibit a strong exchange interaction resulting in thermally populated high-spin states. For the higher generations like **125**, the ESR spectrum gives direct spectral evidence for interaction between end groups from the progressive exchange narrowing.

A couple of disulfide derivative **126** and **127** carrying a NN group at each of the *p*-positions of phenyl rings were prepared (Figure 5.20) and they were found to form a dense self-assembled monolayer (SAM) on a Au surface through a reductive cleavage of the S-S bond [100]. In the case of **127**, π-radical thiol-derivatized nanoparticles with an average diameter of 4.1 nm were isolated and about one hundred π–radical ligands were found to be chemisorbed on the nanoparticle. Magnetic susceptibility and ESR measurement indicated that the unpaired electrons on the π–radical ligands behave paramagnetically but that they interact electronically with the electrons of the gold nanoparticle.

Another series of Au nanoparticles modified with a NR functionalized ligand were prepared by using **128** with a range of spin-label coverage [101]. The X-band ESR spectra of frozen solutions of these nanoparticles showed coverage-dependent line broadening attributed to dipole–dipole interactions between spin labels, and it was found from the analysis of the spectra that if the spin-labeled ligand is substantially longer than the surrounding protecting layer it does not adopt a fully stretched conformation but wraps around the particle immediately above the layer of surrounding ligand.

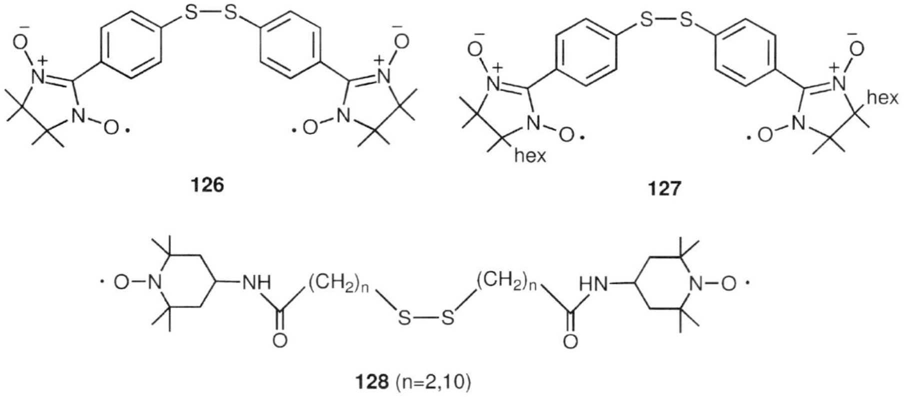

Figure 5.20 Disulfide derivatives carrying NRs for the use of modified Au surfaces or particles.

Organic liquid crystalline radicals also belong to an important class of supramolecular spin systems and the readers are recommended to consult the topics in Chapter 9 of this book.

5.7
Photochromic Spin Systems Carrying NRs

Recently, much attention has been focused on the development of organic photofunctional spin systems and progress started to be made around a decade ago or so in which photochromic spin systems carrying NRs played significant and indispensable roles [102]. Hence, it is thought to be useful for readers to be introduced to recent progress in this field of nitroxide chemistry [103].

A *trans*-azobenzene derivative **129a** carrying NN radicals was first reported by Iwamura, Matsuda *et al.* and its ESR spectra in frozen toluene at 10 K was found to differ before and after irradiation and thus photo-induced change of the magnetic behavior could be observed, though the corresponding *cis*-isomer **129b** was not isolated presumably because of the unstable nature of the *cis*-isomer (Figure 5.21) [104].

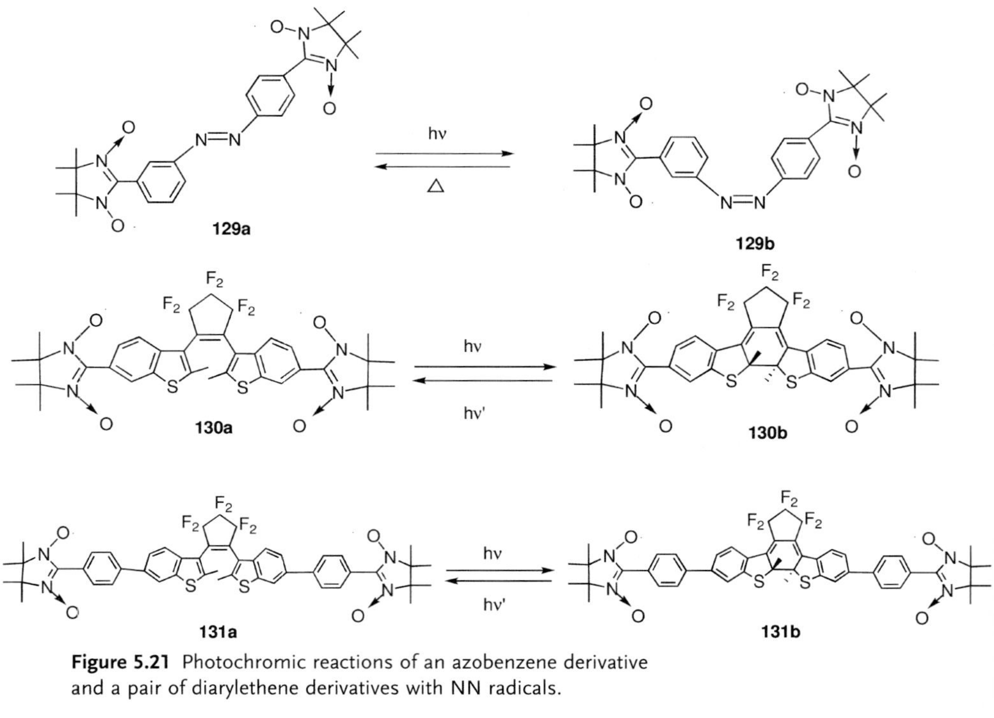

Figure 5.21 Photochromic reactions of an azobenzene derivative and a pair of diarylethene derivatives with NN radicals.

More recently, Matsuda, Irie et al. have prepared a series of diarylethene derivatives carrying NN radicals [105] and a couple of representatives are shown in the same figure.

In these photochromic spin systems, apparent switching of the *intramolecular magnetic interactions* has been realized. For example, the magnitude of the intramolecular antiferromagnetic interaction of $2J/k_B = -2.2$ K (the state of switch off) between the spins in the open-ring isomer **130a** increases appreciably to the lager absolute value of $2J/k_B = -11.6$ K (the state of switch on) in the closed-ring isomer **130b** by photochemical ring closure [106]. A photo-switching of intramolecular magnetic interactions was impressively observed between the spin couples **131a** and **131b** in solution by means of ESR spectroscopy. That is, the initial 15 lines observed for **131a** were found to change to nine lines in benzene solution at ambient temperature. The reason why 15 lines are observed for **131a** is because the exchange interaction between the two radicals is comparable to the hyperfine coupling constant in **131a** by inserting a couple of benzene rings in **130a**, while that of the corresponding closed-ring isomer **131b** is much larger than the hyperfine coupling constant. Thus, the strength of exchange interaction was switched by more than two orders of magnitude in the spin couplers [107].

Two topological isomers of 9,10-diphenylanthracene carrying two IN radicals at *para*-positions (**132a**) and at *meta*-, *para*-positions (**133a**) were designed and synthesized together with another topological isomer with the same radicals at *meta*-positions (Figure 5.22, upper) [108, 109]. Time-resolved ESR (TRESR) spectra of the first excited states with resolved fine structure splittings were observed for both **132a** and **133a** in an EPA or a 2-MTHF rigid matrix and the observed TRESR spectra were unambiguously assigned by the spectral simulation as an excited quintet ($S = 2$) spin state (**132b** from **132a**) and a superimposition of an excited triplet ($S = 1$) and quintet (**133b** from **133a**), respectively, while high-spin excited states were not observed for another topological isomer, showing the crucial role of π–topology in the spin alignment of the excited states.

The clear detection of the excited quintet high-spin state in **132b** shows that the effective exchange coupling between the two radicals through the diphenylanthracene spin coupler changes from antiferromagnetic to ferromagnetic upon excitation. In the case of **133b**, a low-lying quintet state is supposed to locate closely above the photo-excited triplet, being realized by antiferromagnetic spin exchange coupling between two radical spins through the photo-excited triplet spin coupler. Thus photo-induced spin alignments utilizing the excited triplet were realized in these π-conjugated spin systems.

The use of a tetraphenylsilole ring as a central spin coupler showed different behavior from that found for the diphenylanthracene derivatives (Figure 5.22, lower) [110, 111]. In spite of the similarity between the systems **132a** and **134a** as a spin coupler unit, no TRESR signal of the photo-excited species (**134b**) was detected probably because of a shorting of the photo-excited high-spin state or a very low efficiency of the intersystem crossing based on the nitroxide radicals.

Similarly, no TRESR was observed for **135b** upon excitation of **135a**, indicating that the flexibility of the molecule plays a crucial role in the achievement of

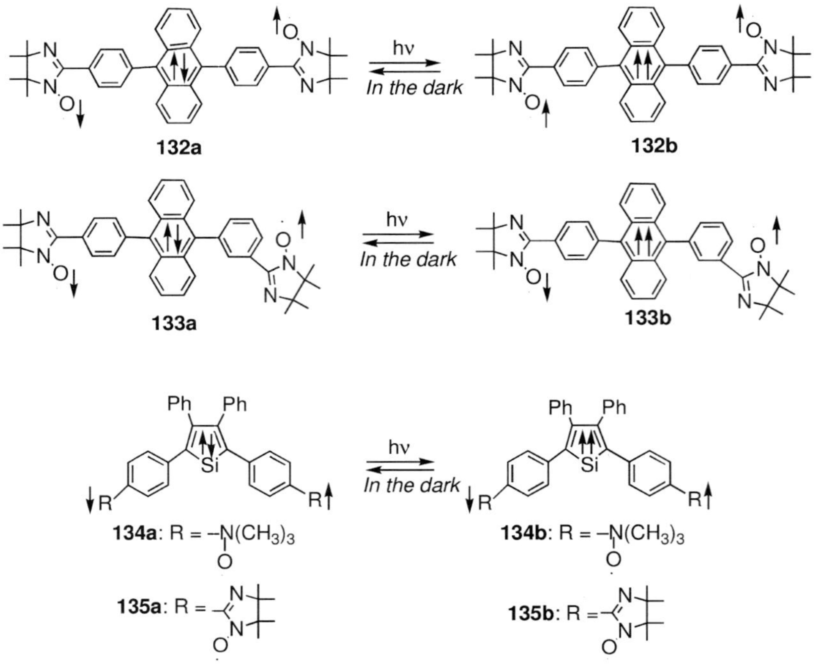

Figure 5.22 Control of spin alignment of 9,10-diphenylanthracenes and tetaphenylsiloles with NRs by light.

photo-induced spin alignment process beside the radical nature, the π–topological requirements and the need of photo-tunable spin states for the coupler.

Preliminary MM calculations for a TEMPO-substituted amide derivative of spiropyran **136a** and the corresponding merocyanine **136b**, in which relatively large differences of their molecular structures were anticipated and such structural change was supposed to cause the differences of their crystal/solid-state structures, which then would give rise to the change of their magnetic properties. Irradiation of the spiropyrans **136a** or **137a** in non-polar solvent such as CCl_4 with light of 365 nm gave the reddish colored precipitates of the corresponding merocyanines **136b** or **137b** and thus these merocyanines could be isolated in the solid state. On standing the merocyanines in polar solvent, such as acetonitrile or MeOH, in the dark the backward reactions were found to occur gradually to give the starting spiropyrans and thus the photochromic properties were actually observed in the systems (Figure 5.23, upper) [112].

As expected previously, clear differences were observed in the magnetic properties in the solid state of spiropyran/merocyanine systems with TEMPO radical, although there were no appreciable differences in their EPR spectra in benzene solution for both couples. Thus, the intermolecular spin–spin interaction of Curie–Weiss (CW) behavior observed in the low temperature region of **136a** was found to be ferromagnetic with Weiss temperature $\theta = 0.38$ K, whereas the interaction in the corresponding merocyanine derivative **136b** was antiferromagnetic with

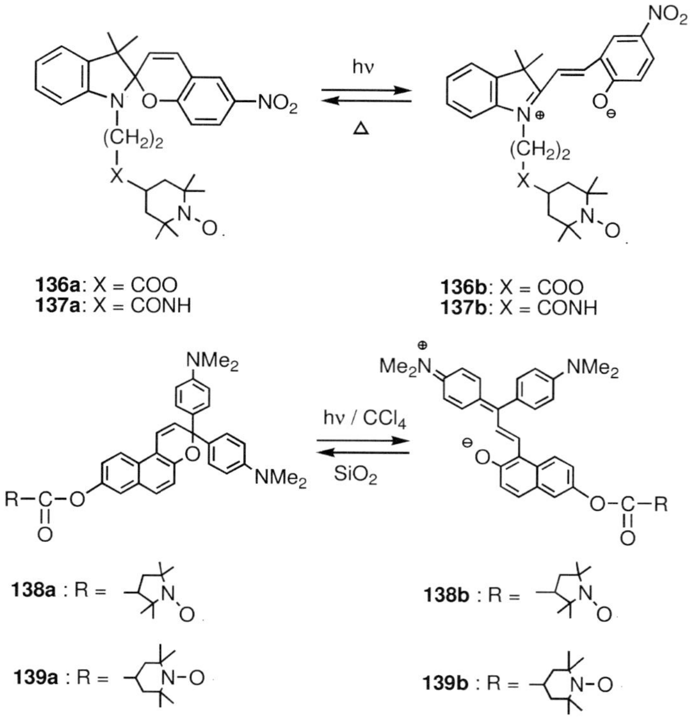

5.7 Photochromic Spin Systems Carrying NRs | 187

Figure 5.23 Photochromic reactions of spiropyrans and naphthopyrans with NRs.

$\theta = -0.82$ K. Similarly, weak ferromagnetic interaction ($\theta = 0.06$ K) found in the spiropyran derivative **137a** turned to antiferromagnetic ($\theta = -0.15$ K) in the corresponding merocyanine **137b**. Thus it is suggested from the results that the switching of the magnetic properties is possible in the spiropyran systems carrying TEMPO radical, although the unstable nature of **136b** and **137b** in solution as well as in the solid state hampers the further repetition of forward and backward reaction. The differences of the magnetic properties observed in the spiropyran/merocyanine systems are supposed to be derived from the differences of their packing features in the crystals due to their molecular structural differences and thus the magnetic interactions can be switched in *intermolecular manner* in this case through the structural changes invoked by light and heat.

A couple of naphthopyran derivatives with a PROXYL-substituent **138a** and a TEMPO-substituent **139a** were prepared from the ethynylated alcohol of Michler's ketone and 2,6-dihydroxynaphthalene. Irradiation of **138a** or **139a** with the light of 365 nm in carbon tetrachloride gave reddish precipitates of **138b** or **139b**, which could be isolated as relatively stable solid materials. The backward reaction was found to occur when the open-formed species was treated with a catalytic amount of SiO_2 in acetonitrile to revert back to the original naphthopyrans in relatively high yields (Figure 5.23, lower) [113].

Antiferromagnetic intermolecular interaction of CW behavior is observed for the PROXYL-couple **138a** and **138b** but the magnitude of the interaction is decreased in **138b** ($\theta = -0.35$ K) compared to that found in **138a** ($\theta = -0.65$ K) as apparently seen in their Weiss temperatures. On the other hand, weak ferromagnetic interaction ($\theta = 0.05$ K) observed in the closed-form species with TEMPO-substituent **139a** is changed to antiferromagnetic interaction ($\theta = -0.20$ K) in the open-formed species **139b**, and thus the signs and the magnitude of their intermolecular magnetic interactions are more distinctly different in this case than those of the former couple. Consequently, apparent tuning of their Curie constants together with their Weiss temperatures has been found to be possible in these two reversible pairs, in principle, by choosing the outer stimuli of light and a catalyst.

Several azobenzene derivatives carrying a NR with some spacer units have been prepared in order to see the possible change of intermolecular magnetic interactions caused by the structural change based on the isomerization (Figure 5.24) [114].

Irradiation of the *trans*-isomer **140a** or **141a** in a dichloromethane solution with light of 365 nm showed apparent absorption spectral change for each case being well ascribed to the photo-isomerization to the corresponding *cis*-isomer **140b** or **141b** and both of them could fortunately be isolated as relatively stable solid substances when kept in the dark and stored in a refrigerator. The backward reaction was found to occur when the *cis*-isomer **140b** or **141b** was exposed to a diffused light or a fluorescent lamp for several hours to revert back to the starting *trans*-isomer. The forward and backward reactions for each photo-isomer pair could be easily monitored by UV/VIS absorption change, although no appreciable change could be seen by ESR spectra in solution at ambient temperature.

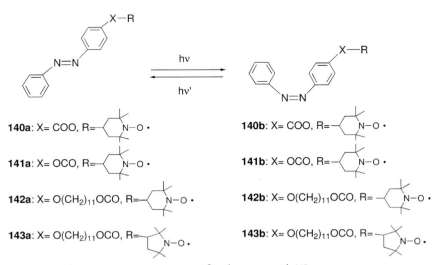

Figure 5.24 Photo-isomerization reactions of azohenzenes with NRs.

An antiferromagnetic interaction being well expressed by singlet–triplet (ST) model was observed in the *trans*-isomer carrying TEMPO-substituent **140a** with a fairly large exchange interaction of $J = -47.6$ K, while a weak ferromagnetic one of Curie–Weiss (CW) model was found in the corresponding *cis*-isomer **140b**, indicating an occurrence of apparent change of intermolecular magnetic interactions between the photo-isomer pair originated possibly from the change of the molecular/crystal structures. Interestingly, quite the reverse magnetic behavior from above mentioned pair **(140a, 140b)** was observed in the pair **(142a, 142b)**. Namely, very weak ferromagnetic interaction of CW behavior observed in the *trans*-isomer **142a** was found to turn to an antiferromagnetic one of ST behavior with a fairly large exchange interaction of $J = -36.7$ K in the corresponding *cis*-isomer **142b**. An antiferromagnetic interaction based on a 1-D Heisenberg model with $J = -1.89$ K was observed in the *trans*-isomer **141a** and the behavior then changed to a ferromagnetic one of CW behavior in the *cis*-isomer **141b**. Concerning the couples with PROXYL-substituent **(143a, 143b)**, the ST behavior with a weak antiferromagnetic interaction ($J = -4.2$ K) in the *trans*-isomer **143a** was turned to the ST behavior with an antiferromagnetic interaction of almost twice the value ($J = -7.7$ K) in the *cis*-isomer **143b**. Thus, a series of photo-responsive spin systems showing different changes to their intermolecular magnetic interactions could successfully be constructed in these cases.

A series of norbornadine derivatives with TEMPO **(144a–146a)** were prepared and their valence isomerization was investigated (Figure 5.25, upper) [112]. Photo-irradiation of the norbornadienes with light of 254 nm in benzene solution for 3–4 days gave the corresponding quadricyclane derivatives **(144b-146b)** in moderate yields. In turn, the quadricyclanes could be reverted to the starting materials by treatment with catalytic amount of 5% Pd-C. However, in so far as these valence isomer pairs are concerned, the tendency of the preservation of intermolecular magnetic interactions was clarified by magnetic measurements, in spite of small differences of their Weiss temperatures.

Being different from the previous cases, intermolecular photochromic reaction is possible in anthracene and the corresponding dimer system. A TEMPO-substituted anthracene derivative **147a** has been prepared for the purpose to see the change in magnetic property based on their light-induced structural changes. The derivative **147a** was found to give corresponding photo-dimer **147b** upon irradiation in benzene solution using a high-pressure Hg-lamp (400 W) and the latter could be reverted to the former by heating in a toluene solution (Figure 5.25, lower) [115]. The distance between intramolecular spins is estimated from the result of point dipole approximation by EPR measurement to be 9.40 Å for **147b**, which is relevant taking *trans*-configuration into account.

A 9-methylanthracene derivative with 4-amino-TEMPO **(148a)** was prepared by the reduction of the Schiff base derived from 9-anthraldehyde and 4-amino-TEMPO. The photo-dimerization of **148a** was carried out by irradiation with a 400 W Hg lamp in benzene solution to give the corresponding dimer **148b**, which, in turn, could be reverted to **148a** by heating in xylene to afford a reversible system in principle. Both forward and backward reactions could be easily monitored by

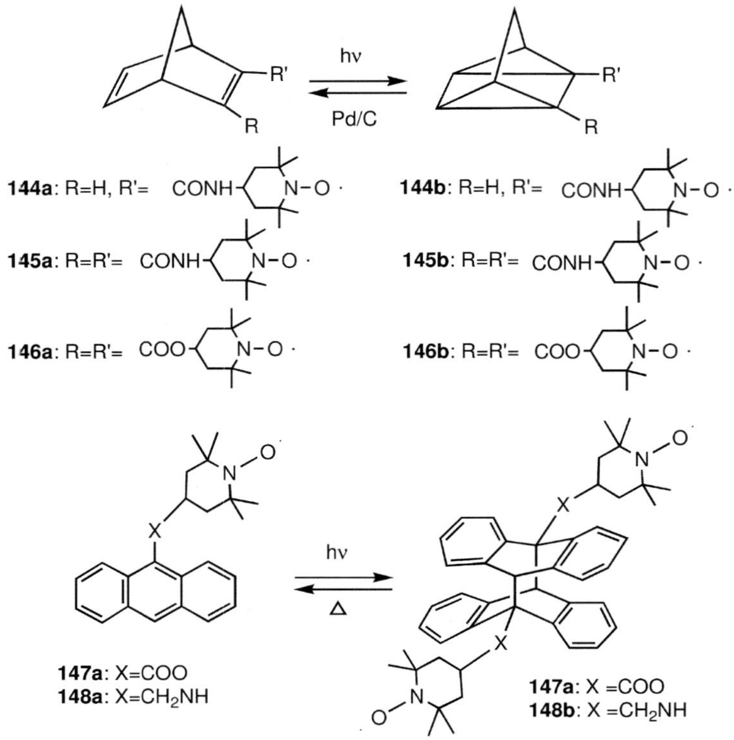

144a: R=H, R'= CONH-⟨N-O·⟩   **144b:** R=H, R'= CONH-⟨N-O·⟩

145a: R=R'= CONH-⟨N-O·⟩   **145b:** R=R'= CONH-⟨N-O·⟩

146a: R=R'= COO-⟨N-O·⟩   **146b:** R=R'= COO-⟨N-O·⟩

147a: X=COO
148a: X=CH$_2$NH

147a: X =COO
148b: X =CH$_2$NH

Figure 5.25 Photochromic reactions of norbornadienes and anthracenes with TEMPO radicals.

electronic spectral change with the indication of the presence or the absence of the anthracene chromophore.

Whereas the tendency to preservation of intermolecular spin–spin interactions was observed in the monomer/dimer pair of **147a**, **147b** from ferromagnetic (θ = 0.21 K) to ferromagnetic (θ = 0.29 K) in spite of the difference of their Weiss temperatures, significant difference was clarified in the pair of **148a**, **148b**.

While a ferromagnetic interaction of CW behavior (θ = −0.97 K) was observed in the spins of **148a**, the solvent molecules used for recrystallization of the dimer **148b** were found to be easily incorporated in the respective crystals to result in different magnetic behaviors. Namely, antiferromagntic interaction (θ = −0.38 K) was observed in the compound obtained from benzene for the recrystallization while ferromagnetic interaction (θ = 0.26 K) was found in the solvated compounds of **148b** with chloroform. Thus, although the efficiency of the reversibility is not so high, it was found to be possible in principle to tune the sign and the magnitude of intermolecular magnetic interactions by choosing *light, heat,* or *appropriate solvent* in this case.

5.8
FNRs for Biomedicinal Applications

The utility as spin probes and the redox activity of NRs, parses the possibility of biomedical applications. Researchers in the field of chemistry, biology, pharmaceutics, and medicine have been pursuing such possibilities for FNRs from a variety of viewpoints. Here, some recent and selected examples of bio-functional NRs will be discussed from a preparative viewpoint.

Considerable attention to the development of new pH-sensitive spin probes has been paid in recent years because they are most promising for pH studies in biological systems, including membranes, proteins, cells and cell organelles. Volodarsky, Grigor'ev et al. of the Novosibirsk school have been actively exploring along this line by using imidazoline and imidazolidine NRs because their stability to acids is far greater than piperidine, pyrroline and pyrrolidine NRs (see Chart 5.2) and their ESR spectra are rather sensitive to the pH in the water, thus providing wide possibilities as "pH probes" in biological systems [116].

Recently, several new pH-sensitive probes have been proposed and among them 4-amino-imidaloline NRs are considered to be the most promising pH-sensitive spin-probes for *in vitro* work due to a relatively large effect on their ESR spectra ($\Delta a_N = 0.7$–$0.9\,G$) and appropriate pK values in the range from 4.5 to 6.6, and it has been shown that substituents at the exocyclic nitrogen atom in the 4-amino NRs may produce a pronounced effect on the pK of the nitroxide. The synthesis of this class of NRs was reported recently [117] and the synthetic scheme is outlined in Figure 5.26; the final step of the synthesis includes a Grignard reagent addition to 5-alkylamino-4H-imidazole 3-oxide **149** to give a series of the corresponding NRs **150**.

Voinov et al. prepared more recently a series of new imidazoline NRs **151** using an approach based on the alkylation of diamagnetic 4-R-amino-1,2,2,5,5-pentamethyl-3-imidazolines and these were shown to have pH-dependent ESR spectra with pK_a ranging from 3.5 to 6.2 (Figure 5.26) [118].

Modification of drugs with NRs gives spin-labeled drugs that can be used to monitor drug–protein, drug–DNA, and other drug–biomaterial interactions. For this purpose Hideg et al. have prepared an interesting series of spin-labeled drugs and related biomolecules [16e, 119].

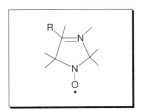
Imidazoline Nitroxide Radicals

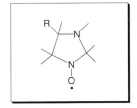
Imidazolidine Nitroxide Radicals

Chart 5.2

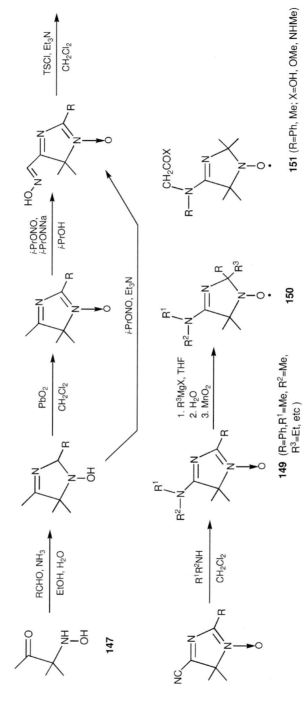

Figure 5.26 Synthesis of pH-sensitive spin probes.

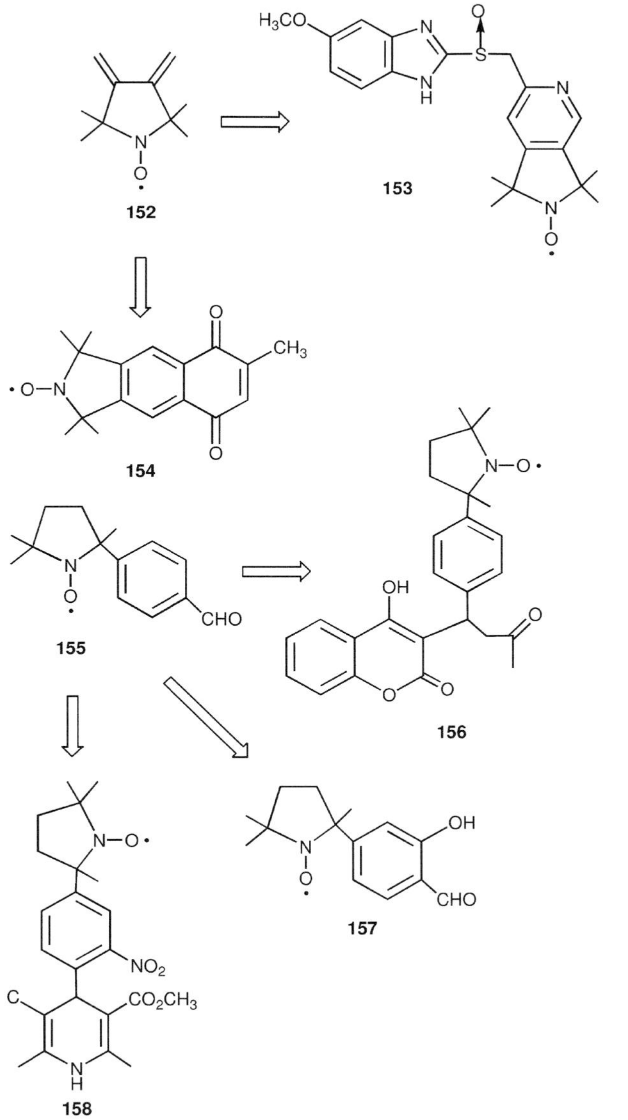

Figure 5.27 Selected examples of spin-labeled drugs.

Some recent examples of this class of compounds were reported by Kálai et al., starting from the paramagnetic pyrrolidine diene **152** [120] and aromatic aldehyde **155** [121]. As illustrated in Figure 5.27, spin-labeled omeprazole (an inhibitor of gastric acid secretion) **153** and vitamin K (one of the blood-clotting vitamins) derivative **154** were prepared by using **152** as the Diel–Alder diene for their construction [120]. A paramagnetic warfarin (the most widely prescribed anticoagulant) **156**, a salicylic acid derivative **157**, and nifedipine (a well-known calcium channel blocker) derivative **158** were prepared from **155** along with a labeled indigo derivative (Figure 5.27) [121].

New platinum (IV) complexes with NRs **(159a,b, 160a,b)** related to cisplatin, a well-known antitumor drug, were prepared as potential antitumor drugs (Figure 5.28) [122a]. Among them, the complex **159b** exhibited high antitumor activity comparable with that of cisplatin against leukemia P388 used as the experimental tumor. Simultaneous administration of low doses of **159b** and cisplatin resulted in synergism of antitumor activity and 100% cure of animals [122b].

Some FNRs could also be applied as spin-trapping agents and/or radical scavengers. For example, the reactivity of the potassium salts of 4-carboxyNN **161** and its homolog **162** with nitric oxide (·NO) was investigated [123]. By using ESR, the NN derivatives **161, 162** were found to react with ·NO in a stoichiometric manner in a neutral solution, resulting in the corresponding Ins **163, 164** and NO_2^-/NO_3^- as shown in Figure 5.29, which is similar to the previous reaction shown in Figure 5.10. Generation of ·NO was identified by reacting the radical **161** in aortic strip and quantitating the reaction product with ESR and it was clarified that the radical **161** effectively antagonized ·NO in biological systems via a unique radical–radical reaction with ·NO.

Marx and Rassat prepared an interesting diradical **165** for spin-labeled spin-trap and applied it to the detection of nitric oxide according to Figure 5.30 [124]. When nitric oxide was reacted with **165**, four new NRs **166a–d** were detected and char-

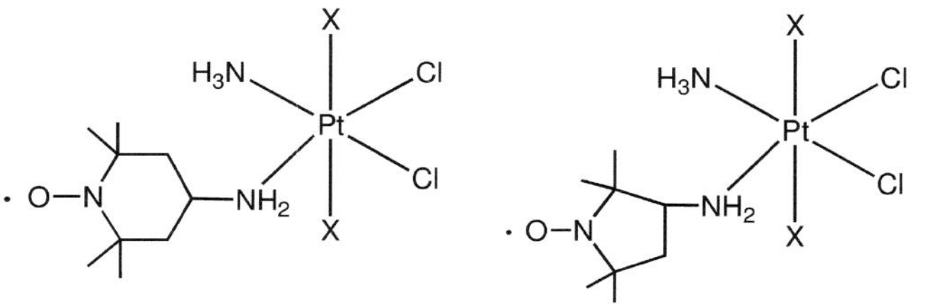

159a (X=OH), 159b (X=OAc)   **160a (X=OH), 160b (X=OAc)**

Figure 5.28 Platinum complexes with NRs.

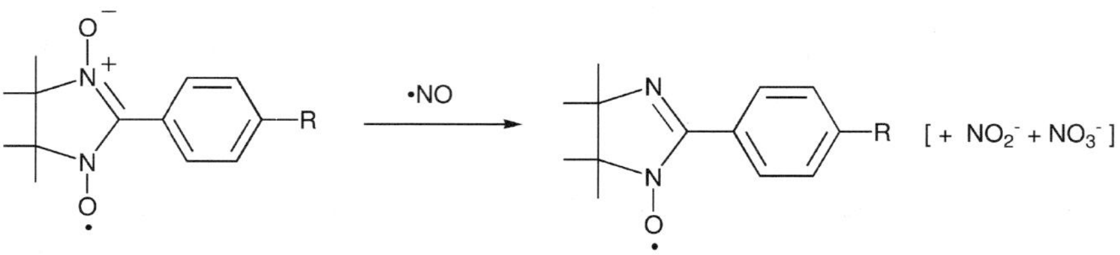

161 R=COO⁻K⁺
162 R=OCH₂COO⁻K⁺

163 R=COO⁻K⁺
164 R=OCH₂COO⁻K⁺

Figure 5.29 Reactions of 4-carboxyNN and its homolog with ·NO.

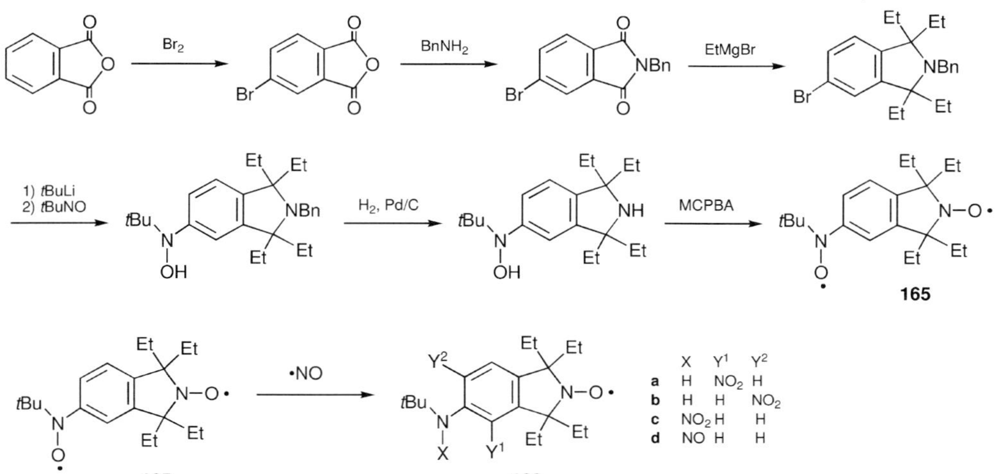

Figure 5.30 Synthesis of a diradical **165** as spin-labeled spin-trap and its use for the detection of ·NO.

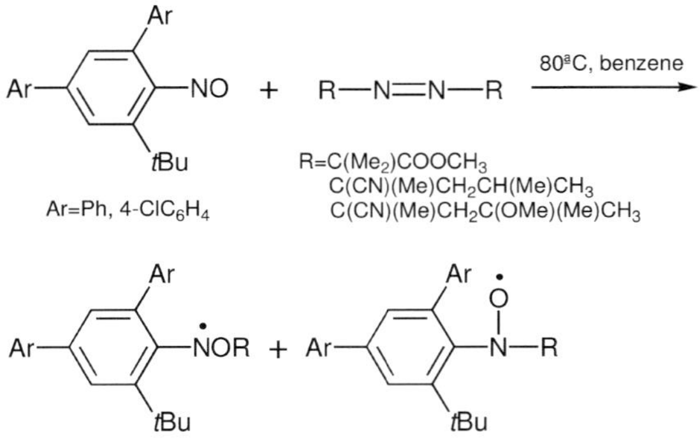

Figure 5.31 Reactions of 2,6-diaryl-6-t-butylinitrsobenzenes with azo compounds.

acterized. They all displayed the same typical isoindolidinyloxyl 3-line ESR spectrum without violation of the radical moiety, thus showing its ability as spin-labeled spin-trap.

A spin-trapping method was used for the preparation of N-t-alkoxyarylaminyl radicals, and their magnetic properties were reported by Miura et al. [125]. As shown in Figure 5.31, the reactions of 2,6-diaryl-6-t-butylnitrsobenzenes with three kinds of azo compounds gave the corresponding oxyaminyl radicals **167** together with fewer isomeric NRs **168**, while the reactions of 2,4,6-triphenylnitrosobenzene with the azo compounds in reflux benzene gave oxyaminyls **167** alone.

5.9
Functional Nitrones

Spin-trapping studies have been widely applied to the detection of reactive radical species in biological processes and, among others, nitrone derivatives have mainly been developed for the purpose [126]. Addition of free radicals to diamagnetic nitrones gives the corresponding paramagnetic NRs to be detected.

As a typical example of nitrone synthesis, the synthesis of 5,5-dimethylpyrroline-N-oxide (DMPO) **169** is illustrated in Figure 5.32 [127]. The reaction of **169** with various free radicals, such as $\cdot O_2^-$ and $\cdot OH$, results in the adduct with spin generation. A variety of nitrones of this class have so far been developed, including **171** and **172** with one or two more methyl groups to enhance the stability of the resulting NRs. Also, acyclic α–aromatic-t-butylnitrones, such as benzyliden-t-butylamine-N-oxide (phenyl t-butyl nitrone, PBN) **173** or its analog **174** (4-POBN), can readily be elicited by heating aromatic aldehyde with t-butylhydroxylamine, although the spin-trapped NRs are relatively unstable compared to the cyclic NRs.

A series of α–aryl-N-adamant-yl nitrones **175** was designed and synthesized to evaluate the stability of the hydroxyl radical adduct (Figure 5.33) [128]. It was found that introduction of an adamantane ring instead of t-butyl group results in reasonably good stability of the hydroxy radical adduct for biological measurements.

In order to improve the stability of spin adducts, and to enhance lipophilicity, recent attention has been addressed to developing new phosphorous-containing nitrones, such as **176** [129], **177** [130], and **178** [131]. Moreover, the ESR phosphorous couplings of their spin-adducts were found to be very sensitive to the ring conformation and could represent a valuable structural probe.

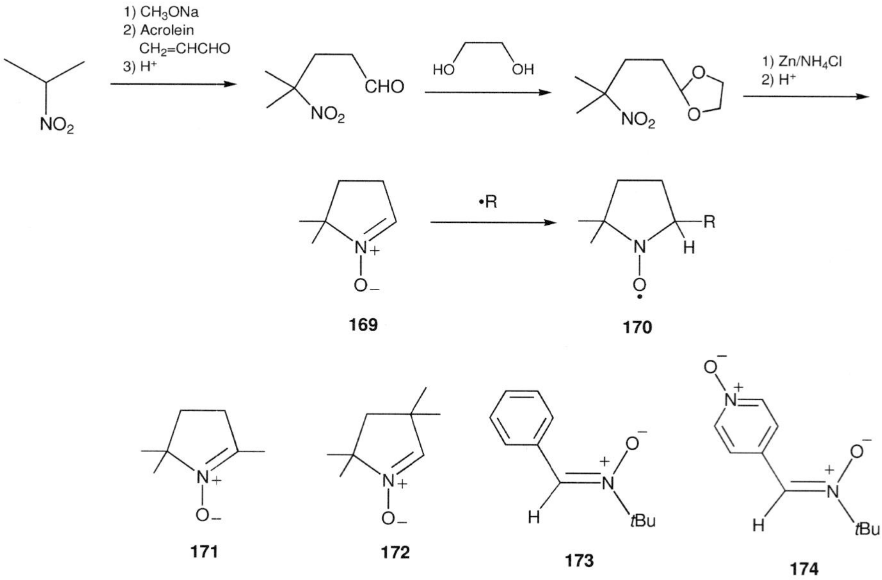

Figure 5.32 A synthetic of nitrone **170** and selected examples of nitrones.

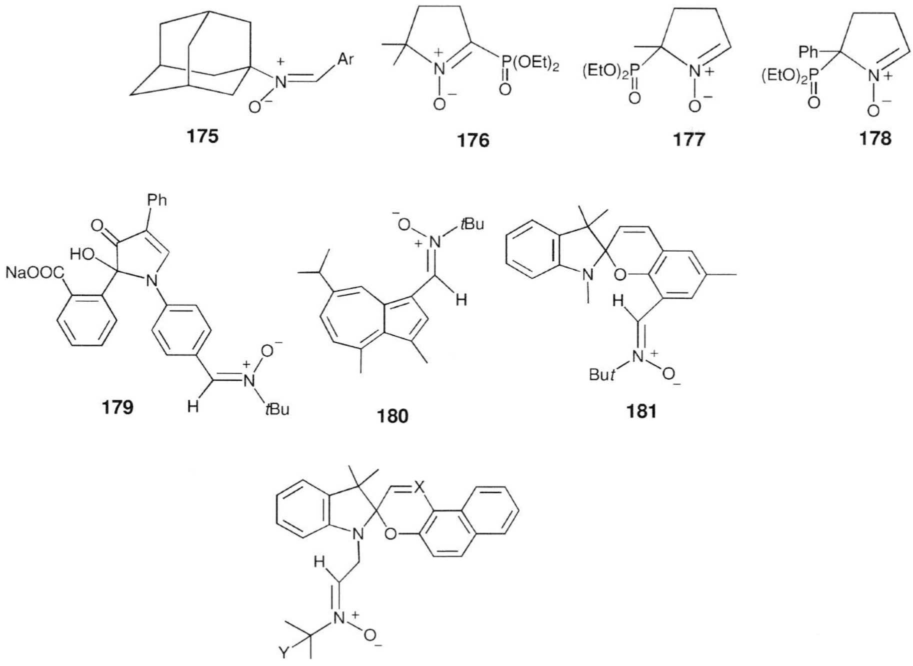

Figure 5.33 Examples of functional nitrones.

Several nitrone derivatives with dual functionality have also been proposed in recent years. As such, a fluorophore-containing spin-trap **179** was designed based on the sensitivity of fluorescence spectroscopy and the specificity of spin-traps toward a variety of free radicals. It was prepared by reacting fluorescamine with N-t-butyl-α-(p-amino)phenylnitrone in aqueous buffer at pH 9 [132]. A nitrone-substituted guaiazulene derivative **180** was prepared by Becker [133] to clarify its chromotropism and reactivity to free radicals at the same time.

Novel photochromic spin traps such as a nitrone-substituted spiropyran **181**, spironaphthopyran **182** (X = CH; Y = CH$_3$) [134], a spiro[indoline-naphthoxazine] **182** (X = N; Y = CH$_3$) [135a], its phosphrylated derivative **182** (X = N; Y = P(O)(OEt)$_2$) [135b], and a 3,3-diphenyl-naphthopyran derivative [136] were synthesized by Campredon et al. and their reactivities toward free radicals together with photochromic properties were investigated.

5.10
Conclusion

We have described in this chapter a wide variety of FNRs ranging from those for magnetic materials and supramolecular spin systems to those with biomedical applications citing relevant and recent references as far as possible and mainly from the viewpoint of preparative chemistry, although we could not give all the details and references to cover this vast area within limited pages. Therefore, readers who want to study FNRs further are recommended to consult relevant papers, reviews and the books cited in the references.

References

1 Fremy, E. (1845) *Annales de Chimie et de Physique, Serie 3*, **15**, 408–88.
2 Zimmer, H., Lankin, D.C. and Horgan, S.W. (1971) *Chemical Reviews*, **71**, 229–46.
3 Teuber, H.-J. and Benz, S. (1967) *Chemische Berichte*, **100**, 2918–29.
4 Piloty, O. and Schwerin, B.G. (1901) *Berichte der Deutschen Chemischen Gesellschaft*, **34**, 1870–87. Cf. Tidwell, T.T. (2001) *Angewandte Chemie (International Ed. in English)*, **40**, 331–7.
5 (a) Gomberg, M. (1900) *Berichte der Deutschen Chemischen Gesellschaft*, **33**, 3150–63. (b) Gomberg, M. (1900) *Journal of the American Chemical Society*, **22**, 757–71. Cf. Rüchardt, C. (2000) *Nachrichten aus der Chemie*, **48**, 904–10.
6 Holden, A.N., Yager, W.A. and Merrity, F.R. (1951) *The Journal of Chemical Physics*, **19**, 1319.
7 (a) Wieland, H. and Offenbächer, M. (1914) *Berichte der Deutschen Chemischen Gesellschaft*, **47**, 2111–15. (b) Wieland, H. and Roth, K. (1920) *Berichte der Deutschen Chemischen Gesellschaft*, **53**, 210–30.
8 Meyer, K.H. and Reppe, W. (1921) *Berichte der Deutschen Chemischen Gesellschaft*, **54**, 327–37.
9 (a) Lebedev, O.L. and Kazarnovsky, S.N. (1959) *Treatises on Chemistry and Chemical Technology*, Gorky, **3**, 649 (in Russian). (b) Lebedev, O.L., Khidekel, M.L. and Razuvaev, G.A. (1961) *Doklady Akademii Nauk SSSR*, **140**, 1327 (in Russian).
10 Neiman, M.B., Rozantsev, E.G. and Mamedova Yu G. (1962) *Nature*, **196**, 472–4.
11 (a) Rozantsev, E.G. and Sholle, V.D. (1971) *Synthesis*, 190–202. (b) Rozantsev, E.G. and Sholle, V.D. (1971) *Synthesis*, 401–14. (c) Keana, J.F.W. (1978) *Chemical Reviews*, **78**, 35–64. (d) Dogonneau, M., Kagan, E.S., Mikhailov, V.I., Rozantsev, E.G. and Sholl, V.D. (1984) *Synthesis*, 895–916.
12 (a) Rozantsev, E.G. (1970) *Free Nitroxyl Radicals*, Plenum Press, New York. (b) Vordarsky, L.B. (1988) *Imidazoline Nitroxides*, CRC Press, Boca Raton, Florida. (c) Volodarsky, L.B., Reznikov, V.A. and Ovcharenko, V.I. (1994) *Synthetic Chemistry of Stable Nitroxides*, CRC Press, Boca Raton, FL.
13 (a) Berliner, L.J. (1976), (1979) *Spin Labeling: Theory and Applications*, Vol. 1, 1976 and Vol. 2, 1979, Academic Press, New York. (b) Likhtenstein, G.I. (1976) *Spin Labeling Methods in Molecular Biology*, Wiley-Interscience, New York. (c) Likhtenstein, G.I. (1993) *Biophysical Labeling Methods in Molecular Biology*, Cambridge University Press, Cambridge.
14 (a) Osiecki, J.H. and Ullman, E.F. (1968) *Journal of the American Chemical Society*, **90**, 1078–9. (b) Boocock, D.G., Darcy, R. and Ullman, E.F. (1968) *Journal of the American Chemical Society*, **90**, 5945–6.
15 (a) Nakatsuji, S. and Anzai, H. (1997) *Journal of Materials Chemistry*, **7**, 2161–2174. (b) Amabilino, D.B. and Veciana, V. (2001) *Magnetism: Molecules to Materials II*, (eds J.S. Miller and M. Drillon), Wiley-VCH, Weinheim.
16 (a) Forrester, A.R., Hay, J.M. and Thomson, R.H. (1968) *Organic Chemistry of Stable Free Radicals*, Academic Press, New York. (b) Aulich, H.G. (1989) *Nitrones, Nitronates and Nitroxides* (eds S. Patai and Z. Rappoprt), John Wiley & Sons, Ltd, New York, pp. 313–99. (c) Birk, M.E. (1995) *Heterocyles*, **41**, 2827–73. (d) Banerjee, S. and Trivedi, G.K. (1995) *Journal of Scientific and Industrial Research*, **54**, 623–36.

(e) Hideg, K., Kálai, T. and Sár, C.P. (2005) *Journal of Heterocyclic Chemistry*, **42**, 437–50.

17 For recent reviews on molecular-based magnetic materials: (a) Lahti, P.M. (1999) *Magnetic Properties of Organic Materials*, Marcel Dekker, Inc, New York, Basel. (b) Itoh, K. and Kinoshita, M. (2000) *Molecular Magnetism*, Kodansha/Gordon and Breach Science Publishers, Tokyo. (c) Veciana, J. (2001) *Structure and Bonding, Vol. 100, π–Electron Magnetism: From Molecule to Magnetic Materials*, Springer Verlag, Berlin. (d) Miller, J.S. and Drillon, M. (2001–2005) *Magnetism: Molecules to Materials*, Vol. I–V, Wiley-VCH, Weinheim.

18 (a) Blaise, A., Lemaire, H., Pilon, J. and Veyret, C. (1972) *Les Comptes Rendus de l'Académie des Sciences Paris, Serie B*, **274**, 157–60. (b) Veret, C. and Blaise, A. (1973) *Molecular Physics*, **25**, 873–82.

19 (a) Korshak, Y.V., Ovchinnikov, A.A., Shapiro, A.M., Medvedeva, T.M. and Spector, V.N. (1986) *Journal of Experimental and Theoretical Physics Letters*, **43**, 309–11 (in Russian). (b) Korshak, Y.V., Medveva, T.V., Ovchinnikov, A.A. and Spector, V.N. (1987) *Nature*, **326**, 370–2.

20 Miller, J.S., Epstein, A.J. and Reiff, W.M. (1988) *Chemical Reviews*, **88**, 201–20.

21 (a) Tamura, M., Nakazawa, Y., Shiomi, D., Nozawa, K., Hosokoshi, Y., Ishikawa, M., Takahashi, M. and Kinoshita, M. (1991) *Chemical Physics Letters*, **186**, 401–4. (b) Nakazawa, Y., Tamura, M., Shirakawa, N., Shiomi, D., Takahashi, M., Kinoshita, M. and Ishikawa, M. (1992) *Physical Review. B, Condensed Matter*, **46**, 8906–14.

22 Awaga, K. and Maruyama, Y. (1989) *Chemical Physics Letters*, **158**, 556–8.

23 Awaga, K., Inabe, T., Nagashima, U. and Maruyama, Y. (1989) *Journal of the Chemical Society. Chemical Communications*, 1617–18.

24 Kinoshita, M. (1997) *Organic Molecular Solids*, Chapter 12, (ed. W. Jones), CRC Press, Boca Raton, FL.

25 (a) Awaga, K., Inabe, T. and Maruyama, Y. (1992) *Chemical Physics Letters*, **190**, 349–52. (b) Blundell, S.J., Pattenden, P.A., Valladares, R.M., Pratt, F.L., Sugano, T. and Hyes, W. (1994) *Solid State Communications*, **92**, 569–72.

26 (a) Sugano, T., Tamura, M., Kinoshita, M., Sakai, Y. and Ohashi, Y. (1992) *Chemical Physics Letters*, **200**, 235–40. (b) Pratt, F.L., Valladares, R., Caulfield, J., Deckers, I., Singleton, J., Fisher, A.J., Hayes, W., Kurmoo, M., Day, P. and Sugano, T. (1993) *Synthetic Metals*, **61**, 171–5. (c) Pattenden, P.A., Valladares, R. M., Pratt, F.L., Blundell, S.J., Fisher, A.J., Hayes, W., Day, P. and Sugano, T. (1995) *Synthetic Metals*, **71**, 1823–4. (d) Sugano, T., Kurmoo, M., Day, P., Pratt, F.L., Blundell, S.J., Hayes, W., Ishikawa, M., Kinoshita, M. and Ohashi, Y. (1995) *Molecular Crystals and Liquid Crystals*, **271**, 107–14.

27 (a) Matsushita, M.M., Izuoka, A., Sugawara, T., Kobayashi, T., Wada, N., Takeda, N. and Ishikawa, M. (1994) *Journal of the Chemical Society. Chemical Communications*, 1723–4. (b) Sugawara, T., Matsushita, M.M., Izuoka, A., Wada, N., Takeda, N. and Ishikawa, M. (1997) *Journal of the American Chemical Society*, **119**, 4369–79.

28 (a) Cirujeda, J., Mas, M., Molins, E., Lanfranc de Panthou, F., Laugier, J., Park, J.G., Paulsen, C., Rey, P., Rovira, C., and Veciana, J. (1995) *Journal of the Chemical Society. Chemical Communications*, 709–10. (b) Garcia-Munoz, J.L., Cirujeda, J., Veciana, J. and Cox, S.F.J. (1998) *Chemical Physics Letters*, **293**, 160–6.

29 Caneschi, A., Ferraro, F., Gatteschi, D., Le Lirzin, A., Novak, M.A., Rentschler, E. and Sessoli, R. (1995) *Advanced Materials*, **7**, 476–8.

30 (a) Nakatsuji, S., Saiga, M., Haga, N., Naito, A., Nakagawa, M., Oda, M., Suzuki, K., Enoki, T. and Anzai, H. (1997) *Molecular Crystals and Liquid Crystals*, **306**, 279–84. (b) Nakatsuji, S., Saiga, M., Haga, N., Naito, A., Hirayama, T., Nakagawa, M., Oda, M., Anzai, H., Suzuki, K., Enoki, T., Mito, M. and Takeda, K. (1998) *New Journal of Chemistry*, **22**, 275–80.

31 Nakatsuji, S., Morimoto, H., Anzai, H., Kawashima, J., Maeda, K., Mito, M. and Takeda, K. (1998) *Chemical Physics Letters*, **296**, 159–66.

32 Chiba, J. and Oda, Y. et al., unpublished results.
33 (a) Nogami, T., Tomioka, K., Ishida, T., Yoshikawa, H., Yasui, M., Iwasaki, F., Iwamura, H., Takeda, N. and Ishikawa, M. (1994) *Chemistry Letters*, 29–32. (b) Ishida, T., Tsuboi, H., Nogami, T., Yoshikawa, H., Yasui, M., Iwasaki, F., Iwamura, H., Takeda, N. and Ishikawa, M. (1994) *Chemistry Letters*, 919–22. (c) Nogami, T., Ishida, T., Tsuboi, H., Yoshikawa, H., Yamamoto, H., Tomioka, K., Ishida, T., Yoshikawa, H., Yasui, M., Iwasaki, F., Iwamura, H., Takeda, N. and Ishikawa, M. (1995) *Chemistry Letters*, 635–6. (d) Nogami, T., Togashi, K., Tsuboi, H., Ishida, T., Yoshikawa, H., Yasui, M., Iwasaki, F., Iwamura, H., Takeda, N. and Ishikawa, M. (1995) *Synthetic Metals*, **71**, 1813–14. (e) Togashi, K., Imachi, R., Tomioka, K., Tsuboi, H., Ishida, T., Nogami, T., Takeda, N. and Ishikawa, M. (1996) *Bulletin of the Chemical Society of Japan*, **69**, 2821–30. (f) Nogami, T., Ishida, T., Yasui, M., Iwasaki, F., Takeda, N., Ishikawa, M., Kawakami, T. and Yamaguchi, K. (1996) *Bulletin of the Chemical Society of Japan*, **69**, 1841–8.
34 Chiarelli, R., Rassat, A. and Rey, P. (1992) *Chemical Communications*, 1081–2.
35 Chiarelli, A., Novak, M.A., Rassat, A. and Tholence, J.L. (1993) *Nature*, **363**, 147–9.
36 Bernstein, J. (2002) *Polymorphism in Molecular Crystals*, Chapter 6, Clarendon Press, Oxford.
37 (a) Rozantsev, E.G. and Scholle, V.D. (1971) *Synthesis*, 401–14. (b) Aulich, H. G. (1989) *Nitrones, Nitronates and Nitroxides*, (eds S. Patai and Z. Rappoport), John Wiley & Sons, New York, p. 607.
38 Bobbitt, J.M. and Flores, M.C.L. (1988) *Heterocycles*, **27**, 509–33.
39 de Nooy, A.E.J., Besemer, A.C. and van Bekkum, H. (1996) *Synthesis*, 1153–75.
40 Nakatsuji, S., Takai, A., Nishikawa, K., Morimoto, Y., Yasuoka, N., Suzuki, K., Enoki, T. and Anzai, H. (1999) *Journal of Materials Chemistry*, **9**, 1747–54.
41 Nakatsuji, S., Mizumoto, M., Takai, A., Akutsu, H., Yamada, J., Kawamura, H., Schmitt, S. and Hafner, K. (2000) *Molecular Crystals and Liquid Crystals*, **348**, 1–6.
42 Sugawara, T., Sakurai, H. and Izuoka A. (2000) *Hyper-Structured Molecules II* (ed. H. Sasabe), Gordon and Breach Publishers, Amsterdam, pp. 35–58.
43 Sakurai, H., Kumai, R., Izuoka, A. and Sugawara, T. (1996) *Chemistry Letters*, 879–80.
44 Kumai, R., Sakurai, H., Izuoka, A. and Sugawara, T. (1996) *Molecular Crystals and Liquid Crystals*, **279**, 133 7–8.
45 Izuoka, A. and Sugawara, T. (1997) *Molecular Crystals and Liquid Crystals*, **305**, 41–54.
46 Nakatsuji, S., Takai, A., Ojima, T. and Anzai, H. (1999) *Journal of Chemical Research (S)*, 93–4.
47 Sugano, T., Fukasawa, T. and Kinoshita, M. (1991) *Synthetic Metals*, 41–43, 3281–4.
48 (a) Sugimoto, T., Tsujii, M., Suga, T., Hosoito, N., Ishikawa, M., Takeda, N. and Shiro, M. (1995) *Molecular Crystals and Liquid Crystals*, **272**, 183–94. (b) Ueda, K., Tsujii, M., Suga, T., Sugimoto, T., Kanehisa, N., Kai, Y. and Hosoito, N. (1996) *Chemical Physics Letters*, **253**, 355–60.
49 (a) Awaga, K., Yamaguchi, A., Okuno, T., Inabe, T., Nakamura, T., Matsumoto, M. and Maruyama, Y. (1994) *Journal of Materials Chemistry*, **4**, 1377–85. (b) Yamaguchi, A., Okuno, T. and Awaga, K. (1996) *Bulletin of the Chemical Society of Japan*, **69**, 875–82.
50 (a) Imai, H., Inabe, T., Otsuka, T., Okuno, T. and Awaga, K. (1996) *Physical Review. B, Condensed Matter*, **54**, R6838–40. (b) Imai, H., Otsuka, T., Naito, T., Awaga, K. and Inabe, T. (1999) *Journal of the American Chemical Society*, **121**, 8098–103.
51 Awaga, K., Okuno, T., Yamaguchi, A., Hasegawa, M., Inabe, T., Maruyama, Y. and Wada, N. (1994) *Physical Review. B, Condensed Matter*, **49**, 3975–81.
52 Aonuma, S., Casellas, H., Faulmann, C., de Bonneval, B.G., Malfant, I., Cassoux, P., Laroix, P.G., Hosokoshi, Y. and Inoue,

K. (2001) *Journal of Materials Chemistry*, **11**, 337–45.

53 Kanbara, K., Akutsu, H., Yamada, J. and Nakatsuji, S. (2005) *Chemistry Letters*, **34**, 306–7.

54 Rahman, B., Akutsu, H., Yamada, J. and Nakatsuji, S. (2007) *Polyhedron* **26**, 2287–90.

55 (a) Akutsu, H., Yamada, J. and Nakatsuji, S. (2001) *Chemistry Letters*, 208–9. (b) Akutsu, H.J. and Yamada, N.S. (2001) *Synthetic Metals*, **120**, 871–2.

56 (a) Akutsu, H., Yamada, J. and Nakatsuji, S. (2003) *Chemistry Letters*, **32**, 1118–19. (b) Akutsu, H., Yamada, J. and Nakatsuji, S. (2005) *Synthetic Metals*, **152**, 377–80.

57 Akutsu, H., Yamada, J., Nakatsuji, S. and Turner, S.S. (2006) *Solid State Communications*, **140**, 256–60.

58 (a) Akutsu, H., Masaki, K., Mori, K., Yamada, J. and Nakatsuji, S. (2005) *Polyhedron*, **24**, 2126–32. (b) Yamashita, A., Akutsu, H., Yamada, J. and Nakatsuji, S. (2005) *Polyhedron*, **24**, 2796–802.

59 (a) Yamaguchi, K., Namimoto, H., Fueno, T., Nogami, T. and Shirota, Y. (1990) *Chemical Physics Letters*, **166**, 408–14. (b) Yamaguchi, K., Okumura, M., Fueno, T. and Nakasuji, K. (1991) *Synthetic Metals*, 41–3, 3631.

60 Nakamura, Y., Koga, N. and Iwamura, H. (1991) *Chemistry Letters*, 69–72.

61 Izuoka, A. and Nakazaki, J. (2004) *TTF Chemistry*, (eds J. Yamada and T. Sugimoto), Kodansha-Springer, Tokyo, Berlin, pp. 137–53.

62 (a) Sugimoto, T., Yamaga, S., Nakai, M., Ohmori, K., Tsujii, M., Nakatsuji, H., Fujita, H. and Yamauchi, J. (1993) *Chemistry Letters*, 1361–4. (b) Sugimoto, T., Yamaga, S., Nakai, M., Ohmori, K., Tsujii, M., Nakatsuji, H. and Hosoito, H. (1993) *Chemistry Letters*, 1817–20.

63 Nakatsuji, S., Hirai, A., Yamada, J., Suzuki, K., Enoki, T. and Anzai, H. (1997) *Molecular Crystals and Liquid Crystals*, **306**, 409–14.

64 Harada, G., Jin, T., Izuoka, A., Matsushita, M.M. and Sugawara, T. (2003) *Tetrahedron Letters*, **44**, 4415–18.

65 Kumai, R., Matsushita, M.M., Izuoka, A. and Sugawara, T. (1994) *Journal of the American Chemical Society*, **116**, 4523–4.

66 Kumai, R., Izuoka, A. and Sugawara, T. (1993) *Molecular Crystals and Liquid Crystals*, **232**, 151–4.

67 (a) Nakatsuji, S., Satoki, S., Suzuki, K., Enoki, T., Kinoshita, N. and Anzai, H. (1995) *Synthetic Metals*, **71**, 1819–20. (b) Nakatsuji, S., Akashi, N., Suzuki, K., Enoki, T., Kinoshita, N. and Anzai, H. (1995) *Molecular Crystals and Liquid Crystals*, **268**, 153–9. (c) Nakatsuji, S., Akashi, N., Suzuki, K., Enoki, T. and Anzai, H. (1996) *Journal of the Chemical Society, Perkin Transactions 2*, 2555–9.

68 Fujiwara, H. and Kobayashi, H. (1999) *Chemical Communications*, 2417–18.

69 Fujiwara, H., Fujiwara, E. and Kobayashi, H. (2002) *Chemistry Letters*, 1048–9.

70 Fujiwara, H., Lee, H.-J., Kobayashi, H., Fujiwara, E. and Kobayashi, A. (2003) *Chemistry Letters*, 482–3.

71 Nakazaki, J., Matsushita, M.M., Izuoka, A. and Sugawara, T. (1999) *Tetrahedron Letters*, **40**, 4523–4.

72 Nakazaki, J., Ishikawa, Y., Izuoka, A., Sugawara, T. and Kawada, Y. (2000) *Chemical Physics Letters*, **319**, 385–90.

73 Matsushita, M.M., Kawakami, H., Kawada, Y. and Sugawara, T. (2007) *Chemistry Letters*, **36**, 110–11.

74 Hiraoka, S., Okamoto, T., Kozaki, M., Shiomi, D., Sato, K., Takui, T. and Okada, K. (2004) *Journal of the American Chemical Society*, **126**, 58–9.

75 Ishizu, K., Kohama, H. and Mukai, K. (1978) *Chemistry Letters*, 227–30.

76 Eastman, M.P., Patterson, D.E., Bartsch, R.A., Liu, Y. and Eller, P.G. (1982) *Journal of Physical Chemistry*, **86**, 2052–8.

77 Dugas, H. and Ptak, M. (1982) *Journal of the Chemical Society. Chemical Communications*, 710–12.

78 Keana, J.F., Cuomo, J., Lex, L. and Seyedrezai, S.E. (1983) *Journal of Organic Chemistry*, **48**, 2647–54.

79 Araki, K., Nakamura, R., Otsuka, H. and Shinkai, S. (1995) *Journal of the Chemical Society. Chemical Communications*, 2121–2.

80 Ulrich, G., Turek, P. and Ziessel, R. (1996) *Tetrahedron Letters*, 8755–8.

81 Wang, Q., Li, Y. and Wu, G. (2002) *Chemical Communications*, 1268–9.
82 Rajca, A., Pink, M., Rojsajjakul, T., Lu, K., Wang, H. and Rajca, S. (2003) *Journal of the American Chemical Society*, **125**, 8534–8.
83 Krock, L., Shivanyuk, A., Goodin, D.B. and Rebek, J., Jr. (2004) *Chemical Communications*, 272–3.
84 Feher, R., Amabilino, D.B., Wurst, K. and Vaciana, J. (1999) *Molecular Crystals and Liquid Crystals*, **334**, 333–45.
85 Taylor, P., Serwinski, P.R. and Lahti, P.M. (2003) *Chemical Communications*, 1400–1.
86 Taylor, P., Lahti, P.M., Carrol, J.B. and Rotello, V.M. (2005) *Chemical Communications*, 895–7.
87 Shiomi, D., Nozaki, M., Ise, T., Sato, K. and Takui, T. (2004) *Journal of Physical Chemistry. B*, **108**, 16606–8.
88 Ise, T., Shiomi, D., Sato, K. and Takui, T. (2006) *Chemical Communications*, 4832–4.
89 a. Aso, M., Norihisa, K., Tanaka, M., Koga, N. and Suemune, H. (2000) *Journal of the Chemical Society, Perkin Transactions*, **2**, 1637. b. Aso, M., Norihisa, K., Tanaka, M., Koga, N. and Suemune, H. (2001) *Journal of Organic Chemistry*, **66**, 3513–20.
90 Okamoto, A., Inasaki, T. and Saito, I. (2005) *Tetrahedron Letters*, **46**, 791–5.
91 Das, K., Pink, M., Rajca, S. and Rajca, A. (2006) *Journal of the American Chemical Society*, **128**, 5334–5.
92 a. Gagnaire, G., Jeunet, A. and Pierre, J.-L. (1989) *Tetrahedron Letters*, **30**, 6507–10. b. Gagnaire, G., Jeunet, A. and Pierre, J.-L. (1991) *Tetrahedron Letters*, **32**, 2021–4.
93 Matsuda, K., Stone, M.T. and Moore, J.M. (2002) *Journal of the American Chemical Society*, **124**, 11836–7.
94 Martin, R.E., Pannier, M., Diederich, F., Gramlich, V., Hubrich, M. and Spiess, H.W. (1998) *Angewandte Chemie (International Ed. in English)*, **37**, 2834–7.
95 Jeschke, G., Pannier, M., Godt, A. and Spiess, H.W. (2000) *Chemical Physics Letters*, **331**, 243–52.
96 Godt, A., Schulte, M., Zimmermann, H. and Jeschke, G. (2006) *Angewandte Chemie (International Ed. in English)*, **45**, 756–7564.
97 (a) Amano, T., Akutsu, H., Yamada, J. and Nakatsuji, S. (2004) *Chemistry Letters*, **33**, 382–3. (b) Nakatsuji, S., Amano, T., Akutsu, H. and Yamada, J. (2006) *Journal of Physical Organic Chemistry*, **19**, 333–40.
98 Matyjaszewski, K., Shigemoto, T., Fréchet, J.M. and Leduc, M. (1996) *Macromolecules*, **29**, 4167–71.
99 Bosman, A.W., Janssen, R.A.J. and Meijer, E.W. (1997) *Macromolecules*, **30**, 3606–11.
100 (a) Matsushita, M.M., Ozaki, N., Sugawara, T., Nakamura, F. and Hara, M. (2002) *Chemistry Letters*, 596–7. (b) Harada, G., Sakurai, H., Matsushita, M.M., Izuoka, A. and Sugawara, T. (2002) *Chemistry Letters*, 1030–1.
101 Ionita, P., Charagheorgheopol, A., Gilbert, B.C. and Chechik, V. (2005) *Journal of Physical Chemistry. B*, **109**, 3734–42.
102 (a) Matsuda, K. and Irie, M. (2004) *Journal of Photochemistry and Photobiology C: Photochemistry Reviews*, **5**, 169–82. (b) Nakatsuji, S. (2004) *Chemical Society Reviews*, **33**, 348–53.
103 Likhtenstein, G., Ishii, K. and Nakatsuji, S. (2007) *Photochemistry and Photobiology*, **83**, 871–81.
104 Hamachi, K., Matsuda, K., Itoh, T. and Iwamura, H. (1998) *Bulletin of the Chemical Society of Japan*, **71**, 2937–43.
105 Matsuda, K. and Irie, M. (2006) *Chemistry Letters*, **35**, 1204–9.
106 (a) Matsuda, K. and Irie, M. (2000) *Chemistry Letters*, 16–17. (b) Matsuda, K. and Irie, M. (2000) *Journal of the American Chemical Society*, **122**, 7195–201.
107 (a) Matsuda, K. and Irie, M. (2000) *Journal of the American Chemical Society*, **122**, 8309–10. (b) Matsuda, K. and Irie, M. (2001) *Chemistry–A European Journal*, **7**, 3466–73.
108 (a) Teki, Y., Miyamoto, S., Imura, M., Nakatsuji, M. and Miura, Y. (2000) *Journal of the American Chemical Society*, **122**, 984–5. (b) Teki, Y., Miyamoto, S., Nakatsuji, M. and Miura, Y. (2001) *Journal of the American Chemical Society*, **123**, 294–305.
109 (a) Teki, Y. and Nakajima, S. (2004) *Chemistry Letters*, **33**, 1500–1. (b) Teki, Y.,

Toichi, T. and Nakajima, S. (2006) *Chemistry – A European Journal*, **12**, 2329–36.

110 Roques, N., Gerbier, P., Nakajima, S., Teki, Y.Guerin, C. (2004) *The Journal of Physics and Chemistry of Solids*, **65**, 757–62.

111 Roques, N., Gerbier, P., Teki, Y., Choua, S., Lesniakova, P., Sutter, J.-P., Guionneau, P. and Guerin, C. (2006) *New Journal of Chemistry*, **65**, 1319–26.

112 Nakatsuji, S., Ogawa, Y., Takeuchi, S., Akutsu, H., Yamada, J., Naito, A., Sudo, K. and Yasuoka, N. (2000) *Journal of the Chemical Society, Perkin Transactions 2*, 1969–75.

113 Kaneko, T., Amano, T., Akutsu, H., Yamada, J. and Nakatsuji, S. (2003) *Organic Letters*, **5**, 2127–9.

114 (a) Fujino, M., Amano, T., Akutsu, H., Yamada, J. and Nakatsuji, S. (2004) *Chemical Communications*, 2310–11. (b) Amano, T., Fujino, M., Akutsu, H., Yamada, J. and Nakatsuji, S. (2005) *Polyhedron*, **24**, 2614–17. (c) Nakatsuji, S., Fujino, M., Hasegawa, S., Akutsu, H., Yamada, J., Gurman, V.S. and Vorobiev, A.K. (2007) *Journal of Organic Chemistry*, **72**, 2021–9.

115 (a) Ojima, T., Akutsu, H., Yamada, J. and Nakatsuji, S. (2000) *Chemistry Letters*, 918–19. (b) Nakatsuji, S., Ojima, T., Akutsu, H. and Yamada, J. (2002) *Journal of Organic Chemistry*, **67**, 916–21.

116 (a) Volodarsky, L.B., Grigoriev, L.A. and Sagdeev, R.Z. (1980) *Biological Magnetic Resonance*, Vol. 2, (eds L.J. Berliner and J. Reubin), Plenum Press, New York, p. 169. (b) Khramtsov, V.V., Weiner, L.M., Grigoriev, L.A. and Volodarsky, L.B. (1982) *Chemical Physics Letters*, **91**, 69–72.

117 Kirilyuk, I.A., Shevelev, T.G., Morozov, D.A., Khromovskih, E.L., Shuridin, N.G., Khramtsov, V.V. and Grigor'ev, I.A. (2003) *Synthesis*, 871–8.

118 Voinov, M.A., Polienko, J.F., Schanding, T., Bobko, A.A., Khramtsov, V.V., Gatilov, Y.V., Rybalova, T.V., Smirnov, A.I. and Grigor'ev, I.A. (2005) *Journal of Organic Chemistry*, **70**, 9702–11.

119 Hideg, K. (1990) *Pure and Applied Chemistry. Chimie Pure et Appliquee*, **62**, 207–12.

120 Kálai, T., Jeko, E. and Hideg, K. (2000) *Synthesis*, 831–7.

121 Kálai, T., Kulcsár, G., Jeko, J., Osy, E. and Hideg, K. (2004) *Synthesis*, 2115–20.

122 (a) Sen', V.D., Tkachev, V.V., Volkova, L.M., Goncharova, S.A., Raevskaya, T.A. and Konovalova, N.P. (2003) *Russian Chemical Bulletin, International Edition*, **52**, 421–6. (b) Sen', V.D., Golubev, V.A., Lugovskaya, N.Y., Saschekova, T.E. and Konovalova, N.P. (2006) *Russian Chemical Bulletin, International Edition*, **55**, 62–5.

123 Akaike, T., Yoshida, M., Miyamoto, Y., Sato, K., Kohno, M., Sasamoto, K., Miyazaki, K., Ueda, S. and Maeda, H. (1993) *Biochemistry*, **32**, 827–32.

124 Marx, L. and Rassat, A. (2000) *Angewandte Chemie (International Ed. in English)*, **39**, 4494–6.

125 (a) Miura, Y. and Muranaka, Y. (2005) *Chemistry Letters*, **34**, 480. (b) Miura, Y., Muranaka, Y. and Teki, Y. (2006) *Journal of Organic Chemistry*, **71**, 4786–94.

126 (a) Rosen, G.M., Britigan, B., Halpern, H.L. and Pou, S. (eds) (1999) *Free Radicals, Biology and Detection by Spin Trapping*, Oxford University Press, New York, Oxford. (b) Janzen, E.G. (1971) *Accounts of Chemical Research*, **4**, 31–40. (c) Lagererantz, C. (1971) *Journal of Physical Chemistry*, **75**, 3466–75. (d) Perkins, M.J. (1980) *Advances in Physical Organic Chemistry*, 1–64.

127 (a) Bonnett, R., Brown, R.F.C., Clark, V.M., Sutherland, I.O. and Todd, A. (1959) *Journal of the Chemical Society*, 2094–102. (b) Rosen, G.M. and Rauckman, E.J. (1984) *Methods in Enzymology*, **105**, 198–209. (c) Janzen, E.G., Jandrisits, L.T., Shetty, R.V., Haire, D.L. and Hilborn, J.W. (1989) *Chemico-Biological Interactions*, **70**, 167–72.

128 Sár, C.P., Hideg, E., Vass, I. and Hideg, K. (1998) *Bioorganic and Medicinal Chemistry Letters*, **8**, 379–84.

129 Janzen, E.G. and Zhang, Y.-K. (1995) *Journal of Organic Chemistry*, **60**, 5441–5.

130 Tuccio, B., Zeghdaoui, A., Finet, J.-P., Cerri, V. and Tordo, P. (1996) *Research on Chemical Intermediates*, **22**, 393–404.

131 Karoui, H., Nsanzumuhire, C., Le Moigne, F. and Tordo, P. (1999) *Journal of Organic Chemistry*, **64**, 1471–7.

132 Pou, S., Bha, A., Bhadti, V.S., Wu, S.Y., Hosmane, R.S. and Rosen, G.M. (1995) *The FASEB Journal: Official Publication of the Federation of American Societies for Experimental Biology*, **9**, 1085–90.

133 Becker, D.A. (1996) *Journal of the American Chemical Society*, **118**, 905–6.

134 Luccioni-Houzé, B., Nakache, P., Campredon, M., Guglielmetti, R. and Giusti, G. (1996) *Research on Chemical Intermediates*, **22**, 449–57.

135 (a) Campredon, M., Guglielmetti, R., Luccioni-Houzé, B., Pèpe, G., Alberti, A. and Macciantelli, D. (1997) *Free Radical Research*, **26**, 529–36. (b) Campredon, M., Luccioni-Houzé, B., Giusti, G., Lauricella, R., Alberti, A. and Macciantelli, D. (1997) *Journal of the Chemical Society, Perkin Transactions 2*, 2559–61.

136 Alberti, A., Campredon, M., Giusti, G., Luccioni-Houzé, B. and Macciantelli, D. (2000) *Magnetic Resonance in Chemistry*, **38**, 775–81.

6
Nitroxide Spin Probes for Studies of Molecular Dynamics and Microstructure
Gertz I. Likhtenshtein

6.1
Nitroxide Molecular Dynamics

6.1.1
Introduction

Molecular dynamic properties of a vast number of materials, including liquids, polymers, organic and inorganic materials, and biological systems, in particular, are a corner stone for their functional activity, technological characteristics, stability, compatibility, and so forth. Nitroxide spin-labeling methods have proved to be a powerful tool for the investigation of the molecular mobility and structure of various materials [1–8].

The basic idea underlying the spin-labeling approach is modification of the chosen sites of the material in question by specific compounds, commonly a nitroxide (NRO), which is bound covalently (labels) and/or non-covalently (probes), and whose properties make it possible to trace the state of the surrounding biological matrix by appropriate physical methods, commonly ESR. The principle advantage of the spin-labeling method is the possibility of gaining direct information about local structure, mobility, micropolarity, acidity, redox status, and electrostatic potential of certain parts of a molecular object of any molecular mass or optical density. Developments in synthetic chemistry, biochemistry, and site-directed mutagenesis in particular, have provided researchers with a wide assortment of labels and probes and have paved the way for the specific modification of molecular materials.

According to theory and a large body of experimental data, the rotation and intramolecular motion of a molecule in a condensed phase are modulated to a great extent by the molecular dynamics of the surrounding molecules. This phenomenon is caused by the relatively tight packing of molecules of liquids and solids, on the one hand, and the existence of static and dynamic defects in these systems, on the other.

Modern ESR techniques (Chapters 3 and 4), including continuous wave electron paramagnetic resonance (CW ESR), pulse ESR methods, echo-detected (ED) EPR,

Nitroxides: Applications in Chemistry, Biomedicine, and Materials Science
Gertz I. Likhtenshtein, Jun Yamauchi, Shin'ichi Nakatsuji, Alex I. Smirnov, and Rui Tamura
Copyright © 2008 WILEY-VCH Verlag GmbH & Co. KGaA, Weinheim
ISBN: 978-3-527-31889-6

multifrequency ESR spectroscopy, ESR imaging (ESRI) and magnetic resonance imaging (MRI), and so forth, together with computational approaches, provide researchers with an array of methods for studying in detail the molecular motion of nitroxides. The combination of these techniques allows us to access dynamic processes that are characterized by a wide range of correlation times and to measure the molecular motion of nitroxide with a wide range of correlation times, $\tau_c = 10^2$–10^{-10} s, and amplitude, ending low-amplitude high-frequency vibration. Measurements of spin relaxation parameters allow for investigating the phonon processes in media. Essential knowledge about the molecular dynamic state of the system under investigation may be derived from measurement of dynamic interactions between nitroxides and other paramagnetics that are encountered.

6.1.2
Molecular Dynamics of Surrounding Molecules

6.1.2.1 Microviscosity and Fluidity

Any motion of a nitroxide radical is greatly influenced by the molecular dynamics of the surrounding molecules. The following ESR parameters, which characterize the molecular motion of nitroxide, are employed: (a) the rotational diffusion correlation time (τ_c); (b) the rotational anisotropy (ε); (c) the angle of restricted motion, wobbling (φ); (d) the rotational diffusion apparent energy (E_{app}) and entropy ($\Delta S^{\#}_{app}$) activation of the Eyring equation; (e) the pre-exponential factor of the Arrhenius equation (ν_0); (f) the electron lattice relaxation time (T_{1e}); (g) the electron phase memory time T_{2e}; (h) the spin-packet width ($\Delta H_{1/2}$); and (i) the anisotropic hyperfine splitting (A_{aniso}) (see Chapter 3). These parameters are sensitive to different dynamic modes, which, in turn, are attributed to the microviscosity (fluidity) and microstructure of the environment and their changes.

The detail of the dynamic theory of the nitroxide ESR spectra for *nitroxide motion in the fast* ($\tau_c = 10^{-9}$–10^{-10} s) *and slow regions* ($\tau_c = 10^{-7}$–10^{-8} s) was developed by D. Kivelson [9] and J.H. Freed [10, 11] and references therein) using the stochastic Liouville equation. For the estimation of apparent rotation correlation time the following equations can be used [2, 7, 9] in the region of the fast motion:

$$\frac{1}{\tau_c} = \frac{3.6 \times 10^9}{\left(\sqrt{\frac{h_0}{h_{-1}}} - 1\right) \Delta H_0} \tag{6.1}$$

where h_0, h_{-1} are the heights of the ESR spectrum hyperfine components, respectively, and ΔH_0 is the line-width of the middle hyperfine component. In the region of slow motion the following equation may be used:

$$\tau_c = ax\left(1 - \frac{A_{zz}}{A^0_{zz}}\right)^b \tag{6.2}$$

where A_{zz}^0 and A_{zz} are the z-components of A-tensor for immobilized (determined from the rigid limit spectrum) and mobile nitroxide, respectively. The coefficient a was found to be 5.4×10^{-10} and 2.6×10^{-10} s for systems modeling isotropic and anisotropic Brownian diffusion, respectively, and b was found to be -1.36 and -1.39 for the aforementioned models.

Many systems, including nitroxide radicals in different solvents and at different pressures, display a functional correlation between τ_c and the viscosity of the solution (μ) that follows from the Stokes–Einstein equation [2]. The values of E_{app} in pure liquids and water–glycerol mixtures are approximately equal to the values of activation energy for viscosity in these systems. A correlation between apparent activation energy of rotational diffusion of nitroxide probes, E_{app}, and the pre-exponential factor, υ_0, on one side, and the corresponding parameters of liquids and polymers obtained by independent methods, including NMR, on another side, has been found [2]. For example, in "soft" polymers there is a satisfactory quantitative correlation between E_{app} and υ_0 for nitroxide rotational diffusion and the corresponding rotational parameters of the polymer segmental motion determined by NMR. In very rigid cross-linking polymers a probe in a microcavity of the polymer, being non-ideally parked (having a free space), undergoes rotational diffusion overcoming a relatively low barrier of "friction" against the wall of a cage, the structure of which is modulated by the cage low-amplitude high-frequency motion. In such a case τ_c is only slightly dependent on temperature. As the polymer is softened its structure became increasingly sensitive to heating. The cage walls acquire segmental mobility, and the probe motion begins to reflect the motion of the polymer cage. The softening is accompanied by a parallel increase of E_{app} and $\Delta S^{\#}$ [2].

In recent work [12] the nitrogen isotropic hyperfine coupling constant (A_{iso}) and the g tensor of a spin probe (di-tert-butyl nitroxide) in aqueous solution have been investigated by means of an integrated computational approach; Car–Parrinello molecular dynamics and quantum mechanical calculations involving a discrete-continuum embedding were performed. Decoupling of the structural, dynamical, and environmental contributions to the nitroxide ESR spectra allowed us to elicit the role played by different dynamic effects and to stress the importance specific vibration modes.

A procedure to measure a radical *super slow rotation* ($\tau_c \geq 10\text{–}13\,\text{s}$) has been developed by Ya. S. Lebedev and coworkers [13]. In the procedure, randomly oriented radicals with anisotropic magnetic parameters are irradiated with intensive, photochemically active polarized light. The polarized radiation destroys radicals whose optical electron transition moment coincides with the polarized vector of the incident light. The light thus causes a dip in the radical ESR spectrum. The recovery time of the dip corresponds approximately to the radical rotation correlation time. Photoactive dual fluorophore–nitroxide molecules being chemically stable would be ideal probes for such studies [7].

6.1.2.2 Motion of Macromoles as a Whole and Segmental Dynamics

In the case of spin-labeled macromolecules (polymers, proteins, and nuclear acids) it is possible to determine the value of the correlation time of the global tumbling of a macromolecule or one of its large segments as a whole (τ_R) using the Stokes–Einstein equation followed by estimation of its molecular volume. The procedure involves an analysis of the experimental dependence of the A_{zz} on temperature and viscosity [14]. The situation appears to be more complicated when the nitroxide segment can be involved at least in two types of motions: high-amplitude low-frequency ($\tau_c \geq 10^{-8}$ s) motion and low-amplitude high-frequency wobbling ($\tau_w \leq 10^{-9}$ s).

To distinguish these motions in macromolecules and membranes, special methods have been developed. According to model I [15], extrapolation of the experimental curve A_{zz} versus T/μ to infinite viscosity allows for estimating the order parameter S, which characterized the nitroxide motion relative to the macromolecule. If the value of the fast nitroxide precession is significantly large, and the wobbling angle is low ($\varphi = 0$–$50°$), the ESR spectrum will exhibit an internal extremum, $A_\perp$ [15]. An alternative model II [9, 10] also suggests that the radical is involved in two types of molecular tumbling, though it is assumed that both motions occur within a slow-motion regimen. In such a case parameter S is related to the hindered rotation of the nitroxide segment with respect to the macromolecule. To distinguish it from the abovementioned model a method based on an analysis of the correlation between the values of the A_{zz} shift and the line-width induced by temperature and viscosity change was proposed [16]. In the case of model I the A_{zz} shift without a marked change of the line-width is theoretically expected. According to the Freed model of anisotropic slow rotation, changes in both the A_{zz} shift and broadening of the spectral lines parallel with each other should occur. Comparison of experimental data for a series of spin-labeled bio-objects, including biomembranes and proteins, indicated that the rotation of the nitroxyl fragments of the spin-labels at an ambient temperature can be described by the model of slow anisotropic rotation with correlation time 10^{-7}–10^{-8} s under conditions where the rotation of the macromolecules and membranes is slow in the ESR timescale and can be neglected [16].

Other approaches to the problem involve the measurement of the angular dependence of the spectrum on ESR upon orientation of microscopically aligned biomaterial as anisotropic objects, such membranes [17] taking separate ESR spectra of the nitroxide under the condition of microwave saturation [18], and the use of ESR high-frequency spectroscopy [19–23]. In [24] continuous-wave EPR at conventional (9.4 GHz) and high (94.2 GHz) frequencies was applied to characterize molecular dynamics in spin-labeled catenanes composed of macrocycles with rigid phenylene–ethynylene and flexible alkyl chain building blocks. By using a set of compounds with increasing complexity, that were all labeled at the center of a rigid building block, it was possible to find a condition under which the spectral line-shapes were either dominated by local motion of the spin-label or those that contained information on tumbling of the building blocks.

6.1.2.3 Low-Amplitude High-Frequency Motion and Phonon Dynamics

Molecular-dynamical processes at sub-zero temperatures appear interesting for a number of reasons: (1) some biological reactions, including electron transfer, were found to occur at low temperatures; (2) cryoprotection of proteins, enzymes and membranes against denaturation and deactivation is an important process in biotechnology and in the investigation of enzymatic mechanisms; and (3) comparison of data for molecular dynamics and enzymatic functions at sub-zero and ambient temperatures paves the way for elucidating which dynamical modes may be responsible for the enzymatic activity and stability of the material under investigation [6, 7, 25–27].

The following parameters of spin probes in liquid glasses and proteins are sensitive to low-amplitude high-frequency and phonon dynamics: (1) line-width of a radical pair (RP); (2) RP ESR signal splitting; (3) amplitude of RP ESR signal; (4) spin packet of a nitroxide probe $\Delta H_{1/2}$; (5) nitroxides' spin-phase ($1/T_{1e} \sim \Delta H_{1/2}$) and spin-lattice relaxation rate ($1/T_{1e}$); (6) ESR spectra of nitroxides' line-width δ; and (7) nitroxides' hyperfine splitting A_{aniso}. Atomic vibration and molecular motions of a nitroxide affect the spin relaxation parameters and line-shapes of the nitroxide ESR spectra. One such effect is narrowing of the spectral line that is initially broadened as a result of inhomogeneous spin dipole–dipole interaction between the unpaired electron and protons of the nitroxides and the medium. Low amplitude of local wobbling of the nitroxide fragment with $\tau_c \leq 10^{-7}$ s averages the latter interaction and leads to narrowing of the inhomogeneous ESR line. According to theory [28, 29], spin electron relaxation rates $1/T_{1e}$ and $1/T_{2e}$ of radicals were drastically affected by the intensity of phonon dynamics in the medium. The phonon oscillations are accompanied by electric field oscillations over a wide range of frequencies. If the distribution of the electric oscillation frequency partially overlaps the microwave resonance frequency, the radical magnetization energy will be transferred to the energy of the phonon oscillation through the spin–orbital interaction. The spin relaxation rate depends on the temperature:

$$1/T_{1e} \sim T^n \tag{6.3}$$

where $n = 1$ for one-quantum direct transfer, $n = 2.63$ and $n = 4.63$ for two-fraction three-dimensional Kramer and non-Kramer transfers, correspondingly, and $n = 7$ and $n = 9$ for the Raman three-dimensional non-Kramer and Kramer transfers, respectively. Thus the fine mechanisms of phonon processes can be studied by measuring the temperature dependence of the spin-relaxation rate of spin probes.

Previously, ultra-slow motion has been studied in frozen 95% ethanol, glycerol, 1:3 water/glycerol, and serum albumin model systems with the use of a series of nitroxides and the diphenyl amine radical pairs (RP) [30–32]. The amplitude of the EPR spectrum of RP, and the spin-packet width ($\Delta H_{1/2}$) of the nitroxides were measured as a function of temperature. In ethanol at 30–60 K, a decrease in the magnitude of doublet splitting, specifically, and a relative increase in spectral intensity were observed for RP. This effect was attributed to a 0.8 nm displacement of the radicals. At 60–95 K some further changes, indicating an intensification of molecular motion, have been observed, and the RP radicals eventually were recombined. At 95–110 K, a linear increase in $1/T_{1e}$ and $\Delta H_{1/2}$, which is related to $1/T_{2e}$, occurred, most likely due to an animation of the phonon processes and the increased mobility of protons. Probe behavior similar to that in ethanol solutions was observed in 75% glycerol and a protein bovin serum albumin (BSA) solutions. Another effect observed was a small (0.1–0.2 mT) gradual A_{zz} shift in the nitroxide ESR spectra in glasses and proteins as the temperature increases from 77 to 200 K [32], followed by a sharp increase in A_{zz} at temperature 200–230 K, depending on the object of investigation. The cause of the low-temperature shift appears to be high-frequency low-amplitude wobbling of the radical on an angle of several degrees. The sharp A_{zz} increase is attributed to intensification of high-amplitude relaxation processes in the media in the slow-motion temporal scale. The occurrence of changes in probe behavior shows the importance of the free volume of the media on their dynamic behavior.

A number of independent physical approaches, including Mössbauer, fluorescence, and phosphorescence spectroscopic methods and NMR, confirmed the abovementioned conclusions regarding low temperature dynamic behavior of glassy materials obtained by the spin probe method ([6, 7, 27] and references therein). Recent spin-probe investigations of glassy systems by an advanced ESR technique provided additional support for this model of low-temperature dynamic behavior of nitroxides, and also elucidated new important dynamic effects.

Pulsed multi-frequency EPR was used to investigate orientational molecular motion of the nitroxide spin-probe ([(SO$_3$)$_2$NO·], Fremy's salt) in glycerol glass near the glass transition temperature [33–35]. Measuring echo-detected EPR spectra at different pulse separation times at resonance frequencies of 3, 9.5, 95 and 180 GHz of Fremy's salt, [(SO$_3$)$_2$NO·], in glycerol glass near the glass transition temperature allowed for estimating parameters of transverse relaxation tensor and thus for discriminating between different relaxation mechanisms and characterizing the timescale of molecular reorientations (10^{-7}–10^{-10} s) [33]. It was found that close to the glass transition temperature, orientation-dependent transverse relaxation is dominated by fast low-amplitude reorientational fluctuations (wobbling), which may be overlapped with fast modulations of the canonical g-matrix values. The data were interpreted using a new simulation program for the orientation-dependent transverse relaxation rate $1/T_{2e}$ of nitroxides based on different models for the molecular motion. The authors suggested that, in frozen solution, orientation-dependent relaxation may be induced by collective motion of the solvent cage, reorientational motion within the cage, or dynamics of A- and g-strains. These processes are discriminated from orientation-independent relaxation mechanisms as methyl group rotation and nuclear spin dynamics. A quantitative comparison of the orientation dependence of transverse relaxation rate with independent measurements of the longitudinal relaxation rates by X-band ESR showed that the increased relaxation rate at the g_X position can be explained by the isotropic dynamics of the g-strain. Another tumbling mode is the orientational motion about all three canonical orientations with a larger excursion about of the molecular y-axis of the nitroxide at $T = 185$ K.

Echo detected (ED) ESR study of orientational molecular motion in glassy solvent was performed for evaluation of anisotropic $1/T_{2e}$ at different positions of

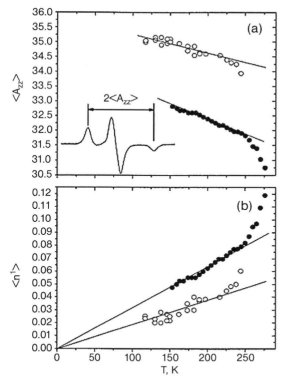

Figure 6.1 (a) Temperature dependence of splitting between two outer peaks (shown in the insert) for TEMPONE in glycerol (open circles) and in o-terphenyl (filled circles). The straight lines are drawn through low-temperature points. (b) The mean-squared amplitudes of angular motion, obtained from data (a) [35]. (Reproduced with permission).

the ESR spectra [34]. Experiments with the use of 13 nitroxides of different sizes and shapes showed no effect of temperature, solvent glasses of different polarity, and the use of deuterated solvents. The solvent glass had a much stronger impact on the relaxation rate than do the size and shape of the nitroxide. These observations are most probably due to phonon processes in the vicinity of the probe nitroxide fragment.

Continuous wave (CW) EPR when applied to nitroxide spin-probes in glassy media showed a linear decrease of A_{zz} with increasing temperature (Figure 6.1) [35]. Above some temperature points the temperature dependence becomes sharper. Within the model of molecular librations, this behavior is in quantitative agreement with the numerical data on neutron scattering and Mössbauer absorption for molecular glasses and biomolecules. The departure from linear temperature dependence in CW EPR was ascribed to the transition from harmonic to anharmonic modes. Experiments showed that ED EPR spectra change drastically above 195 K in glycerol and above 245 K in o-terphenyl (Figure 6.2), indicating the appearance of anisotropic transverse spin relaxation caused by the phase

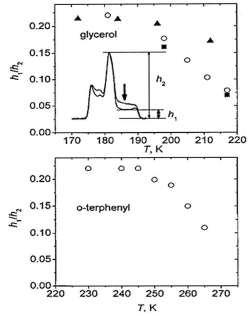

Figure 6.2 The relative intensities of the lines in the ED EPR spectra taken at different temperatures for the field position indicated by an arrow in the insert. (Two different spectra in the insert belong to two different temperatures). Circles: TEMPON, triangles: Fremy's salt, squares: 3-Carboxy-proxyl [35]. (Reproduced with permission).

transition. The authors concluded that the low sensitivity of ED EPR to harmonic motion and its high sensitivity to the anharmonic motion suggest that ED EPR may serve as a sensitive tool to detect dynamical transition in glasses and biomolecules.

All in all, the full information on multimodal behavior of a spin-label may be derived using modern multi-frequency and pulse ESR techniques. Moreover, these methods provide identification of distribution functions for all dynamic parameters used in the description of a physical model of a spin-label motion. The possibility of investigating the molecular dynamics of chemical and biological systems is one of the most important advantages of the spin-labeling method.

6.2
The Spin Label–Spin Probe Methods

This method is based on an interaction of stable radicals with spin probes, which are chemically inert paramagnetic species capable of diffusing freely in solution [2, 7, 8, 36–39]. This method allows for investigating the microstructure and electrostatic potential in the vicinity of a spin-label attached to the object of interest (Figure 6.3).

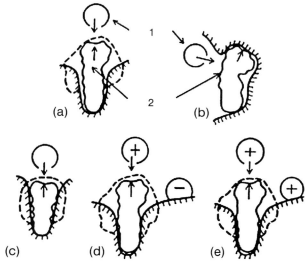

Figure 6.3 The spin-probe (1)–spin-label (2) approach for investigating micro-relief (a, c) and charge distribution (d, e) on the nature of the label-probe interaction [2]. (Reproduced from Likhtenshtein, G.I. (1993) *Biophysical Labeling Methods in Molecular Biology*, Cambridge University Press, with permission) [7].

When a nitroxide radical in solution encounters a paramagnetic species, say, a paramagnetic complex of a transition metal, that is within the concentration range C from 5×10^{-3} to 10^{-1} M (depending on the nature of the probe and nitroxide radical), the magnitude of broadening of the Lorentzian line (ΔH_L) of the nitroxide ESR spectrum is proportional to the radical concentration.

The rate constant of the exchange relaxation (k_{ex}) can be determined by the equation:

$$1/T_{2e} \sim \Delta H_L = 6.5 \times 10^{-9} k_{ex} C \tag{6.4}$$

where peak-to-peak ΔH_L is expressed in mT, C in M, and k_{ex} in $M^{-1}s^{-1}$. In the case of inhomogeneous broadening of the initial line, adequate correction should be made for the value of the experimentally observed broadening [40].

For NRO in solution, the first effect of the exchange interaction with a paramagnetic species is narrowing of the radical spectrum line because of averaging of the dipole contribution from protons. In liquids with dissolved dioxygen the exchange narrowing is usually complete. For a nitroxide at the rigid limit of dynamics, the theoretically calculated width of the central component of the ESR spectrum is also proportional to ΔH_L [41]. This theoretical prediction was supported experimentally [42]. A method has been worked out for determination of k_{ex} from the extent of the change in the rate spin–lattice relaxation ($1/T_{1e}$) and from related saturation curves of the radical in the presence of paramagnetic species [43–45]. Since the relaxation rate $1/T_{1e}$ is more sensitive to the spin interactions than $1/T_{2e}$, in particular in the region of slow rotation, this method allows us to widen markedly the range of the k_{ex} values that could be accessed experimentally. Another method of particular interest is an elegant approach based on ENDOR using two types of nitroxides, ^{15}N and ^{14}N, characterized by non-overlapping positions of ESR lines [46].

In principle, both spin–spin dipole–dipole and exchange interactions can occur in solution. The authors who have studied the behavior of paramagnetic particles in detail ([2, 37] and references therein) have come to the conclusion that, in a medium of relatively low viscosity, the major contribution to ESR line-width for a certain pair spin-label–spin-probe is made by exchange interactions during molecular encounters.

The k_{ex} value depends on the probability of spin relaxation during the encounter (P_{ex}), therefore:

$$k_{ex} = P_{ex} k_d \tag{6.5}$$

where k_d is the rate constant of encounters in solution. According to reference [47]:

$$P_{ex} = \frac{f_g f_{ns} J^2 \tau_{en}^2}{1 + J^2 \tau_{en}^2} \tag{6.6}$$

where f_g is the geometric steric factor, f_{ns} is the nuclear statistical factor, equal to 2/3 for nitroxides, τ_{en} is the lifetime of the encounter complex depending upon viscosity and equal in low-viscosity solutions to about 10^{-10} s, J is the exchange integral depending on the paramagnetic species structure. According to available data [37, 41], J depends on the densities of the unpaired electron at the site of direct contact between the nitroxide N–O· fragment and the probe in the encounter complex. In the case of the long-distance spin exchange interactions the relationship between J and the distance (r) between the label and the probe is described by the empirical equation [7, 26]:

$$J \approx 10^{15} \exp(-\beta r), \text{ s}^{-1} \tag{6.7}$$

where $\beta = 1.3$ Å^{-1} for systems in which the exchanging centers are separated by a medium of molecules or groups with saturated chemical bonds and $\beta = 0.16$ Å^{-1} for systems in which the centers are linked by conjugating bonds. Equation 6.6 is valid if $J \gg \delta$ (the difference between the resonance frequencies of the spins), $\delta \tau_{en} \gg 1$, and the ion spin $S = 1/2$. If $J^2 \tau_{en}^2 \gg 1$. When k_{ex} is independent upon J (strong exchange condition and)

$$k_{ex} = f_g f_{ns} k_d \tag{6.8}$$

In the case of $J^2 \tau_{en}^2 \ll 1$ (weak exchange):

$$k_{ex} = f_g f_{ns} k_d J^2 \tau_c^2 \tag{6.9}$$

A more general theory has been developed for exchange relaxation during encounters in solution [47]. This theory takes into account the influence of factors such as the probe-spin electron relaxation time (T_{1e}), the values of spin numbers or the paramagnetic particles (S_1 and S_2), and the differences between the resonance frequencies of the interacting spins (δ).

The most suitable probes for studies in aqueous solution are found to be potassium ferricyanide and dibenzene chromium iodide [2, 7]. They have a high relaxation probability at the encounter ($P_{ex} \approx 0.2$), form very stable, chemically inert ligand shells, and have short electron relaxation times ($T_{1e} < \tau_{en}$). As was shown, the value of k_{ex} for interaction of both probes with nitroxides in water at $T = 293$ K is approximately equal, $2 \times 10^9 \, \text{M}^{-1}\text{s}^{-1}$, and is inversely proportional to the viscosity of a water–glycerol mixture, indicating that the processes occur in the strong exchange regimen.

For a particular pair of paramagnetic species, the value of k_{ex} depends on microviscosity, steric hindrances, and distribution of electrostatic charges in the region of encounters. Hence, after preliminary empirical calibration, the method of examining of k_{ex} and its variations can be used for experimental study of these factors in biological and other materials.

The translation diffusion coefficient (D_{tr}) is inversely proportional to the hydrodynamic radius R of a particle. Therefore, in the case of labeled macromolecules and a low-molecular weight probe we could neglect the translation diffusion of the macromolecule. Then the accessibility parameter is the value $\alpha = k_w/2k_M$, where k_w and k_M are the k_{ex} constants for the free radical in water and the spin-label on a macromolecule [2]. The parameter α efficiently characterizes an increase of local viscosity and steric hindrance in the region of the encounter between the radical and the probe.

Examples of application of the spin-label–spin-probe methods are given in the following sections.

6.3
Spin Oximetry

Dioxygen is a paramagnetic molecule, and therefore it can affect the spin-lattice ($1/T_{1e}$) and transverse ($1/T_{2e}$) relaxation rates of a NRO (6.4). This property has been used for determination of the $k_{ex}[O_2]$ product, which is an important characteristic of the oxidative capacity of molecular oxygen at the location of the spin-label. In water solution dioxygen is a strong quencher of triplet-excited state of chromophores ($k_{TT} = (3–5) \times 10^9 \, \text{M}^{-1}\text{s}^{-1}$) [48], that is, close to the diffusion limit. If k_{ex} is known or a system is previously calibrated in similar conditions, the dioxygen concentration can be evaluated [7, 49–60].

Several versions of oxymetry have been developed. One of them is based on analysis of the line-shape of the nitroxide or other paramagnetic species, which is sensitive to the presence of dioxygen when the value of $k_{ex}[O_2] \geq 2 \times 10^6 \, \text{s}^{-1}$. Paramagnetic O_2 blurs the superfine structure of inhomogeneously broadening ESR lines and increases the line-width. In water solution the linear relationship between the Lorentzian component of the line width (ΔH_L) of the three hyperfine lines of the nitroxide ESR spectrum (related to $1/T_{2e}$) and the O_2 concentration was proven to hold even when the nitrogen hyperfine splitting is unresolved [58]. Methods of oximetry based on the effect of dioxygen on the spin–lattice relaxation parameters of NRO allow for determination of the product as approximately equal to $k_{ex}[O_2] = 10^5\,s^{-1}$ and, therefore, to expand markedly the O_2 concentration scale available experimentally for the measurement [50–53]. This effect is due to the fact that, for nitroxides, $1/T_{2e} > 1/T_{1e}$. Techniques such as wave power continuous saturation, ELDOR, and magnetization recovery can be used to quantitatively estimate the dioxygen contribution to the value of $1/T_{1e}$.

All of the above-mentioned methods have advantages as well as limitations and drawbacks. The special advantage of the method based on measurement of the nitroxide ESR line-broadening is its relative simplicity and the possibility of the specific introduction to a molecular target of interest. Many of the difficulties of such a method are a lack of sensitivity in viscose media (*in vivo*, for example) and the reducing power of cells. Therefore, several chemically stable paramagnetic probes have been proposed as oxygen sensors: miniature crystals of lithium phthalocyanine [59, 60], activated charcoal [61], and synthetic carbohydrate chars [62]. These probes are more suitable for experiments *in vivo* but their reliability is limited in investigation in cells. For such an approach, noninvasiveness, and biocompatibility of the probes *in vivo* remain one of the most challenging problems.

A problem of the method *in vivo* in particular, is an uncertainty in the k_{ex} value. In the case of the nitroxide sensor, it is possible to overcome this uncertainty by the evaluation of microviscosity in the vicinity of the probe by ESR spectrum analysis that allows us to estimate the k_{ex} value from the Stokes–Einstein equation. Within the last few years, remarkable progress has been achieved in EPR oximetry by developing instrumentation and sensors capable of measuring $k_{ex}[O_2]$ in tissues with sufficient accuracy and sensitivity. Though the above-mentioned measurements, based on detecting changes in electronic spin lattice relaxation time, $1/T_{1e}$, provide more direct evidence of the local concentration in oxygen the ESR instrumentation has to be more elaborate.

The $k_{ex}[O_2]$ product can be determined over a wide range of its values by means of the measurement of the nitroxide power saturation curves [7, 53]. For example, the difference in the amplitude of the saturation curves of nitroxides in the lecithin liposomes with and without air was found to be about 50% while in these conditions no marked change in the probe ESR spectrum was detected. For such a measurement it is enough to detect amplitude of the nitroxide CW ESR spectra in the maximum of the saturation curve. The CW ESR technique is widely available from many biochemical and biophysical laboratories.

At present, spin oximetry has been successfully employed for solving a large number of biomedical and clinical problems [54–62]. *In vivo* measurement can also be performed with low-frequency (0.5 and 2.0 GHz) EPR spectrometers. At low magnetic field, the microwave wavelength becomes compatible with deep penetration depth in tissues, making such an approach suitable for studying animals and/or viable organs by ESR. ESR oximetry has found an important application in the study of ischemia-reperfusion injury and therapeutic agents that are affected by oxygen levels, and in oxidative damage as a reactant and because oxygenation can be affected by NO and other reactive species of interest (see Chapter 11).

Within the last few years, a remarkable advantage has been achieved using EPR oximetry, which has resulted in the use of instrumentation and sensors capable of measuring $k_{ex}[O_2]$ in tissues with sufficient accuracy and sensitivity. It is necessary to emphasize that these advantages in applications have become possible by progress in low-frequency ESR oximetry and experimental methology ([63–67] and references therein). Considering that the reaction of O_2 evolution and consumption are processes of paramount importance in chemistry, biophysics, and medicine, it is difficult to overestimate the potential of the new techniques described here.

6.4
Determination of the Immersion Depth of Radical and Flourescent Centers

In solving problems of polymers, inorganic materials, enzyme and chemical catalysis, molecular biophysics of proteins, biomembranes, nucleic acids, and so forth it is necessary to know the spatial disposition of individual molecules/domains. We must also know the depth of immersion of paramagnetic and luminescence centers in a biological matrix, that is, the availability of enzyme sites to substrates, the distance of electron tunneling between a donor and an acceptor group, the position of a spin-label in a lipid bilayer and in a protein globule, the distribution of the electrostatic field around the paramagnetic center, and so forth.

Here, we describe two approaches to estimating the depth of the immersion based on the measurement of the interactions between nitroxide and a center of interest.

6.4.1
Analysis of Power Saturation Curves in Solids by CW ESR

The effect of spin–spin interactions between paramagnetic molecules is dependent on the distance of the closest approach of the centers and, therefore, on the depth of immersion of the center in a matrix [2, 7, 43, 68–77]. Under certain conditions, a paramagnetic center can affect nuclear spin relaxation parameters monitored by NMR, which can provide information on the location of a nucleus [78–82].

The first attempts to employ CW ESR techniques for the investigation of location and depth of immersions of radicals (r_{min}) were made in the early 1970s [68–71]. The determination of the distance of the closest approach of one paramagnetic center, a "radical", and a paramagnetic species was performed by examining the effect of the dipole–dipole interaction between the radical and paramagnetic species that are distributed uniformly in a vitrified matrix, on the ESR saturation curves of the radicals [72–76, 83]. A similar approach has been employed to examine the kinetics of spin–lattice relaxation for paramagnetic compounds in solids [83].

Methods for estimating a radical depth of immersion are described in detail in [72–77]. Here, we will restrict ourselves to a short illustration of one of the approaches to the problem. A method was developed for determining the nearest distance (r_{min}) between a stable radical (R·) and an ion of paramagnetic metal, an ion-relaxator (IR), which has effects on the spin–lattice relaxation time of R· and is randomly distributed in the bulk of the vitrified sample. In the case of R· penetration into impermeable matrix (macromolecule, membrane, and so forth), the r_{min} value is equal to the radical immersion depth. If the center resides at a sufficient depth, $r_{min} > r_{av}$, where the latter value is the the average distance of radicals, the contribution of the dipole interaction of IR to the R· spin–relaxation rate is expressed by the equation:

$$\Delta(1/T_{1e}) = \frac{A_d \mu^2 \gamma^2 \tau_{1e} C}{r_{min}^3} \tag{6.10}$$

where C is the IR concentration, μ and τ_{1e} are magnetic moment and the spin relaxation rate of the IR, correspondingly, and A_d is a factor that depends on the geometry of surface. For example, if the surface is flat, $A_d = 0.2$. Equation 6.10 predicts the linear dependences of the enhancement of the spin relaxation rate upon C in the case of NR immersion at different depth. An ion-relaxater can be included into superficial portions of the membrane of interest (that is Co.$^{2+}$ acetylacetonate) or into the surface of the membrane (that is Co.$^{2+}$). Such an approach allows precise determination of the location of the radical in the biological molecular object [75].

The above-mentioned methods have been employed for the measuring of depth of immersion of paramagnetic centers in a number of biomembranes and enzymes [72–77].

6.4.2
Determination of Depth of Immersion of a Luminescent Chromophore and a Radical Using Dynamic Exchange Interactions

Fluorescent and phosphorescent chromophores play an important role in numerous biochemical and biophysical processes. Luminescent centers include tryptophane- and tyrosine-containing proteins, flavines, chlorophylls, pheophytins, rhodopsin, carotenoids, phycocyanin, phycoerythrin, pyridoxyl phosphates, heme proteins with iron substituted by Mg or Zn, fluorescent drugs, and so forth. In solving problems relating to enzyme catalysis, the molecular biophysics of proteins and biomembranes, and molecular biology it is necessary to know the depth of immersion of the luminescent centers in the biological matrix, that is, the availability of enzyme sites to substrates, the distance of electron tunneling between a donor and an acceptor group, or the position of labels and probes in a membrane and in a protein globule.

6.4 Determination of the Immersion Depth of Radical and Flourescent Centers

Over the last few decades nitroxide radicals have been extensively used as quenchers of chromophores in exited states [84–86]. The nitroxides have been shown to be strong quenchers of chromophore fluorescence and phosphorescence with rate constant $k_q = 2$–$8 \times 10^9 \, M^{-1} s^{-1}$, values which are close to the diffusion-controlled rate constant k_{en}. These properties make nitroxides effective as inert and water-soluble quenchers suitable for use in the study of the depth of immersion of luminescent chromophores.

In this section we will discuss briefly the theoretical and experimental grounds of the method for measuring the immersion depth of a fluorescent chromophore in biological and model membranes [77, 87]. The proposed approach is based on quantitative investigation of the dynamic quenching of the fluorescence of such a tag incorporated in an object of interest, by a quencher (stable nitroxide radical) freely diffusing in solution. Using the concept of dynamic exchange interactions and the empirical dependence of parameters of the static exchange interactions on the distances between exchangeable centers whether by mechanisms of intersystem crossing (rate constant k_{IC}) or electron transfer (rate constant k_{ET}), the theoretical bases of a method for determination of the immersion depth of chromophore has been developed. The theory established a quantitative relationship between the ratio of the experimental rate constants of dynamic quenching of fluorophore in solution and after fluorophore ducking in the matrix under investigation, on the one hand, and the depth of immersion of the fluorophore, on the other. According to references [77, 87], in the case of the fluorescence quenching in solution, electron exchange-induced intersystem crossing (IC) between singlet and triplet states and electron transfer (ET) between an excited chromophore and nitroxide mechanisms appear to be the prevailing mechanism. In such a case, the value of the fluorophore immersion (R_{im}) can be derived from the following equation:

$$R_{im} = 0.5 \times \beta^{-1}[\ln(k_q^0/k_q^x) + \ln(\tau_{en}^2 10^a)] \tag{6.11}$$

where k_q^0/k_q^x is the ratio of the rate constants of fluorescence quenching of a fluorophore in solution and in an object of interest, correspondingly, τ_{en} is the lifetime of the encounter complex, that depends on viscosity and is equal to about 10^{-10} s for nonviscous media, and $a = 28$ and 26 for the intersystem crossing and the electron transfer mechanisms, respectively. Equation 6.11 can be used to estimate the depth of immersion of a fluorescent chromophore in a "non-conductive" matrix by measuring the ratio k_q^0/k_q^x experimentally with a reasonable choice of values of an encounter complex lifetime τ_{en}. Application of the proposed method to different membranes (lecithin liposomes, membranes from *B. subtiliz* grown in the absence and presence of chloramphenicol) modified by a fluorescence-photochrome stilbene probe, 4,4′-dimethylaminocyanostilbene and quenched by TEMPOL, led to an estimation of the probe depth of immersion in the membranes [87].

A similar approach was proposed for estimating the immersion depth of a paramagnetic center on the basis of its spin–exchange interaction with a nitroxide probe freely diffused in solution [77].

6.5
Nitroxide as Polarity Probes

The dependence of isotropic and anisoptropic hyperfine splitting (A_{iso} and A_{aniso}) and g-factor for nitroxides on the spin-density on nitrogen atoms and the energy of the n → π* transition in the N–O· fragment makes this probe suitable for assessing local polarity and the availability of hydrogen bond donors to form an H-bond with the N–O· group. The physical reason for such dependence is that the contribution of polarized resonance structure N–O· ↔ N$^+$–O$^-$ increases in a polar media causing an increase in the spin density on the nitrogen atom observed as an increase in the A_{iso} value. At the same time, the A_{aniso} values are larger for the ionic structure because of the reduction of the distance between the electron spin and the nitrogen nucleus. The g-factor values are also reported to be sensitive to the medium. A correlation between a decrease of g_x and an increase in A_z has been observed for nitroxides in simple liquids and labeled proteins [7, 20, 88–90]. A polarity effect (up to $\Delta A_{iso} = 0.13\,\text{mT}$) was observed for the TEMPOL radical and for the formation of an H-bond with the N–O group of the nitroxides ($\Delta A_{iso} = 0.15\,\text{mT}$). ESR spectra of a nitroxide probe in solvent of various polarities similar to those shown in Figure 6.4 may be used for estimation of local apparent dielectric constant in the vicinity of a nitroxide incorporated in systems of interest.

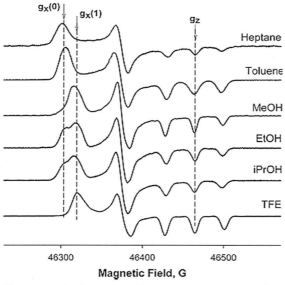

Figure 6.4 Echo detected rigid limit 130 GHz ESR spectra from 5-doxyl SA in series of protic and aprotic solvents of various polarity at $T = 25\,\text{K}$ [88]. (Reproduced with permission).

6.6
Electrostatic Effects in Molecules in Solutions

Methods of determination of the electrostatic potential and local charges in the vicinity of groups bearing magnetic moments are based on the sensitivity of spin-relaxation rates of radicals and nuclei to encounters with charged paramagnetic species in solution [77, 91, 92]. Measurements of the effect of paramagnetics with different charges on the spin relaxation rates of the nuclei or radical allow quantitative characterization of electrostatic interactions.

6.6.1
Effect of Charge on Dipolar Interactions Between Protons and a Paramagnetic Species

According to a number of workers [93–95], the spin–lattice relaxation rate of proton nuclei, $1/T_{1p}$, upon encountering electron spin van der Waals distance, is dominated by dipolar interaction because of low electron spin density on the nuclei. The relaxation rate of dipole interaction can be described by 6.12 and 6.13

$$1/T_{1p} = \left(\frac{4\pi}{9}\right)\left(\frac{\gamma_p^2 \gamma_e^2 \hbar^2}{R_0^6}\right) S(S+1) T_{1e}[R\bullet]f(y) \tag{6.12}$$

$$f(y) = \frac{(4+y)y^2}{9 + 9y + 4y^2 + y^3} \tag{6.13}$$

where γ_p and γ_e are the gyromagnetic ratios of the protons and the electrons, respectively; R_0 is the distance of the closest approach between proton and electron; [R•] is the concentration of the paramagnetic species; $\tau_d = R^2/D$, R is the radius of the radical, $y = (\tau_d/T_{1e})^{1/2}$; T_{1e} is the electron spin–lattice relaxation time; D is the sum of diffusion coefficients of the proton-bearer and radical; and S is the electron spin. Owing to the low value of γ_p, estimates have shown [94, 95] that 6.12 is correct for any reasonable viscosity for the medium.

The slopes, $k^i = d(1/T_{1p})/d[R\bullet]$, of the dependence of proton spin–lattice relaxation rate on the concentration of the nitroxide probes [R•] can characterize the dipole–dipole interaction that dominates in an encounter between a proton and paramagnetic species. Variable k has the dimensions of a second-order rate constant ($M^{-1}s^{-1}$), and can be considered as an apparent relaxation rate constant. The superscript index $i = 0, +1$ or -1 designates data acquired from radical probes with $0, +1$ or -1 charge, respectively. To explore the effect of electrostatic interaction on the dipole–dipole relaxation rate, the ratios of experimental values k^+/k^0 or k^-/k^0 were used to estimate the electrostatic potential, $U(R_0)$, in the vicinity of a proton [96]. The electrostatic effect in the vicinity of a proton coming from charged functional groups (ammonium cation, carboxylate anion, and so forth) in various regions of a target molecule X can be quantitatively characterized in an empirical manner by a relative apparent charge, α^X:

$$\alpha^X = \frac{Z^X_{app}}{Z^0} = \frac{U(R_0)^X_{exp}}{U(R_0)^0_{exp}} \quad (6.14)$$

where α^X is the ratio of experimental log k and indices x or 0 are assigned to a generic target molecule or to a small charged model compound (ethylammonium cation, imidazolium cation, propionate anion, and so forth), respectively. The α^x value depends on the position of the proton in the target molecule relative to the neighboring charged group. Taking $Z_0 = +1$ or -1, we can consider $Z^x_{app} = \alpha^x$ as a parameter indicating an electrostatic effect of the molecule on a charged particle placed in the vicinity of a given proton. Such a parameter can be used in the analysis of electrostatic factors affecting equilibrium and reaction rate constants for nuclei residing in particular local charge environments within the molecule.

Simple organic molecules and amino acids provide well-defined model systems to study the factors that influence electrostatic fields in more complex biologically important molecules. These model systems enable verification of experimental and theoretical approaches to this problem [96]. The only significant difference in the nitroxide spin-probes 1–3 is the difference in the probe charges, Z_p.

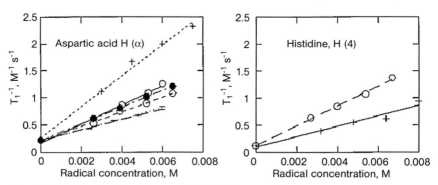

The spin-lattice relaxation rates ($1/T_{1p}$) of protons in small target molecules ethanol, isopropanol, tert-butanol, benzyl alcohol, propionate anion, ethylammonium cation, imidazolium cation) were measured as a function of concentration of spin-probes with different charges (see Figure 6.5 for example). In each case,

Figure 6.5 Spin–lattice relaxation rate of individual protons of small charged molecules as a function of spin probe concentration. Neutral, negative, and charged spin probes (I–III) are respectively denoted as 0, −, and + [96]. (Reproduced with permission).

$1/T_{1p}$ was found to be proportional to the probe concentration. The rate constants, k, (slopes of $d(1/T_{1p})/d[R\bullet]$) were then calculated and used to calculate the electrostatic potential $U(R_0)_{exp}$ (6.14).

The sign and magnitudes of parameters $U(R_0)_{exp}$ corresponded to expected values from simple electrostatic considerations and also to theoretical calculation. The sign is positive for protonated ethylamine and imidazole, and negative for propionate anion. The absolute value of $U(R_0)_{calc}$ is very small for protons in neutral tert-butanol and is markedly larger for small charged molecules (propionate anion, ethylammonium cation and imidazolium cation, and charged amino acids).

6.6.2
Impact of Charge on Spin Exchange Interactions Between Radicals and Paramagnetic Complexes

As discussed in Section **6.2**, the exchange interaction is assumed to dominate over dipolar interaction in fluid solutions. Calculations of local charge Z_x in the vicinity of a paramagnetic particle (such as the active site of metalloprotein or a spin-label) which encounters a nitroxide or metallocomplex with known charge Z_p can be carried out with the use of the Debye equation (6.15) [97].

$$\frac{k^+}{k^0} \text{ or } \frac{k^-}{k^0} = \frac{Z_p Z_x \alpha}{\exp(Z_p Z_x \alpha) - 1} \tag{6.15}$$

where k^+, k^-, and k^0 are the rate constants of encounters for positively charged, negatively charged, and neutral uncharged particles, respectively; and $\alpha = e^2/k_B T \varepsilon_0 r$, where e is the charge of an electron; k_B is the Boltzmann constant; temperature $T = 293$ K, ε_0 is the dielectric constant and r is the distance between the charges in the encounter complex. The values of (k^+/k^0) and (k^-/k^0) are determined by measurement of the rate constants of spin exchange using 6.15, which describes the effect of paramagnetic species on spin-phase and spin–lattice relaxation rates of the radical.

This approach can be applied to two types of problems. First, it can be used to investigate the electrostatic fields in the vicinity of a spin-label or spin-probe using a second paramagnetic species with a different charge, ferricyanide anion, or diphenylchromium cation, for example [2, 9]. The second approach involves monitoring the effect of a paramagnetic species, such as a complex of paramagnetic ion with a protein or the active site of a metalloenzyme, on the spin-relaxation parameters of nitroxide spin-probes of different charges freely diffusing in solution. It can be illustrated by the study of the interaction between ferricyanide anion (FC) and nitroxides of different charges in water solution (Figure 6.6) [77, 91]. The experimental values of k^+, k^-, and k^0 for nitroxide radicals of different charges were found to be 38×10^8, 2×10^8, and $6 \times 10^8 \, M^{-1} s^{-1}$, respectively. From the experimental value of $\log k^+/k = 16$ and the distance between the NO fragment and the ferricyanide ion in the encounter complex ($r = 6$ Å), we can find a product of

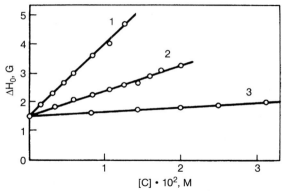

Figure 6.6 Dependence of peak-to-peak line-broadening, $\Delta\Delta H_{pp}$, of EPR signals for positively charged (1), neutral (2), and negatively charged (3). Nitroxide spin-probes on the concentration of $Fe(CN)_6^{3-}$, where $\Delta\Delta H_{pp} = (\Delta H_{pp})_x - (\Delta H_{pp})_0$, and $(\Delta H_{pp})_x$ and $(\Delta H_{pp})_0$ are the line-widths in the presence and absence of $Fe(CN)_6^{3-}$; phosphate buffer, pH 6.9, ionic strength 0.09 M, 20 °C [2]. (Reproduced from Likhtenshtein, G.I. (1993) *Biophysical Labeling Methods in Molecular Biology*, Cambridge University Press, Cambridge, NY, with permission) [7].

charges of radicals (Z_N = +1) and ferricyanide anion (Z_{FC}) $Z_N Z_{FC} = -2.9 \pm 0.1$ and therefore $Z_{FC} = -2.9 \pm 0.1$.

Electron-carrier horse-heat cytochrome c and dioxygen-carrier sperm-whale myoglobin served as models for determination of local electrostatic charges in the vicinity of paramagnetic active sites of metalloenzymes and metalloproteins [77]. The experimental values of the apparent local charge (Z_H) in the vicinity of the heme groups, calculated using 6.15, were found to be equal +0.3 and −0.5 for cytochrome c and myoglobin, respectively. A similar approach was employed in [98] for determination of the electrostatic potential near the surface of calf thymus DNA. Spin–spin interaction between an ^{14}N-nitroxide derivative of 9-aminoacridine attached to DNA and free ^{15}N-labeled nitroxides of different charges was monitored by electron–electron double resonance (ELDOR). The electrostatic potential near the surface of DNA was calculated using a nonlinear Poisson–Boltzman equation. The calculated results were found to be in good agreement with the experimental potentials.

6.7
Spin-Triplet–Fluorescence–Photochrome Method

Measurements of active encounters between molecules especially in native membranes containing ingredients including proteins are of prime importance. Studies of fluorescence quenching of fluorescent probes by nitroxides and exchange interactions between nitroxides have been widely used in applications of fluorescence and ESR to physico-chemical and biophysical problems [99–108]. The principle limitation of fluorescence quenching is the short fluorescent lifetime that causes the need to use relatively high concentrations of quencher (for example, one probe per 10–20 lipids). Such a high dopant concentration could disrupt the membrane integrity. This limitation restricts the use of the standard ESR techniques. The standard ESR technique has its additional drawback: limited sensitivity. Obviously, this technique cannot be applied to objects which require high sensitivity. For quantitative characterization of complicated systems such as biological membranes, analytical techniques over a wide range of distances between encountering molecules and characteristic times of applied methods are required.

For the purpose of the investigation of molecular dynamic processes and the estimation of rare encounters in a high range of rate constants and distances between interacting molecules in membranes in particular, a new expanded cascade method was recently proposed [103, 108–112]. The method combines the three types of probes: (1) photochrome probe, a stilbene derivative, which is fluorescent only in its *trans*-form; (2) a triplet probe, which has a high quantum yield of the triplet excited state and can be used as a sensitizer (E); and (3) a spin-probe, nitroxide radical (R) (Figure 6.7).

These spin-triplet–fluorescence–photochrome techniques are based on the constant-illumination measurements of the *trans-cis* or *cis-trans* photoisomerization kinetics of a photocrome molecule incorporated into an object of interest. The photoisomerization is sensitized by the triplet–triplet energy transfer between a chromophore in its excited-triplet state (sensitizer), which, in turn, is quenched by a nitroxide. Owing to the relatively long lifetime of the sensitizer triplet state,

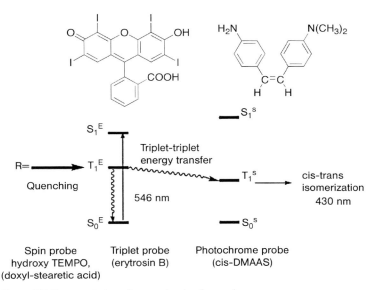

Figure 6.7 Representation of energy levels of cascade reactants and competition between the $T_1^E \rightarrow T_1^S$ and $T_1^E \rightarrow S_0^S$ processes [111]. (Reproduced with permission).

and due to an opportunity to integrate the data on the stilbene photoisomerization, the apparent characteristic time of the cascade method may reach hundreds of seconds. This property of the cascade system allows for investigation of low diffusion processes using very low concentrations of probes.

Kinetic theory (developed in references [109–112]) makes it possible to measure the rate constant of the quenching of the excited-triplet state of the sensitizer with radical (k_q) and the rate constant of the triplet–triplet energy transfer between the sensitizer and the photochrome in the exited-triplet state (k^T).

As an example, the sensitized cascade triplet cis-trans photoisomerization of the excited stilbene includes the triplet sensitizer (Erythrosin B), the photochrome stilbene-derivative probe (4-dimethylamino-4'-aminostilbene) exhibiting the phenomenon of cis-trans photoisomerization and nitroxide radicals (5-doxyl stearic acid) quenching the excited-triplet state of the sensitizer (Figure 6.7) [111]. Measurements of the phosphorescence lifetime of Erythrosin B, and the fluorescence enhancement of the stilbene-derivative photochrome probe at various concentrations of the nitroxide probe made it possible to calculate the quenching rate constant k_q, = 1.05×10^{15} cm^2 mol^{-1} s^{-1} and the rate constant of the triplet–triplet energy transfer between the sensitizer and stilbene probe k^T = 1.2×10^{12} cm^2 mol^{-1} s^{-1}. These values, together with the data on the diffusion rate constant obtained by other methods which cover characteristic times over eight orders of magnitude, were found to be in good agreement with the advanced theory of diffusion-controlled reactions in two dimensions (Figure 6.8) [113].

The relatively long characteristic time of the proposed cascade method (about 0.1 s), makes it possible to follow very rare collisions between molecules using very

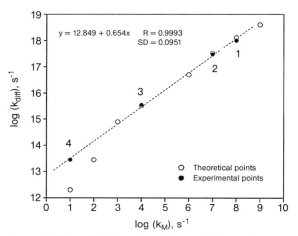

Figure 6.8 Theoretical (○) and experimental (●) dependences of logarithms of the diffusion-controlled rate constants (k_{diff}) on logarithms of the unimolecular decay of the excited species ($k_M = 1/\tau_M$,), which characterizes time scale of different methods [111]. (Reproduced with permission).

low concentrations of free radicals in model and biological membranes, and employing one of the most sensitive and available spectroscopic techniques such as fluorescence spectroscopy. As a result, the membrane properties remain unchanged.

The proficiency of the method can be expanded by choosing cascade participants with the higher efficiency of triplet–triplet energy transfer, higher sensitizer lifetimes, and by an increase of the time of integration of experimental data on a photochrome photoisomerization. It should be noted that *diffusion-controlled rate constants* depending on time and on concentration more adequately depict the actual molecular dynamics in complex systems such as biological membranes. Eventually, for applications in biological systems, it is also important to use photochrome probes with the excitation and emission wavelengths in the near infrared range.

6.8
Dual Fluorophore–Nitroxide as Molecular Dynamics Probes

Over recent decades scientists have faced growing requirements for both novel methods of fast and sensitive analysis of antioxidant properties of biological systems, spin-redox probing and spin-trapping, investigation molecular dynamics, and for convenient models for studies of biophysical and biochemical processes.

In approaching these problems, the use of dual fluorophore–nitroxide compounds (FNRO·) has been suggested. The idea of combining a chromophore and a nitroxide in one molecule for the study of probe mobility was conceived in 1965 by Stryer and Griffith [114]. Starting from the first works of the Likhtenshtein group [30, 102], and the Blough group [115] in 1980, over the last decades, the ability of stable nitroxide radicals to act as quenchers of excited species was intensively exploited as the basis of several molecular probing methodologies [116–125].

Most of the dual fluorophore–nitroxide compounds are based upon the use of two-molecular subfunctionality (a fluorescent chromophore and a stable nitroxide radical) tethered together by a spacer. As was first shown in [102], the nitroxide is a strong intramolecular quencher of fluorescence from the chromophore fragment. When nitroxide moiety is reduced to hydroxylamine or oxidized, or recombined with an active radical, the molecule becomes ESR silent and the fluorescence increases. A reducing substrate (for example, antioxidant, ascorbate, semiquinone and superoxide radicals, nitric oxide) and several other reactions destroying the nitroxide fragment reduce the nitroxide function resulting in a decay of the nitroxide ESR signal intensity plus concomitant enhancement of the chromophore fluorescence. The organic synthetic chemistry allows us to optimize fluorescence, ESR, and the redox properties of such dual molecules by manipulating bridges (spacers), the structure fluorophore and nitroxide fragments, and so forth.

The dual compounds, as well as keeping all the properties of flourescence and nitroxide spin probes, have several new principle advantages.

The most important new properties of nitroxides are the following abilities:

1. Monitoring kinetics of intramolecular fluorescence quenching and electron transfer.
2. Obtaining spatial and temporal information about both fluorescence and ESR behavior of the probe from the same specific part of the system of interest.
3. Estimating the micropolarity and molecular dynamics of media in the vicinity of two different segments (flourophore and nitroxide) of the dual molecule.
4. Monitoring processes in systems of any optical density by ESR technique and of low optical density by fluorescence that is two to three orders of magnitude more sensitive than ESR and optical absorption spectroscopy.

Figure 6.9 shows the chemical structure of the dual fluorophore compounds synthesized by Dr V.V. Martin and A.L. Weis (Lipitek International, Inc.) (private communication). Synthesis of a variety of dual molecules for different purposes was also reported [30, 102, 114, 124].

Recent progress in synthesis of the dual compounds from aminocoumarin, pyrene, and 4-nitrobenzofurazan dyes, and from five- or six-membered nitroxides, has been reported [125]. These new compounds exhibit fluorescence emission between 382 nm and 529 nm, and offer special advantages in studying biological systems, affording various utilization possibilities, especially in biological systems.

In investigating molecular dynamics, dual fluorophore molecules combine all the facilities of a nitroxide probe and fluorescent and phosphorescent chromophores. This advantage has been demonstrated in a series of studies.

As was shown in model liquids (frozen 95% ethanol, 1:3 water/glycerol) and serum albumin systems, phosphorescence, fluorescence, and ESR parameters of the dual molecules (indol–TEMPO and Dansyl–TEMPO) are sensitive to different dynamic modes of media in the vicinity of luminophore and nitroxide segments [30, 102]. In ethanol at 95–110 K, a linear increase in the nitroxide spin-packet ESR line-width $\Delta H_{1/2}$, an abrupt fall in the phosphorescence intensity, a distinct shift in the long-wave phosphorescence maximum (relaxation shift) in the spectrum of the indol–TEMPO, and an abrupt increase in the heat capacity temperature coefficient were detected. This behavior most probably results from some intensification of the phonon dynamics and the polar relaxation of the surrounding molecules on a millisecond timescale. An accompanying increase of the static and dynamic free volume allows for quenching of the indol excited-triplet state by dioxygen. The observed decrease of apparent rotation correlation time of the nitroxyl ring of the Dansyl–TEMPO probe and the relaxation shift in the probe at increased temperature starting above 190–210 K, depending on media, may be explained by the animation nanosecond polar relaxation processes in the medium. Similar dynamic effects were detected for Dansyl–TEMPO and Dansyl–TEMPO modified with a fatty acid moiety embedded in serum albumin and enzymes [6, 107].

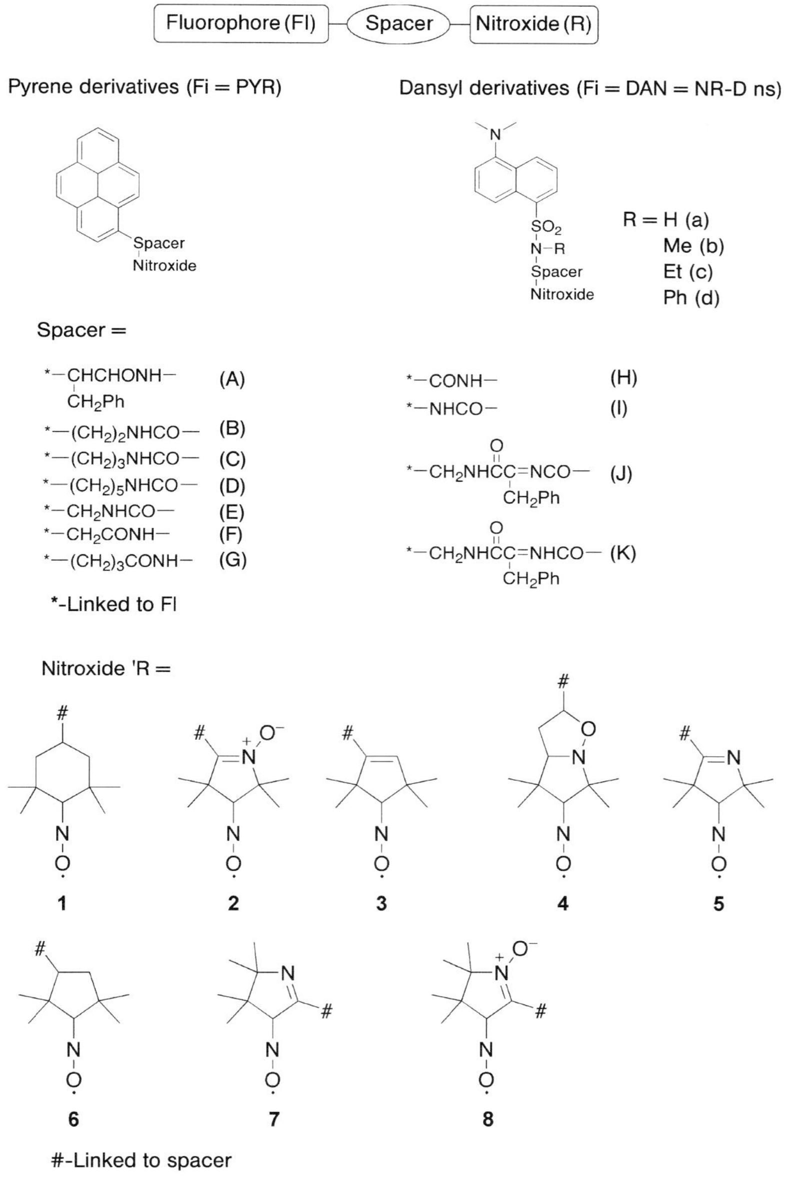

6.8 Dual Fluorophore–Nitroxide as Molecular Dynamics Probes

Pyrene derivatives (Fl = PYR)

Dansyl derivatives (Fl = DAN = NR-D ns)

R = H (a)
Me (b)
Et (c)
Ph (d)

Spacer =

*—CHCHONH— (A)
  |
  CH$_2$Ph

*—(CH$_2$)$_2$NHCO— (B)
*—(CH$_2$)$_3$NHCO— (C)
*—(CH$_2$)$_5$NHCO— (D)
*—CH$_2$NHCO— (E)
*—CH$_2$CONH— (F)
*—(CH$_2$)$_3$CONH— (G)

*-Linked to Fl

*—CONH— (H)
*—NHCO— (I)

*—CH$_2$NHCC=NCO— (J)
        || |
        O   CH$_2$Ph

*—CH$_2$NHCC=NHCO— (K)
        || |
        O   CH$_2$Ph

Nitroxide 'R =

1   2   3   4   5

6   7   8

#-Linked to spacer

Figure 6.9 Chemical structures of dual florophore–nitroxide molecules (private communication of Drs V.V. Martin and A.L. Weiss, Lipitek International, Inc.).

6.9
Nitroxide Spin pH Probes

Local pH is among the most important parameters in chemistry and biochemistry. Stable nitroxides of the imidazoline and imidazolidine types have been shown to be useful spin pH probes for EPR spectroscopy and imaging owing to the large effect of pH on their EPR spectra [126–136]. Figure 6.10 illustrates the chemical origin of the significant pH effect on EPR spectra of the imidazolidine radical 4. Significant changes were observed in both nitrogen hyperfine splitting, A_{iso} ($\Delta a_N \approx 0.8$–1.3 G) and g-factor ($\Delta g \approx 0.002$–000.0003) upon its protonation.

22.5

In practice, two convenient spectral parameters are used as markers of pH. The first is the ratio of peak intensities of RH$^+$ and R spectral components resolved

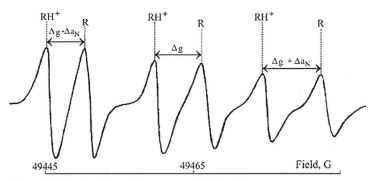

Figure 6.10 The 140 GHz EPR spectrum of a 0.5 mM aqueous solution of the imidazolidine radical **6.4** at pH 4.7. The dotted lines depict the positions of the peaks corresponding to protonated, RH$^+$, and neutral, R, forms of the radical [130]. (Reproduced with permission).

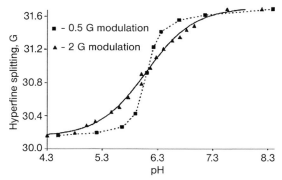

Figure 6.11 The pH dependences of hyperfine splitting, 2 aN, measured as the distance between the low- and high-field components of the L-band EPR spectra of the radical R2 detected at different modulation amplitudes. The EPR spectra of the 0.5 mM aqueous solutions of the radical **5** detected at pH 6.1, and modulation amplitudes 0.5 G (top) and 2 G (bottom) [130]. (Reproduced with permission).

upon detection by high-frequency EPR (Figure 6.10). The second is nitrogen hfs measured as a distance between unresolved spectral components observed at low-field EPR, and is being used almost exclusively as a highly sensitive pH marker in numerous applications (Figure 6.11).

Up to the present time a wide set of pH-sensitive nitroxides of imidazolidine and imidazolidine types with different ranges of pH sensitivity [134], labeling groups, stability to reduction [131, 132] and lipophilicity [133, 135] has been developed. For covalent binding of pH-sensitive nitroxides to the macromolecules, a number of structures with alkylating, iodoacetamido, isothiocyanate, methanethiosulfonate, succinimido, azide, carbodiimide, carboxy, and keto groups were synthesized and are reviewed in reference [131]. Examples of the spin pH probe applications will be given in Section 11.2.1.4.

6.10
Nitroxides as Spin Probes for SH Groups

A direct EPR method for the detection of both low- and high-molecular weight thiols is based on the application of biradical disulfide reagents, RSSR, where R represent imidazoline [136] or imidazolidine [137] radical fragment. These paramagnetic disulfides react with thiols with splitting of the disulfide bond, resulting in characteristic changes of the EPR spectra (Figure 6.12).

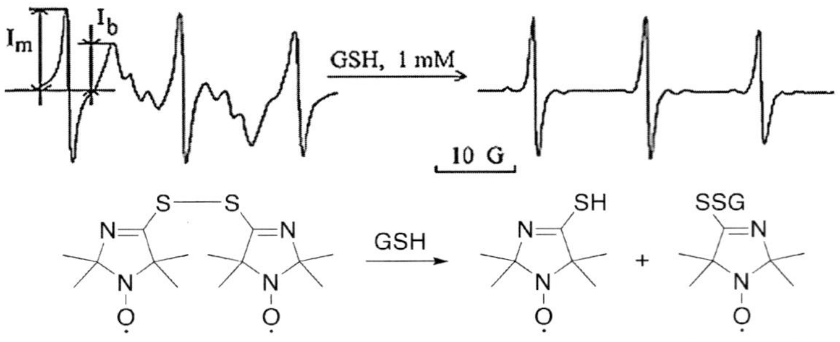

Figure 6.12 X-band EPR spectrum of 0.1 mM solution of the $R_1S\text{-}SR_1$ in 0.1 M PBS buffer, pH 7.4, and its transformation upon treatment with glutathione, GSH, and the scheme of the corresponding reaction of thiol–disulfide exchange responsible for observed spectral changes. The measurement of relative changes of the intensities of monoradical (I_m) or biradical (I_b) components allows quantitative determination of GSH [139]. (Reproduced with permission).

Using imidazolidine disulfide in combination with a "kinetic" EPR approach provides a tool that can be used for real-time measurements of GSH redox state that is not damaging to the tissue being studied [138]. The reasonable value of characteristic time, τ, for the reaction of the imidazolidine biradical $R_2S\text{-}SR_2$ with glutathione at physiological GSH concentrations and neutral pH allows for its application *in vivo*. Recently, low, nontoxic concentrations of $R_2S\text{-}SR_2$ and the kinetic EPR approach have been applied to demonstrate increased GSH oxidation in an animal model of stress-sensitive arterial hypertension [139]. It was found that the concentration of reduced GSH in the blood of the hypertensive rats is significantly lower compared with that of normotensive Wistar rats. These data support an elevated thiol oxidation and free radical mechanism involved in the pathogenesis of this animal model of hypertension.

In conclusion, the spin-label technique has convincingly demonstrated wide possibilities for the study of molecular dynamics over a wide range of frequency and amplitude, and for the quantitative investigation of fundamental properties such as local polarity, acidity, electrostatic potential, and microstructure of various materials. This technique has yielded a whole series of specific important results in many areas of chemistry, biology, and biomedicine (see Chapters 8 to 11).

References

1 McConnell, H.M. and McFarland, B.G. (1970) *Quarterly Reviews of Biophysics*, **3**, 91–136.
2 Likhtenshtein, G.I. (1976) *Spin Labeling Method in Molecular Biology*, Wiley Interscience, New York.
3 Berliner, L. (ed.) (1976) *Spin Labeling. Theory and Applications*, Vol. 1, Academic Press, New York.
4 Berliner, L. (ed.) (1998) *Spin Labeling. The Next Millennium*, Academic Press, New York.

5 Kocherginsky, N. and Swartz, H.M. (1995) *Nitroxide Spin Labels*, CRC Press, Boca Raton.
6 Likhtenshtein, G.I., Febbrario, F. and Nucci, R. (2000) *Spectrochimica Acta, Part A: Molecular and Biomolecular Spectroscopy*, **56**, 2011–31.
7 Likhtenshtein, G.I. (1993) *Biophysical Labeling Methods in Molecular Biology*, Cambridge University Press, Cambridge, New York.
8 Likhtenshtein, G.I. (2000) *Labeling, Biophysical, Encyclopedia of Molecular Biology and Molecular Medicine*, Vol. 7, (ed. R. Meyers), Wiley-VCH, New York, pp. 157–38.
9 Kivelson, D. (1960) *The Journal of Chemical Physics*, **33**, 1094–106.
10 Freed, J.H. (1976) Theory of the ESR spectra of nitroxids, in *Spin Labeling. Theory and Applications*, Vol. 1, (ed. L. Berliner), Academic Press, New York.
11 Tombolato, F., Ferrarini, A. and Freed, J.H. (2006) *Journal of Physical Chemistry B*, **110**, 26260–71.
12 Pavone, M., Cimino, P., De Angelis, F. and Barone, V. (2006) *Journal of the American Chemical Society*, **128**, 4338–47.
13 Lazarev, G.G. and Lebedev, Ya.S. (1981) *Teoretical and Experimental Chemistry*, **17**, 795–805 (in Russian).
14 Shimchick, E.J. and McConnell, H.M. (1972) *Biochemical and Biophysical Research Communications*, **46**, 321–7.
15 Dudich, I.V., Timofeev, V.P., Vol'kenshtein, M.V. and Misharin, A. Yu. (1977) *Molecular Biology (Moscow)*, **11**, 685–7 (in Russian).
16 Antsiferova, L.I., Belonogova, O.V., Kochetkov, V.V. and Likhtenshtein, G.I. (1989) *Izvestiya Akademii Nauk SSSR, Seriya Biologicheskaya*, 494–501.
17 Marsh, D. (1990) *Pure and Applied Chemistry. Chimie Pure et Appliquee*, **62**, 265–70.
18 Squier, T.S. and Thomas, D.D. (1989) *Biophysical Journal*, **56**, 735–48.
19 Krinichny, V.L., Grinberg, O.Ya., Belonogova, O.V., Lyubashevskaya, E. B., Anziferova, L.I., Likhteshtein, G.I. and Lebedev, Ya.S. (1988) *Biofizika*, **32**, 215–22.
20 Krinichnyi, V.I. (1995) *2-mm Wave Band EPR Spectroscopy of Condensed Systems*, CRC Press, Boca Raton, IL.
21 Earle, K.A. and Smirnov, A.I. (2004) *Biological Magnetic Resonance (Very High Frequency (VHF) ESR/EPR)*, **22**, 95–143.
22 Kirilina, E.P., Prisner, T.F., Bennati, M., Endeward, B., Dzuba, S.A., Fuchs, M.R., Mobius, K. and Schnegg, A. (2005) *Magnetic Resonance in Chemistry: MRC*, **43** (special Issue), S119–29.
23 Earle, K.A., Dzikovski, B., Hofbauer, W., Moscicki, J.K. and Freed, J.H. (2005) *Magnetic Resonance in Chemistry: MRC*, **43** (Special Issue), S256–66.
24 Godt, A. and Jeschke, G. (2005) *Magnetic Resonance in Chemistry: MRC*, **43** (Special Issue), S110–18.
25 Likhtenshtein, G.I. (1976) Water and protein dynamics, in *L'eau et les systemes bioologiques*, (eds A. Alfsen and A.J. Berntrand), CNRS, Paris.
26 Likhtenshtein, G.I. (1996) *Journal of Photochemistry and Photobiology A: Chemistry*, **96**, 79–92.
27 Likhtenshtein, G.I. (2003) *New Trends in Enzyme Catalysis and Mimicking Chemical Reactions*, N.Y. Kluwer Academic/Plenum Publishers, New York.
28 Lebedev, A.Ya. and Muromtsev, V.I. (1972) *ESR and Relaxation of Stabilized Radicals*, Khimiya, Moscow. (in Russian).
29 Alexander, S., Entin-Wohlman, O. and Orbach, R. (1986) *Physical Review*, **33**, 3935–46.
30 Likhtenshtein, G.I., Bogatyrenko, V.R., Kulikov, A.V., Hideg, K., Khankovskaya, G.O., Hankovsky, H.O., Lukoyanov, N.V., Kotel'nikov, A.I. and Tanaseichuk, B.S. (1980) *Doklady Akademii Nauk SSSR*, **253**, 481–4.
31 Likhtenshtein, G.I., Bogatyrenko, V.R. and Kulikov, A.V. (1993) *Applied Magnetic Resonance*, **4**, 513–21.
32 Belonogova, O.V., Frolov, E.N., Illyustrov, N.V. and Likhtenshtein, G.I. (1979) *Molecular Biology (Moscow)*, **13**, 567–76.
33 Kirilina, E.P., Dzuba, S.A., Maryasov, S.A., Tsvetkov, A.G. and Yu, D. (2001) *Applied Magnetic Resonance*, **21**, 203–21.
34 Kirilina, E.P., Grigoriev, I.A. and Dzuba, S.A. (2004) *The Journal of Chemical Physics*, **121**, 12465–71.

35 Dzuba, S.A., Kirilina, E.P. and Salnikov, E.S. (2006) *The Journal of Chemical Physics*, **125**, 054502/1–054502/5.

36 Likhtenshtein, G.I., Grebenshchikov, Y.B. and Avilova, T.V. (1972) *Molecular Biology (Moscow)*, **6**, 52–60.

37 Zamaraev, K.I., Molin, Yu.N. and Salikhov, K.M. (1981) *Spin Exchange. Theory and Physicochemical Application*, Springer-Verlag, Heidelberg.

38 Likhtenstein, G.I., Grebentchikov, Yu.B., Bobodzhanov, P.Kh., Kokhanov, Yu.V. (1970) *Molecular Biology (Moscow)*, **4**, 782–9.

39 Hyde, J.S., Swartz, H.M. and Antholine, W.E. (1976) The spin probe-spin label methods, in *Spin Labeling. Theory and Application*, Vol. 2, (eds L. Berliner), Academic Press, New York, pp. 72–113.

40 March, D. (1990) *Pure and Applied Chemistry*, **62**, 265–70.

41 Parmon, V.P., Kokorin, A.I. and Zhidomirov, G.M. (1980) *Stable Biradicals*, Nauka, Moscow (in Russian).

42 Frolov, E.N., Kharakhonicheva, N.V. and Likhtenshtin, G.I. (1974) *Molecular Biology (Moscow)*, **8**, 886–93.

43 Kulikov, A.I. and Likhtenshtein, G.I. (1977) *Advanced Molecular Relaxation Processes*, **10**, 47–78.

44 Kulikov, A.I., Yudanova, E.I. and Likhtenshtein, G.I. (1983) *Journal of Physical Chemistry (Moscow)*, **56**, 2982–7.

45 Hyde, J.S. and Dalton, L.R. (1979) Saturation-transfer spectroscopy, in *Spin Labeling. Theory and Application*, Vol. 3, (ed. L. Berliner), Academic Press, New York.

46 Hyde, J.S. and Feix, J.B. (1989) Electron-electron double resonance, in *Biological Magnetic Resonance, V 8, Spin Labeling. Theory and Application*, Vol. 1, (eds L. Berliner and J. Reuben), Plenum Press, New York, pp. 305–339.

47 Salikhov, K.M., Doctorov, A.B., Molin, Yu. N., and Zamaraev, K.I. (1971) *Journal of Magnetic Resonance*, **5**, 189–205.

48 Demas, J.N., Diemante, D. and Harris, E.V. (1973) *Journal of the American Chemical Society*, **95**, 6864–5.

49 Backer, J.M., Budker, V.A., Eremenko, I., Molin, Yu. N. (1977) *Biochimica et Biophysica Acta*, **460**, 152–6.

50 Subszynski, W.K. and Hyde, J.S. (1981) *Biochimica et Biophysica Acta*, **643**, 283–91.

51 Subczynski, W.K., Felix, C.C., Klug, C.S. and Hyde, J.S. (2005) *Journal of Magnetic Resonance*, **176**, 244–8.

52 Hyde, J.S. and Subczynski, W.K. (1989) Spin label oximetry, in *Biological Magnetic Resonance. Spin Labeling: Theory and Applications*, (eds L.J. Berliner and J. Reuben) Plenum Press, New York, pp. 399–425.

53 Yudanova, E.I. and Kulikov, A.V. (1984) *Biofizika*, **29**, 925–9.

54 Halpern, H.J., Yu, C., Peric, M., Barth, E., Grdina, D.J. and Teicher, B.A. (1994) *Proceedings of the National Academy of Sciences of the United States of America*, **91**, 13047–51.

55 Swartz, H.M. and Glockner, J.F. (1991) *EPR Imaging and in vivo EPR*, (eds G.R. Eaton, S.S. Eaton and K. Ohno), CRC Press, Boca Raton, pp. 261–90.

56 Swartz, H.M. and Dunn, J.F. (2003) *Advances in Experimental Medicine and Biology*, **530**, 1–12.

57 Hou, H., Khan, N., O'Hara, J.A., Grinberg, O.Y., Dunn, J.F., Abajian, M.A., Wilmot, C.M., Demidenko, E., Lu, S., Steffen, R.P. and Swartz, H.M. (2005) *International Journal of Radiation Oncology, Biology, Physics*, **57**, 1503–9.

58 Moscatelli, A., Chen, T.K., Jockusch, S., Forbes, M.D.E., Turro, N.J. and Ottaviani, M.F. (2006) *Journal of Physical Chemistry. B*, **110**, 7574–8.

59 Liu, K.J., Gast, P., Moussavi, M., Norby, S.W., Vahidi, N., Walczak, T., Wu, M. and Swartz, H.M. (1993) *Proceedings of the National Academy of Sciences of the United States of America*, **90**, 5438–42.

60 Vahidi, N., Clarkson, R.B., Liu, K.J., Norby, S.W., Wu, M. and Swartz, H.M. (1994) *Magnetic Resonance in Medicine*, **31**, 139–46.

61 Jordan, B., Baudelet, C. and Gallez, B. (1998) *Magma (New York, NY)*, **7**, 121–9.

62 Clarkson, R.B., Odintsov, B., Ceroke, P., Ardenkjaer-Larsen, J.P., Fruianu, M. and Belford, R.L. (1998) *Physics in Medicine and Biology*, **43**, 1907–20.

63 Dinguizli, M., Jeumont, S., Beghein, N., He, J., Walczak, T., Lesniewski, P.N., Hou, H., Grinberg, O.Y., Sucheta, A., Swartz, H.M. and Gallez, B. (2006) *Biosensors and Bioelectronics*, **21**, 1015–22.

64 Grinberg, V.O., Smirnov, A.I., Grinberg, O.Y., Grinberg, S.A., O'Hara, J.A. and Swartz, H.M. (2005) *Applied Magnetic Resonance*, **28**, 69–78.

65 Swartz, H.M. (2002) *Biochemical Society Transactions*, **30**, 248–52.

66 Matsumoto, K.B., Chandrika, B., Lohman, J.A., Mitchell, J.B., Krishna, M.C. and Subramanian, S. (2003) *Magnetic Resonance in Medicine*, **50**, 865–74.

67 Deng, Y., Pandian, R.P., Rizwan, A.R., Kuppusamy, P. and Zweier, J.L. (2006) *Journal of Magnetic Resonance*, **2**, 254–61.

68 Kotel'nikov, A.I., Likhtenshtein, G.I. and Gvozdev, R.I. (1975) *Studia Biophysica*, **49**, 215–21.

69 Kulikov, A.V. (1976) *Molecular Biology (Moscow)*, **10**, 109–16.

70 Kulikov, A.V. and Likhtenstein, G.I. (1974) *Biofizika*, **19**, 420–4.

71 Case, G.D. and Leigh, J.S., Jr. (1976) *The Biochemical Journal*, **160**, 769–83.

72 Kulikov, A.V., Cherepanova, E.S. and Bogatyrenko, V.R. (1981) *Theoretical and Experimental Chemistry (Moscow)*, **17**, 618–26.

73 Kulikov, A.V., Cherepanova, E.S., Bogatyrenko, V.R., Nasonova, T.A., Fisher, V.R. and Yakubov, H.M. (1987) *Biology Bulletin of the Academy of Sciences of the USSR, Division of Biological Science*, **5**, 7762–9.

74 Kulikov, A.V., Cherepanova, E.S., Likhtenshtein, G.I., Uvarov, V.Yu. and Archakov, A.I. (1989) *Biologicheskie Membrany*, **6**, 1085–94.

75 Cherepanova, E.S., Kulikov, A.V. and Likhtenstein, G.I. (1990) *Biological Membranes (Moscow)*, **77**, 51–6.

76 Likhtenshtein, G.I. (1988) *Chemical Physics of Redox Metalloenzymes*. Springer-Verlag, Heidelberg.

77 Likhtenshtein, G.I. (2000) *Magnetic Resonance in Biology*, Vol. 19, (eds L. Berliner, S. Eaton and G. Eaton) Kluwer Academic Publishers, Dordrecht, pp. 309–45.

78 Alexandrov, I.V. (1975) *Theory of Magnetic Relaxation*, Nauka, Moscow (in Russian).

79 Hwang, L.P. and Freed, J.H. (1975) *The Journal of Chemical Physics*, **63**, 4017–25.

80 Dwek, R.A. (1977) *NMR in Biology*, Academic Press, New York.

81 Berdnikov, V.M., Doktorov, A.B. and Makarshin, L.L. (1980) *Theoretical and Experimental Chemistry (Kiev)*, **16**, 765–71.

82 Borah, B. and Bryant, R.G. (1981) *The Journal of Chemical Physics*, **75**, 3297–300.

83 Bowman, M.K. and Norris, J.R. (1982) *Journal of Physical Chemistry*, **86**, 3385–90.

84 London, E. (1982) *Molecular and Cellular Biochemistry*, **45**, 181–8.

85 Kuzmin, V.A. and Tatikolov, A.S. (1977) *Chemical Physics Letters*, **51**, 45–7.

86 Kuzmin, V.A. and Tatikolov, A.S. (1978) *Journal Chemical Physics Letters*, **53**, 606–10.

87 Strashnikova, N.V., Medvedeva, N. and Likhtenshtein, G.I. (2001) *Journal of Biochemical and Biophysical Methods*, **48**, 43–60.

88 Smirnova, T.I., Chadwick, T.G., Voinov, M.A., Poluektov, O., van Tol, J., Ozarowski, A., Schaaf, G., Ryan, M.M. and Bankaitis, V.A. (2007) *Biophysical Journal*, **92**, 3686–95.

89 Grinberg, O.Y, Dubinsky, A.A. and Lebedev, Y.S. (1983) *Uspekhi Khimmii*, **52**, 1490–513.

90 Griffith, O.N. and Jost, P.C. (1976) *Spin Labeling*. (ed. L. Berliner), Academic Press, New York, pp. 488–59.

91 Likhtenstein, G.I., Grebentchikov, Yu. B., Rosantev, E.G. and Ivanov, V.P. (1972) *Molecularnaya Biologiya (Moscow)*, **6**, 498–507.

92 Hecht, J.L., Honig, B., Shin, Y. and Hubbell, W.L. (1995) *Journal of Physical Chemistry*, **99**, 7782–6.

93 Hwang, L.P. and Freed, J.H. (1975) *The Journal of Chemical Physics*, **63**, 4017–25.

94 Alexandrov, I.V. (1975) *Theory of Magnetic Relaxation*, Nauka, Moscow.

95 Berdnikov, B.M., Doktorov, A.B. and Makarshin, L.L. (1980) *Theoretical and Experimental Chemistry (Kiev)*, **16**, 765–71.

96 Likhtenstein, G.I., Adin, I., Krasnoselsky, A., Vaisbuch, I., Shames, A. and

Glaser, R. (1999) *Biophysical Journal*, **77**, 443–53.
97 Debye, P. (1942) *Journal of the Electrochemical Society*, **82**, 265–72.
98 Hecht, J.L., Honig, B., Shin, Y. and Hubbell, W.L. (1995) *Journal of Physical Chemistry*, **99**, 7782–6.
99 Buchachenko, A.L., Khloplyankina, M.S. and Dobryakov, S.N. (1967) *Optika i Spektroskopiya*, **22**, 554–9.
100 Watkins, A.R. (1974) *Chemical Physics Letters*, **4**, 526–8.
101 Papper, P., Likhtenshtein, G.I., Medvedeva, N. and Khoudyakov, D.V. (1999) *Journal of Photochemistry and Photobiology A: Chemistry*, **122**, 79–85.
102 Bystryak, I.M., Likhtenshtein, G.I., Kotelnikov, A.I., Hankovsky, O.H. and Hideg, K. (1986) *Russian Journal of Physical Chemistry*, **60**, 1679–983.
103 Mekler, V.M. and Likhtenshtein, G.I. (1986) *Biofizika*, **31**, 568–71.
104 Lakowicz, X. (ed.) (1991) *Topics in Fluorescence Spectroscopy*, Plenum Press, New York.
105 Ranganathan, R., Vautier-Ciongo, C. and Bales, B.L. (2003) *Journal of Physical Chemistry. B*, **107**, 10312–18.
106 Green, S.A., Simpson, D.J., Zhou, G., Ho, P.S. and Blough, N.V. (1990) *Journal of the American Chemical Society*, **112**, 7337–46.
107 Fogel, V.R., Rubtsova, E.T., Likhtenshtein, G.I. and Hideg, K. (1994) *Journal of Photochemistry and Photobiology A: Chemistry*, **83**, 229–36.
108 Likhtenshtein, G.I., Bishara, R., Papper, P., Uzan, B., Fishov, I., Gill, D. and Parola, A.H. (1996) *Journal of Biochemical and Biophysical Methods*, **339** (33), 117–33.
109 Papper, P., Likhtenshtein, G.I., Medvedeva, N. and Khoudyakov, D.V. (1999) *Journal of Photochemistry and Photobiology A: Chemistry*, **122**, 79–85.
110 Papper, V., Medvedeva, N., Fishov, I. and Likhtenshtein, G.I. (2000) *J. Appl. Biotech. Biophys*, **89**, 231–47.
111 Medvedeva, N., Papper, V. and Likhtenshten, G.I. (2005) *Physical Chemistry Chemical Physics: PCCP*, **7**, 3368–74.
112 Papper, V. and Likhtenshtein, G.I. (2001) *Journal of Photochemistry and Photobiology A: Chemistry*, **140**, 39–52.
113 Razi Naqvi, K., Martins, J. and Melo, E. (2000) *Journal of Physical Chemistry B*, **104**, 12035–8.
114 Stryer, L. and Griffith, H.O. (1965) *Proceedings of the National Academy of Sciences of the United States of America*, **54**, 1785–90.
115 Blough, N.V. and Simpson, D.J. (1988) *Journal of the American Chemical Society*, **110**, 1915–17.
116 Herbelin, S.E. and Blough, N.V. (1998) *Journal of Physical Chemistry. B*, **102**, 8170–6.
117 Rubtsova, E.T., Fogel, V.R., Khudyakov, D.V., Kotel'nikov, A.I. and Likhtenshtein, G.I. (1993) *Biofizika*, **38**, 211–21.
118 Pou, S., Huang, Y.I., Bhan, A., Bhadti, V.S., Hosmane, R.S., Wu, S.Y., Cao, G.L. and Rosen, G.M. (1993) *Analytical Biochemistry*, **212**, 85–90.
119 Fogel, V.R., Rubtsova, E.T., Likhtenshtein, G.I. and Hideg, K. (1994) *Journal of Photochemistry and Photobiology A: Chemistry*, **83**, 229–36.
120 Lozinsky, E.M., Martin, V., Berezina, T., Shames, A.I., Weism, A. and Likhtenshtein, G.I. (1999) *Journal of Biochemical and Biophysical Methods*, **38**, 29–42.
121 Lozinsky, E., Novoselsky, A., Shames, A.I., Saphier, O., Likhtenshtein, G.I. and Meyerstein, D. (2001) *Biochimica et Biophysica Acta*, **1526**, 53–60.
122 Lozinsky, E., Novoselsky, A., Glaser, R., Shames, A.I., Likhtenshtein, G.I. and Meyerstein, D. (2002) *Biochimica et Biophysica Acta*, **1571**, 239–44.
123 Medvedeva, N., Martin, V.V. and Likhtenshten, G.I. (2004) *Journal of Photochemistry and Photobiology A: Chemistry*, **163**, 45–51.
124 Lozinsky, E.M., Martina, L.V., Shames, A.I., Uzlaner, N., Masarwa, A., Likhtenshtein, G.I., Meyerstein, D., Martin, V.V. and Priel, Z. (2004) *Analytical Biochemistry*, **326**, 139–45.
125 Bognar, B., Osz, E., Hideg, K. and Kalai, T. (2006) *Journal of Heterocyclic Chemistry*, **43**, 81–6.
126 Khramtsov, V.V., Grigor'ev, I.A., Foster, M.A., Lurie, D.J. and Nicholson, I. (2000) *Cell and Molecular Biology (Noisy-le-grand)*, **46**, 1361–74.
127 Keana, J.F.W., Acarregui, M.J. and Boyle, S.L.M. (1982) *Journal of the American Chemical Society*, **104**, 887.

128 Khramtsov, V.V., Weiner, L.M., Grigor'ev, I.A. and Volodarsky, L.B. (1982) *Chemical Physics Letters*, **91**, 69–72.

129 Khramtsov, V.V., Weiner, L.M., Eremenko, S.I., Belchenko, O.I., Schastnev, P.V., Grigor'ev, I.A. and Reznikov, V.A. (1985) *Journal of Magnetic Resonance*, **61**, 397–408.

130 Khramtsov, V.V., Grigor'ev, I.A., Foster, M.A., Lurie, D.J., Zweier, J.L. and Kuppusamy, P. (2004) *Biospectroscopy*, **18**, 213–25.

131 Khramtsov, V.V. and Volodarsky, L.B. (1998) *Spin Labeling. The next Millennium*, Vol 14 (ed. L.J. Berliner), Plenum Press, New York, pp. 109–80.

132 Kirilyuk, I.A., Bobko, A.A., Grigor'ev, I.A. and Khramtsov, V.V. (2004) *Organic and Biomolecular Chemistry*, **2**, 1025–30.

133 Voinov, M.A., Polienko, J.F., Schanding, T., Bobko, A.A., Khramtsov, V.V., Gatilov, Y.V., Rybalova, T.V., Smirnov, A.I. and Grigor'ev, I.A. (2005) *Journal of Organic Chemistry*, **70**, 9702–11.

134 Kirilyuk, I.A., Bobko, A.A., Khramtsov, V.V. and Grigor'ev, I.A. (2005) *Organic and Biomolecular Chemistry*, **3**, 1269–74.

135 Reznikov, V.A., Skuridin, N.G., Khromovskih, E.L. and Khramtsov, V.V. (2003) *Russian Chemical Bulletin*, **52**, 2052–6.

136 Khramtsov, V.V. (2005) *Current Organic Chemistry*, **9**, 909–23.

137 Khramtsov, V.V., Yelinova, V.I., Weiner, L.M., Berezina, T.A., Martin, V.V. and Volodarsky, L.B. (1989) *Analytical Biochemistry*, **182**, 58–63.

138 Khramtsov, V.V., Yelinova, V.I., Glazachev, Yu.I., Reznikov, V.A. and Zimmer, G. (1997) *Journal of Biochemical and Biophysical Methods*, **35**, 115–28.

139 Khramtsov, V.V., Grigor'ev, I.A., Foster, M.A. and Lurie, D.J. (2004) *Antioxidants and Redox Signaling*, **6**, 667–76.

7
Nitroxide Redox Probes and Traps, Nitron Spin Traps
Gertz I. Likhtenshtein

7.1
Nitroxide Redox Probes

7.1.1
Introduction

Redox reactions play a key role in fundamental chemical, photochemical, and biological processes and in the provision of energy in living organisms, in particular. Many destructive processes also involve oxidation and reduction. The nitroxide radicals (NRO) are widely used for quantitative characterization of redox processes and protection from radical damage [1–10]. The spin-redox probe techniques utilize the ability of nitroxides and corresponding hydroxyl amines to proceed in the following chemical reactions (Figure 7.1), which include: (1) reduction of a nitroxide with a reducing agent to a corresponding hydroxyl amine; (2) oxidation of a nitroxide to oxoammonium cation; and (3) oxidation of hydroxyl amine with an oxidant to corresponding nitroxide. All these reactions run as simple one-electron processes and can readily be followed by the standard ESR technique.

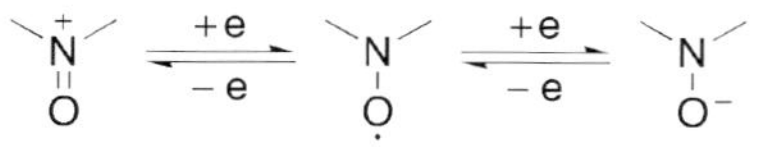

Figure 7.1 One electron redox processes of nitroxides.

Nitroxides are relatively stable towards oxidation but they can be reduced to the corresponding hydroxyl amines. The redox potential of piperidine derivative nitroxide [7] is high enough to oxidize biological compounds such as ascorbic acid, semiquinones, and superoxide radicals [2]. The hydroxylamine derivatives of nitroxides are oxidized directly by superoxide radicals and catalytically by dioxygen in the presence of transition metal ions (Fe^{3+}, Cu^{2+}) [3].

The above-mentioned abilities allow for the use of nitroxides as spin-redox probes. The requirements for efficient spin-redox probes may be formulated as follows: (1) intensive, easily readable, and interpreted ESR signal; (2) fast reaction with a corresponding partner; (3) specificity with regard to a certain redox agent;

Nitroxides: Applications in Chemistry, Biomedicine, and Materials Science
Gertz I. Likhtenshtein, Jun Yamauchi, Shin'ichi Nakatsuji, Alex I. Smirnov, and Rui Tamura
Copyright © 2008 WILEY-VCH Verlag GmbH & Co. KGaA, Weinheim
ISBN: 978-3-527-31889-6

(4) use of a set of nitroxides covering a wide range of redox potentials; (5) chemical stability with regard to nonradical reagents; (6) compatibility to specific properties of the object of interest, bearing suitable groups of different chemical reactivity, hydrophobicity, electrostatic potential, so forth; (7) noninvasiveness *in vivo*; and (8) commercial availability or simplicity of synthesis.

The values of nitroxide redox potential depend on their chemical structures. Thus, pyrrolidine nitroxides have been found to be more stable towards reduction by ascorbate in neutral solution than are piperidine nitroxides. Doxyl radicals are reduced markedly faster than proxyl and acetoxyl radicals, but much more slowly than four-ring nitroxides. Special experiments on the electrochemical reduction of nitroxides have shown that the reduction rate decreases in the following series: imidazolines < imidazoline oxides piperidines [7].

As long ago as 1968, reduction of nitroxides was observed in electron transport in mitochondria [11]. Later, such processes were found to be typical of various biological systems [1–20].

7.1.2
Quantitative Characterization of Antioxidant Status

The parameters which quantitatively characterize the antioxidant ability (status) of nitroxides are the redox potential ($E_{1/2}$), the equilibrium constant (K_{red}), the dissociation energy of the NO–H bond (D_{NO-H}), and the rate constant (k_{red}) for the reaction of nitroxide with a system of interest. The parameter K_{eq}, which obviously relates to the redox potential of the nitroxide fragment, was proposed [21]. This parameter is defined as a relative value of the equilibrium constant for the following exchange reaction (Figure 7.2).

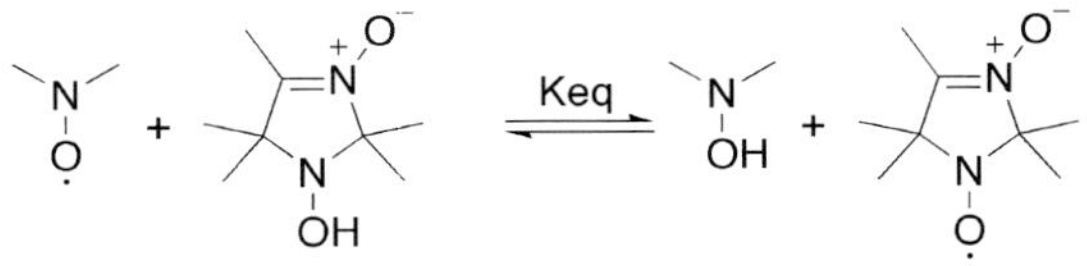

Figure 7.2 Equilibrium of the exchange reaction between nitroxide [21].

According to [21] the K_{eq} value for different nitroxides ranges from 0.02 to 0.23. An important characteristic of nitroxide redox reactions is the energy of dissociation of the N–O bond in nitroxides, which is (in kJ mol^{-1}) 290 ± 1, 296 ± 1, 303 ± 1, and 310 ± 1 for pyrrolidine, imidazoline, pipiredine, and imidazolineoxide derivatives of nitroxides, respectively [22]. The electrochemical reduction of piperidine and pirrolodine nitroxide series is characterized by voltamper curves with the mid-point potential $E_{1/2} = (0.17–0.42)$ V (relative to AgCl electrode, pH 6.8) depending on the nitroxide structure [16, 23, 24]. The rate constant values of the nitroxide reduction (k_{red}) are markedly dependent on the chemical structure of the nitroxide and its reducing partner and on the reaction conditions.

A typical example is the reduction of nitroxides of different structure by ascorbic acid. According to [8, 13, 19, 25] the reaction occurs in the following stepwise fashion:

$$NRO + AH_2 \rightarrow NROH + DH$$
$$NRO + AH^\bullet \rightarrow NROH + A$$
$$2AH^\bullet \rightarrow AH_2 + A$$
$$A \rightarrow DKGA$$
$$DKGA \rightarrow NROH + products$$

where NRO· is the nitroxide radical, NROH is its hydroxyl amine derivative, AH_2 is ascorbic acid and AH· is the ascorbate radical, A is the dehydroascorbate, and DKGA is 2,3 diketogulonic acid, the product of A hydrolysis. The results of comparative studies of antioxidant abilities of nitroxides, hydroxylamines, nitrones, and phenols have been reported [26]. Data on the kinetics of different nitroxide reactions are presented in Figure 7.3.

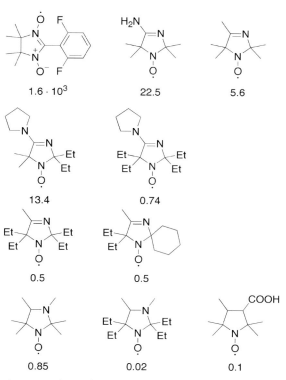

Figure 7.3 Chemical structures and the rate constants (in $M^{-1} s^{-1}$) of nitroxide reduction by ascorbic acid (k_{red}) in phosphate buffer pH 7.5 [19]. (Reproduced with permission).

As we can see from the Figure 7.3, the radical structure markedly affects the rate constants, k_{red}, of ascorbate oxidation. Thus the nitronyl radical reduction is three orders of magnitude faster than the reduction of pyrrolidine nitroxides. The replacement of methyl groups by ethyl moiety leads to a decrease of the k_{red} value by five to eight times. The redox potential of piperidine derivatives of nitroxides are suitable for analysis of ascorbic acid and semiquinone, whereas nitroxides of more positive redox potential (nitronyls and nitroxides with polar substituents) may be used for characterization of antioxidant status of reducing agents, such as flavonols, flavones, and catechins, owing to the greater stability of these antioxidants [27, 28].

The reversible reaction between NR and ascorbate was observed for various types of nitroxides, and the rate constants for direct and reverse reactions were detected [19]. The equilibrium constants for one-electron reduction of the tetraethyl-substituted NR by ascorbate were found to be in the range from 2.65×10^{-6} to 10^{-5}; that is, 40 times lower than corresponding values for the tetramethyl-substituted nitroxides. The redox reactions were found to be significantly affected by glutathione.

7.1.3
Antioxidant Activity

The following mechanisms are responsible for antioxidant effects of nitroxides: (1) superoxide dismutase (SOD) mimetic activity; (2) recombination with free radicals, including radicals in lipid peroxidation; and (3) inducing catalase-like activity in heme proteins [28–31].

7.1.3.1 SOD Mimetic Activity

A dismutation between superoxide radicals ($O_2^{\cdot-}$)

$$O_2^{\cdot-} + O_2^{\cdot-} + 2H^+ \rightarrow H_2O_2 + O_2$$

catalysed in biological systems by enzyme (superoxide dismutase, SOD) proceeds very fast with the rate constant $k_{sod} = \sim 10^9 \, M^{-1} S^{-1}$ [25]. Nitroxides (NRO·) may be involved in the SOD mimetic activity by several mechanisms, which include a series of individual reactions. We performed a comparative theoretical analysis of these reactions using the Marcus equation [32–34]:

$$k_{ET} = \nu_0 \exp\left[-\frac{(\lambda + \Delta G_0)^2}{4\lambda RT}\right] \tag{7.1}$$

where λ is the reorganization energy and ΔG_0 is the standard Gibbs energy. For an aqueous solution a value was estimated within $\lambda = 1.0–1.3 \, eV$ for the electron donor and electron acceptor centers with effective radius $r_{ef} = 0.3–0.4 \, nm$ [32–34]. Taking into consideration that, for $O_2^{\cdot-}$ and the potential intermediate product oxoammonium cation (NRO$^+$), the r_{ef} value is less than 0.3 nm, we suggested that for these centers $\lambda > 1.3$ eV. For a rough estimation of k_{ET} we used the following variables of the Equation 7.1: $\lambda = 1.5$ eV, $\upsilon_0 = 10^9 \mathrm{M^{-1}s^{-1}}$. ΔG_0 was calculated using the following values of redox potentials (E_0, NHE): (HO$_2^•$/HO$_2^-$) = 0.75 eV; (H+, HO$_2^•$/H$_2$O$_2$) = 1.44 eV [35]; (O$_2$/O$_2^-$) = −0.33 eV (pH7) [36]; (H$^+$, NRO·/NROH) = 0.2 eV [16, 23, 24]. The calculated values were compared with available experimental data on piperidinyl nitroxides.

Mechanism I

$$NRO + HO_2^• \leftrightarrow NRO^+ + HO_2^- \tag{1}$$
$$(\Delta G_0 \approx 0, k_{ET} \approx 10^2 \mathrm{M^{-1}s^{-1}})$$
$$HO_2^- + H^+ \leftrightarrow H_2O_2 \tag{2}$$

$$NRO + HO_2^• + H+ \rightarrow NRO^+ + H_2O_2 \tag{3}$$
$$(\Delta G_0 \approx 0.7 \mathrm{eV}, k_{ET} \approx 10^7 \mathrm{M^{-1}s^{-1}})$$

$$NRO^+ + HO_2^• = NRO\cdot + O_2 + H^+ \tag{4}$$
$$(\Delta G_0 \approx -0.9 \mathrm{eV}, k_{ET} \approx 10^8 \mathrm{M^{-1}s^{-1}})$$

$$NRO^+ + O_2^- = NRO\cdot + O_2 \tag{4a}$$
$$\Delta G_0 \approx -1.1 \mathrm{eV}, k_{ET} \approx 2 \times 10^8 \mathrm{M^{-1}s^{-1}}$$

As we can see the estimation predicted very slow rate for reaction 1, while a concerted reaction 3 is expected to proceed relatively fast. Oxidation of a nitroxide (NRO) by HO$_2^•$ radical to corresponding oxoammonium cation at pH ≤ 4 with the rate constant $k = 4.2 \cdot 10^7 \mathrm{M^{-1}s^{-1}}$ (that does not depend on pH) followed by the oxidation of HO$_2^•$ by oxoammonium cation was first reported in [28, 31]. This observation was later confirmed [25, 29, 31]. According to our estimation, reactions 4 and 4a can proceed at a rate that is one or two orders of magnitude more than reaction 3. The experimental value of reaction 4a was reported as $k_{ET} = 3 \cdot 10^9 \mathrm{M^{-1}s^{-1}}$.

Mechanism II

$$NRO\cdot + O_2^- + H^+ \rightarrow NROH + O_2 \tag{5}$$
$$\Delta G_0 \approx -0.5 \mathrm{eV}, k_{ET} \approx 5 \times 10^4 \mathrm{M^{-1}s^{-1}}$$

$$NROH + O_2^- + H^+ \rightarrow NRO\cdot + H_2O_2 \tag{6}$$
$$\Delta G_0 \approx -1.6 \mathrm{eV}, k_{ET} \approx 10^9 \mathrm{M^{-1}s^{-1}}$$

Mechanism II appears to be a non-realistic one because reaction 5, a key process of Mechanism II, is predicted to run several orders of magnitude slower than the key reaction 3 of Mechanism I. As was shown experimentally, superoxide reacts very slowly with nitroxides ($k_{ET} \leq 10^3 \, M^{-1} s^{-1}$) [25, 28, 29, 31] and very fast with NROH (reaction 6) [2, 25]. The relatively low reactivity of superoxide may be explained by a high value of reorganization energy for this tiny radical. The high rate of reactions of dioxygen with hydrogen atom and active radicals [37] provides evidence that the effect of spin change in highly exothermic reactions with participation of superoxide is not predominant as well.

Mechanism III

$$O_2^{\bullet -} + H^+ \leftrightarrow HO_2^{\bullet} \, (pK \, 4.8) \tag{7}$$

$$NRO \cdot + HO_2^{\bullet} \rightarrow NROH + O_2 \tag{8}$$

$$NROH + HO_2^{\bullet} + H^+ \rightarrow NRO \cdot + H_2O_2 \tag{9}$$

Taking in account the values of the following bond energies, in kJ mol^{-1}: (D_{H-O_2H} = 369.4; D_{H-O_2} = 203.6 [38], D_{NO-H} = 303, for TEMPO [22], we calculated the enthalpy change ($\Delta\Delta H_0$) for reactions 8 and 9 as −100 and −66 kJ mol^{-1}, correspondingly. For similar reactions experimental values of the reactions energy activation were found as $E_a \leq 12$ kJ mol^{-1} and estimated rate constants exceed $10^8 \, M^{-1} s^{-1}$ which does not depend on pH [39]. The theoretical estimation of aforementioned redox-reactions rate constants composed the SOD-mimetic process is a relatively crude approximation. Nevethless, this estimation showed rational tendency and even absolute values of the reactions rate constants which colse by order of magnitude to the experimental data.

Both Mechanisms I and III appear to be kinetically effective. Nevertheless, attention is also called to the high possibility of concurrent side reactions of NRO$^+$ in a cell with the other metabolic components such as NADH ($k_{ET} > 10^8 \, M^{-1} s^{-1}$ [31]), ascorbate, and some other antioxidants, semiquinone radical, nucleophilic molecules at high concentration, and so forth, whereas hydroxyl amine NROH is relatively chemically stable with regard to weak oxidants, including O_2. In addition, a very fast enzymatic SOD reaction can successfully compete with the nitroxide SOD mimetic activity. Thus the analysis of available thermodynamic and kinetic data has led to the conclusion that, at least in cells, the nitroxides SOD-mimetic activity most probably involves reactions 7–9 (Mechanism III).

7.1.3.2 Spin Trapping by Nitroxides

7.1.3.2.1 Reaction Nitroxides with Free Radicals

The first evidence about a possibility of reactions of nitroxides with free radicals was obtained by the M.B. Neiman group in 1966 [39]. Nitroxides were found to be effective inhibitors of radical polymerization. The most typical reaction of nitroxides with free radicals such as carbon, sulfur, and nitrogen-centered radicals is a recombination with formation of stable products. These reactive species are formed in processes of the oxidation of methabolites, drugs, and other ingredients of a cell, and as a result of radiation damage. In the absence of serious steric hindrances or electrostatic effects, the recombination is expected to run very fast in the diffusion control limit [37].

SH-containing compounds (glutathione, cysteine, serum albumins, and so forth) make an important contribution to maintaining the antioxidant status of biological systems. Though these compounds do not react directly with nitroxides, nevertheless, in some conditions (the presence of iron-ions and molecular oxygen) the formation of hydroxyl amine and amine from NO and disulphide and sulphonic acid from SH compounds occurs. The $RS^{\bullet}$, $RSOO^{\bullet}$, and $HO^{\bullet}$ radicals are the reaction intermediates. These radicals can, in turn, cause generation of other reactive species [19, 20, 40–42]. Cyclic nitroxides were found to be efficient scavengers of NO_2 ($k = (3-9) \times 10^8 \, M^{-1} s^{-1}$) and CO_3 ($k = (2-6) \times 10^8 \, M^{-1} s^{-1}$) radicals [31]. Derivatives of 2,2,6,6-tetramethylpiperidine-N-oxyl (TPO) react very fast with hydroxyl radical with rate constant $k = (1.0-4.5) \times 10^9 \, M^{-1} s^{-1}$. The kinetic isotope effect suggested the occurrence of $HO^{\bullet}$ in addition to the aminoxyl moiety of 4-O-TPO and H-atom abstraction from the 2- or 6-Me groups or from the 3- and 5-methylene positions [43].

7.1.3.2.2 Reaction Nitroxyl Amines and Nitronyls with Superoxide

The superoxide anion radical (a product of a one-electron reduction of molecular oxygen) has received great attention in the context of oxygen toxicity in biological systems [44–48]. Superoxide can damage biological membranes and tissues directly or it can act as a precursor of more active oxygen species. To understand the role of superoxide, and also to account for its toxic effects, it is necessary to know both how rapidly superoxide reacts with biological molecules and how rapidly it is scavenged by antioxidants.

Nitron spin-trapping is widely used for unambiguous detection of the superoxide radical (Section 7.2). Nevertheless, EPR detection of the O_2^- radicals in biological systems is limited by the slow kinetics of O_2^- spin-trapping and biodegradation of the radical adducts. It was shown [3] that nitroxyl amines are readily oxidized to a corresponding nitroxide. Cyclic hydroxylamines react with O_2^- 100 times faster than spin traps, which allow hydroxylamines to compete with cellular antioxidants and react with intracellular O_2^-. Use of superoxide dismutase or inhibitors of O_2^- production can overcome the lack of specificity of hydroxylamines.

In work [49] a set of cyclic hydroxylamines of various charges, lipophilicity, and cell permeability was synthesized and used to study extra- and intracellular O_2^- in human neutrophils, leucocytes, bovine aortic endothelial cells, and isolated mitochondria:

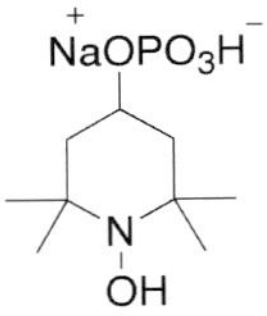

[Structures: PP-H, CAT1-H, TM-H, TMT-H, CM-H, EMPO]

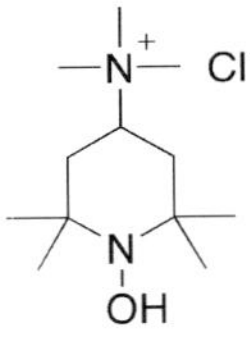

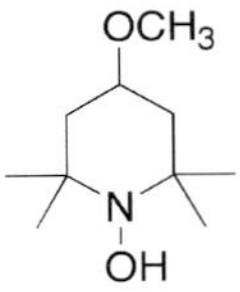

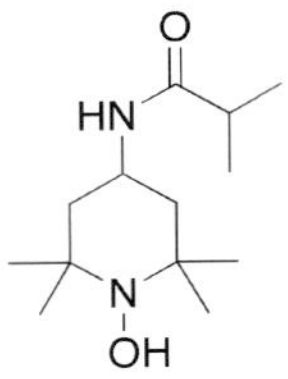

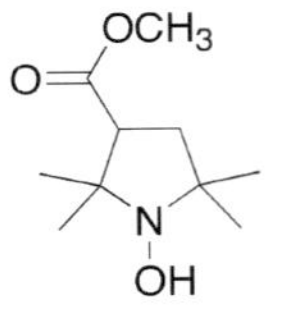

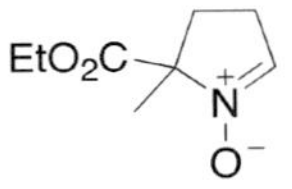

Detection of $O_2^{\cdot-}$ by cyclic hydroxylamines was compared to spin-trapping with DEPMPO and EMPO (Section 7.2). It was demonstrated that these probes allow site-specific $O_2^{\cdot-}$ detection with higher sensitivity than the nitrone spin traps.

The reaction of nitronyls with superoxide occurs by the following scheme (Figure 7.4) [42]. The values of the reaction rate constant were found to be $2 \times 10^3 - 10^3 \, M^{-1} s^{-1}$ depending on the nitronyl structure [29, 42, 50]. A new fluorescence – nitroxide method for the analysis of superoxide radicals in the nanomolar concentration scale is described in Section 7.3.

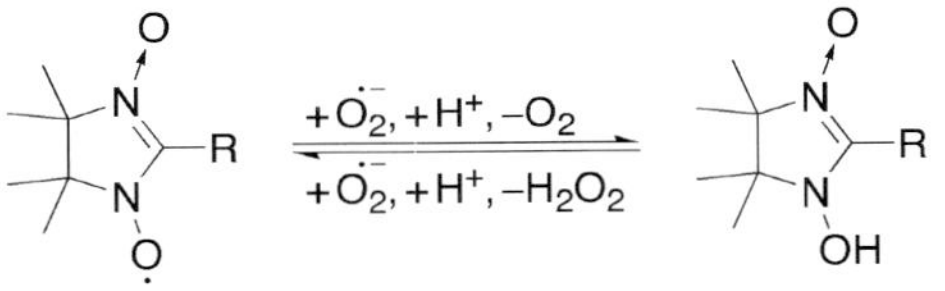

Figure 7.4 Nitronyl nitroxide reaction with superoxide.

7.1.3.2.3 Nitronyl Reactions with Nitric Oxide The widespread physiological and pathophysiological activities of nitric oxide (NO$^{\bullet}$) have led to the development of a variety of NO$^{\bullet}$ donors and scavengers. Among these are nitronyl nitroxides (NNRO), which specifically react with NO$^{\bullet}$ to form imino nitroxides (INRO), with comparatively high reaction constants of about $10^4 \, M^{-1} s^{-1}$ [50–53]. Nitric oxide reduces nitronyl according to the following scheme (Figure 7.5).

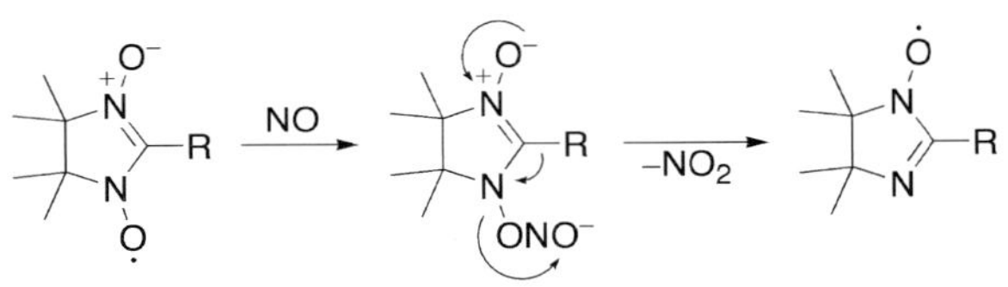

Figure 7.5 Nitronyl nitroxide reaction with nitric oxide [51]. (Reproduced with permission).

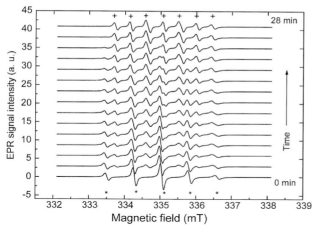

Figure 7.6 Evolution of the EPR spectrum of the pyrene-imino-nitroxide (0.1 mM) after the addition of 0.1 mM SNAP (NO• donor) to a solution of 0.1mM PN in Ringer solution (pH 7.2); $T = 25$ C. Crosses indicate the EPR signals of PI after the reaction is over. Stars indicate the EPR signals of PN before addition of SNAP [54]. (Reprinted from Lozinsky et al. (2004) Analytical biochemistry, detection of NO from pig trachea by a fluorescence method, *Analytical Biochemistry*, 139–145, with permission from Elsevier).

The reaction causes drastic change to the ESR spectrum (Figure 7.6). The nitronyls spectra show five lines with the amplitude ratio 1:2:3:2:1, as a result of hyperfine interaction of two-nitrogen nucleus ($A_{N1} = A_{N3} \approx 8\,G$), whereas iminonitroxyles are characterized by seven lines the interaction with two non-equivalent nitrogen nucleus ($A_{N1} \approx 9\,G$, $A_{N3} \approx 4\,G$). A combination of EPR and NMR spin-trapping to study the mechanisms of the reaction of nitronyl nitroxide with NO• was employed [55]. The NO-induced transformation of the paramagnetic trap was observed by ESR, whereas EPR-invisible diamagnetic products (the hydroxylamines) were detected by ^{19}F-NMR by use of newly synthesized fluorinated traps. The synthesis, kinetics, and stability of new dendrimeric-containing nitronyl nitroxides for analysis of nitric oxides were reported in reference [56].

The above-mentioned nitroxyl reactions can be used for quantitative analysis of nitric oxide. Nevertheless, in biological systems competitive reactions such as the reduction, oxidation, and other processes of chemically active compounds such as nytronyls should be taken into consideration. The application of dual fluorphore – nitroxide as probes for antioxidants, superoxide, and NO• is described in Section 7.3.

7.2
Nitron Spin Trapping

7.2.1
Introduction

Spin-trapping is designed for the investigation of processes with the participation of short-lived particles, reactive free radicals, bearing unpaired electrons. The technique is based on reaction with a molecule (spin trap) with formation of a sufficiently persistent paramagnetic species called the spin adduct, most commonly a nitroxide. The spin adduct can be detected by EPR spectroscopy. From the EPR spectrum of the nitroxide it is usually possible to identify the trapped species. In addition, a variety of physico-chemical methods such as mass spectrometry, HLPC, NMR, time-resolved Chemically Induced Dynamic Nuclear Polarization (CIDNP) and so forth may be employed.

Spin traps were originally utilized in measuring free radical activity. At present they are frequently used as antioxidants, providing unique protection against free radical damage, and for measuring the efficiency of other antioxidants, such as vitamin C, vitamin E, glutathione, R-lipoic acid, and so on. Spin traps are currently being explored as potential therapeutic agents and as a tool for the investigation of molecular mechanisms of a variety of physiological and pathophysiological processes. An "ideal" spin trap is characterized by the following features: (1) intensive, easily readable, and interpreted ESR signal; (2) fast reaction with reactive radicals to win a competition against other reactive species; (3) specificity with regard to certain radicals; (4) chemical stability; (5) compatibility with specific properties of an object of interest bearing a reactive functional group, hydrophility, hydrophobicity, electrostatic potential, and so forth; and (6) commercial availability or simplicity of a synthesis. A series of compounds such as nitroso, pyrollidine-1-oxide derivatives, and various nitrones was proposed as potential spin traps. But, currently, the most commonly used spin traps are analogs of *tert*-butylnitrone, α-phenyl-N-*tert* butyl nitrone (PBN), for example, and cyclic nitrons.

The pioneering works in which the principle possibility of the effectiveness of spin-trapping was demonstrated came to the light in the period 1968–1969 [57, 58]. Since then it has been customary to use this technique for investigating radical processes in chemistry and biology, and many chemical, instrumental, and biochemical efforts have been undertaken to achieve "the ideal".

7.2.2
Chemical Structure and Reactions of Nitrones with Radicals

In the last decade significant progress has been made in the development of new spin traps [59–74]. Several series of spin traps have been synthesized with stable adducts. These series include the nitrones: 5-(diethoxyphosphoryl)-5-methyl-1-pyrroline N-oxide (DEPMPO) [64]; 2-ethoxycarbonyl-2-methyl-3,4-dihydro-2H-pyrrole N-oxide (EMPO) [65, 66]; 5-*tert*-butoxycarbonyl-5-methyl-1-pyrroline N-oxide (BMPO) and 5-carboxy-5-methyl-1-pyrroline N-oxide [67]; 5-carbamoyl-5-methyl-1-pyrroline N-oxide (AMPO) [68]; and 1,1,3-trimethylisoindole N-oxide (TMINO) [69]. The synthesis and application of lipophylic nitrons as PPN, 5-Diisopropoxyphosphoryl-5-methyl-1-pyrroline N-oxide (DIPPMPO), and diethyl-(2-methyl-1-oxido-3,4-dihydro-2H-pyrrol-2-yl)phosphonate (DEPMPO) were described [70–72]. A new set of cationic, anionic, and neutral spin cyclic hydroxylamines with various lipophilicity and cell permeability allow site-specific $O_2^{\cdot-}$ detection with higher sensitivity than nitrone spin-traps [73].

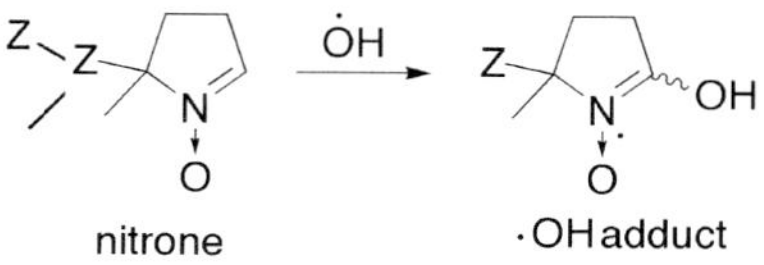

nitrone	Z
DMPO	-Me
EMPO	-CO$_2$Et
BocMPO	-CO$_2$t-Bu
DEPMPO	-P(O)(OEt)$_2$
DIPPMPO	-P(O)(Oi-Pr)$_2$

Figure 7.7 Spin-trapping of hydroxyl radical and chemical formulas of nitrons [73]. (Reproduced with permission).

The typical reaction of reactive free radicals with nitrons is addition to the nitron double bond with formation of a relatively stable radical. Figure 7.7 shows chemical formulas of typical nitron spin-traps and typical reaction of a nitron with a hydroxyl radical. Nevertheless, reactive radicals such as hydroxyl can be involved in different reactions, including the abstraction of hydrogen atoms from a nitron.

The structure of spin-trap conformers was calculated in [75, 76] using density functional theory (DFT) [77, 78]. EPR spectra of EMPO–O$_2$H (a) and DEPMPO–O$_2$H (b) adducts after irradiation using a light-riboflavin system to generate O$_2^-$ is shown in Figure 7.8 [77]. The mechanisms of chemical degradation of spin adducts was investigated in [78]. Figure 7.8 shows examples of ESR spectra of EMPO–O$_2$H (a) and DEPMPO–O$_2$H (b) adducts.

In the Rockenbauer group new approaches in the computer simulation of ESR spectra for nitroxide radicals were proposed and applied to the nitroxide spectra of a number of spin adducts [79–86]. In this approach, called *the two-dimensional simulation technique,* instead of individual curve fitting, a surface fitting procedure is used, when a set of spectra recorded under different conditions are simulated by using a unique parameter set. In ESR spectroscopy the first dimension is the usual magnetic field that is swept to record the spectra. The second dimension could be an external parameter, like the pH or concentration, when different species could be formed from the constituents of a complex system (i.e. ligands coordinated to metal ions) or the temperature, when intra- or intermolecular motions exert a dominant effect on the shape of spectra, or the time when different radicals could be formed, decayed, and converted. This method utilizes the fact that the total number of independent parameters is much less if we analyse the entire surface in a single fitting procedure compared to the case when we adjust all parameters for each spectrum one by one. The molecular dynamics, like chemical exchange, could produce complicated line-forms, which can be demonstrated by the superoxy adduct of DEPMPO [86]. When the exchange phenomenon is combined with molecular association with the solvents, the spectrum variation could be excessively complex for a unique interpretation of the spectra. In this

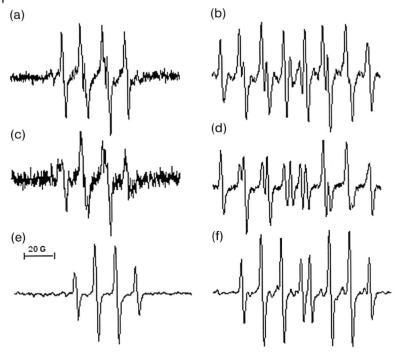

Figure 7.8 Spectral profiles of (a) EMBO/OH, (b) BocMPO/OH, DEMPO/OH, and DIPPMPO/OH. Spectra were taken from 25 mM, 3 mM H_2O_2 with UV radiation [75]. (Reproduced with permission).

case the thermodynamic equations to describe the spectral variations as a function of temperature were used, which allowed us to determine the thermodynamic parameters (enthalpies, entropies, free energies) as well as the spectroscopic parameters from the surface fitting procedure. Figure 7.9 demonstrates an agreement of theoretical and experimental ESR spectra of (5-diethoxyphosphoryl-5-methyl-1-pyrroline N-oxide) superoxy adducts.

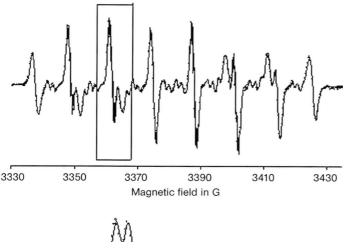

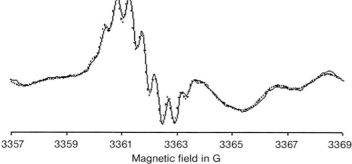

Figure 7.9 The ESR spectra of DEPMPO (5-diethoxy-phosphoryl-5-methyl-1-pyrroline N-oxide) superoxy adducts. Top: full spectra, bottom: excerpts. Solid line: experimental, dotted line [85]. (Reproduced with permission).

In reference [87] a comparative study is made of inhibitor activity of nitrons **1–5** and nitroxide **6** in respect of free radical oxidation of hydrocarbons on model reactions of 2,2′-azobis(2-methylpropionitrile)-initiated oxidation of cimene and of Cu^{2+}-induced oxidation. It has been shown that antioxidant activity of compounds **1–6** corresponds to the activity of alkylphenols (k from $(2-3) \cdot 10^4 \, M^{-1} s^{-1}$ for sterically hindered *p*-alkylphenols to $2 \cdot 10^5 \, M^{-1} s^{-1}$ for α-tocopherol).

In [88, 89] a combination technique consisting of dual spin-trapping (free radicals trapped by both regular and deuterated α-[4-pyridyl 1]-*N*-*tert*-butyl nitrone, POBN), followed by liquid chromatography, ESR, and mass-spectrometry was used to characterize and quantify POBN-trapped free radicals ($\cdot CH_3$, $\cdot OCH_3$, $\cdot CH_2OH$, $\cdot \cdot CH_2S(O)CH_3$) generated from the interaction of HO· and DMSO. This combined approach allowed us to identify and quantify all redox forms, including the ESR-active radical adduct and two ESR-silent forms, the nitrone adduct (oxidized adduct) and the hydroxylamine (reduced adduct).

Modern research trends in spin-trapping, besides the above-mentioned novel synthesis of nitrons and nitroxides, are the stabilization of free radical adducts in dextrin and liposomes, the immunospin-trapping technique, use of modern physico-chemical methods (mass spectrometry, HLPC, NMR) and dual spin-trapping.

To overcome a problem of stability of nitron adducts special efforts have been made in three directions: (1) the chemical synthesis of new nitrones; (2) the inclusion to dextrins; and (3) incorporation to liposomes.

Recently, stable superoxide radical adducts of 5-ethoxycarbonyl-3,5-dimethyl-pyrroline N-oxide (3,5-EDPO) and its derivatives ([90], phosphates containing traps such as products of the 1,3-additions of silicon-phosphorus reagents to nitrones PBN and DEMPO, diisopropoxyphosphoryl-5-methyl-1-pyrroline N-oxide (DIPPMPO) [73] and piro[pyrrolidine-2,2'-adamantane nitrones and nitroxides [91]) have been synthesized.

With the purpose of increasing the stability of adducts and to assign a position of spin-trapping, the spin-trap (N-[4-dodecyloxy-2-(7'-carboxyhept-1'-yloxy) benzylidene]-N-tert-butylamine N-oxide, DOD-8C) was designed [92]. This compound was tested in a model system consisting of large unilamellar vesicles in the presence of ethanol exposed to a copper CuI –1,10-phenanthroline system generating the tert-butoxyl and ·CH$_2$OH radicals from tert-butylhydroperoxide. In an elegant work by the P. Tordo group [84], synthesis, NMR, and EPR investigations of inclusion complexes of EMPO derivatives with 2,6-di-O-methyl-β–cyclodextrin (Figure 7.10) for superoxide detection were performed. The association constants of the complexes of (PBN) analogs and their superoxide spin adducts with ran-

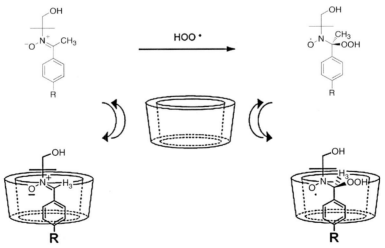

Figure 7.10 Schematic presentation of double superoxide trapping by nitron and dextrin [83]. (Reproduced with permission).

domly methylated β-cyclodextrin and 2,6-di-O-methyl-β-cyclodextrin the complex stoichiometries were measured using parallel ESR and ^{1}H-NMR techniques [93]. The work [94] is an example of the application of the spin-trapping technique coupled with mass spectrometry to the detection of free radicals using a nitrone. Diethyl-(2-methyl-1-oxido-3,4-dihydro-2H-pyrrol-2-yl)phosphonate (DEPMPO), was used to trap the free radicals ·OH, ·CH$_2$OH, and ·CH$_3$, and structure elucidation of the adducts obtained was performed by tandem mass spectrometry.

A new approach to the detection of protein and DNA free radicals using the specific free radical reactivity of nitrone spin-traps in conjunction with a nitrone – antibody sensitivity and specificity was developed [95]. The immuno-spin-trapping technique involves the following steps: (1) the trapping of a free radical by appropriate nitron; (2) preparing an antibody for the free radical adduct; (3) capturing the radical adduct by the antibody; (4) the isolation of captured adduct by HPLC technique; and (5) the analysis of the final product by appropriate physico-chemical methods, including ESR and NMR [96]. The immuno-spin-trapping approach greatly expands the utility of the spin-trapping technique. As an example, polyclonal antibodies were developed. The antibodies bind to protein adducts of the DMPO produced on myoglobin and hemoglobin by self-peroxidation with H_2O_2.

7.2.3
Non-Radical Reactions of Nitrones

In the spin-trapping approach application, potential non-radical reactions which may imitate free radical formation should be taken into consideration [97–105]. The most probable non-radical reaction appears to be the addition of a negatively charged nucleophile, Nu$^-$, to a nitrone, Nn:

$$NuH \leftrightarrow Nu^- + H^+$$

$$Nu^- + Nn \leftrightarrow N^- - Nu$$

$$Nn\text{-}Nu + H^+ \leftrightarrow Nn(H)Nu$$

where Nn(H)Nu and Nn$^-$Nu are the corresponding hydroxylamine and its deprotonated form.

Forrester and Hepburn first suggested the nucleophilic addition of a number of anions. Nitromethane, phthalamide, hydrogen and benzyl cyanide, and acetate added to PBN with formation of the hydroxylamine derivatives followed by their oxidation to EPR-detectable nitroxides [97]. Later the Forrester – Hepburn mechanism in the formation of EPR-detectable adducts of spin-traps was confirmed [101, 102]. Nucleophilic addition reactions of O-centered anions ($^-$OH, $^-$OCH$_3$) [103, 104] and a range of heterocyclic N–H bases toward PBN and DMPO were reported [105].

As an illustration, in an aqueous DMPO solution containing ferric ions (Fe^{3+}), DMPO–OH was produced ([106] and references therein). The addition of methanol, a good scavenger for $\cdot OH$, to this solution led to an aminoxyl radical, DMPO-OCH_3. EPR measurements at 77 K indicated the formation of a chelate between DMPO and Fe^{3+}. Using ^{17}O-enriched water it was demonstrated that under certain conditions in the presence of Fe(III) and Cu(II). Nucleophilic addition of water is a major pathway to the formation of the adduct DMPO–OH [103]. The reaction of the nitrone spin-trap 5,5-dimethylpyrroline-N-oxide (DMPO) with sodium bisulfite in aqueous solution was investigated using NMR and EPR techniques [100]. A reversible nucleophilic addition of (bi)sulfite anions to the double bond of DMPO was observed, resulting in the formation of the hydroxylamine derivative, 1-hydroxy-5,5-dimethylpyrrolidine-2-sulfonic acid. Further oxidation of hydroxylamines results in the formation of a corresponding radical adduct with sulfite radical anion or with the thiyl radicals.

7.2.4
Nitric Oxide Trapping

According to theoretical calculations [107] additions of nitric oxides to nitrons are thermodynamically unfavorable. This conclusion is supported by experimental data. Nevertheless, NO readily reacts with nitronyl nitroxides, Fe(II) hemin and hemin containing proteins (hemoglobin and myoglobin, Fe(II) dithiocarbamate and Fe(II) N-methyl-D-glucamine dithiocarbamate; [108, 109] and references therein) and with complex Co(II)-dansyl aminomethylpyridine [110]. A highly sensitive fluorescent method which allows us to monitor real-time out-flux of nitric oxide from biologically active tissues on a nano concentration scale was developed using super molecular systems (fluorescence-labeled myogobin [111] and a complex of fluorescence-labeled BSA with hemin [112]). Nitric oxide release from the unimolecular decomposition of the superoxide radical anion adduct of cyclic nitrones in aqueous medium was reported [80].

7.2.5
Thermodynamics and Kinetics of Nitron Reactions

The efficiency of the spin-trapping technique depends, to a great extent, upon the thermodynamics and kinetics of formation and breakdown of the spin adducts and a contribution of side non-radical reactions. Thus, quantitative investigation in this area appears to be a problem of paramount importance. A complete account of all the findings in the area is beyond the scope of this section, which is limited to a brief outline of current ideas and a few typical examples in this important and rapidly developing area. Nitrons and their adducts with reactive radicals participate in the following reactions: (1) adduct formations with the rate constant k_{ad}; (2) adducts' reversible decomposition (k_{rev}); (3) adducts' non-reversible decomposition to stable non-radical products k_{dc}; and (4) adduct reactions with formation of secondary radicals (k_{sc}). Nitrons can also participate in non-radical reactions.

7.2 Nitron Spin Trapping

It is known that an addition of reactive free radicals to the double chemical bonds is essentially thermodynamically favorable [37]. Recent calculations confirmed these general observations. A density functional theory (DFT) approach was employed in a study of trapping of free radicals, nitic oxide, and superoxide [75, 76]. It was shown that the order of increasing favorability for the free energy ΔG_o (kcal mol^{-1}) of the radical reaction with various cyclic nitrones follows a trend similar to their redox potentials as well as their second-order rate constants in aqueous solution NO < O$_{2-}$ (−7.51) < O$_2$H (−13.92) < ·SH (−16.55) < CH$_3$ (−32.17) < OH (−43.66). The calculation showed relative high thermodynamic (ΔG_o = 14.57 kcal mol^{-1}) and kinetic barriers for NO (free energy of activation $\Delta G_o^{\#} = 17.7 - 20.3$ kcal mol^{-1}).

Owing to the high reactivity of hydroxyl radicals, ·OH trapping by EMPO, BocMPO, DEPMPO, and DIPPMPO and practically any other nitron occur very fast, with the apparent rate constant $k_{ad} = (2.5-4.5) \times 10^9$ M^{-1}s^{-1}. The commonly used spin-trap 5,5-dimethyl-1-pyrroline N-oxide (DMPO) has a relatively lower k_{ad} of 1.9×10^9 M^{-1}s^{-1}. The half-lives of the ·OH adducts of EMPO, DEPMPO, and DIPPMPO are much longer ($t_{1/2}$ = 127–158 minutes) than those of DMPO and BocMPO, with half-lives of only 55 and 37 minutes, respectively [75]. Amido and spiroester nitrones were predicted to be the most suitable nitrones for spin-trapping of ·OH owing to the similarity of their thermodynamic and electronic properties to those of alkoxyphosphoryl nitrones [113]. The nitrone spin-traps 5,5-dimethyl-1-pyrroline N-oxide and 5-diethoxyphosphoryl-5-methyl-1-pyrroline N-oxide (DEPMPO) have significantly longer pseudo-first-order half-lives of about 130–158 minutes as compared with 55 minutes for DMPO and BocMPO (with the exception of EMPO with a $t_{1/2}$ of about 127 minutes). The BocMPO/·OH adduct gave the shortest $t_{1/2}$ at 37 minutes compared with a $t_{1/2} > 2$ hours for the EMPO/·OH.

Oxygen centered radicals bearing localized free valency react with nitrones sufficiently fast. The rate constant for the trapping of tBuO· by DOD-8C (N-[4-dodecyloxy-2-(7'-carboxyhept-1'-yloxy)benzylidene]-N-tert-butylamine N-oxide) is estimated to be 5.5×10^6 M^{-1}s^{-1} [92]. In the superoxide radical the unpaired electron on oxygen is conjugated with the electron pair of the second oxygen atom and the reaction of formation of the O$_2^-$-nitron adducts is markedly slower as compared with the ·OH radical [114–120]. Correspondingly, the following apparent rate constants for the trapping of superoxide radicals were reported: DMPO (60 M^{-1}s^{-1}); DEPMPO (90 M^{-1}s^{-1}) [63, 114, 115]. The values of first-order approximation half-lives of nitrone – superoxide adducts were reported as 1 minute at pH 7 for DMPO and DEPMPO, respectively [115]. It was shown that, at pH 7.2, nitrones DEPO and EMPO trapped superoxide relatively slowly (DEPO, k_{ad} = 31 M^{-1}s^{-1} and EMPO, k_{ad} = 11 M^{-1}s^{-1}). The rate constant values are dependent on pH. For example, for 5-tert-butoxicarboxy-5methyl-1-prroline N-oxide (BMPO) k_{ad} = 75.0 M^{-1}s^{-1} at pH 7 compared with pH 5, (239 M^{-1}s^{-1}). The adduct O$_2^-$–nitron decomposition rate constant is also dependent on the nitron structure: $k_{dc}/10^{-4}$ s are AMPO 14; EMPO 11.6; DEPMPO 7.5; and DMPO 129 [116]. The rate constant k_{ad} for the reaction of nitrone DMPO with O$_2^-$ was detected to be 170 M^{-1}s^{-1} [117]. According to [118] the half-life of the nitrone 5-carbamoyl-5-methyl-1-pyrroline N-oxide adduct (AMPO–O$_2$H) was about 8 minutes, similar to that observed for EMPO but significantly shorter than that of DEPMPO–O$_2$H with a $t_{1/2}$ of 16 minutes [119].

The O$_2^-$–adduct rate constants decomposition (k_{dc}) are essentially dependent on the adduct structure. The following $k_{dc}/10^{-4}$ s values were found: AMPO (14) [119]; EMPO (11.6); DEMPO (7.5); and DMPO (129) [120]. The half-lives of the ·OH adducts of EMPO, DEPMPO [65], and DIPPMPO are much longer ($t_{1/2}$ = 127–158 minutes) than those of DMPO and BocMPO, with half-lives of 55 and 37 minutes, respectively.

The nitrone EPPN

has been found to selectively detect superoxides in the presence of hydroxyl radicals. The kinetics of superoxide adduct decay showed that the EPPN–OOH half-life time was 7 minutes at pH 5.8 and 5.5 minutes at pH 7. The absence of stereoselectivity in spin-trapping superoxide by 5-tert-butoxycarbonyl-5-methyl-1-pyrroline N-Oxide was shown in [63].

Reactive radicals, such as ·C(OH)(CH$_3$)$_2$, ·CH$_2$OH, and ·CH$_3$, bearing free valency localized on carbon atoms react with different nitrons relatively fast (k_{ad} = (0.28–1.4 × 10^8 M^{-1} s^{-1}) [117]. The monomolecular decomposition of the radical adducts DEPMPO/glutathiyl radical (·GS) and thiyl radical back to the nitrone was demonstrated. The rate constants for this reaction were found to be equal to 0.3 s^{-1} and 0.02 s^{-1} for DMPO/·GS and DEPMPO/·GS radical adducts, respectively [100]. The reactions of the reversible addition of thiols and thiyl radicals to the nitrone spin-traps DMPO (5,5-dimethyl-1-pyrroline N-oxide) and DEPMPO (5-diethoxyphosphoryl-5-methyl-1-pyrroline N-oxide) were invesigated using ^{31}P NMR [100]. The equilibrium constant for the reactions of the formation of the product of DEPMPO with S-centered nucleophiles decreases in the series: sulfite > thioglycolic acid > cysteine > glutathione.

Numerous principle advantages in the investigation of radical processes in chemistry, biochemistry, and biomedical research have been made through the use of the spin-trapping technique (see Chapters 10 and 11).

7.3
Dual Fluorophore–Nitroxides (FNRO·) as Redox Sensors and Spin Traps

7.3.1
Introduction

The use of dual fluorophore–nitroxide compounds (FNRO·) as redox probes and spin-traps is based on the first observations [121] that, in the dual molecule, the

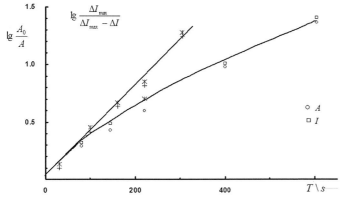

Figure 7.11 Kinetics of the change in amplitude A of the ESR spectrum of the nitroxyl segment and of the intensity I of the fluorescence of the fluorophore of FNRO as a result of photochemical reactions in the presence (1) and absence (2) of oxygen at 293 K in a solution (glycerol, 75%, water 20%, ethanol 5%) for specimens irradiated under identical conditions [121].

nitroxide acts as a quencher of the fluorescence of the chromophore fragment. Photo- and chemical reductions of the nitroxide fragment result in decay of the ESR signal and enhancement of the fluorescence. Experiments with Dansyl – TEMPO were the first examples of intramolecular quenching and the ability of FNRO· to monitor redox processes using two independent methods, ESR and more sensitive fluorescence (Figure 7.11) [121]. A similar effect was observed after the reduction of the dual molecules.

These observations were later confirmed in a series of works [54, 122–132]. For example, it was shown that in paramagnetic nitroxide–naphthalene dual molecules the fluorescence is quenched by 40–60-fold as compared to its diamagnetic hydroxylamine analogues [122]. The residual emission that is observed from these adducts arises from the locally excited singlet of the fluorophore, not from charge recombination.

7.3.2
Analysis of Antioxidant Status

On the basis of the above-mentioned unique properties of the dual molecules, a method for the quantitative analysis of vitamin C in biological and chemical liquids

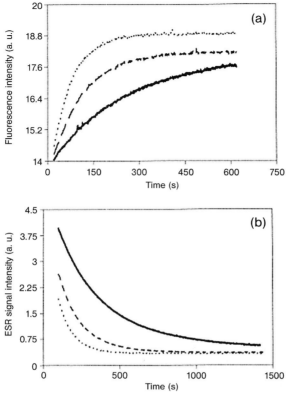

Figure 7.12 Fluorescence enhancement (a) and ESR decay (b) of the probe (1) caused by excess of ascorbic acid. Solid line, 0.1 mM; dashed line, 0.4 mM; dotted line, 0.8 mM. Aquisition parameters of fluorescence: excitation slit 1 nm, emission slit 16 nm, voltage 480 V, $T = 25\,°C$, phosphate buffer (pH 7)[127]. (Reprinted from Lozinsky et al. (1999) Dual fluorophore – nitroxide probes for analysis of vitamin C in biological liquids Elsevier, *J. Biophys. Biochem. Methods*, 29–42, with permission from Elsevier).

was proposed [127]. It was shown that in the presence of ascorbic acid the increase of the fluorescence intensity and the decay of the EPR signal of a dual probe occurred with the same rate constant (Figures 7.12 and 7.13). By performing a series of pseudo-first-order reactions between the dual molecule and ascorbic acid, and consequently plotting rate constants versus ascorbic acid concentrations, the calibration curves for the vitamin C analysis were obtained (Figure 7.13). Variations of the chemical structure of fluorophore and nitroxide fragments allow for regulating fluorescent properties and redox potentials of the dual molecules.

The use of the FNRO· probes for the determination of the ascorbate concentration in biological systems, which, for example, contain proteins, is hindered due to the protein concentration-dependent interactions between the proteins and/or the ascorbate and the probe. It was shown that the addition of BSA to a mixture of (FNRO·) and ascorbate affected the rate of the nitroxide reduction. This reac-

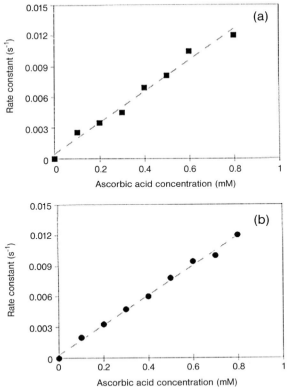

Figure 7.13 Dependence of rate constant of reduction of dansyl – TEMPOL on ascorbic acid concentration. Fluorescence (a) and ESR (b) measurements; $T = 25\,°C$, phosphate buffer (pH 7) [127]. (Reprinted from Lozinsky et al. (1999) Dual fluorophore–nitroxide probes for analysis of vitamin C in biological liquids Elsevier, J. Biophys. Biochem. Methods, 29–42, with permission from Elsevier).

tion was accelerated with increasing BSA concentration [128]. Thus, in complex systems, biological ones in particular, not an antioxidant concentration but *the value of rate constant of the nitroxide reduction may be taken as a quantitative characteristic of the antioxidant status of system of interest*. The proposed method was used for the analysis of the antioxidant status of commercial juices and human blood [127, 130].

A bifunctional stilbene – nitroxide label (BFL1) was applied for the parallel determination of the antioxidant status and measurement of micro- and macro-viscosity of the media [131]. The synthesized stilbene – nitroxide label (BFL1) was immobilized on the surface of a quartz plate. The enhancement of the fluorescence of the photocrome segment of the label indicated the nitroxide reduction by ascorbic acid. At the same time the stilbene moiety photoisomerization kinetics is strongly dependent upon the viscosity of a media. It was demonstrated that the label nitroxide moiety allows for measuring the concentration of ascorbic acid in the solution in a range of $(1–9) \times 10^{-4}$ M. On the other hand, monitoring the rate constant of the stilbene fragment photoisomerization permits the measurement of the viscosity of a medium in a range 1–500 cP. This is done by using the calibration plot (photoisomerization rate constant versus the viscosity of model water – glycerol mixture). All the above-mentioned experiments were performed with the use of a very sensitive fluorescence technique by means of a commercial fluorimeter and of modified quartz plates in a single fast measurement.

The covalent immobilization of the label enables multiple use of the same plate without the need to prepare and calibrate additional solutions – an important advantage for possible industrial use. Replacing the quartz plates by quartz optical fibers will facilitate the use of the given methods for the continuous monitoring of chemical and biological processes.

Six member pyridinyl nitroxides are reduced by reducing agents with relatively low negative redox potential as ascorbic acid and semiquinones but are not able to react with antioxidants of higher redox potential. To overcome this limitation, the dual pyren-nitronyl probe (PNRO) was employed [132].

It was shown that the fluorescence PNRO· assay is suitable as an analysis of antioxidants such as quercetin and galangin the submicro concentration scale. The reaction occurred in excess over the antioxidants, and kinetic analysis suggests a pseudo-first-order process. The linear dependencies of the experimental pseudo-first-order rate constant on concentration of antioxidants in the range $5 \times 10^{-8}–10^{-6}$ M in the presence of catalase gave the rate constant values equal to 7.6×10^{-1} M^{-1}s^{-1} and 1.8×10^{-2} M^{-1}s^{-1} for quercetin and galangin, respectively. These values are close to those obtained independently by following the decay of the PNRO· ESR signal. Catalase essentially affects the reaction rate.

7.3.3
Analysis of Superoxide and Nitric Oxide by Pyren-Nitronyl

7.3.3.1 Superoxide Analysis

A new, fast, and highly sensitive method for superoxide analysis, based on a reaction of fluorophore – nitronyl with $O_2^{-\cdot}$, has been developed [132]. According to [132] the reaction of PNRO·.with superoxide radicals generated by the xanthine/xanthine oxidase system offered a drastic increase (about 2000 times) in the fluorescence intensity and a decrease of the ESR signal. When the rate of superoxide production is slow ($\omega_i < 2 \times 10^{-7}$ M min^{-1}), the reaction followed the zero-order law. Under these conditions, the rate of fluorescence increase (dI/dt) is proportional to the rate of superoxide production (ω_i). The dependence dI/dt on ω_i was used as a calibration curve in subsequent experiments (Figure 7.14). At higher concentrations, the kinetics of fluorescence change takes a first-order character. The kinetic analysis allowed calculation of the rate constant of the PNRO· reduction by superoxide $k_1 = 1.54 \times 10^6$ M^{-1} s^{-1} (pH 7.4, $T = 300$ K).

Owing to its very high sensitivity and simplicity, the proposed method will, under certain conditions, have advantages over conventional light absorption, ESR, and chemiluminescence techniques, and will be able to monitor biological processes *in real time*. The contribution of other reactive oxygen species could be estimated quantitatively by previous treatment with appropriative inhibitors. Thus, the contribution of superoxide radical can be estimated in the presence of superoxide dismutase and catalase.

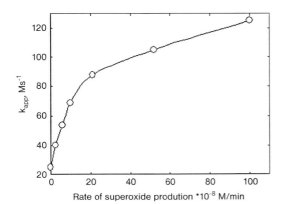

Figure 7.14 Dependence of the apparent rate constant of FNNRO reduction on the rate of superoxide production; PBS (pH 7.4, T=300 K). The insert shows the linear segment of the curve for low rates of superoxide production corresponding to the zero order process ($\omega_i < 2 \times 10^{-7}$ M min^{-1}) [132]. (Reprinted from Medvedeva et al. (2004) Dual fluorophore – nitronyl probe for investigation of superoxide dynamics and antioxidant status of biological systems, *J. Photochem. Photobiol. A: Chem.*, 45–51, with permission from Elsevier).

7.3.3.2 Nitric Oxide Analysis

Real-time monitoring of NO· dynamics under physiological conditions of nanomolar concentrations is a difficult analytical problem [52, 54, 133–135] and references therein). This has been attributed to the labile nature of NO·, which rapidly reacts with many scavenger targets resulting in very rapid diffusion throughout the medium. These scavengers include oxygen, superoxide, amino and mercapto compounds, hemeproteins, and so forth. Although a variety of methods for NO· detection have been proposed, they are not generally amenable to *real-time in-situ* measurement of endogenous NO·.

A new method for nitroxide analysis in nanoconcentration scale and real-time monitoring NO·outflax from tissues has been developed [54]. It was shown that the pyrene – nitronyl (PNRO)· reacts with NO· to yield a pyrene–imino nitroxide radical (PI) and NO_2. Conversion of PNRO· to PI is accompanied by changes in the EPR spectrum from a five-line pattern (two equivalent N nuclei) into a seven-line pattern (two non-equivalent N nuclei) (Figure 7.6). The transformation of the EPR signal is accompanied by a drastic increase in the fluorescence intensity since the imino nitroxide radical is a weaker quencher than the nitronyl. The ESR method was employed for the calibration of NO· on the micromole concentration scale (Figure 7.6), while the essentially more sensitive fluorescence technique allowed us to calibrate NO· on a nanomolar concentration scale (Figure 7.15).

The method was applied to the determination of NO and S-nitroso compounds in tissue from pig trachea epithelia. The measured basal flux of S-nitroso compounds obtained from the tissues was about $1.2\,nmol\,g^{-1}\,min^{-1}$, and NO··-synthase stimulated by extracellular adenosine 5′-triphosphate produced NO· flux of $0.9\,nmol\,g^{-1}\,min^{-1}$ [54].

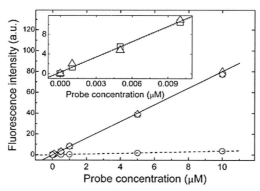

Figure 7.15 Integrated fluorescence measured before (circles) and after 5-hour incubation of PN with excess SNAP (10 µM) (squares) and 10-min incubation with NO (~10 µM) (diamonds). The fluorescence of the pyrene–imino (triangles) is measured for comparison [54]. (Reprinted from *Analytical Biochemistry*, Lozinsky et al., Detection of NO from pig trachea by a fluorescence method, Analytical Biochemistry, 139–145, with permission from Elsevier).

7.3.4
Dual Molecules as Spin Traps

The first reaction of a dual fluorophore–nitroxyl compound with a carbon-centered methyl radical was reported by the N. Blough group [122]. Using ESR and HPLC chromatography the authors showed the formation of the corresponding alkoxy derivative of FNRO·. An ability of dual fluorophore–nitroxide molecules to detect superoxide and hydroxyl radicals was demonstrated in work [125]. The developed method was used for the monitoring dynamics of these reactive radicals generated by stimulated neutrophils.

It was reported [136] that the glutathion thyil radical GS· reacted irreversibly with dual probe Acridin–TEMPO to produce a secondary amine. As expected, during the reaction of Ac–TEMPO with GS, the TEMPO EPR signals decayed and acridine fluorescence concurrently increased. Using combined HPLC and mass spectrometry it was determined that the probe was converted into fluorescent acridine (Ac)piperidine while GSH was primarily oxidized into sulfonic acid. The naphthalene–TEMPO (NTEMPO) was used for probing the hydroxyl radical-mediated reactivity of peroxynitrite [137]. The decomposition of peroxynitrite yielded a hydroxyl radical, which reacted rapidly with DMSO to produce a methyl radical, which was then trapped by the dual probe. The NTEMPO was transformed to a stable diamagnetic o-alkoxyl derivative.

Dual fluorescence – nitroxide probes and photoinitiators were employed to detect free radical generation in polymer films with photoinitiation and to study photoacid-generator behavior using fluorescence spectroscopy and microscopy [138]. The dual probes showed a fluorescence increase after recombination with active radicals. According to the article authors "this system may serve both as a mechanistic tool in the study of photoinitiated radical processes in polymer films and in the preparation of functional fluorescent images". Additionally, a pH-sensitive fluorescence – nitroxide compound was employed as a dual probe for both photogenerated acid and free radicals.

Also demonstrated was the possibility of using PNRO for the detection of hydroxyl radicals generated from hydrogen peroxide by the Fenton mechanism in a less than micromole concentration scale [132]. This method can be used for the analysis of hydrogen peroxide.

Application of dual fluorophore – nitroxide methodology can afford insights into the biochemical significance of reactive free radicals, superoxides, nitric oxides, and hydrogen peroxide, and may create a basis for the development of new methods for biomedical research and medical diagnostics. Several examples of such an application of the dual molecules to investigate biochemical and biophysical processes, spin-trapping, and radical reactions in polymer are described in Chapter 8.

In this chapter we have presented evidence that nitroxide in the free state and as adducts of nitrons with reactive radicals exhibit unique features which allow us to solve a number difficult experimental tasks in practically any system, including living organisms: (1) to characterize kinetically redox properties and antioxidant ability of compounds of a wide range of redox potentials; (2) to detect reactive radicals and to decipher their chemical structure; (3) to serve as antioxidants; and (4) to monitor real-time dynamics of two tiny species of powerful biological impact – nitric oxide and superoxide. Thus, methods based on nitroxide spin-probing and nitron spin-trapping have formed a basis to effectively attack many important chemical, biochemical, and biomedical problems. The effectiveness of these methods and applications will be described in Chapters 10 and 11.

References

1 Keana, J.F.W. (1979) *Spin Labeling*, Vol. 2, Academic, New York, pp. 215–72.
2 Meisel, D. and Czapski, G. (1975) *Journal of Physical Chemistry*, **79**, 1503–9.
3 Rozen, G.D., Finkelshtein, E. and Rauchman, E.J. (1982) *Archives of Biochemistry and Biophysics*, **215**, 367–78.
4 Likhtenshtein, G.I. (1976) *Spin Labeling Method in Molecular Biology*, Wiley Interscience, New York.
5 Likhtenshtein, G.I. (1993) *Biophysical Labeling Methods in Molecular Biology*, Cambridge University Press, Cambridge, NY.
6 Likhtenshtein, G.I. (2005) Labeling, biophysical, in *Encyclopedia of Molecular Biology and Molecular Medecine*, Vol. 7 (ed. R. Meyers), VCH, New York, pp. 157–38.
7 Volodarsky, L.B. (ed.) (1988) *Imidazolin Radicals*, CRC Press, Bosa Raton.
8 Kocherginsky, N. and Swarts, H.M. (1995) *Nitroxide Spin Labels. Reactions in Biology and Chemistry*, CRC Press, Bosa Raton.
9 Rozantsev, E.G., Neiman, M.B. and Likhtenshtein, G.I. (1964), SU 166133 1964110, Application SU 19620529.
10 Keana, J.F.W., Bernard, E.M. and Roman, R.B. (1978) *Synthetic Communications*, **8**, 169–73.
11 Gendel, L.Ya., Goldfeld, M.A., Koltover, V.K., Rozanzev, E.A. and Suskina, V.I. (1968) *Biofizika*, **13**, 1114–15.
12 Mehlhorn, R.J. (1991) *The Journal of Biological Chemistry*, **266**, 2724–31.
13 Van Der Zee, J. and Van Den Broek, P.J. (1998) *Free Radical Biology and Medicine*, **25**, 282–866.
14 Shiga, M.Y., Miyazono, Y., Ishiyama, M., Sasamoto, K. and Ueno, K. (1997) *Analytical Communications*, **34**, 115–17.
15 Likhtenshtein, G.I. (1996) *Journal of Photochemistry and Photobiology A: Chemistry*, **96**, 79–92.
16 Chabita (Saha), K. and Mandal, P.C. (2002) *Indian Journal of Chemistry*, **41A**, 2231–7.
17 Sen, V.D. and Golubev, V.A. (1993) *Izvestiya Akademii Nauk, Seriya Khimicheskaya*, **3**, 542–7.
18 Golubev, V.A., Sen, V.D., Kulyk, I.V. and Aleksandrov, A.L. (1975) *Izvestiya Akademii Nauk SSSR, Seriya Khimicheskaya*, **10**, 2235–43.
19 Bobko, A.A., Kirilyuk, I.A., Grigor'ev, I.A., Zweier, J.L. and Khramtsov, V.V. (2007) *Free Radical Biology and Medicine*, **42**, 404–12.
20 Potapenko, D.I., Bagryanskaya, E.G., Tsentalovich, Y.P., Reznikov, V.A., Clanton, T.L. and Khramtsov, V.V. (2004) *Journal of Physical Chemistry. B*, **108**, 9315–24.
21 Volodarsky, L.A., Grigor'ev, I.A.V. and Dikanov, V.A. (1988) *Imidazoline Nitroxide Radicals*, Nauka (Siberian branch), Novosibirsk, pp. 188–93.
22 Denisov, E.T. and Tumanov, V.E. (2005) *Russian Chemical Reviews*, **74**, 825–58.
23 Fish, J.R., Swarts, S.G., Sevilla, M.D. and Malinski, T. (1988) *Journal of Physical Chemistry*, **92**, 3745–51.
24 Kato, Y., Shimizu, Y., Lin, Y.J., Unoura, K., Utsumi, H. and Ogata, T. (1995) *Electrochimica Acta*, **40**, 2799–802.
25 Czapski, G., Samuni, A. and Goldstein, S. (2002) Superoxide dismutase

mimics: antioxidative and adverse effects, in *Methods in Enzymology (Superoxide Dismutase)*, Vol. 349, Elsevier, pp. 234–42.
26. Rosen, G.M., Cohen, M.S., Britigan, B.E. and Pou, S. (1990) *Free Radical Research Communications*, **9**, 187–95.
27. Magnani, L., Gaydou, E.M. and Hubaud, J.C. (2000) *Analytica Chimica Acta*, **411**, 209–16.
28. Sen', V.D., Golubev, V.A., Kulyk, I.V. and Rozantsev, E.G. (1976) *Russian Chemical Bulletin*, **25**, 1647–54.
29. Krishna, M.C., Samuni, A., Taira, J., Goldstein, S., Mitchell, J.B. and Russo, A. (1996) *The Journal of Biological Chemistry*, **271**, 26018–25.
30. Hideg, K., Kalai, T. and Sar, C. (2005) *Journal of Heterocyclic Chemistry*, **42**, 437–50.
31. Goldstein, S., Samuni, A., Hideg, K. and Merenyi, G. (2006) *Journal of Physical Chemistry. A*, **110**, 3679–85.
32. Likhtenshtein, G.I. (2003) *New Trends In Enzyme Catalysis and Mimicking Chemical Reactions*, Kluwer Academic/Plenum Publishers, New York.
33. Marcus, R.A. and Sutin, N. (1985) *Biochimica et Biophysica Acta*, **811**, 265–322.
34. Gray, G.H. and Winkler, J.R. (1996) *Annual Review of Biochemistry*, **65**, 537–61.
35. Stabury, D.M. (1989) *Advances in Inorganic Chemistry*, **33**, 69–138.
36. Wardman, P. (1989) *Journal of Physical Chemistry*, **18**, 1637–755.
37. Denisov, E.T., Sarkisov, O.M. and Likhtenshtein, G.I. (2003) *Chemical Kinetics. Fundamentals and Recent Developments*, Elsevier Science.
38. Lide, D.R. (ed.) (2004–2005) *Handbook of Chemistry and Physics*, CRC Press, $D_{NO-H} = 303$, for TEMPO.
39. Ruban, L.V., Buchachenko, A.L., Neiman, M.B. and Kokhanov, Yu.V. (1966) *Vysokomolekulyarnye Soedineniya*, **8**, 1642–6.
40. Morrisett, J.D. and Drott, H.R. (1969) *The Journal of Biological Chemistry*, **244**, 5083–4.
41. Glebska, J., Skolimowski, J., Kudzin, Z., Gwozdzinski, K., Grzelak, A., Bartosz, G. and Kirilyuk, I.A. (2003) *Free Radical Biology and Medicine*, A. **35**, 310–16.
42. Goldstein, S., Russo, A. and Samuni, A. (2003) *The Journal of Biological Chemistry*, **278**, 50949–9055.
43. Samuni, A., Krishna, C.M., Riesz, P., Finkelstein, E. and Russo, A. (1988) *The Journal of Biological Chemistry*, **263**, 17921–4.
44. Samuni, A., Goldstein, S., Mitchell, J.B., Krishna, M.C. and Neta, P. (2002) *Journal of the American Chemical Society*, **124**, 8719–24.
45. Halliwell, B. and Gutteridge, J.M.C. (eds) (1989) *Free Radicals in Biology and Medicine*, 2nd edn, Oxford University Press, New York.
46. Fridovich, I. (1983) *Annual Review of Pharmacology and Toxicology*, **23**, 239–57.
47. McCord, J. (1987) *Gastroenterology*, **92**, 2026–8.
48. Fridovich, I. (1986) *Archives of Biochemistry and Biophysics*, **247**, 1–11.
49. Pou, S., Huang, Y., Bhan, A., Bhadti, V.S., Hosmane, R.S., Wu, S.Y., Cao, G. and Rosen, G.M. (1993) *Analytical Biochemistry*, **212**, 85–90.
50. Dikalov, S.I., Kirilyuk, I.A., Voinov, M.A. and Grigor'ev, I.A. (2005) *Abstracts of Conference SPIN-2005*, September 20–24, Novosibirsk, Russia, p. 58.
51. Haseloff, R.F., Zollner, S., Kirilyuk, I.A., Grigor'ev, I.A., Reszka, R., Bernhardt, R., Mertsch, K., Roloff, B. and Blasig, I.E. (1997) *Free Radical Research*, **26**, 7–17.
52. Joseph, J., Kalyanaraman, B. and Hyde, J.S. (1993) *Biochemical and Biophysical Research Communications*, **192**, 926–34.
53. Woldman, Y. Yu., Khramtsov, V.V., Grigor'ev, I.A., Kirilyuk, I.A. and Utepbergenov, D.I. (1994) *Biochemical and Biophysical Research Communications*, **202**, 195–203.
54. Lozinsky, E.M., Martina, L.V., Shames, A.I., Uzlaner, N., Masarwa, A., Likhtenshtein, G.I., Meyerstein, D., Martin, V.V. and Priel, Z. (2004) *Analytical Biochemistry*, **326**, 139–45.
55. Bobko, A.A., Bagryanskaya, E.G., Reznikov, V.A., Kolosova, N.G., Clanton, T.L. and Khramtsov, V.V. (2004) *Free Radical Biology and Medicine*, **36**, 248–58.

56 Rosen, G.M., Porasuphatana, S., Tsai, P., Ambulos, N.P., Galtsev, V.E., Ichikawa, K. and Halpern, H.J. (2003) *Macromolecules*, **36**, 1021–7.

57 Janzen, E.G. and Blackburn, B.J. (1968) *Journal of the American Chemical Society*, **90**, 5909–10.

58 Janzen, E.G. and Blackburn, B.J. (1969) *Journal of the American Chemical Society*, **91**, 4481–90.

59 Villamena, F.A., Locigno, E.J., Rockenbauer, A., Hadad, C.M. and Zweier, J.L. (2007) *Journal of Physical Chemistry. A*, **111**, 384–91.

60 Davies, M.J. (2002) *Electron Paramagnetic Resonance, Specialist Periodical Reports*, Vol. 18 (eds B.C. Gilbert, M.J. Davies and D.M. Murphy), The Royal Society of Chemistry, Cambridge, pp. 47–73.

61 Samouilov, A., Roubaud, V., Kuppusamy, P. and Zweier, J.L. (2004) *Analytical Biochemistry*, **334**, 145–54.

62 Hideg, K., Kalai, T. and Sár, S.P. (2005) *Journal of Heterocyclic Chemistry*, **42**, 437–50.

63 Roubaud, V., Lauricella, R., Tuccio, B., Bouteiller, J.C. and Tordo, P. (1996) *Research on Chemical Intermediates*, **22**, 405–16.

64 Frejaville, C., Karoui, H., Tuccio, B., Le Moigne, F., Calcasi, M., Pietri, S., Lauricella, R. and Tordo, P. (1995) *Journal of Medicinal Chemistry*, **38**, 258–65.

65 Allouch, A., Roubaud, V., Lauricella, R., Bouteiller, J.-C. and Tuccio, B. (2005) *Organic and Biomolecular Chemistry*, **3**, 2458–62.

66 Lauricella, R., Allouch, A., Roubaud, V., Bouteiller, J.-C. and Tuccio, B. (2004) *Organic and Biomolecular Chemistry*, **2**, 1304–9.

67 Olive, G., Mercier, A., Le Moigne, F., Rockenbauer, A. and Tordo, P. (2000) *Free Radical Biology and Medicine*, **28**, 403–8.

68 Villamena, F.A. and Zweier, J.L. (2002) *Journal of the Chemical Society, Perkin Transactions*, RCS, 1340–4.

69 Villamena, F.A., Rockenbauer, A., Gallucci, J., Velayutham, M., Hadad, C., Zweier, J. (2004) *Journal of Organic Chemistry*, **69**, 7994–8004.

70 Bottle, S.E., Hanson, G.R. and Aaron, S. (2003) *Organic and Biomolecular Chemistry*, **14**, 2585–9.

71 Hay, A., Burkitt, M.J., Jones, C.M. and Hartley, R.C. (2005) *Archives of Biochemistry and Biophysics*, **435**, 336–46.

72 Tuccio, B., Lauricella, R. and Charles, L. (2006) *International Journal of Mass Spectrometry*, **252**, 47–53.

73 Chalier, F., Ouari, O. and Tordo, P. (2004) *Organic and Biomolecular Chemistry*, **2**, 927–34.

74 Dikalov, S.I., Dikalova, A.E. and Mason, R.P. (2002) *Archives of Biochemistry and Biophysics*, **402**, 218–26.

75 Villamena, F.A., Hadad, C.M. and Zweier, J.L. (2003) *Journal of Physical Chemistry. A*, **107**, 4407–14.

76 Villamena, F.A.V., Frederick, A., Hadad, C.M. and Zweier, J.L. (2005) *Journal of Physical Chemistry. A*, **109**, 1662–74.

77 Labanowski, J.W. and Andzelm, J. (1991) *Density Functional Methods in Chemistry*, Springer, New York.

78 Parr, R.G. and Yang, W. (1989) *Density Functional Theory in Atoms and Molecules*, Oxford University Press, New York.

79 Villamena, F.A., Rockenbauer, A., Gallucci, J., Velayutham, M., Hadad, C.M. and Zweier, J.L. (2005) *Journal of Physical Chemistry. A*, **109**, 1662–74.

80 Locigno, E.J., Zweier, J.L. and Villamena, F.A. (2005) *Organic and Biomolecular Chemistry*, **3**, 3220–7.

81 Rockenbauer, A., Szabó-Plánka, T., Árkosi, Z. and Korecz, L. (2001) *Journal of the American Chemical Society*, **123**, 7646–54.

82 Karoui, H., Rockenbauer, A., Pietri, S. and Tordo, P. (2002) *Chemical Communications*, **24**, 3031–2.

83 Bardelang, D., Rockenbauer, A., Karoui, H., Finet, J.P. and Tordo, P. (2005) *Journal of Physical Chemistry. B*, **109**, 10521–30.

84 Bardelang, D., Rockenbauer, A., Jicsinszky, L., Finet, J.P., Karoui, H., Lambert, S., Marque, R.A. and Tordo, P. (2006) *Journal of Organic Chemistry*, **71**, 7657–67.

85 Rockenbauer, A. and Korecz, L. (1996) *Applied Magnetic Resonance*, **10**, 29–43.

86 Rockenbauer, A., Nagy, N.V., François Le Moigne, F., Gigmes, D. and Tordo, P.

(2006) *Journal of Physical Chemistry. A*, **110**, 9542–8.
87 Ivanova, A.S., Terakh, E.I., Kndalintseva, N.V. and Grigor'ev, I.A. (2005) Abstracts of Conference SPIN-2005, September 20–24, Novosibirsk, Russia, p. 68.
88 Yue, S., Qian, S.Y., Kadiiska, M.B., Guo, O. and Mason, R.P. (2004) *Free Radical Biology and Medicine*, **36**, 1224–32.
89 Ehrenshaft, M. and Mason, R.P. (2005) *Free Radical Biology and Medicine*, **38**, 125–35.
90 Stolze, K., Rohr-Udilova, N., Rosenau, T., Roswitha, S. and Nohl, H. (2005) *Biochemical Pharmacology*, **69**, 1351–61.
91 Sar, S., Osz, E., József Jeko, J. and Hideg, K. (2005) *Synthesis*, **2**, 255–9.
92 Hay, A., Burkitt, M.J., Jones, C.M. and Hartley, R.C. (2005) *Archives of Biochemistry and Biophysics*, **435**, 336–46.
93 Bardelang, D., Rockenbauer, A., Karoui, H., Finet, J.-P. and Tordo, P. (2005) *Journal of Physical Chemistry. B*, **109**, 10521–30.
94 Tuccio, B., Lauricella, R. and Charles, L. (2006) *International Journal of Mass Spectrometry*, **252**, 47–53.
95 Mason, R.P. (2004) *Free Radical Biology and Medicine*, **36**, 1214–23.
96 Khramtsov, V.V., Berliner, L.J. and Clanton, T.L. (2000) Supramolecular structure and function 7, *Proceedings of the International Summer School on Biophysics*, 7th (ed. G. Pifat-Mrzljak), Rovinj, Croatia, pp. 14–25.
97 Forrester, A.R. and Hepburn, S.P. (1971) *Journal of the Chemical Society. Chemical Communications*, **4**, 701–3.
98 Hanna, P.M., Chamulitrat, W. and Mason, R.P. (1992) *Archives of Biochemistry and Biophysics*, **296**, 640–4.
99 Janzen, E.G., Wang, Y.Y. and Shetty, R.V. (1978) *Journal of the American Chemical Society*, **100**, 2923–5.
100 Potapenko, D.I., Bagryanskaya, E.G., Tsentalovich, Vladimir A., Reznikov, V.A., Clanton, T.L. and Khramtsov, V.V. (2004) *Journal of Physical Chemistry. B*, **108**, 9315–24.
101 Sang, H., Janzen, E.G. and Lewis, B.H. (1996) *Journal of Organic Chemistry*, **61**, 2358–63.
102 Eberson, L. (2000) Spin trapping: problems and artefacts, in *Toxicology of the Human Environment. The Critical Role of Free Radicals* (ed. C.J. Rhodes), Taylor and Francis, New York, pp. 25–47.
103 Hanna, P.M., Chamulitrat, W. and Mason, R.P. (1992) *Archives of Biochemistry and Biophysics*, **296**, 640–4.
104 Janzen, E.G., Wang, Y.Y. and Shetty, R.V. (1978) *Journal of the American Chemical Society*, **100**, 2923–5.
105 Alberti, A., Carloni, P., Eberson, L., Greci, L. and Stipa, P. (1997) *Journal of the Chemical Society, Perkin Transactions 2*, **2**, 887–92.
106 Buettner, G.R., Mason, R. and Ronald, P. (2003) Spin-trapping methods for detecting superoxide and hydroxyl free radicals *in vitro* and *in vivo*, in *Critical Reviews of Oxidative Stress and Aging* (eds R.G. Cutler and H. Rodriguez), pp. 27–38.
107 Villamena, F.A., Hadad, C.M. and Zweier, J.L. (2005) *Journal of Physical Chemistry. A*, **109**, 1662–74.
108 Vanin, A.F., Huisman, A. and Van Faassen, E.E. (2002) *Methods in Enzymology*, (Nitric Oxide, Part D), Elsevier, **359**, 27–42.
109 Dikalov, S.F. (2005) *Methods in Enzymology*, (Nitric Oxide, Part E), **396**, 597–610.
110 Lim, M.H., Kuang, C. and Lippard, S.J. (2006) *Advances in Carbohydrate Chemistry and Biochemistry*, **7**, 1571–6.
111 Hen, O., Uzlander, N., Priel, Z. and Likhtenshtein, G.I., *Journal of Biochemical and Biophysical Methods* (in press).
112 Parkhomyuk-Ben Arye, P. (2004) *Development of multifunctional fluorescent proes for analysis of nitric oxide and antioxidants*, Doctor of Phylosophy Thesis, Ben-Gurion University of the Negev, Beer-Sheva, Israel (in English).
113 Villamena, F.A., Merle, J.K., Hadad, C.M. and Zweier, Jay L. (2005) *Journal of Physical Chemistry. A*, **109**, 6083–8.
114 Finkelstein, E., Rosen, G. and Rauckman, E. (1980) *Journal of the American Chemical Society*, **102**, 4995–5002.
115 Villamena, F.A. and Zweier, J.L. (2002) *Journal of the Chemical Society, Perkin Transactions*, **2**, 1340–4.

116 Villamena, F.A., Xia, S., Merle, J.V., Lauricela, R., Tuccio, B., Habad, C., Zweiz, J.L. (2007) *Journal of the American Chemical Society*, **129**, 8177–91.

117 Goldstein, S., Rosen, G.M., Russo, A. and Samuni, A. (2004) *Journal of Physical Chemistry. A*, **108**, 6679–85.

118 Samouilov, A., Roubaud, V., Kuppusami, P. and Zweier, J.L. (2004) *Analytical Biochemistry*, **334**, 145–54.

119 Villamena, F.A., Rockenbauer, A., Gallucci, J., Velayutham, M., Hadad, C.M. and Zweier, J.L. (2004) *Journal of Organic Chemistry*, **69**, 7994–8004.

120 Allouch, A., Roubaud, V., Lauricella, R., Bouteiller, J.-C. and Tuccio, B. (2005) *Organic and Biomolecular Chemistry*, **3**, 2458–62.

121 Bystryak, I.M., Likhtenshtein, G.I., Kotelnikov, A.I., Hankovsky, O.H. and Hideg, K. (1986) *Russian Journal of Physical Chemistry*, **60**, 1679–983.

122 Blough, N.V. and Simpson, D.J. (1988) *Journal of the American Chemical Society*, **110**, 1915–17.

123 Herbelin, S.E. and Blough, N.V. (1998) *Journal of Physical Chemistry. B*, **102**, 8170–6.

124 Rubtsova, E.T., Fogel, V.R., Khudyakov, D.V., Kotel'nikov, A.I. and Likhtenshtein, G.I. (1993) *Biofizika*, **38**, 211–21.

125 Pou, S., Huang, Y.I., Bhan, A., Bhadti, V.S., Hosmane, R.S., Wu, S.Y., Cao, G.L. and Rosen, G.M. (1993) *Analytical Biochemistry*, **212**, 85–90.

126 Fogel, V.R., Rubtsova, E.T., Likhtenshtein, G.I. and Hideg, K. (1994) *Journal of Photochemistry and Photobiology A: Chemistry*, **83**, 229–36.

127 Lozinsky, E.M., Martin, V., Berezina, T., Shames, A.I., Weism, A. and Likhtenshtein, G.I. (1999) *Journal of Biochemical and Biophysical Methods*, **38**, 29–42.

128 Lozinsky, E., Novoselsky, A., Shames, A.I., Saphier, O., Likhtenshtein, G.I. and Meyerstein, D. (2001) *Biochimica et Biophysica Acta*, **1526**, 53–60.

129 Lozinsky, E., Novoselsky, A., Glaser, R., Shames, A.I., Likhtenshtein, G.I. and Meyerstein, D. (2002) *Biochimica et Biophysica Acta*, **1571**, 239–44, 29.

130 Silberstein, T., Shames, A.I., Likhtenshtein, G.I., Maimon, E., Mankuta, D., Mazor, M., Katz, M., Meyerstein, D. and Meyerstein, N. Saphier, O., (2003) *Free Radical Research*, **37**, 301–8.

131 Parkhomyuk-Ben Arye, P., Strashnikova, N. and Likhtenshtein, G.I. (2002) *Journal of Biochemical and Biophysical Methods*, **51**, 1–15.

132 Medvedeva, N., Martin, V.V. and Likhtenshtein, G.I. (2004) *Journal of Photochemistry and Photobiology A: Chemistry*, **163**, 45–51.

133 Packer, L. (1994) Nitric oxide, in *Methods in Enzymology*, Academic Press, San Diego, pp. 739–49.

134 Stuehr, D.J. and Ghosh, S. (2000) Enzymology of nitric oxide synthases, in *Handbook of Experimental Pharmacology*, 143 (Nitric Oxide), Springer, Berlin, pp. 33–70.

135 Utepbergenov, D.I., Khramtsov, V.V., Vlassenko, L.P., Markel, A.L., Mazhukin, D.G., Tikhonov, A.Y. and Volodarsky L.B. (1995) *Biochemical and Biophysical Research Communications*, **214**, 1023–32.

136 Borisenko, G., Martin, I., Zhao, Q. and Kagan, V.E. (2004) *Journal of the American Chemical Society*, **126**, 9221–32.

137 Bian, Z.-Y., Guo, X.-Q., Zhao, Y.-B. and Du, J.-O. (2005) *Analytical Sciences: The International Journal of the Japan Society for Analytical Chemistry*, **21**, 553–9.

138 Coenjarts, C., García, O., Llauger, L., Palfreyman, J., Vinette, A.L. and Scaiano, J.C. (2003) *Journal of the American Chemical Society*, **125**, 620–1.

8
Nitroxides in Physicochemistry
Gertz I. Likhtenshtein

8.1
Polymers

8.1.1
Introduction

Polymer materials are widely employed in industry, technology, medicine, and people's everyday life. The spin-labeling method has proved to be one of the most effective tools for the study of molecular dynamics and microstructure of polymers. Nitroxides are effective regulators of radical polymerization, and modern chemistry allows us to modify any macromolecules by chemical covalent bonds, absorption, or solid-phase mechanochemical incorporation, and provides a wide arsenal of nitroxides and related compounds for polymerization. In this section we will briefly describe nitroxide applications in the following areas of polymer science: (1) polymerization and its regulation; (2) intermolecular dynamics; (3) microstructure and heterogeneity; (4) polymer complexes; (5) chemical and photodegradation; and (6) miscellaneous aspects.

8.1.2
Polymerization: Nitroxides Mediated Living Polymerization (NMLP)

The first observation of inhibition of radical polymerization with nitroxide mono- and biradicals was made in 1966 by the M.B. Neiman group [1]. The inhibiting effect of nitroxides on styrene polymerization at 50 °C, initiated by azodiisobutyronitrile, was interpreted as a recombination of nitroxides and polymer radicals, which led to the termination of the process. The linear termination rate constant for different radicals was found to be $2.1–3.2\,10^4 \times M^{-1}s^{-1}$. The application of nitroxides as inhibitors and regulators of polymerization provides an opportunity for process monitoring by ESR techniques and its regulation by the use of nitroxides of various structures. In this section we will concentrate on recent advances in this field which are important in basic and applied aspects.

Nitroxides: Applications in Chemistry, Biomedicine, and Materials Science
Gertz I. Likhtenshtein, Jun Yamauchi, Shin'ichi Nakatsuji, Alex I. Smirnov, and Rui Tamura
Copyright © 2008 WILEY-VCH Verlag GmbH & Co. KGaA, Weinheim
ISBN: 978-3-527-31889-6

8.1.2.1 Phenomenon of and Chemistry of Nitroxide Mediated Living Polymerization

A new era in nitroxide-mediated polymerization started when Solomon, Richardo, Griffiths, and Cacioli [2, 3] showed that it was possible to prepare well-controlled and living (homo-, co)polymer by radical polymerization in the presence of a nitroxyl radical as controlling agent. According to the authors, besides the "regular" steps of the radical polymerization inhibited by nitroxide (NRO·), that is, initiation, polymeric chain prolongation, polymeric radical (PR·) recombination or disproportion, and the formation of polymeric alkoxyamine (PR–NRO), a new key reaction observed is the *reversible* thermal C–O bond homolysis of polymeric alkoxyamines PR–NRO **(R–Y)** with a regeneratrion of PR· **(R·)** and NRO· **(Y)** included (Figure 8.1). This phenomenon was named nitroxide-mediated living polymerization (NMLP) [4–6].

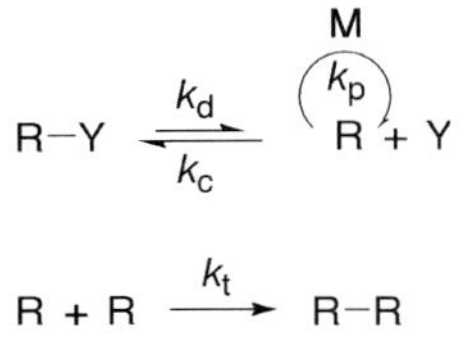

$$R-Y \underset{k_c}{\overset{k_d}{\rightleftharpoons}} \overset{M\ (k_p)}{R\cdot + Y}$$

$$R + R \xrightarrow{k_t} R-R$$

Figure 8.1 Principle reactions in nitroxide mediated living polymerization [5]. (Reproduced with permission).

where k_d and k_c are the rate constants of decomposition and composition of the compounds, respectively; k_t is rate constant of the polymer radical's recombination or disproportion.

If the equilibrium is shifted to alkoxyamine, it leads to a low concentration of free radicals and ensures a low fraction of irreversible termination via polymer radical dimerization/disproportionation processes. This eventually leads to prolongation of the polymeric chains and low polydispersities. The special advantages of NMLP are that it opens wide an opportunity to control polymerization by affecting the rate of formation and dissociation of the polymeric alkoxyamine.

During the last two decades nitroxide-mediated controlled free radical polymerizations have been intensively investigated [6–26].

Several series of compounds are currently employed in NMLP; that is, nitroxides, amioxylamins, nitrons, aromatic and aliphatic nitroso derivatives, and aliphatic amines. Figure 8.2 gives examples of the chemical structures of nitroxides and alkoxyamines reproduced from [21]. The aim of investigations in this field is to produce polymeric products possessing optimal properties such as homogeneity, definite chain length and texture at relatively low temperature, and production in short time and low mediator concentrations. Reducing temperature is also important for preventing side oxidative processes. To advance this aim it is necessary to have an optimal combination of rate constants for the stages shown in the Figure 8.1, and formation and dissociation of alkylamines in particular. The commonly accepted effective means for controlling the process is the use of sterically

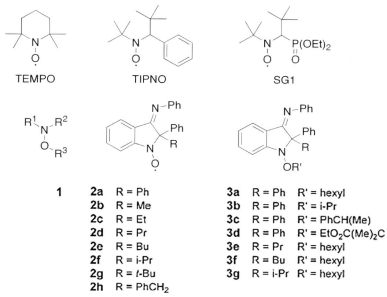

Figure 8.2 Chemical structures of nitroxides and aminoxylamines [21]. (Reproduced with permission).

hindered mediators and phosphorous-containing nitroxides, which react sufficiently fast with polymeric radicals but at the same time do not form very stable alkoxyamines. Other variable parameters of NMLP are temperature, solvent, and specific technological conditions.

Each mediator is characterized by a specific temperature range for the definite NMLP process. For example, TEMPOL in sterene polymerization below 100 °C operates as an inhibitor, in the range 100–150 °C, as a mediator, and above 160 °C the process acquires an uncontrolled-free-radical character [14, 15].

The effectivness of alkoxyamine-mediated steren polymerization essentially depends on its structure. For example, it was found [16] that for sterically highly hindered alkoxyamines such as 2,6-bis(*tert*-butyldimethylsilanyloxy methyl)-2,6-diethyl-1-(1-phenylethoxy)piperidine, 2,2,6,6-tetraethyl-4-methoxy-1-(1-phenylethoxy)piperidine, and the open-chain alkoxyamine *N*-*tert*-butyl-*N*-(2-methyl-1-phenylpropyl)-*O*-(1-phenylethyl)hydroxylamine (styryl-TIPN), the equilibrium constants K are three orders of magnitude larger than the K of styryl–TEMPO. As a result, increase of the polymerization conversion by about a factor of 2 was achieved with only 25% of the alkoxyamine additive with respect to styryl–TEMPO. The use of a new tertiary SG1-based alkoxyamine instead of the secondary SG1-based alkoxyamine improved the polymerization of both styrene and butyl acrylate [9]. The polymerization of butyl acrylate can be carried out at 90 °C as compared with $T > 120$ °C for TEMPOL. Use of SG1-based alkoxyamine offered a lowering of the methyl methacrylate polymerization temperature to 45 °C. In an attempt to optimize the processes of radical polymerization, besides cyclic aliphatic nitroxides, several series of nitrogen-containing compounds such as nitrons, nitroso derivatives, aliphatic and aromatic amins, and nitronyls have been suggested.

Nitrons were suggested and employed as new regulation agents for NMLP [8, 15]. The application of nitrons expands the arsenal of potential mediators and allows us to perform polymerization at relatively low temperatures, and to monitor the process of additive formation by ESR spectroscopy. C-phenyl-N-tert-butylnitrone promoted butyl acetate and methacrylate polymerization at 50–60 °C, that is a much milder regimen compared to TEMPO and other analog nitroxides. Synthesis of poly(N-vinylpyrrolidone) and N-vinylpyrrolidone-methyl methacrylate copolymers was also accomplished in the presence of C-phenyl-N-tert-butylnitrone at 60 °C [8]. It was possible to synthesize poly(N-vinylpyrrolidone) and its copolymers without any gel effect and to control molecular-mass characteristics of homo- and copolymers.

It was shown that N-tert-butyl-α-isopropylnitrone pre-reacted with the radical initiator, and mediated the radical polymerization of styrene, styrene/acrylonitrile mixtures, and dienes [17]. Benzoyl peroxide-initiated polymerization of styrene mediated by phenyl-tert-butylnitrone was studied by isothermal calorimetry, ESR, and gel permeation chromatograph [18]. At 120 °C the reaction follows the mechanism of reversible inhibition of growing radicals by macromolecular nitroxides resulting from the trapping of primary growing radicals by the nitrone. In a mixture of styrene and butyl acrylate the corresponding block copolymer was obtained. Copolymerization of styrene with acrylonitrile at 100 °C in the presence of 2-methyl-2-nitrosopropane was characterized by the absence of the gel effect, relatively low polydispersity of the reaction product, and a shift in molecular mass distribution curves to high molecular masses [19].

8.1.2.2 Thermodynamic and Kinetics

Numerous studies on the mechanism and the kinetics of nitroxide-mediated polymerization (NMP) have been carried out after Fischer's group works which showed that the kinetics of NMLP were based on the persistent radical effect (PRE) (Figure 8.1). The proposed kinetic model formed a basis for the development of NMLP theory.

Several works were related to the theoretical calculation of the bond energy of dissociation (BDE) of (C)O–N and (N)O–C bonds in alkoxyamine compounds. Figure 8.2 shows the chemical formulas of a series of PR–NROs. According to [20], the structures of the nitroxide ande of the alkoxy fragment of the polymer radicals markedly affect BDE, and the main factor affecting BDE is most probably steric hindrances in the alkoxy compounds. In [21] the lability of the N–O(C) and (N)O–C bonds in a series of stable 2,2-disubstituted 3-(phenylimino)indol-1-oxyls, the alkoxyamines, was compared. According to theoretical calculation, alkoxyamines with a primary- or secondary-alkyl group bound to the O-atom of the nitroxide function (hexyl and i-Pr) mainly underwent N–O bond homolysis. When the O-alkyl, radical was a tertiary or a benzyl group (crotonyl or styryl), O–C bond cleavage occurred as the main process. The authors suggested that alkoxyamine decomposition may occur in two different ways as a result of a competition between (N)O–C (BDE 175–102 kcal mol^{-1}) and (C)N–O (136–126 kcal mol^{-1}). As expected, the steric effect is stronger in C–O bond cleveage. In reference [22] it was found that calculated BDE of C–O bonds decreased in the following sequence: 2,2,6,6 tetra-methyl-4-oxopiperidinyloxy (TEMOPO) > polyvinyl chloride > polybutyl acetate > polysteryne > polymethyl methacrylate. BDE for the C–O bond between sterene and nitroxide in corresponding derivatives gave a sequence as follow: TEMPO > 2-methyl-2-nitroso propane > C-phenyl-tret-butyl nitron.

Multiparameter analysis of stereoelectronic effects of Me (17 kJ mol^{-1}) in alkoxyamine C–ON bond homolysis in the TEMPO-based alkoxyamine and SG1 [N-tert-butyl-N-(1-diethoxyphosphoryl-2,2-dimethyl-propyl)aminoxyl] series was performed [23]. The analysis showed that the difference in reactivity between diastereoisomers is caused by an $n\sigma \rightarrow \sigma$ interaction between the $n\sigma$ lone pair of the oxygen atom of the ester bond and the σ orbital of the cleaved O–C bond. The following values of the rate and equilibrium constants in the process for sterine mediated by TEMPO were reported [24]: $k_d = 5.5 \times 10^{-4}$ s^{-1} at 114 °C, $k_c = 7.6 \times 10^7$ M^{-1} s^{-1} at 125 °C, $K = 7.8 \times 10^{-12}$ M at 114 °C, $K = 2.1 \times 10^{-11}$ M at 125 °C, the rate constant of polymeric charge prolongation $k_p = 1740$ M^{-1} s^{-1}, and termination $k_t = 5 \times 10^8$ M^{-1} s^{-1} [24]. Kinetic investigation of nitroxide-mediated living polymerization of various polymers and additives [10] gave the following rate constants for reaction polymer radicals, k_c, with nitroxides at 12 °C: polysteryl radical and TEMPO (1.4×10^7 M^{-1} s^{-1}), methyl acrylate and TEMPO (2.5×10^7 M^{-1} s^{-1}), polysteryl radical and 2-methyl-2-nitroso propan (0.2×10^7 M^{-1} s^{-1}), PMMA radical and TEMPO (0.16×10^7 M^{-1} s^{-1}), polysterene substituted macronitroxyl with tret-butyl nitroso terminal group (0.07×10^7 M^{-1} s^{-1}), and 4-linoleoamido TEMPO (0.4×10^7 M^{-1} s^{-1}). For the copolymerization process of sterene with other monomers the following values of k_c were found for the formation alkoxyamines: methyl acrylate (0.8×10^7 M^{-1} s^{-1}), vinyl acetate (1.4×10^7 M^{-1} s^{-1}), methacrylo nitryl 0.15×10^7 M^{-1} s^{-1}), butylacrylate 0.8×10^7 M^{-1} s^{-1}.

The classic Fischer equations [5] are not precise for systems when the amount of persistent radical produced exceeds ~10–15%. In the work [25], an analysis was made of NMLP systems, to provide deeper insight into the operation of PRE by use of Predici simulations. The NMLP kinetic process was divided into three stages, that is, preequilibrium, transition period, and quasi-equilibrium, which were analysed separately. The kinetics of NMLP of styrene and methyl methacrylate mediated by N-tert-butyl-N-[1-diethylphosphono-(2,2-dimethylpropyl)] nitroxide (SG-10) were evaluated. It was shown that the Fisher equations are fulfilled for $k_c = 5.0 \times 10^5$ M^{-1} s^{-1} and $k_t = 2.5 \times 10^9$ M^{-1} s^{-1}; that is, $K_{eq} = 2.0 \times 10^{-8}$ M $< I_0 k_c / 16 k_t$ $= 6.2 \times 10^{-7}$ M. A difference in the plots for the classic Fischer's equation and a new equation for the system was stressed. After equilibrium is established, new equations were suggested which are applicable up to very high conversion and essentially infinite reaction time.

In the reference [26], technological aspects of NMLP were discussed. In particular, optimal operating policies for the industrial-level semibatch living free radical polymerization of styrene were proposed. The optimization performed involved the determination of design parameters such as the initial holdup, feed stream temperature, and concentrations required for a large-scale nonlinear program. The resulting optimal operating conditions led to a polymer featuring living characteristics matching monomer conversion, molecular-weight distribution, and polydispersity index target values.

Strategies to improve the performance of living NMLP were discussed [24]. The following ways to enhance the process rate were kinetically analysed: (1) an additional initiation; (2) enhanced monomer conversion rates; and (3) increase of the initial excess of the persistent species.

8.1.3
Molecular Dynamics and Microstructure of Polymers

8.1.3.1 Introduction

A wide diversity of chemical, physical, and technological characteristics of polymers is affected, to a great extent, by molecular dynamic behavior. As early as 1968 Neiman, Buchachenko, and Wasserman first demonstrated the powerful potential of nitroxide spin-label methods when investigating molecular dynamics and the microstructure of high molecular mass compounds [27–29]. Since then nitroxides have found great application in investigations of polymers. These investigations involve the modification of a polymer of interest by nitroxides and measurement of the nitroxide's molecular dynamic parameters by ESR techniques in various conditions. An entire arsenal of experimental and theoretical means of modern ESR and related methods (Chapters 3, 4, and 6) have been employed to do this.

The first studies of the dynamic properties of a series of polymers (polystyrene, atactic and isotactic polypropylene, polybutadiene rubber, polyisobutylene, natural rubber, and butadiene rubber, etc.) using nitroxide probes (3,3,5,5-tetramethyl-4-oxopiperidinooxy and 1,1,3,3,5-pentamethyl-1,3,4,4a,5,9b-hexahydro-2H-pyridol[4,3-b]indol-2-yloxy) indicated a correlation between the apparent activation parameters of the nitroxide rotational diffusion and the corresponding parameters of the polymers' segmental motion detected by NMR [27]. Further investigations elicited a number of structural and dynamic properties for the polymer and its complexes, such as heterogeneity, local polarity and acidity, conformational and phase transitions, and so forth.

Both methods of polymer modification by nitroxides, covalent linking and adsorption, have been employed. A few typical examples of hydrophobic spin-probes and modified polymers [30] are presented here.

4: n = 5 (5DSA)
5: n = 16 (16DSA)

PVP*-

PVP*-

PMAA

8.1.3.2 Polymers Segmental Dynamics

In the investigation of a polymers' molecular dynamics, a set of dynamic parameters such as rotational correlation time, rotation energy, and entropy activation and its distribution and spin relaxation times are commonly measured. After the pioneering works of the Neiman group [27–29], multiple results of investigations of segmental dynamics of various homo- and copolymers were reported. In this section we describe several typical recent results in polymer segmental dynamics.

Segmental dynamics of the vesicle and aqueous phases of mixed membranes made from poly(oxyethylene) hydrogenated castor oil and hexadecane labeled with TEMPO and di-*tert*-Bu nitroxide (DTBN) were investigated by ESR spectroscopy [31]. Detection of partition coefficients, rotational correlation times, and the rotation activation energies (E_a) of the spin-probes in the vesicle phase indicated a similarity in dynamic behavior of both nitroxides. The E_a values obtained by simulation for DTBN and TEMPO were found to be 21 and 20 kJ mol^{-1}, respectively. Dynamic heterogeneity in the interfacial region of microphase-separated polystyrene-block-poly(methyl acrylate) (PS-block-PMA) was studied by the spin-label technique [32]. The distribution of the motional correlation time (τ_c) in the interfacial region was much broader than that in the homopolymers. The glass transition temperature, T_g, in PS-block-PMA, detected by appearance in label ESR spectra hyperfine splitting of 5 mT, was found to be almost the mean value of those of the spin-labeled PS and PMA homopolymers. The polymer's molecular mass strongly influenced the interfacial thickness, the segmental mobility, the width of the distribution of τ_c in the interfacial region, and the T_g of the microdomains.

Molecular mobility in regular and irregular polycarbosylan dendritic macromolecules of various generations was studied by the spin-probe technique [33]. It was shown that an increase in the molecular mass of regular polycarbosilane dendrimers is accompanied by a stepwise reduction in the rotational mobility of spin-probes introduced into these polymers. Within the temperature interval from −60 to +30 °C, a correlation was observed between a change in the mobility of a spin-probe incorporated into dendritic macromolecules and coefficients of permeability of simple gases (O_2, CO_2) through these polymers.

To understand the self-assembling structure and dynamics in poly(ethylene-co-methacrylic acid) (EMAA) ionomers in membranes and water dispersions, the polymer was modified by doxyl steric acids (nDSA) with the doxyl group attached to carbon atoms at different positions relative to the head group, and a probe with the nitroxide group attached to a nonadecane backbone [34]. The ESR spectra of the modified EMMA (Figure 8.3) exhibited inner and outer extremes that indicated two types of probe motion; rotational diffusion in a slow region and fast small-angle wobbling (see Chapter 6). By examination of ESR spectra, detailed information on the local environment of the probes, chain mobility, and structural heterogenity in the vicinity of the ionic aggregates in the polymer was obtained.

EPR-spectroscopy of spin-labels has turned out to be very informative when investigating self-associating polymer systems, in particular complexes of polyelectrolytes with surface active substances (SAS) [30]. Application of this method

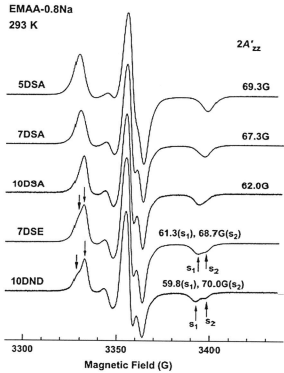

Figure 8.3 X-band ESR spectra at 293 K for the indicated spin probes in EMAA-0.8Na. Spectra were normalized to a common microwave frequency (9.43 GHz). The extreme separation, $2A'_{zz}$, is shown on the right. Thin and thick arrows point to the s_1 and s_2 spectral components for 7DSE and 10DND [34]. (Reproduced from Kutsumizu, S. and Schlick, S. (2005) Structure and dynamics of ionic aggregates in ethylene ionomer membranes: recent electron spin resonance (ESR) studies. *J. Mol. Struct.* **739**, 191–198, with permission of Elsevier).

allowed us to establish both particularities of molecular dynamics and the local organization of SAS molecules, and also to segment the mobility of macromolecules in such complexes [30]. The effect of formation of complexes of poly-N-ethyl-4-vinylpyridinium bromide (PEVP) with sodium dodecylsulfate (SDD) at various mole ratios $Z = [SDD]/[PEVP]$ is shown in Figure 8.4.

As we can see in Figure 8.4, an increase in the ratio $Z = [SDD]/[PEVP]$, where SDD is sodium dodecylsulfate and PEVP is poly-N-ethyl-4-vinylpyridinium bromide, leads to marked immobilization of the probe [30].

Several publications relate to the investigation of the interaction between organic polymers and inorganic materials. The structural–dynamic relationships of interphase layers in particles of mineral filler due to chemical bonds or adsorptive forces, was investigated by the spin-probe method ([35] and references therein). A

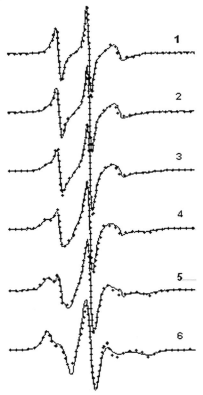

Figure 8.4 Experimental (firm line) and calculated (dotted line) EPR-spectra of spin-marked PEVP* (P_w = 1000) in solution (1) and in complexes poly-N-ethyl-4-vinyl-pyridinium bromide (PEVP) with sodium dodecylsulfate (SDD) at various mole ratios Z = [SDD]/[PEVP]: Z = 0,1 (**2**); Z = 0,2 (**3**); Z = 0,3 (**4**); Z = 0,4 (**5**); Z = 1,0 (**6**) [30]. (Reproduced with permission).

correlation between vitrification temperatures of the polymer matrix and the ratio between segment mobility of chains in the boundary layer and in the bulk of the polymer has been established. Nanocomposites of poly(Me acrylate) (PMA) with synthetic fluoromica (Somasif) as the inorganic component were studied as a function of clay content using a spin-label ESR (ESR) technique, X-ray diffraction (XRD), and differential scanning calorimetry (DSC) [36]. The spin-probe's ESR spectra indicated that the mobility of PMA chains in the nanocomposites is constrained due to interactions in the interface region. On the basis of the deconvolution of ESR spectra measured as a function of temperature into slow (S) and fast (F) components, the average thickness of the rigid interface in the nanocomposites was established to be in the range 5–15 nm. The average interface thickness decreased with increasing Somasif content.

8.1.3.3 Spatial and Orientational Distribution of Nitroxides

A method of EPR-tomography, developed in [37–39], allows both detection of molecular mobility and its change due to thermo- or photo-destruction of the polymer at various points and registration of the distribution of oxidation active sites through the sample. This method made possible identification of polymer parts in which the destruction process proceeds. Solution of this problem is of great importance for the selection of conditions for exploitation of polymer materials. In work [39] the thermo- and photo-oxidation of poly(acrylonitrile-butadiene)styrene (ABS) copolymer was studied by EPR-tomography. It was shown that in a given case the process of destruction proceeds mainly near the surface; in the middle of the sample the process of oxidative destruction practically does not proceed. The spatial distribution of stabilizer-derived nitroxide radicals during thermal degradation of poly(acrylonitrile-butadiene-styrene) copolymers was investigated by pulsed Double Electron-Electron Resonance (DEER) and ESR imaging (ESRI) [40]. DEER provided information on the spatial distribution of radicals on the length-scale of a few nanometers, whereas ESRI showed a length-scale of millimetres with a resolution of about 100 µm. The Hindered Amine Stabilizer (HAS) Tinuvin 770 was incorporated in poly(acrylonitrile-butadiene-styrene) (ABS) copolymers and the distribution of local concentrations of the monoradical derived from bifunctional HAS was measured.

A method based on computational spectra simulation has been developed to determine the orientation distribution function (ODF) of anisotropic paramagnetic species by analysis of the angular dependence of the ESR spectra [41]. The method was used for determination of the orientation distribution functions for radical probe 2,2,6,6-tetramethyl-4-ol-piperidinooxyl in 4-n-amyl-4′-cyanobiphenyl aligned by magnetic field and 2-septadecyl-2,3,4,5,5-pentamethylimidazolidine in polyethylene stretched films (Figure 8.5).

8.2
Nitroxides in Photochemistry and Photophysics

The main application trends for nitroxides and nitrons in photochemistry and photophysics are the following: (1) studies of fluorescence quenching and electron transfer mechanisms; (2) the use of nitroxides and nitrons as spin-traps in photochemical processes and photoswitching magnetic systems; and (3) modeling multispin processes consisting of a fragment of metal complexes or highly conjugated compounds.

8.2.1
Fluoresence Quanching, Photoelectron Transfer and Photoreduction

8.2.1.1 Duel Fluorophore–Nitroxide Compounds

For the study of mechanisms of intermolecular fluorescence quenching, electron transfer, and nitroxide photoreduction a method based upon the use of dual

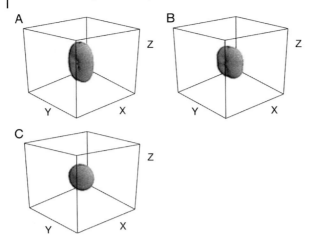

Figure 8.5 The orientation distribution functions of TEMPOL in aligned 4-n-amyl-4′-cyanobiphenyl determined for unannealed sample (−196 °C) (A) and for samples annealed at temperatures of −78 °C (B) and −37 °C (C) [41]. (Reprinted from Vorobiev, A. Kh. and Chumakova, N. A. (2005) Orientation distribution of nitroxides in polymers, *J. Magn. Reson.*, 175, 146–157, with permission from Elsevier).

fluorophore–nitroxide compounds (FNRO) has been suggested. A FNRO· consists of two-molecular subfunctionalities (a fluorescent chromophore and a stable nitroxide radical) tethered together by a spacer. Starting from the first work of the G.I. Likhtenshtein group [42], the ability of stable nitroxide radicals to act as quenchers of excited species was intensively exploited as the basis of several methodologies which include redoxprobing, spin-trapping, modeling intramolecular photochemical and photophysical processes, and the construction of new magnetic materials [43–51] (see also Chapters 5 to 8).

Probing biological and nonbiological environments with dual fluorophore–nitroxide molecules in which a fluorophore is tethered with nitroxide, a fluorescence quencher, opens unique opportunities to study molecular dynamics and the micropolarity of the medium which affects intramolecular fluorescence quenching (IFQ), electron transfer (ET), and photoreduction. In such molecules, the excited fragment of the chromophore can serve as an electron donor, and the nitroxide fragment as an electron acceptor. The same groups allow the monitoring of molecular dynamics and also make it possible to measure the micropolarity of the medium in the vicinity of the donor (by a fluorescence technique) and acceptor (by electron spin resonance) moieties. Thus, the dual compounds, having the properties of flourescence and nitroxide spin-probe at the same time have new principle advantages in the investigation of photochemical and photophysical processes. The organic synthetic chemistry allows us to play with the chemical structure of the dual molecules to give different absorption, fluorescence and ESR spectra, and redox properties with a variety of bridges (spacers) tethered to the chromophore and nitroxide segments (see Chapter 5).

As was first shown in [42], in a dual fluorophore–nitroxide compound (FNRO), the nitroxide is a strong fluorescent quencher. Irradiation of the chromophore segment of a (FNRO) Dansyl–TEMPO in a glassy liquid (glycerol, 75%, water 20%, ethanol 5%) invoked production of the hydroxylamine derivative accompanying a decay of the nitroxide ESR signal and a parallel eightfold increase of fluorescence (Figure 7.11). Both processes run with the same rate constant the nitroxide segment photoreduction k_{pred} under identical conditions. The temperature dependence of k_{pred} was investigated in the range 77–320 K. Experiments have shown that the k_{pred} values drastically increased when the temperature was increased, starting from 210 K. The k_{pred} change correlated with an animation of the nanosecond relaxation dynamics in media monitored by the fluorescence and ESR techniques, while the rate constant of the intramolecular fluorescence quenching, k_q, was found to be temperature-independent. Similar kinetic and molecular dynamic behavior was observed in bovine serum albumin (BSA) and human serum albumin (HSA) in which the hydrophobic site was modified with Dansyl–TEMPO and its fatty acid derivative [43–47]. In the case of the Dansyl–TEMPO fatty acid derivative, nanosecond scale relaxation of the protein around excited dansyl segments incorporated in the site was directly monitored by a picosecond time-resolved fluorescence technique [47].

In order to establish a mechanism of intramolecular fluorescence quenching (IFQ) and photoreduction of the nitroxide segment in dual molecules, a series of dansyl–nitroxides of different structures and flexibility of spacer group, and different redox potential of nitroxide was synthesized and investigated (Figure 6.9) [45].

The rate constants of IFQ (k_q), quantum yields of paramagnetic and diamagnetic forms, and rate constant of photoreduction (k_{pred}) in various solvents for a series of dansyl compounds containing various nitroxides were measured. Figure 8.6a and b shows the positive correlation between the rate constant of the nitroxide fragment photoreduction k_{pred} and the equilibrium constant K_{eq} for the chemical exchange reaction between different nitroxides depending on the nitroxide redox potential established. Nevertheless, the k_q values were found not to be dependent on K_{eq}.

On the basis of these and other available data, two mechanisms of IFQ were proposed: the major mechanism, Intersystem Crossing (IC) and the minor mechanism, reversible intramolecular electron transfer (ET) from the excited singlet of the fluorophore (donor D) to nitroxide (acceptor A) followed by fluorophore segment regeneration and hydroxyl formation [45]. The latter mechanism is responsible for photoreduction. The authors of work [48] also came to the conclusion that the fluorescence-quenching in a series of dual compounds arises through electron exchange which causes relaxation of the singlet state to the triplet and/or ground state of the fluorophore due to the effects of nitroxide segments.

Based on these experimental data on local polarity and molecular dynamics in the vicinity of donors and acceptors in dual molecules incorporated in BSA and HSA, the following parameters of the Marcus theory of electron transfer were estimated for electron transfer at $T = 300$ K: the Gibbs energy $\Delta G_0 = -1.75$ eV, the reorganization energy $E_r = 0.9$ eV, and activation energy $E_a \approx 0.25$ eV [42–44, 46,

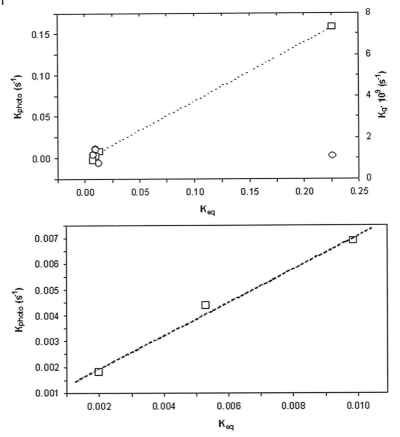

Figure 8.6 a,b Dependence of rate constant of photoreduction, k_{photo}, of compound **4** on redox power of nitroxide fragment, K_{eq}, (squares) contrary to independence of rate constant of IFQ, k_q, (circles) (G. I. Likhtenshtein, private communication).

47]. Such a set of parameters is closely related to the inverted Marcus region and should run with a rate which is essentially lower than that expected from experimental data. Therefore, the reversible electron transfer in the dual molecules cannot be a major mechanism of fluorescence-quenching. However, charge photoseparation is still a necessary step in nitroxide radical photoreduction. Thus it was suggested ET occurs at the expense of vibrational stabilization of the photoseparated ionic pair (D^+A^-) and at the expense of partial stabilization owing to fast polar relaxation modes.

For the problem of utilization of light energy, donor–acceptor structures should be capable of retaining the photoseparated state long enough for the occurrence of secondary chemical reactions of those charges of special interest. The above-mentioned data demonstrated that nitroxide photoreduction in dual probes occurs without violation of the fluorophore structure. This process is, in fact, photoinduced electron transfer from molecules which are very weak reducing agents: glycerol, for example, transfers an electron to nitroxide with the formation of hydroxylamine derivatives (FNROH) with a moderate reducing power. Therefore, photochemical reactions in the dual molecules may be considered to be a process of light energy transfer.

It was demonstrated that a nitroxide (TEMPO) and oxygen in the ground triplet state inhibited back electron transfer in photoseparated donor–acceptor pairs in dual compounds by enhancing intersystem crossing of a singlet radical ion pair into its triplet state [49]. Paramagnetic species can induce intersystem crossing from a singlet state into the triplet state. Because triplet pair recombination to the ground singlet state is a spin-forbidden process, the lifetime of such an ion pair state is substantially longer. The effect can be observed only when the energy of the charge-separated state is lower than that of the locally excited triplet states. Because of the spin statistics, the reverse intersystem crossing is less efficient, allowing use of oxygen and other paramagnetic species to impede charge recombination in various electron-transfer systems. The phenomenon was demonstrated on a series of dual compounds (porphyrin/chlorine and fullerene C60). In the case of the system TEMPO–ZnChl-C_{60}, the intersystem crossing rate constant k_{isc} = $4.4 \times 10^9 \,M^{-1}\,s^{-1}$ and the rate constant of the complex TEMPO–ZnChl-C_{60} dissociation $k_{-d} = 7 \times 10^8\,s^{-1}$ were found. The authors suggested that similar mechanism may play a role in naturally occurring photoinduced electron-transfer reactions where oxygen or other paramagnetic species are involved. It can also be applied in manipulating outcomes in artificial photoinduced electron-transfer processes.

Photochemical stability of dual chromophore–imidazoline compounds was investigated by ESR and magnetic susceptibility [50]. The substances under investigation are divided into three groups by their properties and the way nitroxyl radicals are added: (1) rhodium complexes bearing two or three imidazoline ligands; (2) derivatives of fullerenes C_{60} and C_{70}; and (3) copper complexes with imidazoline moieties as bidentate ligands. Irradiation of rhodium imidazoline without oxygen complexes led to the reversible conversion of nitroxyl radicals into corresponding hydroxylamino derivatives. When oxygen presents in this system, hydroxylamino groups are transformed into nitroxyl groups. A similar effect of oxygen has been noted for nitroxyl derivatives of fullerenes as well. A similar effect of oxygen has also been noted for nitroxyl derivatives of fullerenes, C_{60} and C_{70}, with one, two or three imidazoline nitroxyl groups. On the contrary, copper complexes with nitroxyl ligands attached in a bidentate manner were found to be stable both with and without oxygen. Such behavior is attributed to the quenching effect of paramagnetic Cu^{2+} on the complex's exited states.

Photophysical and photochemical processes in fixed distance electron donor-chromophore–acceptor-TEMPO molecules were investigated in detail in work [51]. A donor-chromophore–acceptor (DCA) system, MeOAn-6ANI-Ph$_n$-A-NRO (where MeOAn = p-methoxyaniline, 6ANI = 4-(N-piperidinyl)naphthalene-l,8-dicarboximide, Ph = 2,5-dimethylphenyl (n = 0,1), and A = naphthalene-

1,8:4,5-bis(dicarboximide) (NI) or pyromellitimide (PI)) were synthesized. After DCA excitation in the triplet-exited state, processes in the photogenerated triradical system (MeOAn$^{+\cdot}$-6ANI-Ph$_n$-A$^{-\cdot}$- NRO·) were monitored using both time-resolved optical and EPR spectroscopy. It was shown that the lifetimes of charge recombination, τ_{CS}, and the coupling constant ($2J$) drastically depended on the compound's structure. Variation of the spacer between the donor and acceptor groups led to changing τ_{CS2} (ps) from 1.3 to 5000, τ_{CR} (ns) from 10.5 to 506, and the coupling constant $2J$ (mT) from <1 to 66. The NRO· moiety accelerates the radical pair intersystem crossing and therefore modulates the charge recombination rate within the triradical compared with the corresponding biradical lacking nitroxide. The radical markedly (15–2.5 times) enhanced the lifetime of charge recombination but only slightly affected charge separation and the coupling constant because the exchange interaction is strong only with the neighboring A$^{-\cdot}$ group.

8.2.1.2 Nitroxides in Multispin Systems

Electron exchange interactions with paramagnetic species generate the excited state relaxation of chromophores. The quenching mechanism originates from changes in the spin multiplicity of the electronic states [43, 52–64]. The singlet ground state (S_0) and the lowest excited singlet state (S_1) of the chromophore become the doublet (D_0 and D_n, respectively) states because of the unpaired electron spin of the doublet nitroxide radical (NRO). The lowest excited doublet (D_1) and quartet (QA_1) states are generated by an interaction between the NRO· and the T_1 chromophore. Thus, the spin-forbidden transitions of the chromophore, that is, $S_1 \to T_1$ and $T_1 \to S_0$, partially transform into the $D_n \to D_1$ and $D_1 \to D_0$ transitions, respectively (Figure 8.7).

A series of chromophore–nitroxide compounds has been synthesized and investigated (see, for example, Figure 8.8) The series includes metalloporphyrins and metallophthalocyanines coordinated or linked to nitroxide radicals [54], Iron(II) complexes [55], Prussian blue analogs [56], an azobenzene derivative bearing two stable nitronyl nitroxide radicals [57], and diaryethene as a photoswitching unit [57, 58], and azobenzene derivatives with a long alkyl chain and aminoxyls [59].

Chromophores bonded to nitroxides have been studied by time-resolved ESR (TRESR) spectroscopy in order to directly investigate the excited multiplet states consisting of a photoexcited triplet chromophore and doublet nitroxides in C_{60} and phthalocyaninatosilicon derivatives covalently linked to one nitroxide radical [60–64]. The D_1 and QA_1 states, which consist of a T_1 chromophore and doublet NRO·,

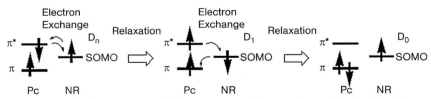

Figure 8.7 Molecular structures and electronic states of SiPc covalently linked to NRs [43].

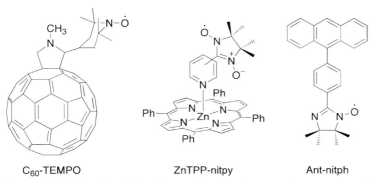

Figure 8.8 The $D_n \rightarrow D_1$ and $D_1 \rightarrow D_0$ transitions via electron exchange processes [43].

have been observed in solution and in frozen solutions for several systems. The TRESR spectra of ZnTPP-nitpy and ZnTPP-py consisting of sharp signals at $g = 2.00$ (attributed to a doublet state) and pair of signals (attributed to a quartet state) indicated the magnetic interactions between the T_1 ZnTPP and the nitroxide doublet. The photophysical properties of silicon phthalocyanine (SiPc) covalently linked to one or two NRO· have been systematically studied by fluorescence, transient absorption, and TRESR spectroscopies [63]. *In vitro* photodynamic effects of phthalocyaninatosilicon covalently linked to 2,2,6,6-tetramethyl-1-piperidinyloxy radicals on cancer cells was investigated [64]. It was shown that the decay rate of the $D_1 \rightarrow D_0$ transition is well correlated with the magnitude of the electron exchange interaction between the T_1 chromophore and the doublet NR.

Excited multiplet states consisting of the T_1 chromophore and doublet NRs are promising in terms of their magnetic properties and applications in photophysics, photochemistry, and photobiology.

8.2.1.3 Spin Trapping in Photochemical Reactions

In a series of elegant works, nitric oxide was suggested as a new spin-trapping agent [65, 66]. A suggested mechanism implies a two-step reaction of a reactive radical R with NO

$$NO + R^{\bullet} \rightarrow RNO \quad (1)$$

$$R^{\bullet} + NO \rightarrow R_2NO^{\bullet} \quad (2)$$

Radicals such as acyl, hydrated acyl, alkyl and ketyl produced from formaldehyde, acetaldehyde, acetaldehyde-d4, propionaldehyde, isobutyraldehyde, isopentanal, and tert-pentanal were detected by ESR spectroscopy. Photochemically generated 2-amido-2-Pr radical generated photochemically in aqueous solution attracted hydrogenatoms from aldehydes with the formation of a corresponding radical. The structure of radicals was identified using the deuterium substitution effect in the substrates on the radical EPR spectra. Similar results were obtained in NO spin-trapping and EPR studies on the photochemistry of aliphatic amides (formamide, acetamide, and their N-methyl- or deuterium-substituted derivatives).

The spin-trapping method was employed for the investigation of photoinduced electron transfer in ruthenium bipyridyl complexes [67]. Ruthenium complexes with three bipyridyl ligands, one of which is modified by attaching one or two hydroxamic acid groups (R_1 and R_2) were synthesized and investigated. The concentration of photochemically produced nitroxide radicals in aerobic conditions markedly increased in the presence of spin-traps DMPO (5,5′-dimethyl-1-pyrroline-N-oxide) and PBN (N-tert-butyl-α-phenylnitrone) which are characterized by strong affinity to superoxide radicals. It was suggested that these radicals are produced in a complex R_2–O_2. This suggestion was supported by the observation of shifts in the ^1H NMR spectra of R_2 in the presence of O_2.

The kinetics of photolysis and photonucleation of benzaldehyde (BA) vapor in air and in an inert gas were investigated using the spin-trapping method [68]. To identify short-lived free radicals accompanying BA photolysis, acyclic α-phenyl nitron (PN), 1,2,2,5,5-pentamethyl-3-imidazoline-3-oxide (PMIO), and 3-hydroxy-2,3-dihydro-2,2,5-trimethylpyrazine-1,4-dioxide (HDPDO) were employed. The PN trap was used to establish the presence of short-lived free radicals in the mixture under photolysis. Different free radicals are formed under irradiation with short-wavelength light (<290 nm) in air and in nitrogen. On the basis of ESR hyperfine splitting constants of adducts, it was assumed that the photolysis occurs by simple photochemical reaction with formation of ·C_6H_5 and HCO· radicals. The photogeneration of HCO and phenyl radicals was also observed using other starting compounds (formaldehyde, chlorobenzene).

Time-resolved chemically induced dynamic nuclear polarization (TR-CIDNP) and laser flash photolysis (LFP) techniques have been used to measure rate constants for coupling between acrylate-type radicals and a series of newly synthesized stable imidazolidine N-oxyl radicals [69]. The changes in the coupling rate constants k_c due to the varied steric ($E_{s,n}$) and electronic ($\sigma_{L,n}$) characters of the substituents were described in terms of the Hammett linear free energy relationship: $\log(k_c/M^{-1}s^{-1}) = 3.52\Sigma_{L,n} + 0.47 E_{s,n} + 10.62$.

Radical production by three natural eumelanins exposed to solar levels of UVA (sepia melanin from *Sepia officinalis* and eumelanins isolated from Oriental human and domestic cat hair) was investigated in aerobic conditions by spin-trapping methods [70]. It was shown that UVA irradiation of sepia melanin in solution at pH 4.5 in the presence of the spin trap 5,5-dimethyl-1-pyrroline N-oxide (DMPO) gave hydroperoxyl and hydroxyl radical-adducts. Dual fluorescence–nitroxide probes, derivatives of quinoline–TEMPO, were employed for mapping photogenerated radicals from hydroxy ketone and diaryl-α-disulfone in thin polymer films of poly(methyl methacrylate) (PMMA) [71]. Additionally, a pH-sensitive prefluorescent radical probe was employed as a dual probe for both photogenerated acids.

8.3
Complexes Transition Metals with Nitroxide Ligands

Complexes of transition metals with nitroxide ligands (CTMN) have been attracting wide attention for several reasons. They can serve as a convenient model for investigating fine intra- and intermolecular quantomechanical effects and for establishing a relationship between structure and physical and chemical properties. The models are well defined structurally and are characterized in detail by a whole arsenal of physical and chemical methods. CTMN are promising molecular materials for molecular magnets and switches (see Chapters 5 and 9). In this section we will focus on the former aspect of the matter.

After the first reported synthesis of the chelate complex Cu^{2+}–Schiff bases ligand derivative of TEMPO [72] and of hexafluoroacetylacetonato copper(II)–free-radical base [73], and subsequent publications in the 1970s [74–76], hundreds of complexes of transition metals with nitroxide ligands were synthetized and investigated (for reviews see [77–89].

To understand the magnetic and optical properties of complexes of transition metals with nitroxide ligands it is important to consider principal quantomechanical effects in these compounds, such as the spin–spin exchange and spin–spin dipole–dipole interactions, spin-lattice and spin-phase relaxations, and the formation of new molecular orbitals in ground and excited states.

The Hamiltonian for the exchange interaction between two paramagnetic centers with spins S_1 and S_2 is represented as $-JS_1S_1$, where J is the exchange integral [77]. Negative and positive values of J indicate an antiferromagnetic and ferromagnetic interaction, respectively. For two centers bearing $S = 1/2$ a negative J value gives a singlet-ground state and triplet-excited state, and J is the magnetic energy separation between the singlet and triplet levels. When the value of J is larger than the energy separation between the ESR transition for two paramagnetic centers (ΔE), the average g-factor and hyperfine splitting is observed. When $J \approx \Delta E$ for two slow relaxation centers, ESR line-splitting occurs. In fluid solution the complex rotation averages partially or completely to the spin–spin dipole–dipole interaction and the J value can be evaluated by the analysis of the observed splitting or by computer examination of ESR spectra. In frozen solutions both dipole and exchange interactions commonly contribute to the splitting. For this system the values of J and r can be obtained by ESR spectra simulation or, in some cases, directly from the intensity of the half-field transition.

The value of the exchange integral is influenced by an overlap of the orbital bearing the unpaired electron and the spin density on it. Spin exchange can occur via direct overlap of orbitals or indirectly by means of intermediate overlaps (super exchange). The super exchange causes delocalization of spin over intermediate ligands and coligands, which can be independently examined by ESR and NMR techniques. The contribution of the spin relaxation processes in ESR behavior of CTMN should also be taken into consideration. When the rate of spin relaxation of a metal unpaired electron increases, the metal ESR signals broaden, and the splitting of the nitroxyl signal eventually collapses.

The exchange interaction between d- or f-orbitals of transition metals and the p-orbital of nitroxide gives new multiplicity and, as a result, formally spin-forbidden transitions become spin-allowed with the breakdown of $\Delta S = 0$ [79]. In such a case complexes exhibit fairly strong formally spin-forbidden absorption bands. Besides the magnetic interaction between the ground states in the spin-coupled systems, there are intramolecular magnetic interactions between the excited state in one moiety of the multi-spin systems and the ground state of the other.

Spin–spin dipole–dipole interactions in CTMN, which generate a splitting in ESR spectra ($\Delta\Delta H_{dd}$), are strongly dependent on the distance between the spin of metal and the spin of nitroxide (R). Another factor affecting the spin dipole–dipole interaction is the spin-relaxation rate. For a slow relaxing spin of metal, the splitting is proportional to $1/R^3$, whereas for the fast relaxing spin it is $\Delta\Delta H_{dd} \sim 1/R^6$). The following are methods of measuring distances in complexes ([77] and references therein) (see also Chapter 5): (1) decrease of amplitude of nitroxide signal; (2) change of electron spin relaxation; (3) change of nitroxyl d_1/d ratio; (4) intensity of half-field transition; (5) analysis of resolved splitting [77, 80].

Over a period from 1970 to 1980, complexes containing piperidine, pyrolidine, and imidazole derivatives of nitroxide and multiple intermediate ligands were synthesized and investigated. The main regularity in this area in this period was summarized in the review [77]. An example of a CMTN with two imino nitroxides coordinated through the diamagnetic Cu(I) is shown in Figure 8.9 [81].

Nitroxide radical ligands have been found to exhibit various coordination effects on the spectroscopic properties of the metal complexes. These effects are strongly dependent on either direct coordination of the nitroxide moiety or indirect coordination through the pyridyl or piperidyl moieties takes place or the magnetic d orbital symmetry or the kinds of coligands occurs. The orthogonality or overlap of the magnetic orbitals, magnetic d orbital symmetry of the transition metal, and type of coligands can significantly affect the magnetic and optical behavior of CTMN.

In complexes in which nitroxyl oxygen was coordinated to the metal, the exchange interaction is large relative to the nuclear hyperfine splittings and the g value difference. As a result, averaging g and reduced A values for the spin-labeled complexes in fluid solution is observed. About 30 spin-labeled Cu(II) or vanadyl

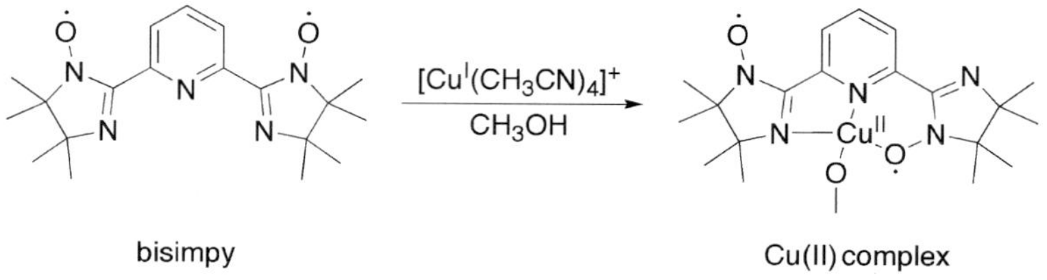

Figure 8.9 Structure of complex Cu(I) with imino nitroxides [81]. (Reproduced from Oshio, H. and Ito, T. (2000) Assembly of imino nitroxides with Ag(I) and Cu(I) ions, *Coord. Chem. Rev.*, 198, 329–346, with permission of Elsevier).

complexes exhibited this effect [77]. In these complexes there is no signal from the singlet state and incomplete motional averaging of the large zero-field splitting of the triplet state causes broadening of the fluid solution spectra. Coordination of the nitroxyl oxygen gives ferromagnetic or antiferromagnetic interaction that is large enough to measure by magnetic susceptibility. Rapid relaxation of the metal decouples the two spins and a triplet from nitroxide is detected in fluid solutions (Fe(III)–nitroxide) complexes. In Ni(II)–nitroxide complexes with intermediate values $1/T_1$, broadening of the signal is observed.

In contrast, when the liganding does not occur via coordination with the nitroxide oxygen, the exchange interaction is relatively weak and is dependent on the linkage between ligand and nitroxide. The magnitude of J decreases in the order: Shiff base > urea > amide > ester, which parallels the decrease of the p contribution to the bonding in the linkage [77]. Decreasing J in the order pyrroline > tetrohydropyridin > pyrolidine > piperidine was observed. For spin-labeled porphyrin, J decreased in the order Ag(II) > Cu(II) > VO(IV), which parallels the extent of spin-delocalization from metal onto porphyrin orbitals. This effect was monitored by the measurement of a nuclear hyperfine coupling constant for the porphyrin nitrogen and the pyrrol hydrogens. For porphyrin–Cu(II) complexes, J decreased in the order *trans*-olefin > saturated > $(CH_2)_2$ > *cis*-olefin. This pattern indicates that there is a significant s contribution to the spin delocalization in the porphyrins. J is dependent on conformation in the greater range for 5-member rings than for 6-member rings. When a nitroxide was attached to one phenyl ring of a tetraphenyl porphyrin, the exchange interaction was several orders of magnitude greater for the *ortho*-isomers then for the meta- or *para*-isomers.

Experimental values of the exhange integrals for a series of complexes of Cu^{2+}, Co^{2+}, Cr^{2+}, and VO^{2+} with a weak exchange interaction beween spins are summarized and discussed [77]. It was shown that the majority of complexes exhibit ferromagnetic properties with $J = 0.0002$–$0080\,cm^{-1}$ dependent on the ligand's structure. Several Cr^{2+} complexes showed antiferromagnetic interaction with $J = -(0.003$–$0.01)\,cm^{-1}$.

In complexes of nitroxides coordinated with hexafluoroacetylacetonatonate $[M(hfac)_2]$ (where M = VO, Mn, Cu) via nitroxide oxygen, strong antiferomagnetic coupling with $J = -75$–$700\,cm^{-1}$ takes place [79]. Strong antiferromagnetic interaction was observed also in $Mn(hfac)_2$ complexes of isoindoline nitroxide radicals [82]. Ferromagnetic interactions in complexes $[CuCl(bisimpy)(MeOH)](PF_6)$, [NiCl(bisimpy)$(H_2O)_2$]Cl$\cdot 2H_2O$ (2), and [$ZnCl_2$ (bisimpy)] (3, bisimpy = 2,6-bis(1'-oxyl-4',4',5',5'-tetramethyl-4',5'-dihydro-1'H-imidazol-2'-yl)pyridine, with imino nitroxyl diradicals were reported [83]. The value of J was significantly dependent on the complex structure and ranged from $13\,cm^{-1}$ to $165\,cm^{-1}$.

A significant contribution of spin relaxation rates to ESR properties of CTMN was demonstrated with the use of time-domain spin-echo techniques [84, 85]. It was shown that for low-spin FeIII(TPP)(MeIm)$_2$ (TPP = tetraphenylporphyrin; MeIm = methylimidazole) at a temperature of about 28 K, the relaxation rates for the Fe(III) drastically increased and the spin–spin splitting collapsed. Above about 10 K, iron relaxation rates increased in the order X = F < Cl < Br, which is the order of

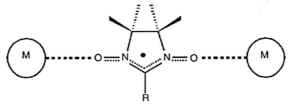

Figure 8.10 Schematic presentation of a binuclear nitronyl complex [79]. (Reproduced from Kaizaki, S. (2006) Coordination effects of nitroxide radicals in transition metal and lanthanide complexes, *Coord. Chem. Rev.*, 250, 1804–1818, with permission of Elsevier).

increasing zero-field splitting. Between 10 and 120 K high-spin Fe(III) (S = 5/2) had a greater impact than low-spin Fe(III) (S = 1/2) caused by differences in values of spin, spin relaxation rate, and the difference in resonance frequency for the fast-relaxing (iron) and slowly-relaxing (nitroxyl) spins in the denominator.

The use of nitronyl and its derivatives as ligands [86] gave new impact to the area ([78, 79, 87–89] and references therein).

One of the principle advantages of nitronyl is its ability to coordinate directly two metal atoms (Figure 8.10). This particular property opened a wide possibility for building up two- and three-dimensional magnetic materials.

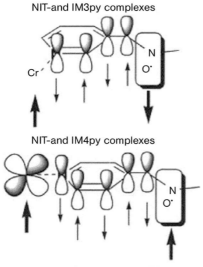

Figure 8.11 Molecular orbitals of ferromagnetic or antiferromagnetic complexes [79]. (Reproduced from Kaizaki, S. (2006) Coordination effects of nitroxide radicals in transition metal and lanthanide complexes, *Coord. Chem. Rev.*, 250, 1804–1818, with permission of Elsevier).

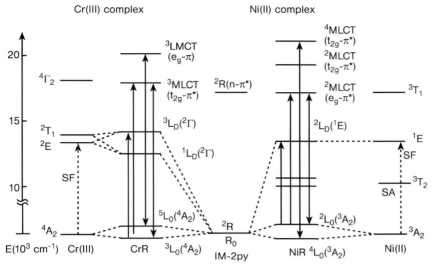

Figure 8.12 Energy levels of the nitroxide radical Cr(III) and Ni(II) complexes together with the non-radical complexes [79] (Reproduced from Kaizaki, S. (2006) Coordination effects of nitroxide radicals in transition metal and lanthanide complexes, *Coord. Chem. Rev.*, 250, 1804–1818, with permission of Elsevier).

Figures 8.11 and 8.12 demonstrate the major quantomechanical effects in metal complexes with nitronyls [79]. As is seen from the figures, the structure of complexes dictates ferromagnetic or antiferromagnetic behavior of the molecular systems. Figure 8.12 shows that exchange coupling in the CTMN $^3A_2{}^2L_0$(Ni) or $^4A_2{}^2L_0$ (Cr) ground state gives a doublet and quartet or triplet and quintet, whereas only the doublet or the singlet and triplet are generated in the CTM excited state $^2(^1E^2L_0)$ or $^1(^2E/^2T_1{}^2L_0)$ or $^3(^2E/^2T_1{}^2L_0)$. Therefore, the spin-forbidden $^2(^3A_2{}^2L_0)$ → $^2(^1E^2L_0)$(Ni) or $^3(^4A_2{}^2L_0)$ → $^3(^2E/^2T_1{}^2L_0)$ transition becomes formally spin-forbidden or actually doublet–doublet or triplet–triplet spin-allowed with the breakdown of the $\Delta S=0$ restriction.

Results of a series of works devoted to magneto-structural correlations inherent to heterospin compounds based on metal hexafluoroacetylacetonate and stable nitronyl nitroxides containing pyrazole substituents were summarized in [88, 89]. The structure of mixed-metal hexafluoroacetylacetonates with nitronyl nitroxide, [Cu$_{1-x}$M$_x$(hfac)$_2$L] (M = Mn, Ni, Co)] and the temperature of its transition in solid solution were established [89]. It was shown that magnetic anomalies induced by the phase transitions originate from specific motions in coordination units containing two types of exchange clusters, Cu(II)–O·–N< or >N–·O–Cu(II)–O·–N<, and are accompanied by significant changes in the crystal volume after multiple cooling/heating cycles ("breathing crystals"). The structure of nitronyl complexes is shown in Figure 8.13.

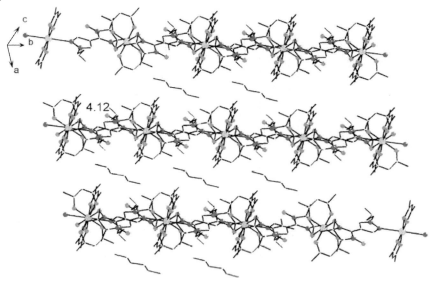

Figure 8.13 Layers of the structure of heterospin complexes of Cu(hfac)2 (hfac, hexafluoroacetylacetonate) with pyrazole-substituted nitronyl nitroxides [89]. (Reproduced with permission).

The aforementioned particular properties of the complexes of transition metals with nitroxide ligands open many possibilities for deeper insight into quantomechanical effects in molecules and for building up two- and three-dimensional magnetic materials.

8.4
Nitroxides in Inorganic Chemistry

The applications of nitroxides in inorganic chemistry and technology are currently not as many as in other areas. Nevertheless, available data indicate wide possibilities or the spin label method for solving a number of molecular dynamic and structural tasks in the investigation of inorganic materials (gold, clay, silica, zeolites, graphite, oxides, alloys, and so forth). Information on microstructure, local polarity, viscosity, electrostatic potential, and surface acidity, and in porous materials, kinetics and the mechanism of adsorption and radical processes, molecular dynamics and structure of surface thin films, surface biovailability, distribution of hydrophobic and hydrophilic sites, and production of surface magnetic materials can be obtained by modifying objects of interest by nitroxides and investigating by ESR technique.

8.4.1
Langmuir–Blodgett (LB) Films on Inorganic Substrates

Self-assembled monolayers as well as Langmuir–Blodgett (LB) films have attracted considerable interest, especially in connection with technological applications including highly specific sensors or devices based on molecular electronics. The spin-labeling method can provide information about molecular motion of the films on solid surfaces. A common problem for the use of ESR spectroscopy in studies of surface processes is a lack of sensitivity. In a series of works this problem has been overcome.

The organization of triazolehemiporphyrazine derivatives bearing crown ether macrocycles within the LB multilayer was characterized by use of IR dichroism, ESR and IR methods [90]. It was shown that aggregates of 50 molecules of these compounds form stacks of macrocycles and are tilted with respect to the normal of the substrate. This configuration is a first step to organizing the crown ether group to form ionic channels in LB films. Mono- and multilayer LB films with long hydrocarbon chain nitroxides (doxyl stearic acid and hexadecanoylpiperidine-oxyl radicals) were built up and investigated by ESR spectroscopy [91]. These films were prepared from (1) single components, (2) nitroxide binary mixtures, and (3) a diluted solution of the paramagnetic components in a diamagnetic (stearicacid) host. The ESR spectra of undiluted nitroxides were characterized by strong spin–spin (dipole–dipole and Heisenberg spin exchange) magnetic interactions whose extent also depended on the nitroxide type and mixture. A degree of order of the spin probes and the chain flexibility, in particular for diamagnetically diluted systems was established.

Self-assembled stearic acid films doped with spin labels at different positions on the alkyl chain adsorbed on an Al_2O_3-film were studied by ESR technique and near-edge X-ray absorption fine structure (NEXAFS) spectroscopy [92]. The line-shape analysis of temperature-dependent EPR spectra showed the spatial distribution and a strong dependence of the rotational motion of the nitroxide moieties on its location along the chain and allowed us to calculate parameters of the probe's rotational motion. At high probe concentrations the spectra indicated a strong exchange and dipole–dipole interactions between nitroxides.

EPR spectroscopy in combination with FTIR and ellipsometery measurements were used for investigating the spatial and temporal organization of self-assembled monolayers of 5- and 16-doxyl stearic acid (5 DSA and 16 DSA, respectively) adsorbed on a GaAs substrate [93]. The examination of ESR spectra of the labeled films indicated different dynamic behavior of the nitroxide fragment of 5 DSA and 16 DSA. For 5DSA, with the spin-label group situated close to the substrate, the EPR spectrum immediately after adsorption showed the distribution of the probe correlation times and the spin-exchange interaction between neighboring molecules. The ESR spectra show spin–spin exchange interaction with frequency from 32 to 102 MHz, which depends on time. Induced ferromagnetic interaction of nitroxides in Langmuir–Blodgett Films on inorganic substrates was reported [94].

8.4.2
Surface Microstructure and Dynamics

The properties of the silica layer during the formation of a mesoporous material were investigated by ESR and ESEEM measurements with the use of organo(trialkoxy)silane spin probes [95]. The spin-probes were co-condensed with the silica source, tetra-Et orthosilicate, in the synthesis of MCM-41 with cetyltrimethylammonium bromide. This approach allowed us to establish the probe location on the MCM-41 and to monitor all stages of material formation.

Adsorption of solutions of spin probes in water, methanol, acetonitrile (ACN), and binary MeOH–water and ACN–water solutions onto porous silica was investigated by ESR spectroscopy, using nitroxide probes of different hydrophobicity, and reversed-phase high-performance liquid chromatography [96]. The silica samples differed from each other with regard to surface area, pore size, particle size, surface functions (NH2, C8, and C18), and percentage of functionalization. The structural and dynamic parameters of spin-probes and their environments were established by an examination of the spectral line-shape. It was shown that an endcapping process of the residual silanols strongly enhanced the surface hydrophobicity tested by the probes. At the highest water content, the adsorption of the polar or charged probes onto the hydrophobic surface is the lowest and self-aggregation occurs. When the probes bear both hydrophilic and hydrophobic moieties, adsorption is enhanced by synergy between hydrophilic and hydrophobic bonds with the surface. This insertion and the response on the formation of the ordered layer were favored in ACN–water with respect to MeOH–water.

A distribution of a hydrophobic spin probe (TEMPO benzoate) in suspensions of humic acids, hectorite, and aluminum hydroxide-humate-hectorite complexes was investigated by ESR spectroscopy [97]. Distribution coefficient (K_d) values for the adsorption of TEMPO benzoate on different substrates were measured. The ability of EPR to provide evidence that hydrophobic molecules in the presence of geosorbents can segregate in multimolecular clusters that are in equilibrium with aqueous probe concentrations was demonstrated. The ESR technique was used to study distribution of 4-phenyl-2,2,5,5-tetramethyl-3-imidazoline-1-oxyl on fluorographite and graphite oxide [97]. Depending on the concentration of the probes in the interlayer, nitroxides exist as isolated ordered species or form antiferromagnetic phases. Formation of one-dimensional Heisenberg antiferromagnets in a synthetic layered silicate (Laponite) with a particle size of 25 nm modified by nitroxides was proved by ESR and X-ray diffraction methods [98]. The analysis of line-shape of the probe's ESR spectra provided evidence for noncovalent assembly of nitroxide radicals in the Laponite film into structures which act as one-dimensional Heisenberg antiferromagnets.

A bifunctional stilbene–nitroxide label (BFL1, Section 7.3) was applied for the dual determination of the antioxidant status of the media and measurement of micro- and macro-viscosity on a silica plate surface [99]. A synthesized stilbene–nitroxide label (BFL1) was immobilized on the surface of a quartz plate. Previously it was shown that the photoisomerization kinetics of a stilbene probe are strongly dependent upon the viscosity of a media. Monitoring the rate constant of photo-isomerization of the BFL firstilbene by a fluorescence technique permitted measurement of the viscosity of a medium on the plate surface in the range 1–500 cP.

The degradation of spin probes (4-hydroxy–TEMPO) located on the surface of clay suspensions and pastes caused by bacteria was investigated [100]. The observed probe ESR signal decay can most probably be explained by nitroxide reduction with the bacteria metabolites.

8.4.3
Nanoparticles

Metal nanoparticles modified by organic ligands including nitroxide derivatives are very promising in the area of building materials, with unusual chemical, catalytic, optical, and electronic properties of these materials which could lead to a range of potential applications. A series of Au nanoparticles modified with nitroxide spin labels (Figure 8.14) has been prepared and molecular dynamics and mutual disposition of the label was studied by EPR spectroscopy [101]. Figure 8.15 shows covalent modification of the particles by the label N4.

Samples with low coverage of the spin labels at different distances from the Au surface were used to estimate its rotational correlation times which vary from 10^{-10} second to more than 3×10^{-9} second. Examination of the ESR spectra lineshape revealed dynamic heterogeneity of the probe EPR signals of isolated radical pairs, which were observed at intermediate coverage. Quantitative analysis of the

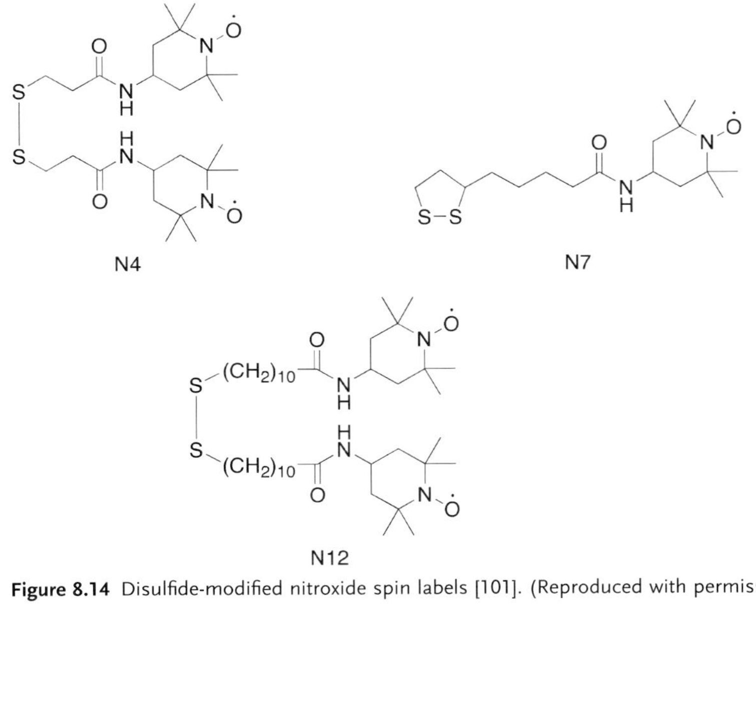

Figure 8.14 Disulfide-modified nitroxide spin labels [101]. (Reproduced with permission).

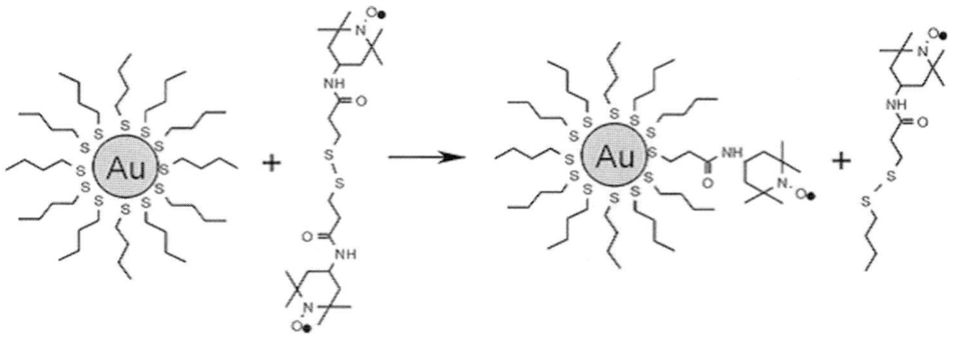

Figure 8.15 Preparation of spin-labeled Au nanoparticles by exchange reaction [101]. (Reproduced with permission).

EPR spectra of samples with higher coverage of the spin-probes suggested a marked contribution of exchange interaction between nitroxides and the presence of non-equivalent binding sites on the surface of Au nanoparticles.

Quenching of fluorescence: fluorescence of CdSe nanoparticles of different sizes (2.4–6.7 nm) and λ_{max} (525–630 nm) was investigated by TEMPO (non-binding) and 4-amino–TEMPO (binding) [102]. Data obtained using fluorescence and ESR techniques suggested that, depending on the sample's structure and the quencher concentration, the process occurs by exchange dynamic interaction at encounter complex or/and static interaction between particles and the binding nitroxide. Possible mechanisms of fluorescence quenching are discussed in Section 7.3.

8.4.4
Local Acidity

Determination of the local pH (pH_{loc}) values for characterizing the acid–base properties of different inorganic materials such as zeolites and related mesoporous materials (MM) is of a great practical and theoretical interest in the fields of heterogeneous catalysis and separation by adsorption. A new method for determining acidity of a medium inside solid-state pores (pH_{loc}) using pH-sensitive nitroxide radicals (NR) as probes and the ESR technique has been elaborated [103, 104]. The method is based on principles described in Section 6.9. Direct measurements of acidity inside micropores of inorganic (hydrogels based on TiO_2, SiO_2, zeolites, and kaolin) sorbents and organic (synthetic ion-exchange resins) were carried out. It was found that local pH values at the site of the label's location in the sorbents differed from the pH values of the solutions by 1–2 units.

Two pH-sensitive parameters of nitroxide ESR spectra have been used for pH measurements in porous solids: the hyperfine coupling constant of isotropic ESR signals of NR (a_N), which allows us to calculate an average pH in liquid phase of a system; and the parameter defined as either a fraction of protonated or unprotonated form for an immobilized nitroxide. The latter was used for determining a local concentration of protons near the surface of MM and surface electrostatic potential (SEP). The method was employed to characterize the acid–base properties of a number of inorganic MM systems and crosslinked polyelectrolytes.

The nitroxide radical 4-dimethylamine-2-ethyl-5,5-dimethyl-2-pyridine-4-yl-2, 5-dihydro-1H-imidazol-1-oxyl, being pH-sensitive in the range from 2.5 to 7 pH units, was used as a pH probe for SiO_2 gel and aluminium oxides modified by incorporating F- and SO_4^{2-} [105]. The local pH values (pH_{loc}) inside sorbent pores (excluding γ-Al_2O_3) were found to differ from the pH-values of external solution by 0.5–1.3 units. It was found that the titration curves for the NR incorporated into γ-Al_2O_3 and in SiO_2 gel are shifted to the left and right sides (respectively) relative to the graduation curve of the same NR in aqueous solution,. The pk_a value for SiO_2 gel silanol groups was found to be 4.25. This result accords with previously obtained data [106].

The formation of radical pairs after adsorption of 2,2′,5,5′-tetramethyl-4-phenyl-3-imidazoline-1-oxyl (**R**) on the acid sites of beta zeolites previously dehydrated by heat treatment in air at 500 °C has been reported [107]. In addition to the conventional EPR spectrum of monomer adsorbed radicals, new lines with $g = 2.005$ and symmetrical splitting with 150 and 300 G appeared in the spectrum. The analysis of this spectrum typical for immobilized radicals made it possible to assign it to paramagnetic species in triplet state (radical pairs) with $g = 2.005$, $D = 150 G$, $E \approx 0$. Suggesting spin–spin dipole–dipole interaction, the distance between the radicals was estimated to be equal to 0.57 nm. This distance is close to the distance between the opposite oxygen atoms in 12-member rings of the sinusoidal channels of zeolites under investigation. The authors suggested that the results obtained clearly indicate that radical pairs can be formed after adsorption of radicals on the active sites of zeolites. The authors suggested that such information is important for heterogeneous catalysis.

Several typical examples described in this chapter have clearly demonstrated the great potential of nitroxide application in various areas of modern physico-chemistry.

References

1. Ruban, L.V., Buchachenko, A.L., Neiman, M.B. and Kokhanov, Yu.V. (1966) *Vysokomol Soed*, **8**, 1642–6.
2. Griffiths, P.G., Rizzardo, E. and Solomon, D.H. (1982) *Journal of Macromolecular Science, Chemistry*, **A17**, 45–50.
3. Solomon, D.H., Rizzardo, E. and Cacioli, P. (1985) U.S. 4,581,429, March 27.
4. Greszta, D. and Matyjaszewski, K. (1996) *Macromolecules*, **29**, 7661–70.
5. Fischer, H. and Souaille, M. (2001) *Chimia*, **55**, 109–13.
6. Guillaneuf, Y., Gigmes, D., Marque, S.R.A., Astolfi, P., Greci, L., Tordo, P. and Bertin, D. (2007) *Macromolecules* (Washington, DC, USA), **40**, 3108–14.
7. Gavranovic, G.T., Csihony, S., Bowden, N.B., Hawker, C.J., Waymouth, R.M., Moerner, W.E. and Fuller, G.G. (2006) *Macromolecules*, **39**, 8121–7.
8. Grishin, D.F., Kolyakina, E.V. and Polyanskova, V.V. (2006) *Vysokomol*

Soed, Seriya A i Seriya B, **48**, 764–70.
9. Chauvin, F., Dufils, P.-E., Gigmes, D., Guillaneuf, Y., Marque, S.R., Tordo, P. and Bertin, D. (2006) *Macromolecules*, **39**, 5238–50.
10. Zaremskii, M.Yu. (2006) *Vysokomol Soed, Seriya A i Seriya B*, **48**, 404–22.
11. Matyjaszewski, K. (2000) *Controlled/Living Radical Polymerization*, Oxford University Press, Oxford.
12. Matyjaszewski, K. and Mueller, A.H.E. (2006) *Progress in Polymer Science*, **31**, 1039–40.
13. Moad, G. and Solomon, D.H. (1995) *The Chemistry of Free Radical Polymerization*, Pergamon, Oxford.
14. Cunningham, M.F. (2003) *Comptes Rendus Chimie*, **6**, 1351–74.
15. Potapov, E.E., Boreiko, N.P. and Strygin, V.D. (2005) *Kauchuk i Rezina*, **3**, 7–11.
16. Che-Chien Chang, Ch-Ch. and Studer, A. (2006) *Macromolecules*, **39**, 4062–8.
17. Detrembleur, C., Sciannamea, V., Koulic, C., Claes, M., Hoebeke, M. and Jerome, R. (2002) *Macromolecules*, **35**, 7214–23.
18. Zaremskii, M.Yu., Orlova, A.P., Garina, E.S., Olenin, A.V., Lachinov, M.B. and Golubev, V.B. (2003) *Vysokomol Soed, Seriya A i Seriya B*, **45**, 871–82.
19. Semenycheva, L.L., Lazarev, M.A. and Grishin, D.F. (2004) *Vysokomol Soed, Seriya A i Seriya B*, **46**, 1225–9.
20. Gigmes, D., Gaudel-Siri, A., Marque, S.R.A., Bertin, D., Tordo, P., Astolfi, P., Greci, L. and Rizzoli, C. (2006) *Helvetica Chimica Acta*, **89** (10), 2312–26. Gaudel-Siri, A., Siri, D. and Tordo, P. (2006) *Chem Phys Chem*, **7**, 430–8.
21. Gigmes, D., Gaudel-Siri, A., Marque, S.R.A., Bertin, D., Tordo, P., Astolfi, P., Greci, L. and Rizzoli, C. (2006) *Helvetica Chimica Acta*, **89**, 2312–26.
22. Grishin, D.F., Ignatov, S.K., Razuvaev, A.G., Kolyakina, E.V., Shchepalov, A.A., Pavlovskaya, M.V. and Semenycheva, L.L. (2001) *Vysokomol Soed, Seriya A i Seriya B*, **43**, 1742–9.
23. Beaudoin, E., Bertin, D., Gigmes, D., Marque, S.R.A., Siri, D. and Tordo, P. (2006) *European Journal of Organic Chemistry*, **7**, 1755–68.
24. Souaille, M. and Fischer, H. (2002) *Macromolecules*, **35**, 248–61.
25. Tang, W., Fukuda, T. and Matyjaszewski, K. (2006) *Macromolecules*, **39**, 4332–7.
26. Lemoine-Nava, R. and Flores-Tlacuahuac, A. (2006) *Industrial and Engineering Chemistry Research*, **45**, 4637–52.
27. Vasserman, A.M., Buchachenko, A.L., Kovarskii, A.L. and Neiman, M.B. (1968) *Vysokomol Soed, Seriya A*, **10**, 1930–6.
28. Vasserman, A.M., Buchachenko, A.L., Kovarskii, A.L. and Neiman, M.B. (1969) *European Polymer Journal*, (Suppl.), 473–8.
29. Buchachenko, A.L., Lebedev, Ya.S. and Neiman, M.B. (1970) *Usp. Khim. Fiz. Polim. "Khimiya"* (ed. Z.A. Rogovin), Xxxx, Moscow, pp. 409–47.
30. Wasserman, A.M. (2005) *Chemical and Biological Kinetics. New Horizons*, Vol. 1, *Chemical Kinetics*. In commemoration of Professor N.M. Emanuel's 90th Anniversary (eds E.B. Burlakova, A.E. Shilov, S.D. Varfolomeev and G.E. Zaikov), Brill Academic Publishers, Leiden (The Netherlands); Boston (USA), pp. 385–404.
31. Nakagawa, K. (2005) *Lipids*, **40**, 745–50.
32. Yamamoto, K., Okamoto, S., Sakaguchi, M., Sakai, M., Makita, S., Sakurai, S. and Shimada, S. (2004) *Macromolecules*, **37**, 3707–16.
33. Krykin, M.A., Wasserman, A.M., Motyakin, M.V., Gorbatsevich, O.B. and Ozerin, A.N. (2002) *Vysokomol. Soed., Seriya A i Seriya B*, **44**, 1325–30.
34. Kutsumizu, S. and Schlick, S. (2005) *Journal of Molecular Structure*, **739**, 191–8.
35. Kovarskii, A.L., and Yushkina, T. V. (2005) *Chemical and Biological Kinetics New Horizons*, **1** (ed. E.B. Burlakova), Academic Publisher, Leiden, The Netherlands, pp. 67–79.
36. Miwa, Y., Yamamoto, K., Tanabe, T., Okamoto, S., Sakaguchi, M., Sakai, M. and Shimada, S. (2006) *Journal of Physical Chemistry. B*, **110**, 4073–82.
37. Yakimchenko, O.E., Smirnov, A.I. and Lebedev, Y.S. (1990) *Applied Magnetic Resonance*, **1**, 1–19.
38. Eaton, G.R., Eaton, S.S. and Ohno, K. (eds) (1991) *EPR Imaging and in vivo EPR*, CRC Press, Boca Raton, FL.
39. Schlick, S., Kruczala, K., Motyakin, M.V. and Gerlock, J.L. (2001) *Polymer Degradation and Stability*, **73**, 471–5.

40 Jeschke, G. and Schlick, S. (2006) *Physical Chemistry Chemical Physics: PCCP*, **8**, 4095–103.

41 Vorobiev, A.K. and Chumakova, N.A. (2005) *Journal of Magnetic Resonance (San Diego, Calif.)*, **175**, 146–57.

42 Bystryak, I.M., Likhtenshtein, G.I., Kotelnikov, A.I., Hankovsky, O.H. and Hideg, K. (1986) *Russian Journal of Physical Chemistry*, **60**, 1679–983.

43 Likhtenshtein, G.I., Nakatsuji, S. and Ishii, K. (2007) *Photochemistry and Photobiology*, **83**, 871–81.

44 Fogel, V.R., Rubtsova, E.T., Likhtenshtein, G.I. and Hideg, K. (1994) *Journal of Photochemistry and Photobiology A: Chemistry*, **83**, 229–3.

45 Lozinsky, E.A., Shames, A. and Likhtenshtein, G.I. (2000) Dual fluorophore-nitroxides: Models for investigation of intramolecular quenching and novel redox probes, in *Recent Research Development in Photochemistry and Photobiology*, Vol. 2 (ed.S.G. Pandalai), Transworld Research Network, Trivandrum, India, pp. 35–48.

46 Likhtenshtein, G.I. (1996) *Journal of Photochemistry and Photobiology A: Chemistry*, **96**, 79–92.

47 Likhtenshtein, G.I., Febrario, F., Nucci, R. and Spectrochem (2000) *Acta Part A. Biomol Spectrosc*, **56**, 2011–31.

48 Green, S.A., Simpson, D.J., Zhou, G., Ho, P.S. and Blough, N.V. (1990) *Journal of the American Chemical Society*, **112**, 7337–46.

49 Vlassiouk, I., Smirnov, S., Kutzki, O., Wedel, M. and Montforts, F.-P. (2002) *Journal of Physical Chemistry. B*, **106**, 8657–66.

50 Ivanova, V.N., Nadolonnyi, V.A. and Grigorie, I.A. (2004) *Journal of Structural Chemistry*, **45** (Suppl.), S71–S75.

51 Qixi Mi, O., Chernick, E.T., McCamant, D.W., Weiss, E.A., Ratner, M.A. and Wasielewski, M.R. (2006) *Journal of Physical Chemistry. A*, **110**, 7323–33.

52 Ishii, K.N. and Kobayashi, N. (2003) *The Porphyrin Handbook*, Vol. 16 (eds K. M. Kadish, R.M. Smith and R. Guilard), Academic Press, New York, pp. 1–42.

53 Ishii, K. and Kobayashi, N. (2000) *Coordination Chemistry Reviews*, **198**, 231–50.

54 Ishii, K., Ishizaki, T. and Kobayashi, N. (1999) *Journal of Physical Chemistry. A*, **103**, 6060–2.

55 Gütlich, P., Hauser, A. and Spiering, H. (1994) *Angewandte Chemie (International Ed. in English)*, **33**, 2024–54.

56 Mizuochi, N.Y., Ohba, Y. and Yamauchi, S. (1999) *Journal of Physical Chemistry. A*, **103**, 7749–52.

57 Matsuda, K. and Irie, K. (2004) *Journal of Photochemistry and Photobiology C: Photochemistry Reviews*, **5**, 169–82.

58 Hamachi, K., Matsuda, K., Itoh, T. and Iwamura, H. (1998) *Bulletin of the Chemical Society of Japan*, **71**, 2937–43.

59 Nobusawa, M., Akutsu, H., Yamada, J.-I., and Nakatsuji, S. (2006) *Letters in Organic Chemistry*, **3**, 685–8.

60 Corvaja, C., Maggini, M., Prato, M., Scorrano, G. and Venzin, M. (1994) *Journal of the American Chemical Society*, **117**, 8857–8.

61 Ishii, K., Takeuchi, S. and Kobayashi, N. (2001) *Journal of Physical Chemistry. A*, **105**, 6794–9.

62 Teki, Y.S., Miyamoto, K., Iimura, M., Nakatsuji, S. and Miura, Y. (2000) *Journal of the American Chemical Society*, **122**, 984–5.

63 Ishii, K., Takayanagi, A., Shimizu, S., Abe, H., Sogawa, K. and Kobayashi, N. (2005) *Free Radical Biology and Medicine*, **38**, 920–7.

64 Ishii, K., Ishizaki, T. and Kobayashi, N. (2001) *Journal of the Chemical Society, Dalton Transactions*, 3227–31.

65 Wang, F., Jin, J. and Wu, L. (2003) *Magnetic Resonance in Chemistry: MRC*, **41**, 647–59.

66 Yang, D., Lei, L., Liu, Z., Wang, F., Lei, L. and Wu, L. (2005) *Magnetic Resonance in Chemistry: MRC*, **43**, 156–65.

67 Yavin, E., Weiner, L., Arad-Yellin, R. and Shanzer, A. (2004) *Journal of Physical Chemistry. A*, **108**, 9274–82.

68 Dubtsov, S.N., Dultseva, G.G., Dultsev, E.N. and Skubnevskay, G.I. (2006) *Journal of Physical Chemistry*, **110**, 645–9.

69 Zubenko, D., Tsentalovich, Yu., Lebedeva, N., Kirilyuk, I., Roshchupkina, G., Zhurko, I., Reznikov, V., Marque, S.

R. and Bagryanskaya, E. (2006) *Journal of Organic Chemistry*, **71**, 6044–52.
70 Haywood, R.M., Lee, M. and Linge, C. (2006) *Journal of Photochemistry and Photobiology B: Biology*, **82**, 224–35.
71 Coenjarts, C., García, O., Llauger, L., Palfreyman, J., Vinette, A.L. and Scaiano, J.S. (2003) *Journal of the American Chemical Society*, **125**, 620–1.
72 Medzhidov, A.A., Kirichenko, L.N. and Likhtenshtein, G.I. (1969) *Izvestiya Akademii Nauk SSSR, Seriya Khimicheskaya*, **000**, 698–700.
73 Lim, Y.Y. and Drago, R.S. (1972) *Inorganic Chemistry*, **11**, 1334–8.
74 Medzhidov, A.A., Shapiro, A.B., Mamedova, P.S., Musaev, A.M. and Rozantsev, E.G. (1977) *Izv. Akad. Nauk SSSR, Ser. Khim.*, 538–43.
75 Larionov, S.V., Ovcharenko, V.I., Sadykov, R.A., Sagdeev, R.Z. and Volodarskii, L.B. (1975) *Koord. Khim.*, **1**, 1312–18.
76 Eaton, D.L., DuBois, P.M., Boymel, G.R. and Eaton, G.R. (1979) *Journal of Physical Chemistry*, **83**, 332.
77 Eaton, S.S. and Eaton, G.R. (1988) *Coordination Chemistry Reviews*, **83**, 29–72.
78 Luneau, D. and Rey, P. (2005) *Coordination Chemistry Reviews*, **249**, 2591–611.
79 Kaizaki, S. (2006) *Coordination Chemistry Reviews*, **250**, 1804–18.
80 Likhtenshtein, G.I. (1993) *Biophysical Labeling Methods in Molecular Biology*, Cambridge University Press, Cambridge, NY, pp. 57–62.
81 Oshio, H. and Ito, T. (2000) *Coordination Chemistry Reviews*, **198**, 329–46.
82 Smith, D., Bottle, S.E., Junk, P.C., Inoue, K. and Markosyan, A.S. (2003) *Synthetic Metals*, **138**, 501–6.
83 Oshio, H., Yamamoto, M., Hoshino, N. and Ito, T. (2001) *Polyhedron*, **20**, 1621–5.
84 Rakowsky, M.H., More, K.M., Kulikov, A.V., Eaton, G.R. and Eaton, S.S. (1995) *Journal of the American Chemical Society.*, **117**, 2049–57.
85 Rakowsky, H., Zecevic, A., Eaton, G.R. and Eaton, S.S. (2001) *Journal of Magnetic Resonance*, **20**, 1621–5.
86 Caneschi, A., Gatteschi, D., Sessoli, R. and Rey, P. (1989) *Accounts of Chemical Research*, **22**, 392
87 Shultz, D.A. (2002) *Comments on Inorganic Chemistry*, **23**, 1–21.
88 Ovcharenko, V.I., Fokin, S.V., Romanenko, G.V., Tretyakov, E.V., Boltacheva, N.S., Filyakova, V.I. and Charushin, V.N. (2006) *Russian Chemical Bulletin*, **55**, 2122–4.
89 Ovcharenko, V.I., Fokin, S.V., Romanenko, G.V., Ikorskii, V.N., Tretyakov, E.V., Vasilevsky, S.F. and Sagdeev, R.Z. (2002) *Molecular Physics*, **100**, 1107–15.
90 Pfeiffer, S., Mingotaud, C., Garrigou-Lagrange, C., Delhaes, P. and Sastre, A.T. (1995) *Langmuir: The ACS Journal of Surfaces and Colloids*, **11**, 2705–12.
91 Martini, G., Bonosi, F., Ottaviani, M.F. and Gabrielli, G. (1989) *Thin Solid Films*, **178**, 271–9.
92 Risse, T., Hill, T., Schmidt, J., Abend, G., Hamann, H. and Freund, H.-J. (1998) *Journal of Physical Chemistry. B*, **102**, 2668–76.
93 Ruthstein, S., Artzi, R., Goldfarb, D. and Naaman, R. (2005) *Physical Chemistry Chemical Physics: PCCP*, **7**, 524–52.
94 Osipov, M.A., Gallani, J.-L. and Guillon, D. (2006) *The European Physical Journal E: Soft Matter*, **19**, 213–21.
95 Baute, D., Frydman, V., Zimmermann, H., Kababya, S. and Goldfarb, D. (2005) *Journal of Physical Chemistry. B*, **109**, 7807–16.
96 Ottaviani, M.F., Cangiotti, M., Famiglini, G. and Cappiello, A. (2006) *Journal of Physical Chemistry. B*, **110**, 10421–9.
97 Danilenko, A.M., Boguslavsky, E.G., Nadolinny, V.A., Gromilov, S.A., Grigor'ev, I.A. and Rejerse, E. (2003) *Applied Magnetic Resonance*, **24**, 225–32.
98 Chavez, L., Bain, E., Eastman, M., Porter, T.L. and Parnell, R. (2003) *Langmuir: The ACS Journal of Surfaces and Colloids*, **19**, 1143–7.
99 Parkhomyuk-Ben Arye, P., Strashnikova, N. and Likhtenshtein, G.I. (2002) *Journal of Biochemical and Biophysical Methods*, **51**, 1–15.
100 Dumestre, A., Spagnuolo, M., Bladon, R., Berthelin, J. and Baveye, P. (2006)

Environmental Pollution (Barking, Essex: 1987), **143**, 73–80.

101 Chechik, V., Wellsted, H.J., Alexander Korte, A., Gilbert, B.C., Caldararu, H., Ionita, P. and Caragheorgheopol, A. (2004) *Faraday Discussions*, **125**, 279–29.

102 Scaiano, J.C., Laferriere, M., Galian, R.E., Raquel, E., Maurel, V. and Billone, P. (2006) *Physica Status Solidi A: Applications and Materials Science*, **203**, 1337–43.

103 Molochnikov, L.S., Kovalyova, E.G., Lipunov, I.N. and Grogor'ev, I.A. (1997) Book of abstracts, 214th ACS National Meeting, September 7–11, Las Vegas, NV.

104 Molochnikov, L.S., Kovalyova, E.G., Grigor'ev, I.A. and Zagorodni, A.A. (2004) *Journal of Physical Chemistry. B*, **108**, 1302–13.

105 Molochnikov, L.S., Kovalyova, E.G., Medyantseva, E.L., Kirilyuk, I.A. and Grigor'ev, I.A. (2005) Abstracts of Conference SPIN-2005, September 20–24, Novosibirsk, Russia.

106 Mendez, A., Bosch, E. and Roses, M. (2003) *Journal of Chromatography. A*, **986**, 33–9.

107 Timofeeva, M.N., Ayupov, A.B., Volodin, A.M., Pak, Yu.R., Volkova, G.G. and Echevskii, G.V. (2005) *Kinetics Catalysis*, **46**, 123–7.

9
Organic Functional Materials Containing Chiral Nitroxide Radical Units
Rui Tamura

9.1
Introduction

Among persistent or stable organic free radicals such as nitroxides [1, 2], verdazyls [3, 4], thioaminyls [5], certain hydrazyl [6], phenoxyls [7, 8], and carbon-centered radicals [9], nitroxide radicals (NRs) show outstanding thermodynamic stability ascribed to the delocalization of the unpaired electron over the N–O bond and hence the lack of the occurrence of dimerization [10]. In fact, sterically protected NRs bearing two quaternary carbon atoms attached to the nitroxyl group show a robust nature and therefore have found various practical applications, including a number of spin labels and probes for EPR spectroscopic studies [1, 11–13], the precursor of organic oxidants (oxoammonium ions) [14–21], the reactant for diastereoselective coupling reactions with prochiral radicals [13, 22–24], the controller for living free-radical polymerization processes [13, 25–27], the spin source for elaboration of purely organic, paramagnetic materials [28–30], and the ligand for transition metal complexes [1]. Furthermore, if the molecular, supramolecular, or superstructural design is appropriate, chiral nitroxides would exhibit unique optoelectronic, magneto-optical, and magnetoelectric properties in the condensed phases owing to: (1) the relatively large electric dipole moment (μ = 3 Debye) of the nitroxyl group; (2) the magnetic moment arising from the unpaired electron; and (3) the molecular chirality giving a helical structural motif. It should also be stressed that, nowadays, the development of the chiral NR chemistry owes a great deal to recent significant advances in HPLC analytical techniques using chiral stationary-phase columns as well as X-ray crystallographic analyses.

Summarized in this chapter are the structural, optical, electric, and magnetic properties of functional chiral NRs, which have mainly emerged during the past decade, together with recent applications of the redox properties of chiral and achiral NRs.

Nitroxides: Applications in Chemistry, Biomedicine, and Materials Science
Gertz I. Likhtenshtein, Jun Yamauchi, Shin'ichi Nakatsuji, Alex I. Smirnov, and Rui Tamura
Copyright © 2008 WILEY-VCH Verlag GmbH & Co. KGaA, Weinheim
ISBN: 978-3-527-31889-6

9.2
Synthesis and Structure of Chiral NRs

Optically active nitroxides have been prepared and utilized for a wide variety of purposes. Their synthetic methods are classified into two categories: one involves the construction of a chiral nitroxide structure in novel skeletal systems, and the other consists of the attachment of a preexisting achiral nitroxide unit to an optically active molecule [13]. The latter protocol was particularly useful for the preparation of biologically active spin-labeled derivatives of carbohydrates, amino acids, and nucleotides, and their synthesis was well surveyed in a review article by Braslau [13]. In this section, we focus on the former strategy by which diverse manipulation of molecular structures is possible.

9.2.1
Chiral Five-Membered Cyclic NRs

α-Nitronyl nitroxides (α-NNs), pyrrolidine-1-oxyls (PROXYLs), and oxazolidine-1-oxyl (DOXYLs) are the representative NRs classified into this category.

9.2.1.1 Chiral α-NNs

A series of stable achiral α-NNs were synthesized by condensation of 2,3-dimethyl-2,3-bis(hydroxyamino)butane sulfuric acid salt (**1**) and achiral aldehydes followed by oxidation with PbO_2, MnO_2, or $NaIO_4$ by Ullman in 1968 [31]. By using chiral aldehydes, α-NNs bearing the chiral component on the C(2) position of the imidazolyl ring were obtained. The optically active histidine analog **2**, an α-NN mimic of the imidazole group, was synthesized by Jorgenson in 1971 (Figure 9.1) [32]. Kahn prepared α-NN-substituted triazole derivatives (*R*)-**3** [33]. Recently, Inoue and Veciana synthesized enantiomerically pure phenyl α-NNs, (*S*)-**4** and (*R*)-**5**, respectively, for a study on magneto-optical phenomena (Figure 9.1, see also Section 9.3.1) [34, 35].

Rey has pointed out that this conventional procedure by Ullman for the preparation of bis(hydroxyamino) compound **1** by reduction of the precursor *vic*-dinitro compound with zinc in an ammonium chloride (Zn/NH_4Cl) buffered solution is less reliable in the case of dissymmetric *vic*-dinitro compounds [36]. This is because the resulting dissymmetric bis(hydroxyamino) compounds are unstable in solution and susceptible to gradual decomposition over the prolonged reaction time. Therefore, rapid reduction of a dinitro compound to the bis(hydroxyamino) derivative, followed by solvent-free condensation with an aldehyde, are necessary to overcome this intractable problem [37].

According to this scenario, the improved synthetic procedure for optically active and racemic α-NNs **6** having a stereogenic center at the C(4)-position of the imidazolyl ring was established with the aim of obtaining liquid crystalline compounds (Figure 9.2) [37]. This procedure consists of: (1) the synthesis of a dissymmetric *vic*-dinitro compound by the Kornblum reaction; (2) the enantiomeric resolution of the racemic dinitro compound by the diastereomer method; (3) the rapid reduction of the dinitro compound to the bis(hydroxyamino) derivative with Al/Hg

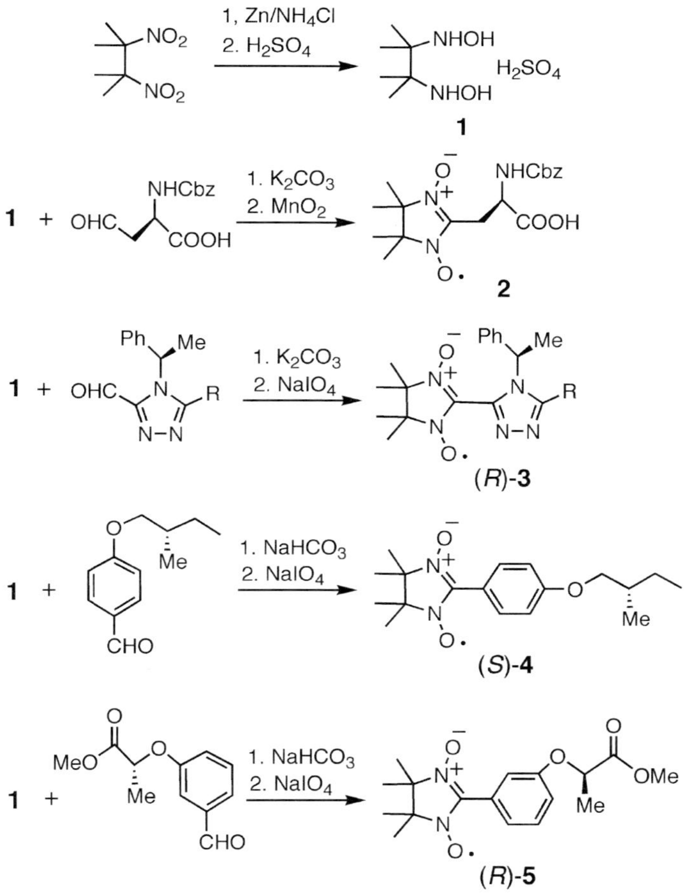

Figure 9.1 Synthesis of chiral α-NNs.

[38, 39]; (iv) the solvent-free condensation of the bis(hydroxyamino) compound with an aldehyde to give 1,3-dihydroximidazolidine; and (v) the final oxidation of the α-NN precursor with aqueous NaIO$_4$.

An alternative synthetic route to chiral α-NNs **8** bearing stereogenic centers at the C(4) and C(5) positions of the imidazolyl ring was proposed by Rey (Figure 9.3) [36]. This procedure involves the key condensation of 2,3-diamino-2,3-dimethylhexane and an aldehyde followed by 3-chloroperbenzoic acid-oxidation of the resulting imidazolidine into α-NN **8**. By using this modified procedure, enantiopure α-NNs **8** were prepared from (4R,5R)- and (4S,5S)-**7**, both of which were obtained by enantiomeric resolution of racemic **7** using the diastereomer method [40].

9.2.1.2 Chiral PROXYLs

In 1983, with a view to preparing a series of spin labels, Keana developed a synthetic method for chiral racemic *trans*-2,5-difunctionalized 2,5-dimethyl-PROXYLs

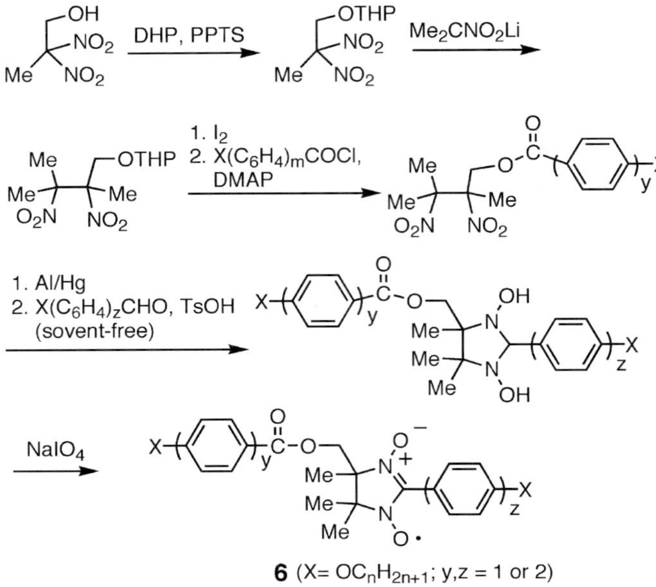

Figure 9.2 Synthesis of chiral α-NNs.

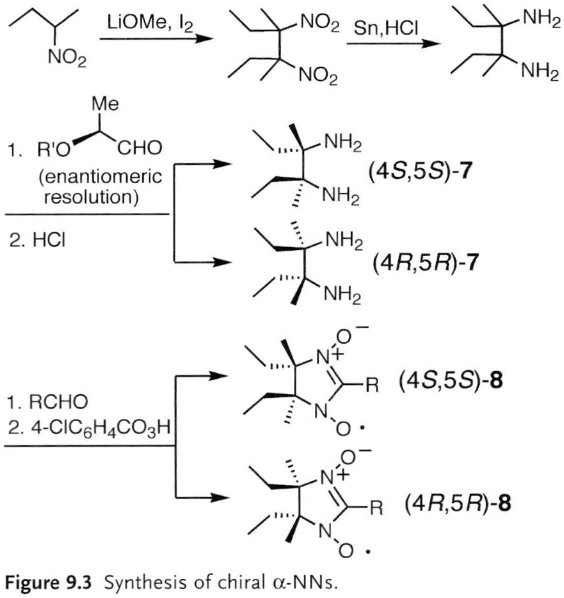

Figure 9.3 Synthesis of chiral α-NNs.

by tandem nucleophilic addition–oxidation sequences on racemic nitrone **9**, which was obtained by the oxidation of 2,5-dimethylpyrrolidine (Figure 9.4a) [41]. Recent progress in organic chemistry using this type of PROXYL derivatives deserves attention.

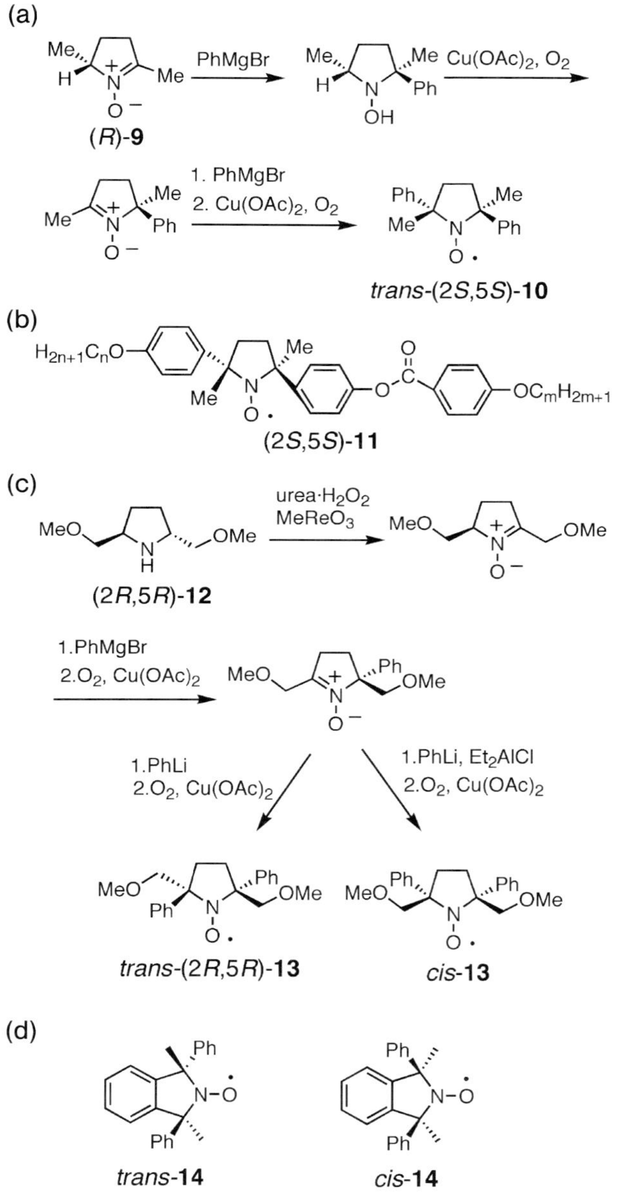

Figure 9.4 Synthesis of chiral PROXYLs.

9.2.1.2.1 Synthesis Selective synthesis of C_2-symmetric *trans*-(2*S*,5*S*)-2,5-dimethyl-2,5-diphenylpyrrolidine-1-oxyl (**10**) was achieved by Einhorn starting from the *R*-enriched nitrone **9** (Figure 9.4a) [42]. Since the *S*-enriched nitrone **9** is also available from *S*-enriched 2,5-dimethylpyrrolidine [43], both enantiomers of **10** and their derivatives can be prepared. Recently, by using this synthetic procedure, 2*S*,5*S*-enriched all-organic liquid crystalline radical compounds **11** were prepared (Figure 9.4b, also see Section 9.4.2) [44–46].

In 2000, Yamamoto reported the synthesis of *trans*-(2*R*,5*R*)-**13** from *trans*-(2*R*,5*R*)-**12** by an analogous procedure (Figure 9.4c) [47]. Interestingly, in the presence of Et_2AlCl as a Lewis acid, *cis*-**13** was selectively obtained. The syntheses of racemic samples of C_2-symmetric isoindoline nitroxide *trans*-**14** and meso *cis*-**14** were achieved by two different routes by Braslau (Figure 9.4d) [48].

9.2.1.2.2 **Conglomerates** A racemic conglomerate, which gives a mixture of enantiopure *R* and *S* chiral crystals by spontaneous resolution from the supersaturated solution, and hence can benefit from enantiomeric resolution by the preferential crystallization method [49], is very rare for chiral nitroxides.

In 1994, a new type of chiral bicyclic PROXYL **16** and **17** was prepared by the reduction of homoallylic nitro enones **15** with SmI_2 in THF and HMPA/THF, respectively (Figure 9.5a) [50–52]. Nitroxides **16** bearing one or two nitro groups on the meta position of the benzene ring were found to exist as a racemic conglomerate. By X-ray crystallographic analysis of **16** (Ar = 3,5-$(NO_2)_2C_6H_3$, R = Me), it was found that the molecules were arranged around a threefold screw axis by the electrostatic intermolecular interactions between the nitroxyl oxygen atom and the nitrogen atom of one of two nitro groups on the benzene ring (O–N distance 2.87 Å) with a space group of $P3_1$ or $P3_2$, eventually resulting in a head-to-tail arrangement around this axis (Figure 9.5b) [50, 53].

More recently, it was found that racemic samples of chiral PROXYLs **18**, *trans*-**19**, and *trans*-**20** bearing one or two 4-hydroxyphenyl groups on the stereogenic centers [C(2), or C(2) and C(5)] attached to the nitroxyl group existed as a racemic conglomerate in the stable crystalline state, whereas **21** and *trans*-**22** having one or two 3-hydroxyphenyl groups at the same positions belonged to a racemic compound (Figure 9.6a). The origin of the conglomerate formation with respect to **18**, *trans*-**19**, and *trans*-**20** can be interpreted in terms of strong intermolecular OH···ON hydrogen bonds, together with weak attractive forces such as $C(sp^n)H···ON$ interactions and dipole–dipole interactions between the NO groups, by X-ray crystallographic analysis and magnetic susceptibility measurements (Figure 9.6b) [54, 55].

9.2.1.2.3 **Racemization** Two examples of racemization of chiral cyclic nitroxides were documented; one was a self-racemization in solution, while the other was induced by chemical oxidation.

Enantiomerically enriched samples of chiral PROXYLs **18**, *trans*-**19**, and *trans*-**20** with one or two 4-hydroxyphenyl groups underwent unprecedented self-racemization and/or epimerization in aprotic solvents, whereas **21** and *trans*-**22** having one or two 3-hydroxyphenyl groups were not isomerized [56]. Racemization proceeded promptly in EtOAc or was dramatically accelerated by irradiating with a 27 W fluorescent lamp in CH_2Cl_2 at 25 °C, implying a radical mechanism. In contrast, racemization was greatly suppressed in protic solvents or hydrogen-bonding acceptor solvents such as EtOH or THF, or at low temperatures such as

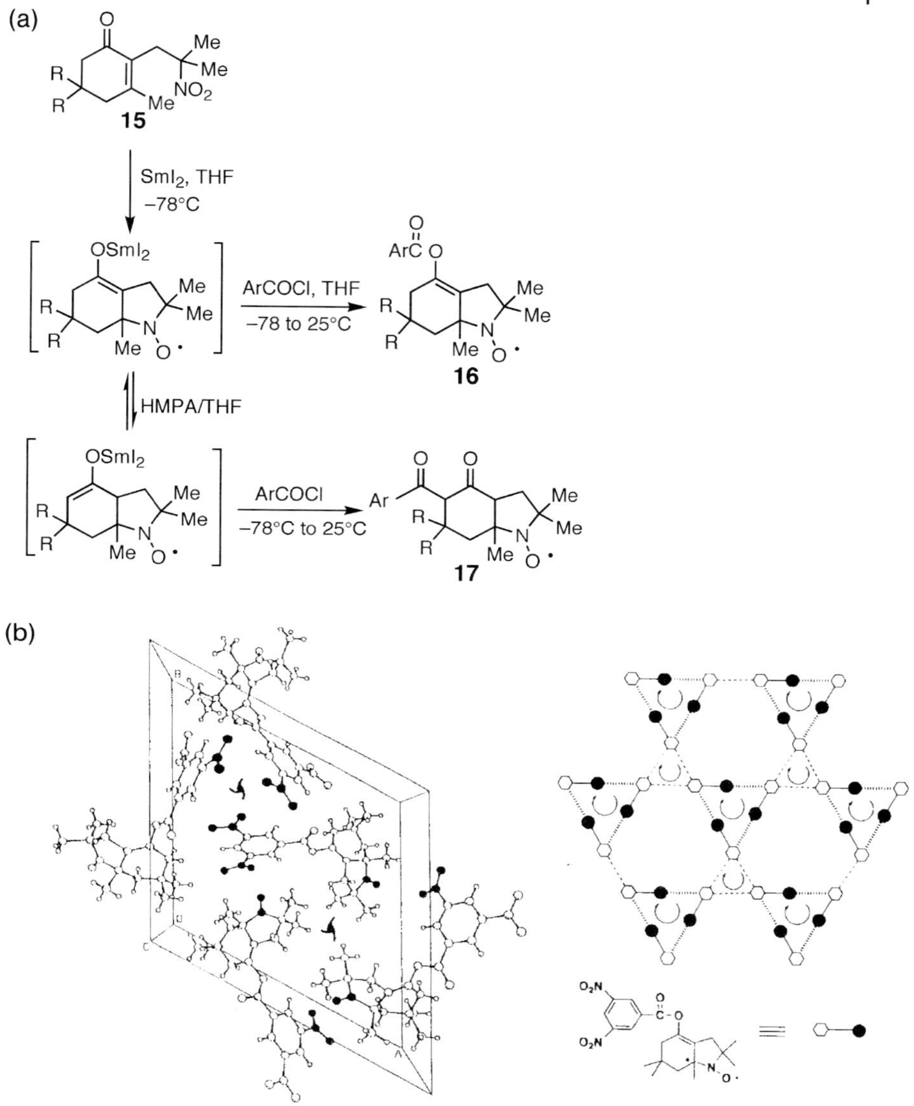

Figure 9.5 Synthesis of PROXYL type of chiral bicyclic nitroxides **16** and **17**, and X-ray crystal structure of conglomerate **16** (Ar = 3,5-$(NO_2)_2C_6H_3$, R = Me).

−20 °C, even in CH_2Cl_2. The mechanism of epimerization of *trans*-(2*S*,5*S*)-**20** could be accounted for in terms of the multi-step equilibrations involving: (a) the intermolecular abstraction of a phenolic hydrogen atom by the neighboring nitroxyl radical to generate the phenoxy radical *trans*-**A** and the corresponding hydroxyamine; (b) the formation of the planar quinoid intermediate **B**; (c) the free rotation

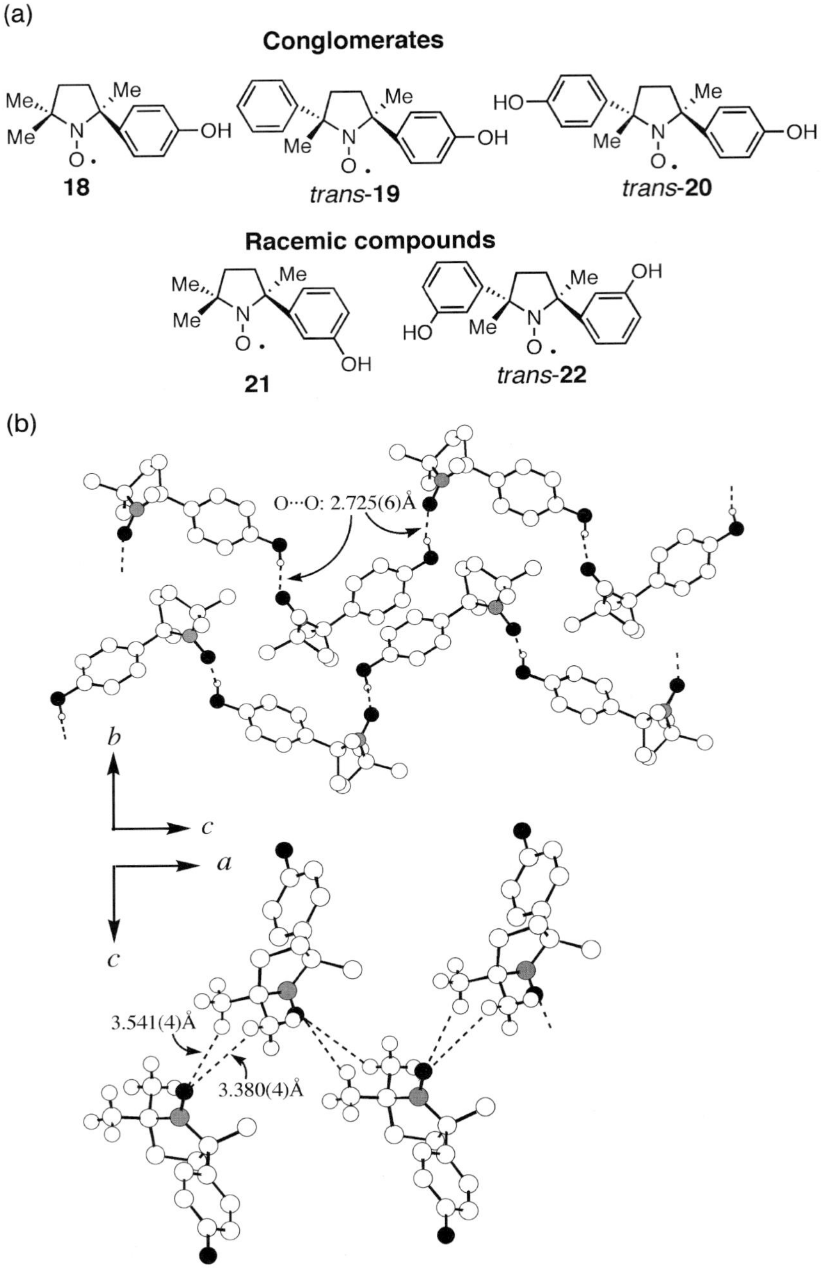

Figure 9.6 Conglomerates and racemic compounds, and X-ray crystal structure of conglomerate **18**.

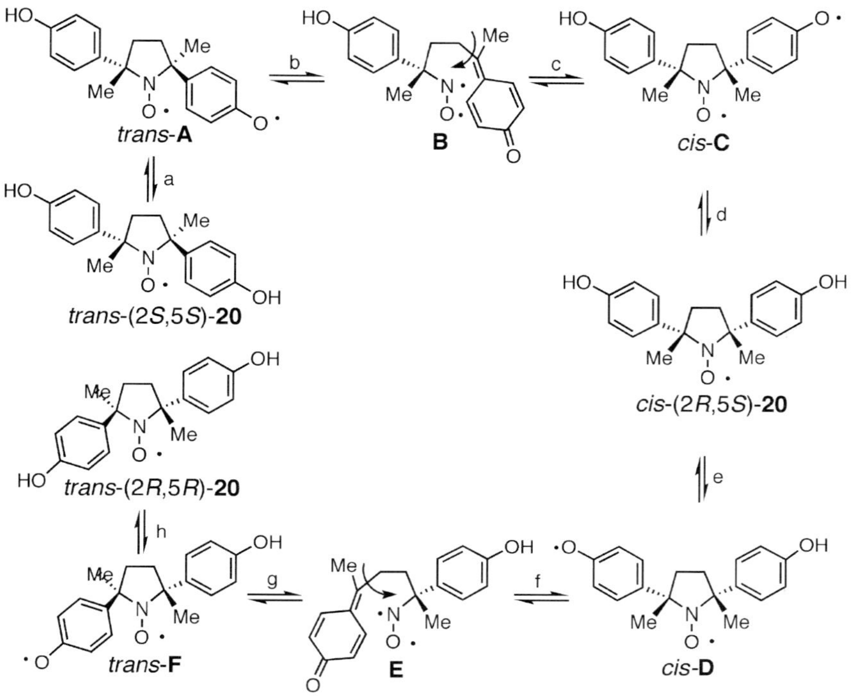

Figure 9.7 Mechanism of epimerization and racemization of trans-(2S,5S)-20.

of the quinoid moiety in **B** and the subsequent intramolecular regeneration of the C–N bond to give cis-**C**; and (d) the intermolecular hydrogen abstraction by the phenoxy radical of cis-**C** from the resulting hydroxyamine or another hydroxy group to give meso cis-(2R,5S)-**20** (Figure 9.7). Additional multi-step equilibrations involving: (e) the other phenoxy radical (cis-**D**) formation; (f) the quinoid intermediate (**E**) formation; (g) the intermolecular recombination to give trans-**F**; and (h) the intermolecular hydrogen abstraction to give trans-(2R,5R)-**20** can clearly explain the mechanism of the eventual racemization of trans-(2S,5S)-**20**.

Optically pure indane nitroxide **23** bearing a spirocarbon center attached to the nitroxyl group was susceptible to racemization under oxidizing conditions with 3-chloroperbenzoic acid in CH_2Cl_2 [57] (Figure 9.8). The mechanism of this racemization involves: (a) the solvolytic ring opening of N-oxoammonium salt **24**, an overoxidation product of the nitroxide **23**, to give the indane cation and nitroso group: and (b) the subsequent ring closure to give **24**.

9.2.1.2.4 Enantiomeric Resolution
Although enantiomeric resolution of the precursor racemic dialkylamines was very often accomplished by diastereomeric salt formation with a resolving acid to obtain enantiomerically enriched nitroxides after oxidation of the resolved amines, only two examples have been reported for the direct resolution of racemic PROXYLs.

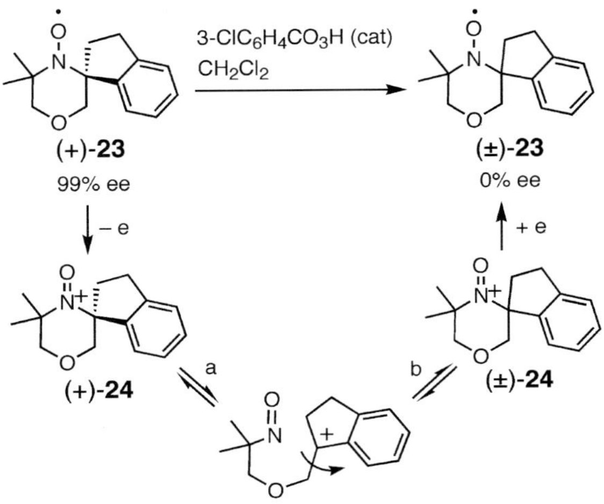

Figure 9.8 Mechanism of racemization of (+)-23.

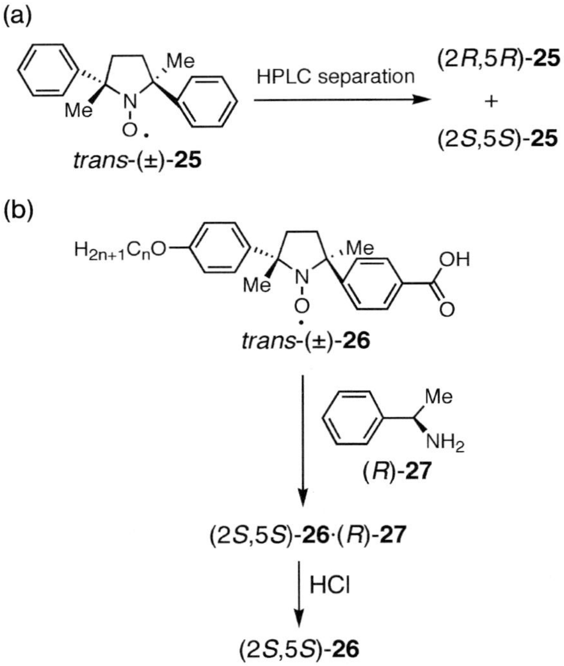

(a)

trans-(±)-25 →(HPLC separation)→ (2R,5R)-25 + (2S,5S)-25

(b)

trans-(±)-26

$H_{2n+1}C_nO$—

↓ (R)-27 (PhCH(Me)NH$_2$)

(2S,5S)-26·(R)-27

↓ HCl

(2S,5S)-26

Figure 9.9 Enantiomeric resolution of trans-(±)-25 and trans-(±)-26.

Separation of the two enantiomers of a racemic compound [(±)-25] was achieved by using a semipreparative chiral HPLC column (Figure 9.9a) [2, 58]. On the other hand, chemical resolution of a series of racemic trans-4-[5-(4-alkoxyphenyl)-2,5-dimethylpyrrolidine-1-oxyl-2-yl]benzoic acids (26), which are the key intermediates for the synthesis of chiral organic liquid crystalline radical compounds and are crystallized to give a racemic compound, was accomplished (Figure 9.9b) [59]. Racemic acid **26** with a long alkyl chain (C7 to C13) could be resolved by conventional diastereomeric salt formation using *(R)*- or *(S)*-1-phenylethylamine **27** as the resolving agent, whereas resolution of (±)-**26** with a short alkyl chain (C4 to C6) was unsuccessful. The use of 6 equiv of *(R)*- or *(S)*-**27** for the initial diastereomeric salt formation of (±)-**26** with a C7–C13 alkyl chain, followed by recrystallization of the resulting salts once or twice, gave 2*S*,5*S* or 2*R*,5*R*-enriched **26**, respectively, in an *ee* range of 75–92% and with an overall recovery of 11–27% based on the original quantity of (±)-**26**.

9.2.1.3 Chiral DOXYLs

The most common optically active nitroxides were the so-called DOXYL compounds, which could be easily derived from optically active ketones and achiral β-aminoalcohols (Figure 9.10a). By this method, a number of steroidal DOXYLs were stereoselectively prepared and used as spin labels (Figure 9.10b) [1, 13, 60, 61]. On the other hand, Braslau showed the opposite combination; camphonyl DOXYLs **28** were prepared from achiral acetal and the optically active β-aminoalcohol derived from (−)-camphene (Figure 9.10c) [13, 62].

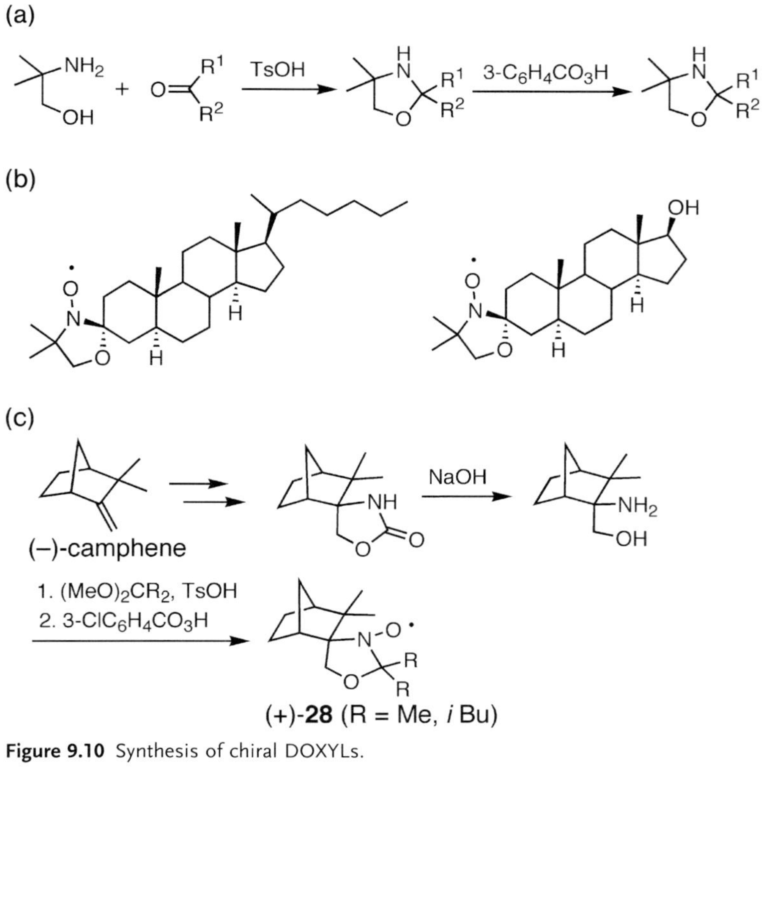

Figure 9.10 Synthesis of chiral DOXYLs.

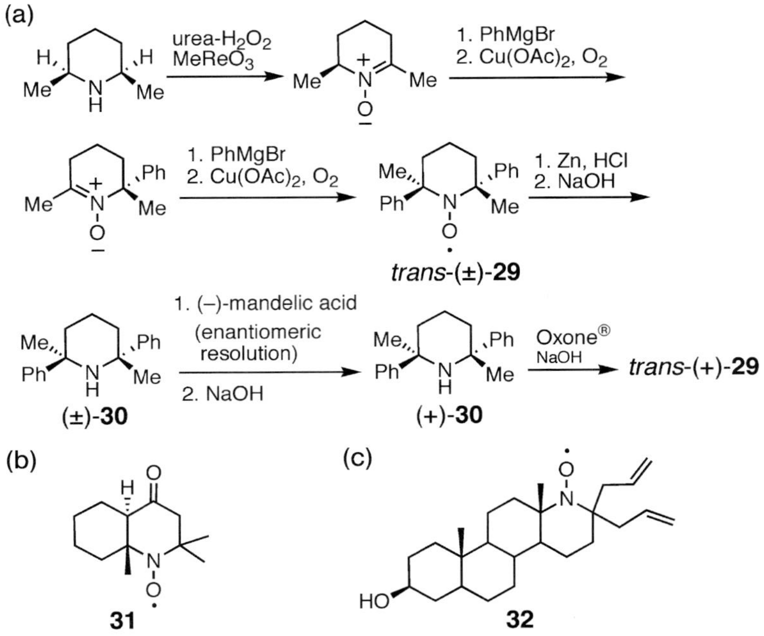

Figure 9.11 Synthesis of chiral six-membered cyclic nitroxides.

9.2.2
Chiral Six-Membered Cyclic NRs

C_2-Symmetric trans-(2R,6R)- and trans-(2S,6S)-2,6-dimethyl-2,6-diphenylpiperidin-1-oxyls (**29**), the piperidine analogs of **25**, were synthesized by the enantiomeric resolution of racemic piperidine **30** with (R)- or (S)-mandelic acid followed by oxidation with Oxone (Figure 9.11a) [63]. Optically active decahydroquinoline nitroxides **31** were obtained from diastereomeric derivatives after enantiomeric resolution (Figure 9.11b) [64]. Rassat synthesized the optically active steroidal nitroxide spin-probe **32**, in which the nitrogen atom was introduced into the steroidal D ring, starting from isoandrolactam acetate (Figure 9.11c) [65].

9.2.3
Miscellaneous Examples

To obtain an excellent catalyst for enantioselective oxidation of chiral secondary alcohols, Rychnovsky synthesized a C_2-symmetric seven-membered cyclic nitroxide **33** with axial chirality from enantiomerically enriched azepine [66] (Figure 9.12a, also see Section 9.5.1.2). Rassat prepared t-butylcamphenylnitroxide **34** in three steps from optically active camphoroxime (Figure 9.12b) [67].

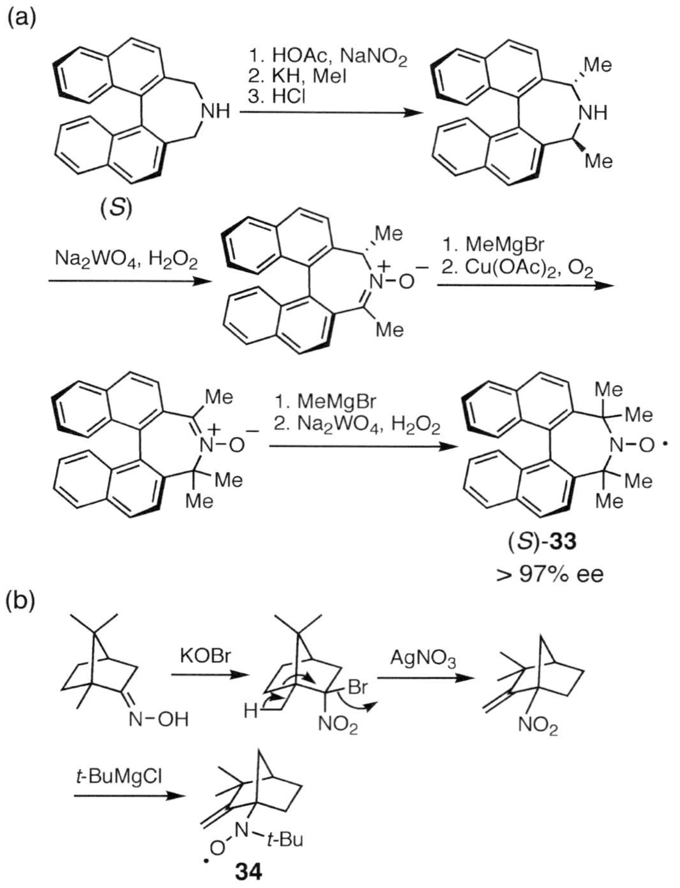

Figure 9.12 Synthesis of chiral nitroxides 33 and 34.

9.3
Magnetic Properties of Chiral NRs in the Solid State

9.3.1
Chiral Nitoxide-Mn²⁺ Complex Magnets

Stimulated by the observation of magneto-chiral dichroism (MChD) in luminescence from an optically active and paramagnetic europium (III) complex in solution [68], complexes of optically pure nitroxides with Mn(hfac)$_2$ were prepared and their magnetic properties in the solid state were fully characterized [69, 70]. Inoue prepared the optically active dinitroxide complex [(S)-35·Mn(hfac)$_2$]$_n$, trinitroxide complex [(R,R)-36·Mn(hfac)$_2$], and α-NN complex [(S)-4·Mn(hfac)$_2$]$_n$, which showed a metamagnetic behavior below 5.4 K, a ferrimagnetic behavior below 33 K, and a ferrimagnetic behavior at 4.6 K, respectively (Figure 9.13) [34, 71, 72].

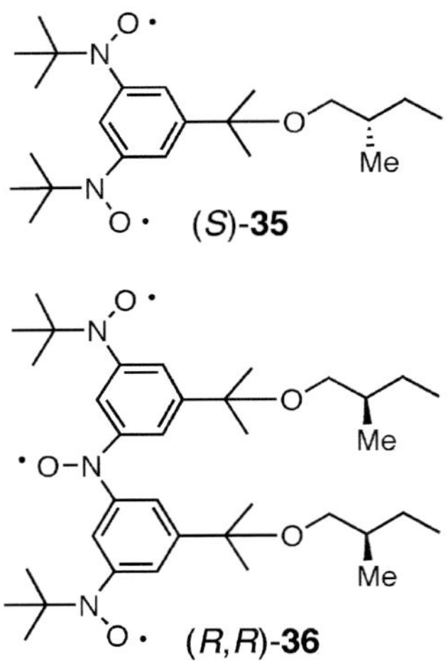

Figure 9.13 (S)-**35** and (R,R)-**36**.

Later, Veciana revealed that the α-NN complex [(R)-5·Mn(hfac)₂]ₙ showed a ferromagnetic phase transition at 3 K (Figure 9.14) [73]. Thus far, however, MChD has not been observed for these complexes.

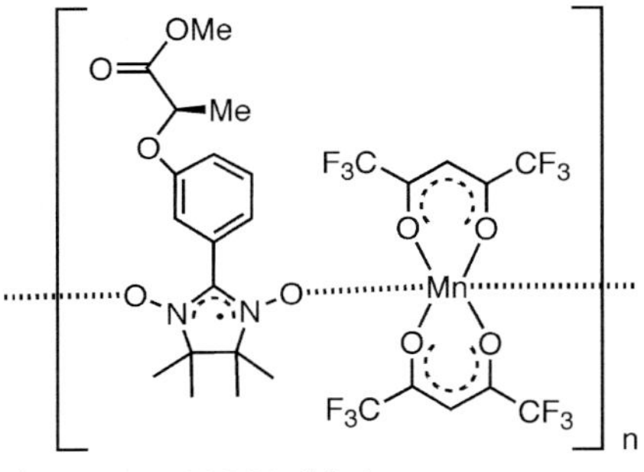

Figure 9.14 (R)-**5**·Mn(hfac)$_2$.

9.3.2
Chiral Multispin System

Chiral mononitroxide-, dinitroxide-, tetranitroxide-, and hexanitroxide-cyclotriphosphazene hybrid compounds (S)-**37a–37d** together with a chiral 2,2′-bridged trinitroxide-bis(cyclotriphosphazene) hybrid compound (S)-**38** were prepared to examine the potentiality for the use of the cyclotriphosphazene framework as a molecular scaffold for elaborating chiral multispin systems (Figure 9.15a) [74, 75]. EPR spectroscopic studies in solution and in frozen solvent matrices indicated that strong intramolecular through-space electron-exchange interactions were observed for the tetranitroxide and hexanitroxide hybrid compounds due to the intramolecu-

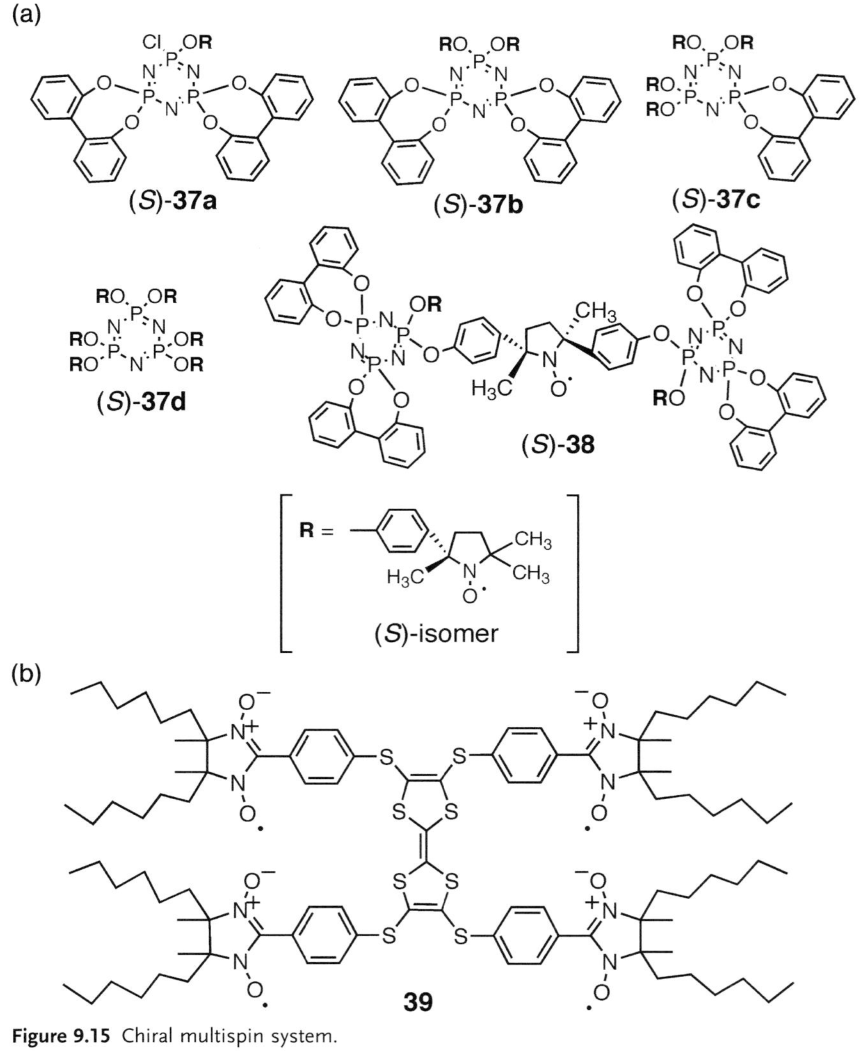

9.3 Magnetic Properties of Chiral NRs in the Solid State

Figure 9.15 Chiral multispin system.

lar electron-exchange interactions between two or three neighboring radical moieties disposed on the same side of the cyclotriphosphazene ring, while magnetic susceptibility measurements showed that weak antiferromagnetic interactions were uniformly recognized in the solid states for all of these. The X-ray crystallographic analysis of the dinitroxide hybrid compound (S)-**37b** ascribed the latter antiferromagnetic behavior to the intermolecular interactions between nitroxyl radicals [74, 75].

To study whether a p-phenylenethio unit serves as an intramolecular ferromagnetic coupler, the chiral TTF-based tetraradical donor **39** was prepared as a mixture of diastereomers (Figure 9.15b) and its electronic structure was characterized by spectroscopic measurements, magnetic susceptibility measurements, and electrochemical analyses [38]. Introduction of the dissymmetric α-NN group increased the solubility of the precursor containing four cyclic bis(hydroxyamino) groups.

9.4
Properties of Chiral NRs in the Liquid Crystalline State

Paramagnetic rod-like liquid crystals (LCs) have attracted great interest as soft materials to enhance the effect of magnetic fields on the optical and electric properties of liquid crystals [76]. These paramagnetic LC compounds are classified into two categories; the majority are the 3d- or 4f-metal-containing liquid crystals [77], while only a few LCs containing an organic spin center have been prepared, most likely because the geometry and bulkiness of the radical-stabilizing substituents are detrimental to the stability of LC phases, which require molecular linearity and planarity [78]. The large paramagnetic anisotropy ($\Delta\chi_{para}$) of the metal-containing LCs due to spin orbital coupling seems advantageous to the orientation control of LC molecules by magnetic fields [77, 79]. However, the intrinsic high viscosity of the ligand-coordinated metal-complex structure frequently renders the response to weak magnetic fields difficult. On the other hand, organic rod-like LCs with a stable nitroxyl group as a spin source can benefit from the low viscosity, although the $\Delta\chi_{para}$ is very small due to the p-orbital origin [78].

Several rod-like organic LCs with a DOXYL or TEMPO group as a spin source were prepared (Figure 9.16). However, their molecular structures were limited to those containing a nitroxyl group within the terminal alkyl chain, away from the rigid core, and hence allowed the free rotation of the nitroxyl moiety inside the molecule, leading to a decrease in the $\Delta\chi_{para}$ of the whole molecule. All attempts to prepare monomeric or polymeric LC compounds by using the nitroxyl group as part of the rigid core had been unsuccessful until the first PROXYL liquid crystalline compounds **11** were prepared by the present authors [44]. The molecular structures and magnetic properties of organic liquid crystalline nitroxyl radicals, together with the aim of each study, are briefly summarized here.

9.4.1
DOXYL and TEMPO Liquid Crystals

Chiral racemic and achiral compounds **40–42** were synthesized by Dvolaitzky to use as a LC spin-probe for EPR spectroscopic studies [80–82]. Racemic **40** showed stable smectic phases such as SmA, SmC, and SmE. The temperature dependence of the magnetic susceptibility (χ) was not measured.

With a view to measuring the magnetic properties of a LC structure at low temperatures, Finkelmann prepared chiral racemic radical polymer LC **43** which can retain the LC structure in the supercooled glassy phase [83]. By means of a Faraday balance, the temperature dependence of the molar magnetic susceptibility (χ_M) was measured in the temperature range from 6 to 350 K, in which the crystal-to-liquid crystal-to-liquid (C-to-LC-to-I) phase transition occurred. Consequently,

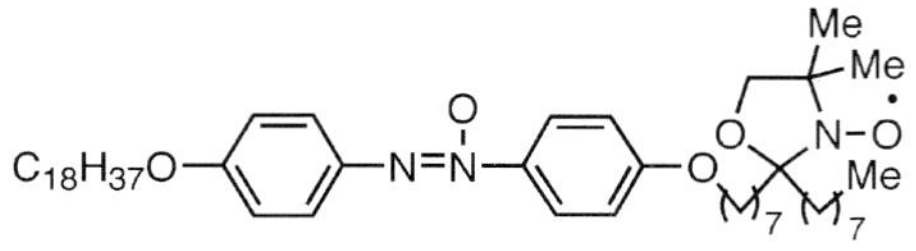

40 C 99 SmA 115 I

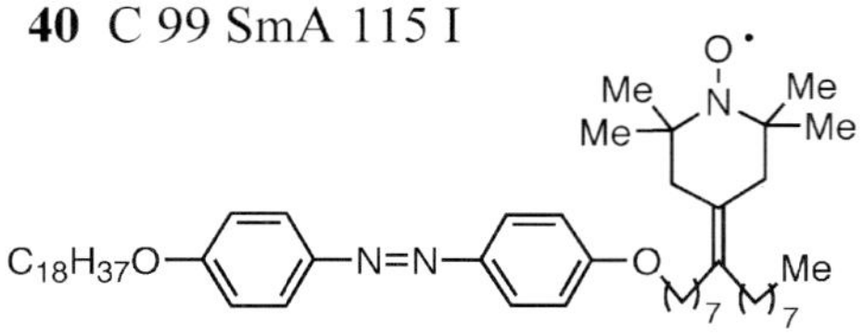

41 C 86 SmB 99 SmC 115 I

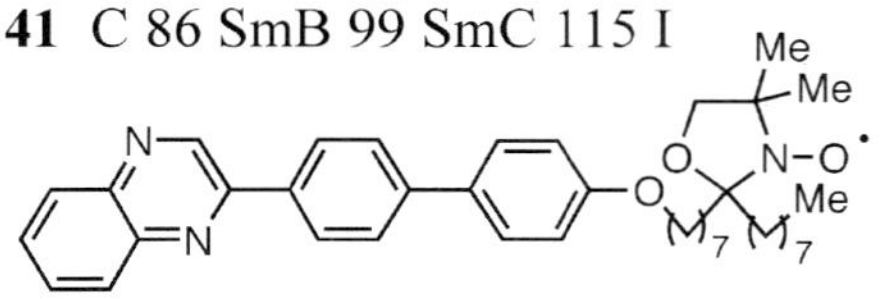

42 C 66-79 SmE 92-99 SmC 117-119 SmA 130-134 I

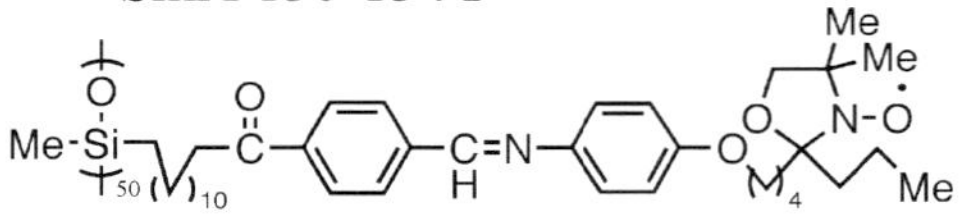

43 C –6 Sm 43 N 50 I

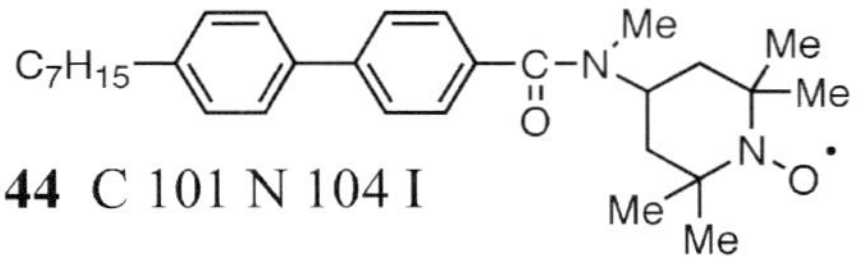

44 C 101 N 104 I

Figure 9.16 DOXYL and TEMPO liquid crystals.

upon the heating run in this temperature range, **43** showed no orientation change; no appreciable change in the χ_M was observed at the C-to-LC phase-transition temperature. This is most likely due to the high viscosity of the polymer material. It is desirable to measure the temperature dependence of the χ_M upon the cooling run from the isotropic phase on a SQUID susceptometer, as carried out for viscous metal-containing liquid crystals.

With the aim of preparing a supercooled glassy material and crystal polymorphs in the applied magnetic fields, and of observing the change in magnetic behavior accompanying the alteration in the solid-state structure, Nakatsuji et al. synthesized the achiral liquid crystalline compound **44** [84]. Achiral **44** showed the N phase within a narrow temperature range of 3 °C, with a distinct increase in χ_M at the C-to-LC transition upon the heating run. These authors also observed the difference in the magnetic behavior between the heating and cooling runs; **44** showed antiferromagnetic interactions based on a singlet–triplet model before the thermal phase transition upon the heating run of the crystals, while the magnetic behavior changed to a Curie–Weiss behavior after the thermal phase transition upon the cooling run from the isotropic phase.

9.4.2
PROXYL Liquid Crystals

It is expected that the hitherto-unknown mesoscopical ferromagnetic spin–spin interactions ($J > 0$) may be operative in a paramagnetic LC domain because of the "complexity" behavior of the molecules in the organic radical LCs, although the possibility of ferromagnetic LC materials has been considered unrealistic due to the inaccessibility of long-range spin–spin interactions between rotating molecules in the LC state [79]. Furthermore, the electric dipole moment (circa 3 Debye) of a nitroxyl group is large enough for the source of the spontaneous polarization (Ps) of ferroelectric liquid crystals. However, no precedent was reported for this use. In this context, of particular interest are the hitherto-unknown magnetic dipole–electric dipole interactions (magnetoelectric interactions) in the ferroelectric LC (FLC) state of organic radical compounds [85, 86], which may lead to an increase in the molecular $\Delta\chi_{para}$ and, thereby, enable the orientation control of organic FLC molecules by weak magnetic fields.

To observe and evaluate these unknown phenomena in the LC state, chiral nitroxides **11** with a negative dielectric anisotropy ($\Delta\varepsilon < 0$) were designed and synthesized [44] (Figure 9.17). Nitroxides **11** can satisfy the following three requirements.

- A nitroxyl radical group with a large electric dipole moment (circa 3 Debye) and known principal g-values (g_{xx}, g_{yy}, g_{zz}) [87] should be used as the spin source, because the dipole moment is enough for the source of the Ps and the known principal g-values are useful to understand the direction of molecular alignment in the LC phase by EPR spectroscopy.
- To avoid the free rotation of the nitroxyl group and maximize the paramagnetic anisotropy, a geometrically fixed chiral cyclic nitroxide unit should be incorporated into the core of LC molecules.
- To obtain a zigzag molecular structure advantageous for the appearance of a SmC phase, a *trans*-2,5-diphenylsubstituted pyrrolidine skeleton is the best choice.
- Since both chiral and achiral liquid crystals are required for comparison in their optical and magnetic properties, the molecule should be chiral and both racemic and enantiomerically enriched samples need to be available.

9.4.2.1 Phase-Transition Behavior

Transition temperatures of (±)-**11** are summarized in Figure 9.17a [44]. All of the racemic samples showed enantiotropic (thermally reversible) LC phases. (±)-**11** with short C4–C10 alkyl chains showed only a nematic phase, and upon the cooling run their supercooled N phases were preserved even at 25 °C. (±)-**11** with long C11–C13 chains showed both SmC and N phases, while (±)-**11** with longer C14 and C15 chains showed only a SmC phase.

9.4 Properties of Chiral NRs in the Liquid Crystalline State | 321

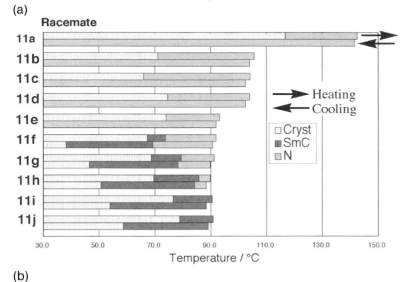

11a: m = n = 4
11b: m = n = 7
11c: m = 8, n = 7
11d: m = n = 8
11e: m = n = 10
11f: m = n = 11
11g: m = n = 12
11h: m = n = 13
11i: m = n = 14
11j: m = n = 15

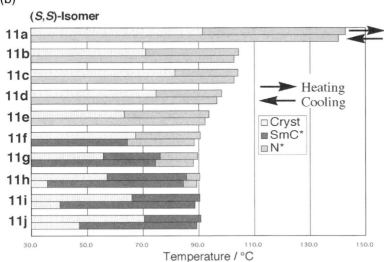

Figure 9.17 Phase-transition temperatures of (±)- and (2S,5S)-**11**.

Transition temperatures of (2S,5S)-**11** are shown in Figure 9.17b [44–46]. Similar to the racemic samples, (2S,5S)-**11** with C4–C10, with C12 and C13, and with C14 and C15 chains showed only a N* phase, both SmC* and N* phases, and only a SmC* phase, respectively. All (2S,5S)-**11** samples showed the liquid crystalline phase over a wider temperature range than the corresponding racemic samples upon the cooling run.

9.4.2.2 Ferroelectric Properties

(2S,5S)-**11** showing a SmC* phase indeed exhibited explicit ferroelectricity; a maximum spontaneous polarization (Ps) value of 24 nC cm^{-2}, an optical response time (τ) in the order of a hundred µs, and a layer tilt angle of 34 ~ 41° were recorded in a 4 µm sandwich cell (Figure 9.18) [45, 46]. Since intermolecular nonlinear

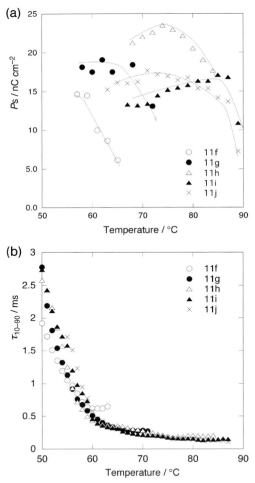

Figure 9.18 Ferroelectric properties of (2S,5S)-**11**.
(a) Spontaneous polarization (Ps). (b) Optical response time (τ).

mesoscopical–magnetic interactions were observed in the SmC phase as shown in Section 9.4.2.3, it is expected that a magnetoelectric effect may be observed in the FLC state of **11** [85, 86].

9.4.2.3 Nonlinear Mesoscopical–Magnetic Interactions

Contrary to general expectations, unique intermolecular nonlinear mesoscopical–magnetic interactions were explicitly observed in the SmC phase of **11** by measuring the temperature dependence of the χ_M on a SQUID magnetometer (Figure 9.19a) (Uchida, Y., Ikuma, N., Tamura, R., et al., manuscript submitted for publication). In fact, the magnetic interactions actually allowed the LC particles and grains on water to be attracted by a weak permanent magnet (<0.5 T) (Figure 9.19c). Further observation of the nonlinear relationship between the applied magnetic field and the observed magnetization (Figure 9.19b) supports the occurrence of weak intermolecular ferromagnetic-like interactions that allow a coherent collective motion of the LC molecules under the influence of weak magnetic fields. Such magnetic interactions may be useful in the development of nonmetallic paramagnetic advanced materials usable at ambient temperature, such as in a drug delivery system.

9.5 Application of Redox Properties of NRs

TEMPO-based nitroxide radicals are very stable and can be reversibly oxidized into the corresponding oxoammonium cations, both chemically and electrochemically ($E_{1/2}$ = 0.27 ~ 0.44 V vs Ag/AgNO$_3$) [2, 88]. Therefore, such redox properties have been utilized for the environmentally benign oxidation of alcohols and the development of an organic cathode-active material for rechargeable batteries.

9.5.1 Oxidation Catalyst

9.5.1.1 Achiral Catalyst

Rozantsev reported that the N-oxoammonium salt **45**, which is obtained by the oxidation of a TEMPO radical, oxidizes primary and secondary alcohols to aldehydes and ketones [15] (Figure 9.20a). Later, Semmelhack showed that the N-oxoammonium ion acts as a mediator for the electrolytic oxidation of alcohols [89], and that the presence of a bulk oxidant transforms nitroxyl radicals to N-oxoammonium salts that, in turn, rapidly oxidize alcohols to aldehydes or ketones. The resulting hydroxyamines are then reoxidized by the bulk oxidant to N-oxoammonium salts to complete the catalytic cycle [90]. Since then, a number of bulk oxidants have been used for the efficient and selective oxidation of alcohols [16–18]. Consequently, environmentally benign and economic catalytic systems using O$_2$ and catalysts such as TEMPO/RuCl$_2$(PPh$_3$)$_3$ and TEMPO/Br$_2$/NaNO$_2$ have been devised (Figure 9.20b) [19–21].

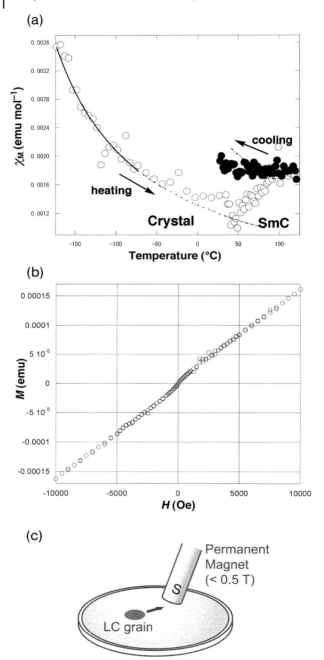

Figure 9.19 Magnetic properties of (±)-**11h** in the SmC phase. (a) Temperature dependence of χ_M. (b) Magnetic field *(H)* dependence of magnetization *(M)*. (c) Attraction by a weak permanent magnet of an LC grain on water.

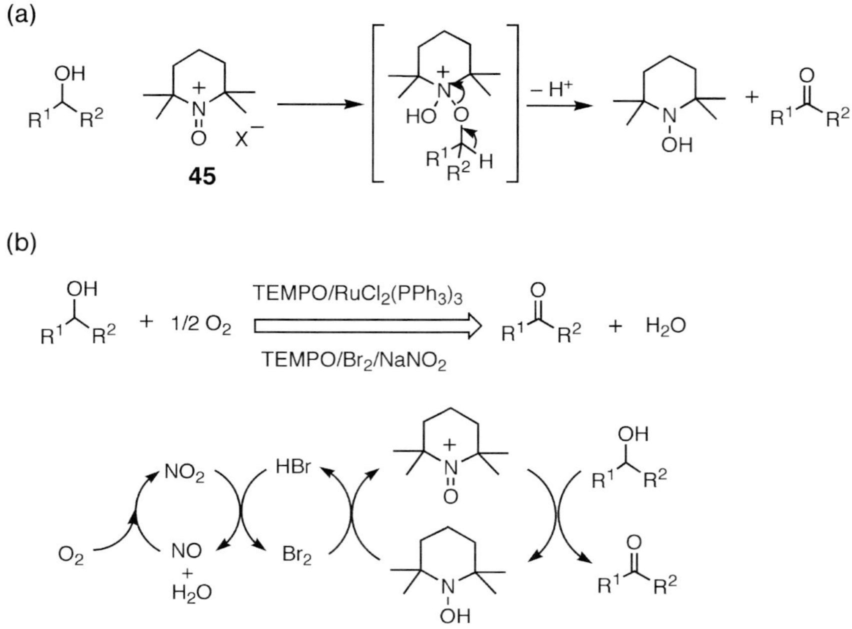

Figure 9.20 Oxidation of alcohols by N-oxoammonium salt. (a) Stoichiometric reaction. (b) Environmentally benign and economic catalytic system.

9.5.1.2 Chiral Catalyst

By analogy based on the resolution of racemic secondary alcohols achieved by enantioselective oxidation using the redox enzyme horse liver alcohol dehydrogenase (HLADH), Bobbitt investigated the use of optically active piperidine nitroxides **46** prepared from commercially available (+)-dihydrocarvone in the enantioselective oxidation of a meso diol, as well as in the kinetic resolution of a racemic alcohol (Figure 9.21a) [91]. Later, Rychnovsky reported the efficient, enantioselective oxidation of secondary alcohols using the C_2-symmetric seven-membered cyclic nitroxide **33** as a catalyst (Figure 9.21b) [66]. Furthermore, Osa and Bobbitt demonstrated efficient kinetic resolution of racemic secondary alcohols by using a TEMPO-modified graphite felt electrode in the presence of (−)-spartane [92, 93].

9.5.2 Radical Battery

With a view to developing environmentally benign organic rechargeable batteries with a high energy-density, Nakahara utilized a stable nitroxyl polyradical, poly(2,2,6,6-tetramethylpiperidinyloxy methacrylate) (PTMA), as a new class of cathode-active material for rechargeable lithium batteries (Figure 9.22) [94, 95]. These fabricated PTMA/Li batteries demonstrated an average discharge voltage of 3.5 V and initial

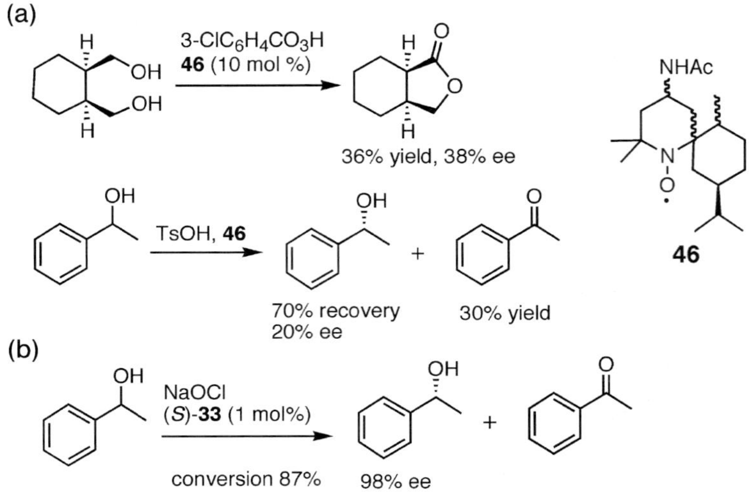

Figure 9.21 Enantioselective oxidation of racemic secondary alcohols.

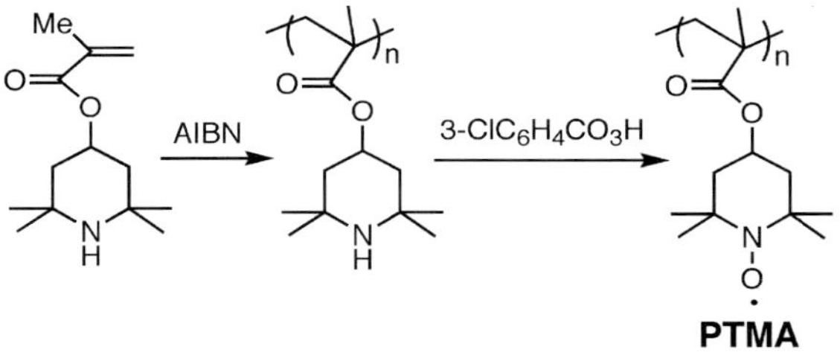

Figure 9.22 Stable achiral nitroxyl polyradical PTMA utilized as a new class of cathode-active materials.

discharge capacity of 77 Ah kg^{-1} of the PTMA weight at a current density of 0.1 mA cm^{-2}, which corresponds to 70% of the theoretical capacity. The capacity remains unchanged for over 500 cycles of charging and discharging at a high current density of 1.0 mA cm^{-2}. The charge–discharge curves indicate that the charge process at the cathode consists of oxidation of PTMA to the oxoammonium salt, while the discharge process involves reduction of the oxoammonium salt to PTMA.

9.6
Conclusion

Chiral nitroxide chemistry has been making gradual progress for the past two decades, as described in this chapter, and will continue to find further fundamental uses and practical applications for NRs in the research fields of both materials and life sciences. It is noteworthy that the introduction of chirality into NRs can lead to structural diversity and uniqueness in their condensed phases, and thereby new optical, electric, and even magnetic properties may emerge. The recent advent of nonlinear mesoscopical–magnetic and ferroelectric organic LC NRs is a good example of this case. Alternatively, the recent development of organic NR rechargeable batteries could open up a new research field for NRs in materials science.

References

1 Volodarsky, L.B., Reznikov, V.A. and Ovcharenko, V.I. (1994) *Synthetic Chemistry of Stable Nitroxides*, CRC Press, Boca Raton.
2 Aurich, H.G. (1989) *Nitrones, Nitronates and Nitroxides* (eds S. Patai and Z. Rappoport), John Wiley, pp. 371–99.
3 Neugebauer, F.A. (1973) *Angewandte Chemie (International Ed. in English)*, **12**, 455–64.
4 Azuma, N., Ishizu, K. and Mukai, K. (1974) *The Journal of Chemical Physics*, **61**, 2294–6.
5 Miura, Y. (1990) *Reviews on Heteroatom Chemistry*, **3**, 211–32.
6 Goldschmidt, S. and Renn, K. (1922) *Chemische Berichte*, **55**, 628–43.
7 Mukai, K., Ueda, K., Ishizu, K. and Deguchi, Y. (1982) *The Journal of Chemical Physics*, **77**, 1606–7.
8 Awaga, K., Sugano, T. and Kinoshita, M. (1986) *The Journal of Chemical Physics*, **85**, 2211–15.
9 Armet, O., Veciana, J., Rovira, C., Riera, J., Castaner, J., Molins, E., Rius, J., Miravitlles, C., Olivella, S. and Brichfeus, J. (1987) *Journal of Physical Chemistry*, **91**, 5608–16.
10 Griller, D. and Ingold, K.U. (1976) *Accounts of Chemical Research*, **8**, 13–19.
11 Likhtenstein, G.I. (1990) *Pure and Applied Chemistry. Chimie Pure et Appliquée*, **62**, 281–8.
12 Marsh, D. (1990) *Pure and Applied Chemistry. Chimie Pure et Appliquée*, **62**, 265–70.
13 Naik, N. and Braslau, R. (1998) *Tetrahedron*, **54**, 667–96.
14 Yamaguchi, M., Miyazawa, T., Takata, T. and Endo, T. (1990) *Pure and Applied Chemistry. Chimie Pure et Appliquée*, **62**, 217–22.
15 Rozantsev, E.G. and Sholle, V.D. (1971) *Synthesis*, 401–14.
16 Bobbitt, J.M. and Flores, M.C.L. (1988) *Heterocycles*, **27**, 509–33.
17 Yamaguchi, M., Miyazawa, T., Takata, T. and Endo, T. (1990) *Pure and Applied Chemistry. Chimie Pure et Appliquée*, **62**, 217–22.
18 de Nooy, A.E.J., Besemer, A.C. and van Bekkum, H. (1996) *Synthesis*, 1153–74.
19 Sheldon, R.A., Arends, I.W.C.E., Brink, G.-J. T. and Dijksman, A. (2002) *Accounts of Chemical Research*, **35**, 774–81.
20 Lenioir, D. (2005) *Angewandte Chemie (International Ed. in English)*, **45**, 3206–10.
21 Liu, R., Liang, X., Dong, C. and Hu, X. (2004) *Journal of the American Chemical Society*, **126**, 4112–13.
22 Braslau, R., Burril, L.C., II, Mahal, L.K. and Wedeking, T. (1997) *Angewandte Chemie (International Ed. in English)*, **36**, 237–8.
23 Braslau, R., Burrill, L.C., II, Chaplinski, V., Howden, R. and Papa, P.W. (1997) *Tetrahedron: Asymmetry*, **8**, 3209–12.
24 Braslau, R., Naik, N. and Zipse, H. (2000) *Journal of the American Chemical Society*, **122**, 8421–34.
25 Puts, R.D. and Sogah, D.Y. (1996) *Macromolecules*, **29**, 3323–5.
26 Ananchenko, G. and Matyjaszewski, K. (2002) *Macromolecules*, **35**, 8323–9.
27 Zhou, N., Xu, W., Zhang, Y., Zhu, J. and Zhu, X. (2006) *Journal of Polymer Science Part A: Polymer Chemistry*, **44**, 1522–8.
28 Tamura, M., Nakazawa, Y., Shiomi, D., Nozawa, K., Hosokoshi, Y., Ishikawa, M., Takahashi, M. and Kinoshita, M. (1991) *Chemical Physics Letters*, **186**, 401–4.
29 Chiarelli, R., Novak, M.A., Rassat, A. and Tholence, J.L. (1993) *Nature*, **363**, 147–9.
30 Nokami, T., Ishida, T., Yasui, M., Iwasaki, F., Takeda, N., Ishikawa, M., Kawakami, T. and Yamaguchi, K. (1996) *Bulletin of the Chemical Society of Japan*, **69**, 1841–8.
31 Ullman, E.F., Osiecki, J.H., Boocock, D.G.B. and Darcy, R. (1972) *Journal of the American Chemical Society*, **94**, 7049–59.
32 Weinkam, R.J. and Jorgensen, E.C. (1971) *Journal of the American Chemical Society*, **93**, 7028–33.

33 Sutter, J.-P., Golhen, S., Ouahab, L. and Kahn, O. (1998) *Comptes Rendus de l'Academie des Sciences Serie IIc: Chemie*, **1**, 63–8.
34 Kumagai, H., Markosyan, A.S. and Inoue, K. (2000) *Molecular Crystals and Liquid Crystals*, **343**, 97–102.
35 Minguet, M., Amabilino, D.B., Cirujeda, J., Wurst, K., Mata, I., Molins, E., Novoa, J.J. and Veciana, J. (2000) *Chemistry – A European Journal*, **6**, 2350–61.
36 Hirel, C., Vostrikova, K.E., Pecaut, J., Ovcharenko, V.I. and Rey, P. (2001) *Chemistry – A European Journal*, **7**, 2007–14.
37 Shimono, S., Tamura, R., Ikuma, N., Takimoto, T., Kawame, N., Tamada, O., Sakai, N., Matsuura, H. and Yamauchi, J. (2004) *Journal of Organic Chemistry*, **69**, 475–81.
38 Harada, G., Jin, T., Izuoka, A., Matsushita, M.M. and Sugawara, T. (2003) *Tetrahedron Letters*, **44**, 4415–18.
39 Weis, C.D. and Newkome, G.R. (1995) *Synthesis*, 1053–65.
40 Hirel, C., Pécaut, J., Choua, S., Turek, P., Amabilino, D.B., Veciana, J. and Rey, P. (2005) *European Journal of Organic Chemistry*, 348–59.
41 Keana, J.F.W., Seyedrezai, S.E. and Gaughan, G. (1983) *Journal of Organic Chemistry*, **48**, 2644–7.
42 Einhorn, J., Einhorn, C., Ratajczak, F., Gautier-Luneau, I. and Pierre, J.-L. (1997) *Journal of Organic Chemistry*, **62**, 9385–8.
43 Beak, P., Kerrick, S.T., Wu, S. and Chu, J. (1994) *Journal of the American Chemical Society*, **116**, 3231–9.
44 Ikuma, N., Tamura, R., Shimono, S., Kawame, N., Tamada, O., Sakai, N., Yamauchi, J. and Yamamoto, Y. (2004) *Angewandte Chemie (International Ed. in English)*, **43**, 3677–82.
45 Ikuma, N., Tamura, R., Shimono, S., Uchida, Y., Masaki, K., Yamauchi, J., Aoki, Y. and Nohira, H. (2006) *Advanced Materials (Deerfield Beach, Fla.)*, **18**, 477–80.
46 Ikuma, N., Tamura, R., Masaki, K., Uchida, Y., Shimono, S., Yamauchi, J., Aoki, Y. and Nohira, H. (2006) *Ferroelectrics*, **343**, 119–25.
47 Shibata, T., Uemae, K. and Yamamoto, Y. (2000) *Tetrahedron: Asymmetry*, **11**, 2339–46.
48 Braslau, R. and Chaplinski, V. (1998) *Journal of Organic Chemistry*, **63**, 9857–64.
49 Jacques, J., Collet, A. and Wilen, S. (1994) *Enantiomers, Racemates, and Resolutions*, Krieger Publishing Co., Malabar.
50 Tamura, R., Susuki, S., Azuma, N., Matsumoto, A., Toda, F., Kamimura, A. and Hori, K. (1994) *Angewandte Chemie (International Ed. in English)*, **33**, 878–80.
51 Tamura, R., Susuki, S., Azuma, N., Matsumoto, A., Toda, F. and Ishii, Y. (1995) *Journal of Organic Chemistry*, **60**, 6820–5.
52 Tamura, R., Shimono, S., Fujita, K. and Hirao, K. (2001) *Heterocycles*, **54**, 217–24.
53 Tamura, R., Susuki, S., Azuma, N., Matsumoto, A., Toda, F., Takui, D., Shiomi, D. and Itoh, K. (1995) *Molecular Crystals and Liquid Crystals*, **271**, 91–6.
54 Ikuma, N., Tamura, R., Shimono, S., Kawame, O., Sakai, N., Yamauchi, J. and Yamamoto, Y. (2003) *Mendeleev Communications*, 109–11.
55 Ikuma, N., Tamura, R., Shimono, S., Kawame, N., Tamada, O., Sakai, N., Yamamoto, Y. and Yamauchi, J. (2005) *Molecular Crystals and Liquid Crystals*, **440**, 23–35.
56 Ikuma, N., Tsue, H., Tsue, N., Shimono, S., Uchida, Y., Masaki, K., Matsuoka, N. and Tamura, R. (2005) *Organic Letters*, **7**, 1797–800.
57 Rychnovsky, S.D., Beauchamp, T., Vaidyanathan, R. and Kwan, T. (1998) *Journal of Organic Chemistry*, **63**, 6363–74.
58 Benfaremo, N., Steenbock, M., Klapper, M., Müllen, K., Enkelmann, V. and Cabrera, K. (1996) *Justus Liebigs Annalen Der Chemie*, 1413–15.
59 Uchida, Y., Uematsu, T., Nakayama, Y., Takahashi, H., Tsue, H., Tanaka, K. and Tamura, R. (2007) *Chirality*, **19** (in press).
60 Keana, J.F., Keana, S.B. and Beetham, D. (1967) *Journal of the American Chemical Society*, **89**, 3055–6.
61 Rassat, A. and Michon, P. (1974) *Journal of Organic Chemistry*, **39**, 2121–4.
62 Braslau, R., Kuhn, H., Burrill, L.C., Lanham, K. and Stenland, C.J. (1996) *Tetrahedron Letters*, **37**, 7933–6.

63 Einhorn, J., Enihorn, C., Ratajczak, F., Durif, A., Averbuch, M.-T. and Pierre, J.-L. (1998) *Tetrahedron Letters*, **39**, 2565–8.
64 Roberts, J.S. and Thomson, C. (1972) *Journal of the Chemical Society, Perkin Transactions*, **2**, 2129–40.
65 Ramasseul, R. and Rassat, A. (1971) *Tetrahedron Letters*, 4623–4.
66 Rychnovsky, S.D., McLernon, T.L. and Rajapakse, H. (1996) *Journal of Organic Chemistry*, **61**, 1194–5.
67 Bunnel, Y., Lemaire, H. and Rassat, A. (1964) *Bulletin de la Societe Chimique de France*, 1895–9.
68 Rikken, G.L.J.A. and Raupach, E. (1997) *Nature*, **390**, 493–4.
69 Inoue, K., Ohkoshi, S. and Imai, H. (2005) *Magmetism: Molecules to Materials V* (eds J.S. Miller and M. Drillon), Wiley-VCH Verlag GmbH, Weinheim, pp. 41–70.
70 Amabilino, D.B. and Veciana, J. (2006) *Topics in Current Chemistry*, **265**, 253–302.
71 Kumagai, H. and Inoue, K. (1999) *Angewandte Chemie (International Ed. in English)*, **38**, 1601–3.
72 Ghalsasi, P.S., Inoue, K., Samant, S.D. and Yakhmi, J.V. (2001) *Polyhedron*, **20**, 1495–8.
73 Minguet, M., Luneau, D., Lhotel, E., Villar, V., Paulsen, C., Amabilino, D.B. and Veciana, J. (2002) *Angewandte Chemie (International Ed. in English)*, **41**, 586–9.
74 Shimono, S., Tamura, R., Ikuma, N., Takahashi, H., Sakai, N. and Yamauchi, J. (2004) *Chemistry Letters*, **33**, 932–3.
75 Shimono, S., Takahashi, H., Sakai, N., Tamura, R., Ikuma, N. and Yamauchi, J. (2005) *Molecular Crystals and Liquid Crystals*, **440**, 37–52.
76 Blinov, L.M. (1983) *Electro-Optical and Magneto-Optical Properties of Liquid Crystals*, John Wiley, New York.
77 Griesar, K. and Haase, W. (1999) *Magnetic Properties of Organic Materials* (ed. P.M. Lahti), Marcel Dekker, New York, pp. 325–44.
78 Kaszynski, P. (1999) *Magnetic Properties of Organic Materials* (ed. P.M. Lahti), Marcel Dekker, New York, pp. 305–24.
79 Dunmur, D. and Toriyama, K. (1999) *Physical Properties of Liquid Crsytals* (eds D. Demus, J. Goodby, G.W. Gray, H.-W. Spies and V. Vill), Wiley-VCH Verlag GmbH, Weinheim, pp. 102–12.
80 Dvolaitzky, M., Billard, J. and Polydy, F. (1974) *Comptes Rendus Hebdomadaire des Seances de l'Academie des Sciences, Serie C*, **279**, 533–5.
81 Dvolaitzky, M., Taupin, C. and Polydy, F. (1976) *Tetrahedron Letters*, 1469–72.
82 Dvolaitzky, M., Billard, J. and Polydy, F. (1976) *Tetrahedron*, **32**, 1835–8.
83 Allgaier, J. and Finkelmann, H. (1994) *Macromolecular Chemistry and Physics*, **195**, 1017–30.
84 Nakatsuji, S., Mizumoto, M., Ikemoto, H., Akutsu, H. and Yamada, J. (2003) *European Journal of Organic Chemistry*, 1912–18.
85 Domracheva, N.E., Ovchinnikov, I.V., Turanov, A.N. and Konstantinov, V.N. (2004) *Journal of Magnetism and Magnetic Materials*, **269**, 385–92.
86 Erenstein, W., Mathur, N.D. and Scott, J.F. (2006) *Nature*, **442**, 759–65.
87 Noda, Y., Shimono, S., Baba, M., Yamauchi, J., Ikuma, N. and Tamura, R. (2006) *Journal of Physical Chemistry. B*, **110**, 23683–7.
88 Sümmermann, W. and Deffner, U. (1975) *Tetrahedron*, **31**, 593–6.
89 Semmelhack, M.F., Chou, C.S. and Cortes, D.A. (1983) *Journal of the American Chemical Society*, **105**, 4492–4.
90 Semmelhack, M.F., Schmid, C.R. and Cortes, D.S. (1986) *Tetrahedron Letters*, **27**, 1119–22.
91 Ma, Z., Huang, Q. and Bobbitt, J.M. (1993) *Journal of Organic Chemistry*, **58**, 4837–43.
92 Kashiwagi, Y., Yanagisawa, Y., Kurashima, F., Anzai, J., Osa, T. and Bobbitt, J.M. (1996) *Chemical Communications*, 2745–6.
93 Osa, T., Kashiwagi, Y., Yanagisawa, Y. and Bobbitt, J.M. (1994) *Chemical Communications*, 2535–7.
94 Nakahara, K., Iwasa, S., Satoh, M., Morioka, Y., Iriyama, J., Sugaro, M. and Hasagawa, E. (2002) *Chemical Physics Letters*, **359**, 351–4.
95 Nishide, H., Iwasa, S., Pu, Y.-J., Suga, T., Nakahara, K. and Satoh, M. (2004) *Electrochimica Acta*, **50**, 827–31.

10
Spin Labeling in Biochemistry and Biophysics
Gertz I. Likhtenshtein

10.1
Proteins and Enzymes

Starting from the pioneering studies of the McConnel group ([1] and references therein), the nitroxide spin-labeling method finds massive application in protein and enzyme chemistry and biophysics. The labels can be incorporated in chosen parts of macromolecules by covalent chemical bonds, physical adsorption, or genetic site-directed incorporation.

The following physical processes involving nitroxides attached to proteins and enzymes, including their binding sites, can be investigated [2–10]: (1) rotational diffusion and wobbling of nitroxide labels and probes; (2) rate constant of dynamic spin exchange between nitroxides and nitroxides and paramagnetic complexes; (3) static spin exchange and dipole–dipole interactions between nitroxides and nitroxide and paramagnetic molecules; (4) dipole–dipole interactions between a nitroxide spin electron and proton nuclear spin; and (5) quenching of singlet- and triplet-excited states of natural and artificial chromophores with nitroxides.

These measurements allow us to study the following structural and dynamic properties of proteins and enzymes: (1) the distances between chosen functional groups; (2) local intramolecular dynamics (molecular "breathing"); (3) the rotational diffusion of macromolecules as a whole; (4) conformational transitions; (5) the distribution of electrostatic potential over molecules; (6) the microstructure (microrelief) of binding sites, including the active centers of enzymes; (7) local pH in different regions of molecules; (8) local polarity (apparent dielectric constant, ε_0); and (9) the kinetics of emzyme reactions.

The effect of any external factors, such as temperature, pH, composition, substrate and inhibitor concentrations, on the above-mentioned properties can be investigated.

Nitroxides: Applications in Chemistry, Biomedicine, and Materials Science
Gertz I. Likhtenshtein, Jun Yamauchi, Shin'ichi Nakatsuji, Alex I. Smirnov, and Rui Tamura
Copyright © 2008 WILEY-VCH Verlag GmbH & Co. KGaA, Weinheim
ISBN: 978-3-527-31889-6

10.1.1
Intramolecular Dynamics and Conformational Transition in Enzymes

10.1.1.1 Introduction

The molecular dynamics of proteins is a key factor governing their most important properties, such as enzyme catalysis, conformational transitions, recognition, and stability. The present concept of the intramolecular dynamics of proteins is based on hypotheses put forward during the 1950s and 1960s [11–13]. Although these hypotheses were based on intuitive considerations and indirect evidences, the authors have formulated a fundamental problem of great importance and thus initiated a major challenge to researchers.

Since the late 1960s, more direct approaches for the investigation of protein dynamics have been intensively developed, and have featured the application of physical methods, such as physical labeling, NMR, optical spectroscopy, fluorescence, differential scanning calorimetry, and X-ray and neutron scattering. The purposeful application of these approaches has allowed detailed information to be obtained on the mobility of the different regions of protein globules, and to compare this mobility with both the functional characteristics and stability of proteins, and with results of the theoretical calculations of protein dynamics.

The essential contribution when tackling the problem, especially at the early stages of a study of protein dynamics, was the development and use of biophysical labeling methods [14–29]. The basic idea underlying this approach to the study of intramolecular dynamics of proteins is the specific incorporation of a nitroxide label into a protein macromolecule, followed by the measurement of parameters of rotational diffusion of the label using ESR spectroscopy; that is, the rotational correlation time (τ_c), energy (E_a), and entropy ($\Delta S^{\#}$) of activation of the rotation. According to the available data (Section 6.1) these parameters are dependent on the molecular dynamics of the media.

The first, direct evidence for a protein's intramolecular dynamics, in nanosecond temporal rank under ambient conditions, was obtained using nitroxide spin-labeling [14] and Mössbauer [15] labeling techniques.

The existing experimental and theoretical data offer evidence that the nitroxide spin-labeling method is suited for the investigation of dynamic processes in proteins: (1) high-frequency, low-amplitude harmonic nuclear vibrations with $\tau_c = 10^{-12}$–10^{-13} s and amplitude $A = 0.001$–0.0005 nm and phonon processes; (2) anharmonic low-frequency ($\tau_c = 10^{-7}$–10^{-9} s) and relatively high-amplitude motions (0.02 nm and more); (3) rotation and wobbling protein side groups ($\tau_c = 10^{-7}$–10^{-11} s); (4) slow motion ($\tau_c = 10^{3}$–10^{-6} s) of protein domains; and (5) the protein globules' folding and unfolding ($\tau_c = 10$–10^{-9} s) and rotation as a whole ($\tau_c = 10^{-7}$–10^{-9} s).

10.1.1.2 Low-Temperature Molecular Dynamics

Molecular-dynamical processes at sub-zero temperatures appear interesting for a number of reasons:

- Some biological reactions, including electron transfer, were found to occur at low temperatures.

- The cryoprotection of proteins and enzymes against denaturation and deactivation is an important method in biotechnology and in the investigation of enzymatic mechanisms.
- A comparison of data of molecular dynamics and enzyme functions at sub-zero and ambient temperatures paves the way for elucidating which dynamical modes can be responsible for the enzymatic activity and stability of the object under investigation.

The first direct experimental evidence of the intramolecular mobility of a protein matrix at low temperatures was obtained using spin and Mössbauer labels during the 1970s and 1980s [6, 14–19]. A hydrophobic aromatic derivative of a nitroxide radical was embedded in the human serum albumin (HSA) binding site, and the mobility of the spin probe was traced using ESR spectroscopy. Several modes of the low-amplitude, high-frequency and phonone dynamics were detected in the temperature range of 40 to 200 K. These effects were similar to those observed in simple glassy solutions (Section 6.1) [9, 19–23]. It was shown that the probe correlation time decreased sharply ($\tau_c < 10^{-7}$ s) at temperatures of 200–230 K (depending on sample humidity). This conclusion was further supported by investigations of the mobility of Mössbauer atoms ^{57}Fe, which were attached as a metal-complex to the surface of a HSA globule and incorporated as a polynuclear serum-iron cluster within the globule. The experiment performed in the temperature range 77–300 K showed a sharp decreased in the Mössbauer spectra intensity (f) and a increased in the spectral line-width at temperatures exceeding 210 K. Such a change is caused by an anharmonic vibration of the Mössbauer atoms of which the correlation time was less than 10^{-7} s and the amplitude (A) was >0.04 nm at $T > 210$ K.

Subsequent systematic studies of the intramolecular mobility of bovine serum albumin (BSA) and HSA, lysozyme, myoglobin, α-chymotrypsin, ferredoxin, and bacterial photosynthetic reactions over a wide temperature range (20–300 K) by the combined use of biophysical labeling methods (radical-pair, spin, fluorescence, phosphorescence, Mössbauer labeling) and NMR, allowing the motion to be studied at a frequency of $\tau_c = 10^2 – 10^{-10}$ s, have been conducted. These studies have revealed a general picture of the dynamic effects in these proteins [8, 9, 22, 24]. Starting from 40 K, the broadening of the width ($\Delta H_{1/2}$) of the ESR signal from the spin label attached to the HSA surface was observed, showing an intensification of vibration processes. At increasing temperatures, the general tendency is as follows: the lower the value of the characteristic frequency of the method, the lower the temperature at which the label mobility can be monitored. Over a wide temperature range (130–300 K), the experimental data for surface labels follow an Arrhenius straight line, with $E_{app} = 16$ kcal mol^{-1} and $\Delta S^{\#}_{app} = 38$ e.u. Thus, the recorded mobility stems from a gradual softening of the water–protein interface, rather than from an individual phase transition. The spin, fluorescence, and Mössbauer label mobility at $\upsilon_c > 10^7$ s^{-1} increases from approximately 200 K upwards and reaches the nanosecond range at physiological temperatures. Recent studies have confirmed the basic inferences highlighted in the studies cited earlier,

10.1.1.3 Protein Dynamics at Ambient Temperature

Direct experimental evidence of *the nanosecond intramolecular mobility* of a protein matrix at ambient temperatures was obtained using spin and Mössbauer labels and probes [14–19, 22, 25, 26]. To illustrate this, a hydrophobic aromatic derivative of a nitroxide radical was embedded in the HAS binding site, and the mobility of the spin-probe traced using ESR spectroscopy. The correlation time ($\tau_c = 10^{-8}$ s) of the probe, which is essentially faster than the protein macromolecular tumbling, was not found to be dependent on viscosity and, therefore, was attributed to the local mobility of the label. The apparent energy ($E_{app} = 2.5$ kcal mol^{-1}) and activation entropy ($\Delta S^{\#}_{app} = -15.0$ e.u.) were determined. Thus, it was concluded that probe mobility follows the mobility of the flexible walls of the protein binding site with a similar frequency. The serum albumin's intramolecular mobility in a nanosecond temporary region at ambient temperatures was confirmed later by a series of independent dynamical methods such as spin and fluorescence labeling, tryptophane fluorescence, and proton NMR [9, 22, 23, 26]. In a series of studies ([20] and references therein), the dynamics of the HSA binding site around the dansyl and nitroxide moieties of the dual fluorophor–nitroxide probe were monitored indirectly, using ESR spectroscopy and by the temperature-dependent relaxation shift $\Delta \lambda^{fl}_{max}(T)$, and directly using the picosecond fluorescent time-resolved technique (Figure 10.1) [9]. All three methods showed that the relaxation of the protein groups in the vicinity of nitroxide and the excited chromophore occurred with a rate constant of approximately 10^9 s^{-1}. This conclusion was confirmed in theoretical investigations [27, 28].

The first direct evidence of the intramolecular mobility of the heme group of hemeproteins was obtained by the spin and Mössbauer labeling methods [25, 29]. These experiments were carried out on dry and moistened powders, which excluded any motion by the macromolecule as a whole in the nanosecond time range. The ^{57}Fe atoms were incorporated into the heme group in myoglobin and hemoglobin. Given that the rigid heme ring is bound to the protein by numerous contacts, it is evident that anharmonic motion of heme above 200 K is related to the intramolecular mobility of the protein globule in the nanosecond time region. This mobility appears only at a critical degree value of hydration. The increase in mobility recorded by the Mössbauer spectroscopy correlated with the data for isotopic H–D, the relatively higher amplitude of which must be accompanied by displacement of the helical polypeptide chain. Such an unharmonic nanosecond motion with $\langle \Delta x \rangle_{nh} > 0.02$ nm also was recorded at temperature $T > 210$ K in myoglobin using spin and fluorescence labeling methods [8, 9]. The flexibility of the cavity of the myoglobin active site is evidenced by the mobility of a spin probe, a derivative of isocyanate attached to the heme group in the single crystal. At room temperature, the mobility parameters were found as follows: correlation frequency τ_c approximately 10^{-8} s; $E_{app} = 2.8$ kcal mol^{-1}; and $\Delta S_{app} = -14$ e.u. High-frequency dynamics in hemoglobin measured by magnetic relaxation dispersion was reported

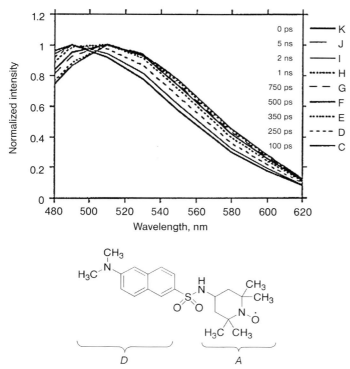

Figure 10.1 Time-resolved fluoresence spectra of dansyl–TEMPO dual probe in human serum albumin [9]. Reproduced from Likhtenshtein, G.I., Febrario, F. and Nucci, R. (2000) Dual fluorophore-nitroxides, Spectrochem. Acta Part A. Biomolecule. Spectroscoscopy, 56, 2011–2031, with permission of Elsevier.

[30]. Paramagnetic contributions from two spin labels attached to two β-93 positions in the protein to the ^1H spin-lattice relaxation rate constant were observed. This effect characterizes fluctuations sensed by dipolar interactions in the time range from tens of microseconds to 1 ps.

Various theoretical and experimental investigations have shown independently the intramolecular nanosecond dynamics of proteins and its key role in the functional activity of proteins and enzymes. A correlation between the dynamic parameters of physical labels and the kinetic parameters of processes in proteins and enzymes has been observed ([8, 9, 24] and references therein). Such a correlation was found for photodecarboxylation and carboxylation in myoglobin [31], the α-chymotrypsin-catalyzed hydrolysis of the cyanomoyl α-chymotrypsin [32], and in a series of photochemical processes.

Photosensitive systems provide convenient methods for analysing a possible correlation between the dynamic and functional properties of proteins. After a short light pulse, it is possible to observe a chemical reaction and to trace the dynamical state of the matrix with the aid of internal and external physical labels. A detailed investigation of the possible role of media (protein and membrane) dynamics in electron transfer was carried out on the reaction center (RC) extracted from *Rhodopseudomonas spheroidas* in the isolated state, and on the composition of the photosynthetic membrane [8, 17, 18, 33]. Spin, Mössbauer, fluorescent, and phosphorescent labels were introduced into the various portions of the system being studied. These were covalently bound to the RC surface groups, adsorbed by the hydrophobic segments of the protein and membrane, after which ^{57}Fe atoms were incorporated by way of biosynthesis into iron-containing proteins. Then, in the same samples, the dependence on temperature, moisture content, and viscosity was measured for the label mobility and the rate constant of electron transfer (ET) between the components of the photosynthetic chain.

The following correlations were found. The emergence of an electron from the primary photosynthetic cell – that is, the transport from the reduced primary acceptor Q_{A-} to the secondary acceptor Q_B – was shown to take place only under conditions in which the labels record the mobility of the protein moiety in the membrane with $\tau_c < 10^{-7}$ s. The rate of another important process, recombination of the primary product of the charge separation – that is, reduced primary acceptor (Q_{A-}) and oxidized primary donor, bacteriochlorophyll dimer (P^+) – falls from 10^2 to $10^3 s^{-1}$ when dynamic processes with $\tau_c = 10^{-3}$ s occur. Very rapid electron transfers from P^+ to bacteriochlorophyll (Bchl) and from (Bchl)- to Q_A do not depend on media dynamics, and occur via conformationally non-equilibrium states.

10.1.2
Conformational Changes in Proteins and Enzymes, and Mechanism of Intramolecular Dynamics

The spin-labeling method has been used successfully for the detection of various types of conformational change, including: (1) large-scale denaturation and unfolding or folding processes; (2) predenaturational phenomena; (3) transglobular allosteric transitions; (4) interglobular allosteric transitions; and (5) conformational transitions in enzyme active sites. Although extensive bibliography has accumulated in this field, in this section we need to restrict this to only a few typical examples.

Allosteric effects were first reported in the pioneering studies of McConnel and associates while investigating spin-labeled horse hemoglobin [34]. According to the results of these studies, when oxygen is bound in the active center of hemoglobin, a splitting of the ESR spectral components is observed. Experiments with the protein individual co-subunits and its combination showed that the main allosteric effects occur on the surface of the β-subunits in the areas of contact with the α-subunits, and this was later confirmed by a series of investigations. (For recent results see [35–40] and references therein.)

Temperature-dependent local conformational changes before denaturation (predenaturational transitions) were detected in a number of proteins (lyzozyme, myoglobin, serum albumins, etc.) by using biophysical labeling methods ([6–9, 41]

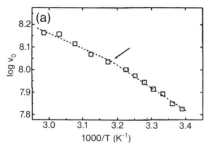 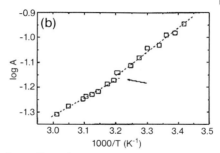

Figure 10.2 Arrhenius plots for (a) the temperature dependence of the rotation frequency $\gamma = 1/\tau_c$) of spin label and (b) of the polarization of FMA in labeled b-glycosidase, respectively. The arrows indicate the inflection points [9]. Reproduced from Likhtenshtein, G.I., Febrario, F. and Nucci, R. (2000) Dual fluorophore–nitroxides, Spectrochem. Acta Part A. Biomolecules. Spectroscoscopy, 56, 2011–2031, with permission of Elsevier.

and references therein). The common feature of these proteins is that the transitions in the mobility of spin and fluorescence labels is more sensitive to temperature increase after the transition than before. Although it might be said that, after the transitions, the protein globules became "softer", for the hyperthermostable β-glycosidase, the Arrhenius plots of the spin-label rotation frequency and polarization of the fluorescent labels, which are attached to the protein SH and NH_2 groups, respectively, exhibit less sensitivity to temperature increase after an inflection point T_{in} at about 314 K (Figure 10.2) [9]. Speculation may be offered as to the reason why there is an increase in protein rigidity in this hyperthermostable enzyme above T_{in} = 312 K. The efficiency of the chemical processes requires optimum flexibility of the enzyme active site, which is enzymatically active at temperatures up to 90 °C. If the low-temperature tendency towards an increase in the enzyme conformational flexibility in the active area continues at higher temperatures, the site optimum dynamic state would be destroyed. Thus, the conformational transition may be necessary to maintain a balance between the activity and stability of the enzyme at high temperatures.

A comparative analysis of the data obtained [7–9, 15, 16, 41] revealed an apparent discrepancy between the physical labeling approach and certain other physical methods. Thus, the methods of spin- and fluorescence-labeling and deuterium exchange showed predenaturational conformational changes in egg lysozyme at temperatures above 30 °C [41], whereas the temperature dependences of the enzyme acivity, heat capacity, and water protons' relaxation change monotonically in this temperature region. Measurements of the temperature dependence of the heat capacity (c_p) of proteins, α-chymotrypsin, lysozyme, myoglobin and collagen at T 180–210 K, at various degrees of hydration, indicated only a monotonic increase of c_p and did not detect any pronounced phase transitions [42, 43]. However, spin-, fluorescence-, and Mössbauer-labeling, H–D exchange, nonelastic neutron scattering, and absorption spectra of heme in heme proteins detected sharp transitions within this temperature range [6, 8, 9].

The above-mentioned discrepancy was easily explained by the following proposal [8, 17]. This type of nanosecond intermolecular mobility consists of movement of the relatively large and rigid parts of the protein macromolecules. Although such hinge-like oscillations of the tightly packed polypeptide blocks do not provide any measurable contribution to the overall heat capacity and the helicity degree of polypeptide chains, they strongly affect the mobility of those Mössbauer labels that are firmly bound to the protein blocks, the mobility of spin and fluorescence labels, and the native chromophores located in cavities between the blocks. At a later stage, the mechanism proposed was confirmed by independent experimental investigation and molecular dynamics simulation [44–49]. Depending on protein specifics, the correlation time of domain motion can vary from milliseconds to nanoseconds.

10.1.3
Structure of the Enzymes' Active Centers

The first study of the active center of an enzyme using a nitroxide spin-label was undertaken by Berliner and McConnel [50]. The nitroxide spin-labeled substrate, DL-2,2,5,5-tetramethyl-3-carboxypyrrolidine p-nitrophenyl ester, was incorporated into the α-chymotrypsin active site. The label mobility, monitored by ESR spectroscopy, was found to occur significantly faster than did the rotation of the enzyme globule as a whole. A significant insight into the mechanism of enzyme catalysis was offered by the group of Mikenen [51–53]. To illustrate, the spin-labeled transition-state analog in the α-chymotrypsin reaction, N-(2,2,5,5-tetramethyl-1-oxypyrrolinyl)-L-phenylalaninal, has been synthesized. The stereoview of this molecule into the active site of α-chymotrypsin obtained by ENDOR spectroscopy is shown in Figure 10.3. The conformation of the acyl moiety of the substrate analog within the active site of the reaction intermediate differs significantly from that of the free substrate in solution. This is strong evidence that torsional alterations are induced in the substrate by binding to the enzymes to form a catalytically active productive pretransition state. A true acylenzyme reaction intermediate of TEM-1, β-lactamase, which is formed with specific spin-labeled substrate 6-N-(2,2,5,5-tetramethyl-1-oxypyrrolinyl-3-carboxyl)-penicillanic acid, was trapped at low temperature (−70 °C and detected using ENDOR spectroscopy. The obtained electron–nucleus distances were applied as constraints to assign the conformation of the substrate in the active site and of amino acid side chains by molecular modeling. The ENDOR results provide experimental evidence of the functional role of glutamate-166 as the general base catalyst in the wild-type enzyme for the hydrolytic breakdown of the acyl–enzyme reaction intermediate of TEM-1 β-lactamase.

A series of the spin-labeled nucleotides was synthesized and used for a study of the active centers of enzymes. A number of probes analogous to NAD and ATP containing ^{2}H and ^{15}N were synthesized and used to study the active center of glyceraldehyde-3-phosphatase [54, 55]. The spin–spin dipole–dipole splitting of the labeled enzyme ESR spectrum corresponded to about a 1.2 nm distance between

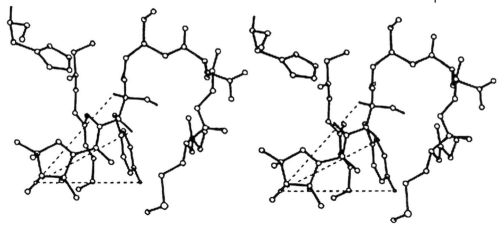

Figure 10.3 Stereo diagram of the acyl moiety of the spin-labeled tryptophanyl-acylenzyme reaction intermediate of α-chymotrypsin. Active site residues close to the acyl moiety are labeled [52]. (Reproduced from Makinen, M.W. (2000) Electron nuclear double resonance determed structure of enzyme reaction intermediate, Spectrochem. Acta Part A. Biomolecules. Spectroscoscopy, 54, 2269–2281, with permission of Elsevier).

nitroxides, which was in agreement with data obtained by X-ray analysis. Spin-labeled adenine nucleotides were employed to investigate nucleotide binding to the 70-kDa heat shock protein, DnaK, from *Escherichia coli,* whereupon it was shown that the spin-labeled analogs function in a similar manner to the natural nucleotide-substrates of the protein.

ESR spectroscopy using spin-labeled ATP was employed to study nucleotide binding to, and structural transitions within, the multidrug resistant P-glycoprotein, wild type, and mutant of P-gp, in which all naturally occurring cysteines were substituted with alanines [56]. 2-Azido-2′,3′-O-(1-oxyl-2,2,5,5-tetramethyl-3-carbonylpyrroline)-ATP (2-N3-SL-ATP), a photoaffinity spin-labeled derivative of ATP with a nitroxide moiety, was attached to the ribose ring and an azido group was attached to C2 of the adenine ring, and these were used to study the nucleotide-binding site stoichiometry of sarcoplasmic reticulum (SR) Ca^{2+}-ATPase. Spin-labeled nucleotide mobility at the boundary of the EcoRI endonuclease binding site in a complex consisting of the EcoRI endonuclease site-specifically bound to spin-labeled DNA 26mers was investigated [57]. No spectral changes were observed which indicated an absence of any significant structural disruption being propagated along the helix as a result of protein binding.

Electron-carrier horse heat cytochrome c and dioxygen-carrier sperm whale myoglobin were used as models for determining local electrostatic charges in the vicinity of paramagnetic active sites of metalloenzymes and metalloproteins [58]. Neutral TEMPOL, positively charged nitroxide, or negatively charged nitroxide were used to probe the local charge in the proteins' paramagnetic heme region (Section 6.6). The apparent local charges (Z_H) in the vicinity of the cytochrome c heme groups of the cytochrome c and myoglobin at pH 7 were found to be +0.3 and −0.3, respectively.

Data on the investigation of the nitrogenase active center by a spin-labeling method have been summarized in [8].

For recent reviews of applications of advanced ESR techniques (high-field, pulsed, multiquantum ESR, and ESR imaging) in the investigation of the structure and dynamics of proteins see Chapter 4.

10.1.4
Site-Directed Spin-Labeling (SDSL)

10.1.4.1 Introduction

The use of site-directed mutagenesis in combination with modern ESR spectroscopy gave new life to the nitroxide spin-labeling method [59–65]. Given the current state of molecular genetics, it is not a problem to modify a selected amino acid in a protein or peptide, or a nucleotide in DNA or RNA, with any other corresponding moiety. Using modern ESR methods such as high-field, pulsed, multiquantum ESR, and ESR imaging allows us to obtain detailed information about the molecular structure and dynamics in the region of labeling. The main advantage of SDSL is the possibility of overcoming the limitations of a choice of amino acids suitable for labeling in native proteins. A widely accepted procedure in protein SDSL is the substitution of a selected amino acid for cystein via a site-directed mutagenesis technique. The resultant cysteine is then chemically modified with a sulfhydryl reactive nitroxide radical (1-Oxy-2,2,5,5-tetramethyl-3-pyrroline-3 methyl) methanethiosulfonate (MTSL) (Figure 10.4). A typical procedure used in site-directed nitroxide spin-labeling includes stages of expression to a cell gene that contains

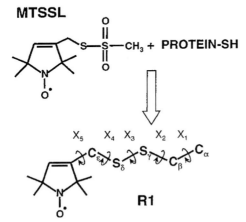

Figure 10.4 Structure of the spin label obtained by reaction of the sulfhydryl group of a cysteine with 1-oxyl-2,2,5,5-tetramethyl-3-pyrroline-3-(methyl)methanethiosulfonate [78].

an amino acid-to-cysteine substitution at a chosen position, purification of the protein, and its specific labeling with MTSL.

The complete arsenal of chemical, biochemical, biophysical, and physical methods (see Chapters 3 to 7) has been applied in SDSL investigations.

10.1.4.2 Soluble Proteins

After the first works concerning spin-labeled lysozyme, published in the 1970s, which clearly demonstrated that this protein is a convenient model for the study of the structure, dynamics, and conformational changes of proteins and enzymes [66–68], more than 130 articles on enzyme spin-labeling have been published. Site-directed spin-labeling has provided a new results in the application of the spin-labeling method in this area.

Modeling of the effects of structure and dynamics of the nitroxide side chain on the ESR spectra of spin-labeled T4 lysozyme has been reported [69]. The 72R1 and 72R2 mutants of the T4 lysozyme, which bear the spin-label at a solvent-exposed helix site, have been considered (Figure 10.5). Two types of nitroxide moiety motion were taken into consideration: slow reorientations of the whole protein and fast chain motions, which have been identified with conformational jumps and fluctuations in the minimum of the chain torsional potential. Fast chain motions were analyzed within the framework of the stochastic Liouville equation (SLE) methodology. It was shown that, for the side chain of a bulky spin label, a

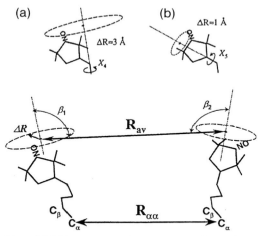

Figure 10.5 The geometrical arrangement of the two R1 side chains and the inter-nitroxide distances consistent with the X4/X5 model. This arrangement is used to calculate distance distributions, $P(r)$ between the nitroxides, where r is the inter-nitroxide distance. Here $R_{\alpha\alpha}$ is the distance between the two C_α carbons located on the backbone. Each circle with radius ΔR represents the possible location of the nitroxide spins due to rotations about either the X4's or the X5's; R_{av} connects the centers of these two circles. The plane of each circle is referred to R_{av} by polar angles β_1 (circle 1) and β_2 (circle 2). Insets more clearly show a rotation about X4 and about X5 [78]. (Reproduced with permission).

few minor conformers are possible, whose mobility is limited to torsional fluctuations. In the case of "more elegant spin labels", ESR spectra result from the simultaneous presence of constrained and mobile chain conformers. The proposed model provides an explanation for the experimentally observed dependence of the spectral line-shapes on temperature, solvent, and substituents in the nitroxide pyrroline ring. The following are values of rotation diffusion parameters: $D_{0,\perp} = 0.8 \times 10^7\,\mathrm{s}^{-1}$ and $D_{0,\parallel} = 1.35 \times 10^7\,\mathrm{s}^{-1}$ ($T = 283\,\mathrm{K}$) and $D_{0,\perp} = 1.2 \times 10^7\,\mathrm{s}^{-1}$ and $D_{0,\parallel} = 1.9 \times 10^7\,\mathrm{s}^{-1}$ ($T = 312\,\mathrm{K}$). These values fitted to the labeled enzyme experimental spectra. The pulsed ESR technique for distance measurement, based on the detection of double quantum coherence (DQC) was used for the measurements of long distances between labeled side groups in eight T4 lysozyme mutants, doubly labeled with methanethiosulfonate spin-label (MTSSL) (Figure 10.4) [70]. The measurements were performed at 9 and 17 GHz. The high quality of the dipolar spectra allowed us to measure distances which span a range from 20 Å for the 65/76 mutant to 47 Å for the 61/135 mutant and to determine the distance distribution. The ribbon structure of T4L, showing positions of the spin-labels, is reproduced in Figure 10.6 [70].

Site-directed spin-labeling is used to investigate the tertiary structure and orientation of T4 lysozyme (T4L), a monolayer of which was adsorbed via a His-group on quartz [71, 72]. Simulations using the stochastic Liouville equation revealed conservation of the secondary and tertiary structures of T4L upon adsorption, although slight conformational changes in the presence of the surface were detected. The orientation of the entire protein was deduced on the basis of an anisotropic motional model for the spin-labeled side chain. At high ionic strength, significant changes of the backbone fold are limited to the region around the enzymic cleft, while at low ionic strength the previously unperturbed parts of the protein interact with the surface [73]. Hydrophobic interactions were shown to play an important role in the partial unfolding at high ionic strength.

A tether-in-a-cone model, in which the nitroxides adopt a range of interprobe distances and orientations, was developed for the simulation of ESR spectra of

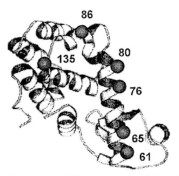

Figure 10.6 Ribbon structure of T4L, showing spin-labeled mutated sites studied in this work [78]. (Reproduced with permission).

dipolar coupled nitroxide spin-labels attached to T4 [74]. Simulations demonstrate the sensitivity of ESR spectra to the orientation of the cones as a function of cone half-width and other parameters. For small cone half-widths (<~40°), simulated spectra are strongly dependent on the relative orientation of the cones. For larger cone half-widths, spectra become independent of cone orientation. In work [74] ESR spectroscopy at 250 GHz and 9 GHz was utilized to study the dynamics and local structure of T4 lysozyme labeled at sites 44 and 35 in aqueous solution from 10 °C to 35 °C. The 250-GHz ESR spectra were described by a model which takes into consideration the influence of the tether connecting the probe to the protein but not the contribution of the protein motion as a whole. For the 9-GHz ESR spectra, the overall rotational diffusion was accounted for in the analyses. It was shown that two distinct motion/ordering modes of the probe exist for both lysozyme derivatives, indicating that the tether exists in two distinct conformations on the ESR timescale [75].

The transglobular conformation changes in egg lysozyme when binding to substrates were first demonstrated in works [66, 67]. Later, the substrate-induced conformational transition in T4 lyzsozyme was studied in details by a site-directed spin-labeling technique [76]. A single or pair of nitroxide side chains were introduced into the protein to monitor tertiary contact interactions and inter-residue distances, respectively, in solution. According to analysis of dipole–dipole interactions, in the absence of a substrate, differences in both inter-residue distances and tertiary contact interactions relative to this reference state are consistent with a hinge-bending motion that opens the active site cleft. According to analysis of the contribution of dipole–dipole interaction in the nitroxide ESR spectra, the substrate binding was accompanied by a change of the distance from 1.3 nm to 2.1 nm. Derivatives of the pyrroline or pyrrolidine series of nitroxides which were 4-substituted were utilized for studies of the molecular motion of spin-labeled side chains in the α-helix surface of T4 lysozyme mutants (D72C and V131C) [77]. The spectral line-shapes were analyzed as a function of side chain structure and temperature using a simulation method with a single-order parameter with diffusion rates about three orthogonal axes as parameters. The results provided strong support for an anisotropic motion model of the side chain in the nanosecond or faster timescale.

The influence of the disulfide bond configuration on the dynamics of the spin-label attached via the methanethiosulfonate moiety to cytochrome c at position C102 was studied [79]. On the basis of a series of multi-nanosecond molecular dynamics (MD) simulations, the authors concluded that the protein secondary structure is slightly changed in the cytochrome c labeled at position C102 as compared with the non-labeled protein. The MD simulations also showed that the disulfide bond tethering a spin-probe to a protein strongly influenced the behavior of the nitroxide group. The DOQ time evolution method was used for determination of distances between labels specifically attached to the 61 and 80 sites of the cytochrome c mutant (Iso-1-cyt c). The Heme iron in the protein was reduced under anaerobic conditions to lengthen the T_2s of nitroxide spin-labels. The detected distance between the mutants 61/80 showed a bimodal distribution. The distribution was characterized by peaks at 2.9 and 3.4 nm, with relative weights of 2 and 1, respectively; the average distance was 3.06 nm.

The effect of temperature on two-component spectra of spin labels attached to cysteine at the 102 and 47 positions of iso-1-cytochrome was studied by CW ESR [79]. The computer simulations revealed that the CW EPR spectrum for each form of cytochrome c consists of at least two components, a fast (F) and a slow (S) component, which differ in the values of the rotational correlation times (longitudinal rotational correlation time) and (transverse rotational correlation time) and in the temperature dependence.

10.1.4.3 Rhodopsin and Bacteriorhodopsin

Site-directed spin-labeling has provided extensive information about the structure and structural changes of different loop regions of rhodopsin and bacteriorhodopsin which is not available by other methods [80–83].

In order to obtain information on the structure and dynamics of the rhodopsin II helix F, EPR experiments have been carried out on a truncated transducer (NpHtrII157) and NpSRII, site-directed spin-labeled and reconstituted into purple membrane lipids [82]. Data on label accessibility and dynamics allowed the authors to identify a helical region up to residue Ala94 in the AS-1 amphipathic sequence, followed by a highly dynamic domain protruding into the water phase. It was shown that light activation of NpSRII leads to a displacement of helix F, which in turn triggers a rotation or screw-like motion of TM2 in NpHtrII. Time-resolved electron paramagnetic resonance spectroscopy in combination with SDSL was utilized to unravel photoexcited conformational changes of bacteriorhodopsin [81, 83]. The C–D loop, positions 100–107, and those of the E–F loop, including the first α-helical turns of helices E and F, positions 154–171, were modified with a methanethiosulfonate spin label followed by an examination of the label's ESR spectra (Figure 10.7). It was shown that a small movement of helix C and an outward tilt of helix F are accompanied by a rearrangement of the E–F loop and of the C-terminal turn of helix E.

10.1.4.4 Muscle Proteins

The dynamic of a hydrophobic loop of actin (residues 262–274) was investigated utilizing four mutants that have cysteine residues [84]. The data obtained indicated that, in solution, the loop position within the filament is able to dynamically populate other conformational states which stabilize or destabilize the filament. Structural and dynamic characterization of a vinculin binding site in the talin rod, which is a key protein involved in linking to the actin cytoskeleton, was investigated [85]. The talin single vinculin binding site (VBS) was labeled by nitroxides via appropriately located cysteine residues. Measurements of inter-nitroxide distances in doubly spin-labeled protein showed that, as the result of the binding to the vinculin Vd1 domain, the protein helical bundle is disrupted and the mobility of the helices, except for the VBS helix, is markedly increased.

Site-directed spin-labeling EPR in combination with dipolar ESR measurements was used to detect the structure of the inhibitory region of troponin I (TpI) [86].

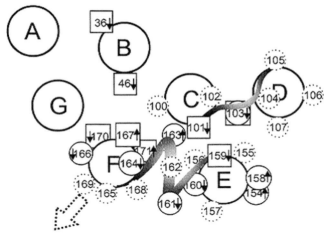

Figure 10.7 Schematic top view of the cytoplasmic structure of bacteriorodopsin. The locations of the spin-labeled side chains are shown. The transient increase and decrease in the nitroxide motional freedom during the photocycle are indicated by small arrows pointing upward and downward, respectively. Squares denote spin-labeled side chain positions, where the rise of the conformational change occurs during the M-to-N transition or later, and circles denote positions where the structural change is monitored during the lifetime of the M intermediate. The shapes of the transient spectral changes are in line with an outward movement of helix F and rearrangements of the A–B, C–D, and E–F loops and the C-terminal turn of helix E. Results on D36R1 and T46R1 [81]. (Reproduced with permission).

The residues 129–145 of cardiac TpI were mutated sequentially to cysteines and labeled with MTSSL. Measurements of four intradomain distances within the inhibitory sequence, using dipolar EPR has confirmed an α-helical structure of the inhibitory region.

10.1.4.5 Membrane Proteins

The arrangement of two β-subunits in the holo-enzyme F_0F_1–ATP synthase from *E. coli*, which couples proton translocation with the synthesis of ATP from ADP and phosphate, was investigated by the SDSL method [87]. Spin labels were introduced in the tether domain of the β-subunit at 40, 51, 53, 62, or 64 sites followed by conventional (9 GHz), high-field (95 GHz), and pulsed EPR investigations. On the basis of the findings it was concluded that the distance between the spin labels at each β-subunit is 2.9 nm in each mutant and the protein helices are separated by a large distance (1.9 nm). A parallel arrangement of the two helices was also revealed. It was shown that the binding of the non-hydrolyzable nucleotide AMPPNP to the spin-labeled enzyme had no significant influence on the distances compared to that in the absence of nucleotides.

Motion restrictions in the bacteriophage M13 major coat protein were investigated by the SDSL method and CD spectroscopy [88]. Twenty-seven single-cysteine mutants were labeled with nitroxide spin labels and incorporated into phospholipid bilayers. On the evidence of the examination of the nitroxide ESR spectra, it was suggested that the coat protein has a large structural flexibility, which facilitates a stable protein-to-membrane association.

The structural properties of a transmembrane helix for proton translocation in vacuolar H$^+$-ATPase in sodium dodecyl sulfate micelles were studied using double site-directed spin-labeling combined with ESR and circular dichroism spectroscopy [89]. A synthetic peptide derived from transmembrane helix 7 of subunit α, from the yeast *Saccharomyces cerevisiae* vacuolar proton-translocating ATPase that contains two natural cysteine residues suitable for spin-labeling, was used. Based on the combined results from ESR, circular dichroism, and molecular dynamic simulations the authors concluded that the peptide forms a dynamic α-helix. The enzyme inhibitor bafilomycin was shown to interact with residues E721, L724, and N725 (13↓), making the helix rigid, thereby blocking proton translocation.

Phosphatidylcholine activation of human heart (R)-3-hydroxybutyrate dehydrogenase (BDH) mutants lacking active center sulfhydryls was investigated [90]. Six cysteine mutants each have an increased K_m (NADH) (2–6-fold) but an unchanged K_m (NAD$^+$). It was shown that the C242 site in the C-terminal lipid binding domain of BDH, which is close to the enzyme active site, located a hydrophobic environment and only indirectly influenced the substrate binding site at the catalytic center of BDH. Site-specific spin-labeling of a single cysteine mutation within a water-soluble mutant of subunit β of the F_0F_1–ATP-synthase and ESR spectroscopy were used to study binding interactions of the b dimer (b_2) with F_1-ATPase [91]. Based on experimental data, the authors suggested that b_2 packs tightly to F_1 between residues 80 and the C terminus. Nevertheless, there are segments of b_2 within that region where packing interactions are quite loose and that the reconstitution of the ATP synthase is not ordered with respect to these subunits. Results of detail EPR studies of the F_0F_1–ATP-synthase structure and function using affinity and site-specific spin labeling were reported in [92].

Recent applications of SDSL in the investigation of protein structure, protein–protein and protein–membrane interactions, and membrane proteins were reviewed in [93]. Data on combining high-field EPR with site-directed spin labeling in active proteins are reviewed in [94, 95].

10.2
Biomembranes

The role of membranes in fundamental biochemical, biophysical, physiological, and biomedical processes cannot be overestimated. The spin-labeling method (Chapters 6 and 7) provides many opportunities for quantitative measurements of chemical, biochemical, and physical processes in biological and model membranes. The principles of spin-labeling for investigating biological membranes and recent results have been reviewed [8, 10, 96–100].

10.2.1
Structure and Dynamics

10.2.1.1 Location of Labels, Water, and Oxygen in Membranes

During analysis of experimental data on the structural and dynamic features of biological and model membranes obtained by spin- and fluorescence-labeling methods, it is important to know a label position in the object of interest. Localization of the nitroxide moiety of spin-probes in biological membranes relative to aqueous and lipid phases was investigated by methods based on the registration of the shift of saturation curves of the EPR signals of paramagnetic centers in the presence of paramagnetic ions (see Section 6.4.1) [101]. The immersion depths of nitroxide groups of fatty acid probes (R_0) in microsome membranes and chromatophores from *Rhodospirillum rubrum*, chloroplasts, and *M. capsulatus* were detected. The investigation revealed a variety of conformations *for the same spin-labels in different biological membranes*. Thus, the R_0 values for spin-probes in *M. capsulatus* membranes were for more extended conformations than those in the chromatophore, chloroplast, and microsome membranes.

A method has been developed for measuring depth of immersion (R_0) of a fluorescent chromophore in biological matrices, such as biomembranes and proteins (see Section 6.4.2) [102]. The method is based on dynamic quenching of chromophore fluorescence by a nitroxide probe freely diffused in solution taking into consideration the long-distance exchange interaction between nitroxide and a fluorophore. The proposed method was applied to investigation of lecithin liposomes and membranes from *Bacillus subtilis* modified by the photochrome-fluorescent probe 4,4'-dimethylaminocyanostilbene. The results indicated a distribution of R_0 values: a part of the fluorophore molecule is located in the liposome in an area about 0.7 nm from the membrane surface while another part is plunged deeper.

Micelles of lysomyristoylphosphatidylcholine (LMPC) and mixed micelles of LMPC with anionic detergent sodium dodecyl sulfate (SDS) have been characterized by spin-probe-partitioning electron paramagnetic resonance (SPPEPR) and time-resolved fluorescence quenching (TRFQ) experiments [103]. The rate of energy transfer from a fluorophore to the spin probe di-*tert*-butyl nitroxide (DTBN) and the activation energy of the transfer process in LMPC and LMPC–SDS micelles have been determined. It was shown that the activation energy of DTBN transfer in pure lysophospholipid micelles does not change with LMPC concentration while it decreases with the increasing molar fraction of SDS in mixed LMPC–SDS micelles.

Oxygen profiles in lipid bilayers were obtained by the spin-label method [104]. It was found that the oxygen transmembrane profiles have a Boltzmann sigmoidal dependence on the depth into each lipid leaflet and correlate with the transmembrane profiles for intramembrane polarity, and for water penetration into the membrane. The association of water (D_2O) in membranes of dipalmitoyl phosphatidylcholine with and without 50 mol% of cholesterol was studied by using pulsed-ESR techniques [105]. The results indicated that, for phosphatidylcholine spin-labeled at different positions down the *sn*-2 chain, the amplitude of the deuterium signal decreases toward the center of the membrane, and is reduced to zero from the C-12 atom position onward. At chain positions C5 and C7 closer to the phospholipid headgroups, the amplitude of the deuterium signal is greater in the presence of cholesterol than in its absence.

Water concentration profiles in bilayer membranes of dipalmitoyl phosphatidylcholine with and without 50 mol % cholesterol were measured by ESEEM of spin-labeled lipids [106]. Quantum chemical calculation (DFT) and ESEEM simulations assigned the broad spectral component to one or two D_2O molecules that are directly hydrogen bonded to the N–O group of the spin label. The result suggested that the midpoint of the water concentration sigmoidal profile is shifted toward the membrane center for membranes without cholesterol, relative to those with cholesterol, and the D_2O–ESEEM amplitude in the outer regions of the chain is greater in the presence of cholesterol than in its absence. Water penetration depth into the lipid bilayer composed of egg-phosphatidylcholine and dicetylphosphate was studied by the electron spin echo envelope modulation technique (Fourier transform–ESEEM) [107]. The stearic acid spin probes with nitroxide moiety at different positions on the acyl chain were dissolved in the bilayer of liposomes prepared in buffer solution with D_2O and the deuterium effect on the ^{14}N-hyperfine splitting constant of the spin probe was examined. The water distribution in phosphatidylcholine vesicles was investigated using the hydrophobic spin probe (TEMPO–stearate) [108]. The effective water concentration (C_w) of the polar shell of dimyristoyl-phosphatidylglycerol vesicles is greater by about 4.0 M than the effective water concentration of the polar shell of dimyristoyl-phosphatidylcholine vesicles. The C_w decreases by about 0.5 M for an increase of two carbons in the chain, and increases noticeably with hydrocarbon unsaturated chain.

10.2.1.2 Membrane Microstructure

A new method for determining tie-lines in coexisting membrane phases using spin-label ESR was applied to 100 different lipid compositions containing the spin-labeled phospholipid 16-PC in or near the two-phase coexistence region of the liquid-disordered and the gel phases of dipalmitoyl-PC/dilauroyl-PC/cholesterol (DPPC/DLPC/Chol) [109]. The coexisting phases in equilibrium were characterized by the partition coefficient, K_p, of the spin-label and its respective dynamic parameters obtained from fitting the ESR spectra to dynamical models. Direct evidence from ESR studies for the presence of two different types of lipid populations in the plasma membranes of live cells from four different cell lines was provided in [110]. The analysis of ESR spectra of spin probes incorporated into the membranes indicated coexisting liquid-ordered (LO) and liquid-disordered (LD) lipid domains with distinct order parameters and rotational diffusion coefficients. The LO region was found to contribute a major component.

The magnetically alignable phospholipid bilayered membranes at different mole ratio q (1,2-dihexanoyl-sn-glycerol phosphatidylcholine/1,2-dimyristoyl-sn-glycerol phosphatidylcholine) and temperature were studied by spin-labeled X-band ESR spectroscopy and solid-state ^2H and ^{31}P NMR spectroscopy [111]. For higher q ratios (≥5.5), bicelles in the gel phase maintain magnetic alignment below the main phase transition temperature of 30°C in the presence of lanthanide cations. The global and the local molecular architecture of supported phospholipid model membranes at solid interfaces was investigated by small- and wide-angle X-ray surface scattering and spin-label methods using 9.6 GHz and 1.2 GHz EPR [112]. The effects of humidity, temperature, and pressure were examined.

10.2.2
Membrane Dynamics

The dynamic behavior of spin-labels with nitroxides located in different positions on aliphatic chains was first investigated with the use high-frequency (2-mm) ESR spectroscopy [113, 114]. Based on examination of the spin probe's temperature dependence over the range 120–260 K (Figure 10.8), it was concluded that in the temperature range 220–260 K the nitroxide rotation is essentially anisotropic with correlation time t_c 10^{-7}–10^{-8} s and about 10^{-9} s at physiological temperatures. Detailed information on translational diffusion of doxyl stearic spin-probes bearing ^{15}N and ^{14}N in nitroxide fragments was obtained by the ELDOR method [115].

Two-pulse, echo-detected (ED) ESR (EPR) spectroscopy was used to study the librational motions of spin-labeled lipids (sn-2 chain (n = 5, 7, 10, 12, and 14)) in membranes of dipalmitoylphosphatidylcholine + 50 mol % cholesterol over the temperature range 77–240 K [116]. The librational amplitude, $\langle\alpha_2\rangle$, was detected using a spin label at the 14-C position of the lipid chain from the partially motionally averaged hyperfine splitting in the conventional EPR spectra. The librational correlation time, τ_c, which is deduced from a combination of the conventional and ED-EPR results, lies in the subnanosecond region and depends only weakly on temperature.

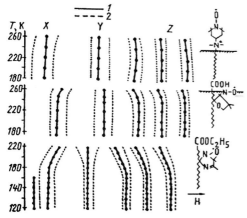

Figure 10.8 Temperature dependences of the position (1) and line-width (2) of the 2-mm ESR spectra of nitroxide spin probes (right) in lecithin lyposomes [114]. (Reproduced from Likhtenshtein G.I. (1993) Biophysical Labeling Methods in Molecular Biology, Cambridge University Press, Cambridge, NY, with permission).

The values of diffusion rate constants in model lipid membranes together with similar data obtained by other methods, that cover characteristic times over eight orders of magnitude, were found to be in good agreement with the advanced theory of diffusion-controlled reactions in two dimensions [117] (Section 6.7, Figure 6.8).

The effects of pressure were examined for the partitioning and rotational correlation time of nitroxide probes, TEMPO and DTBN, in dispersions of a triglyceride membrane [118]. It was shown that as external pressures increased, nitroxide probes shifted from the vesicle phase to the aqueous phase. The values of molar change at the rotation label were found to be $\Delta V = 65.6 \, \text{cm}^3 \, \text{mol}^{-1}$ for TEMPO and $24.4 \, \text{cm}^3 \, \text{mol}^{-1}$ for DTBN.

Results of molecular dynamic simulations of the structural properties of perifosine and its synthetic spin-labeled alkylphospholipid analogs in a DPPC were reported [119]. As expected, alkyl chains tend to insert into the hydrophobic core, whereas charged groups stay at the lipid–water interface. Nevertheless, a doxyl group in the middle of the alkyl chain moves up to the interface region, thus preventing adoption of the extended conformation in agreement with previous experimental data [101]. Characterization of chain motions in saturated phosphatidylcholine/cholesterol mixtures at the high packing densities achieved at low temperatures was reported [120, 121].

10.2.3
Proteins and Peptides in Membranes

A protein–lipid interaction is an important factor in determining a stable thermodynamic association of a membrane protein with the lipids [122]. This interaction provides a number of changes with increasing acyl chain length in the membrane's component structure: (1) the conformational space of the amino acid side chains; (2) backbone conformations; (3) the tilt angles of the transmembrane segments; (4) protein partitioning in the lipid bilayer, or aggregation state in the membrane; and (5) the order of the lipid acyl chains. Also with increasing acyl chain length, phase transition to nonlamellar structures incorporated into phospholipid bilayers occurs.

SDSL is combined with ESR and CD spectroscopy has been used to investigate 27 single cysteine mutants of the bacteriophage M13 major coat protein incorporated into phospholipid bilayers with increasing acyl chain length [122]. The spin-label ESR spectra were analyzed using a new spectral simulation approach based on hybrid evolutionary optimization. The residue-level free rotational space was established and the rotational diffusion constants of the spin-label attached to the protein were measured. It was shown that in thelipid systems investigated, there is a relatively strong restriction for positions 23–29. The most pronounced effect is found for position 25, which is much more constrained than the average for the spin labels in the α-helical transmembrane domain. The authors suggested that it is due to a cluster of bulky amino acid residues (Ile-22, Tyr-21, Tyr-24, and Trp-26). The membrane location at the C-terminus is not affected by the lipid bilayer thickness. It was suggested that this effect can be assigned to a unique concerted action of the C-terminal phenylalanines (Phe-42 and Phe-45) and lysines (Lys-40, Lys-43, and Lys-44) in membrane anchoring. The results indicated that the coat protein has a large structural flexibility, which facilitates a stable protein-to-membrane association in lipid bilayers with various degrees of hydrophobic mismatch.

The topography of the prostaglandin endoperoxide H2 synthase-2 in liposomes was investigated [123]. T4 mutant enzymes, each containing a single free cysteine substituted for an amino acid in the COX-2 membrane binding domain, were labeled and reconstituted into liposomes. EPR power saturation experiments indicated that the labeled protein's side chains have limited accessibility to both polar and nonpolar paramagnetic relaxation agents. It was suggested that COX-2 associated primarily with the interfacial membrane region near the glycerol backbone and phospholipid head groups and that the protein–membrane interaction is promoted by the hydrophobic Phe59, Phe66, Tyr76, and Phe84, and by the charged amino acids, Arg62 and Lys64.

Substrate-dependent unfolding of the energy-coupling motif of a Ton box dependent membrane transport protein (BtuB, the carrier of vitamin B12 across the outer membrane of *Escherichia coli*) was studied by spin-labeling methods in combination with double electron–electron resonance (DEER) [124]. The distances between the Ton box and the periplasmic surface of the transporter with and without substrate were measured using DEER. From these distances, a model for the position of the Ton box was constructed, and this indicates that the N-terminal end of the Ton box extends from 2.0 to 3.0 nm into the periplasm upon the addition of substrate. The authors speculated that this substrate-induced extension provides the signal that initiates interactions between BtuB and the inner membrane protein TonB.

Quenching by nitroxide spin-label-modified phospholipids of the fluorescence of Trp residues substituted into cytochrome P 450 2C2, modified to contain Trp only at position 120, was used to identify regions of P 450 inserted into the lipid core and to established the depth of penetration [125]. The results obtained were consistent with an orientation of cytochrome P 450 2C2 in the membrane in which positions 36, 69, and 380 are inserted into the lipid bilayer and residues 80 and 225 are near or within the phospholipid headgroup region. In this orientation, the F–G loop, which contains residue 225, could form a dimerization interface.

A method based on the measurement of magnetic nuclear relaxation enhancement (PRE) by nitroxide spin-labels was used to provide long-range distance information about the integral membrane protein OmpA [126]. The method is based on parallel spin-labeling of a protein with paramagnetic and diamagnetic labels that allows us to detect distances in the range 1.5–2.4 nm. The OmpA was labeled at 11 water-exposed and lipid-covered sites, and 320 PRE distance restraints were measured. The authors stressed the proposed method has the advantage of calculating protein structures with PRE distance restraints only.

Cooperative interactions of the Synaptotagmin I Tandem C2 domains in phosphatidylcholine and phosphatidylserine membranes were detected by the SDSL method [127]. A water-soluble fragment of synaptotagmin I (C2AB) that contains its two C2 domains (C2A and C2B) was labeled in 19 positions and membrane depth parameters of labeled mutant proteins were measured by ESR spectroscopy. It was shown that both C2A and C2B domains bind to the membrane interface with their first and third Ca^{2+} binding loops penetrating the membrane interface while the polybasic face of C2B does not interact with the membrane lipid. Surface charge and curvature for the binding orientation of *Thermomyces lanuginosus* lipase (TTL) on negatively charged or zwitterionic phospholipid vesicles has been studied by ESR SDSL methods.

Eleven single-cysteine TLL mutants were spin-labeled and studied during membrane binding using the water soluble spin-relaxation agent chromium (III) oxalate [128]. Additional information was derived from the fluorecesnce quenching of dansyl probes located in the membrane by the nitroxide labels. The ESR and fluorescence results indicated that that TLL associates more closely with the negatively charged PG surface than the zwitterionic PC surface, and binds to the membrane predominantly through the concave backside of TLL opposite to its active site.

An elegant method for determining the position of proteins in membranes was proposed in [129]. The method utilizes an oxygen permeability gradient, which is a smooth continuous function of the distance from the center of the membrane. The relaxation gradient of oxygen in zwitterionic and anionic phospholipid membranes was detected by attaching a single nitroxide probe to "a ruler" (the transmembrane α-helical polypeptide) followed by studies of the oxygen effect on spin relaxation parameters of the spin labels. The peptide ruler was used to determine the depth of penetration of the calcium-binding loops of the C2 domain of cytosolic phospholipase A2.

The location and aggregation of the spin-labeled peptide trichogin GA IV in dipalmitoylphosphatidylcholine membrane were studied by ESEEM spectroscopy and PELDOR) [130–132]. The location of TOAC (2,2,6,6-tetramethylpiperidine-1-oxyl-4-amino-4-carboxylic acid), spin-labeled analog of the lipopeptaibol trichogin GA IV peptide, in the membrane was determined by comparing the ESEEM spectra for peptides labeled at different positions along the amino acid sequence. The PELDOR spectroscopy technique was chosen to study peptide aggregation and to determine the mutual distance distribution of the spin-labeled peptides in the membrane [131]. It was shown that the aggregates are characterized by a broad range of intermolecular distances (1.5–4 nm) between the labels at the N-terminal residues. At a low peptide/lipid molar ratio (less than 1:100) the nonaggregated peptide chain of the trichogin molecules lies parallel to the membrane surface, with TOAC at the fourth residue located near the 9th–11th carbon positions of the *sn*-2 lipid chain. Peptide conformations and cholesterol effect on the spatial distribution of spin-labeled analogs of trichogen GA IV in the multilamellar membranes of egg L-α-phosphatidylcholine were studied by PELDOR and CW ESR [132]. The analysis of dipole–dipole interactions between spin-labels for monolabeled analogs of trichogen GA IV bound to the membranes indicated that membrane trichogen molecules are distributed homogeneously and are likely to be located on or near the inner and outer membrane surfaces. Addition of cholesterol leads to an increase of the local concentration of trichogen molecules in the membranes. For the double-labeled trichogen, the intramolecular distance distribution of two main maxima located at distances of 1.3 and 1.8 nm of the intramol was observed. The distance of 1.3 nm is close to that expected for the α-helix structure of the peptide chain. The suggested model of the trichogen GA IV location in a membrane is reproduced in Figure 10.9.

A method to record high-field ESR spectra from different membrane orientations in a magnetic field was developed and applied to spin-labeled Gramicidin A and cholestane in aligned DPPC membranes [133]. A combination of isopotential spin-dry ultracentrifugation (ISDU), spin-labeling, microtome techniques, and use of the Fabry–Perot resonator allowed the collection of 170 GHz ESR spectra corresponding to different orientations of the membrane normal relative to the magnetic field.

Utilizing ESEEM spectroscopy to locate the position of specific regions of the 26-mer bee venom melittin and a *de novo* designed 15-mer D,L-amino acid amphipathic peptide (5D-L9K6C), within model membranes has been reported [134]. A nitroxide spin-label was introduced to the N-terminus of the peptides and measurements were performed either in H_2O solutions with deuterated model membranes or in D_2O solutions with nondeuterated model membranes. The results suggested that the N-terminus of both peptides is situated in the solvent layer in membranes and the peptides change the water distribution in the membrane, making it "flatter" and increasing the penetration depth into the hydrophobic region.

Protein–lipid interactions of *Escherichia coli* outer-membrane protein X (OmpX) in the dihexanoylphosphatidylcholine (DHPC) micelles have been studied by NMR spectroscopy with the use of paramagnetic ions and lipophilic nitroxides [135]. The

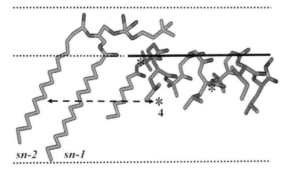

Figure 10.9 Model of the in-plane-bound trichogin GA IV molecule showing its immersion depth in the DPPC gel-state membrane. The fourth TOAC label is located at the same level as that of the tenth carbon atom of the sn-2 fatty acyl chain of the lipid (broken line). The solid line indicates the amphipathic plane of the peptide molecule separating the opposite hydrophilic and hydrophobic faces of the helix. This plane coincides with the border between the polar interface and the hydrocarbon interior of the membrane. Nitroxide-labeled positions (residues 1, 4, and 8) are depicted by asterisks [130].

observation of highly selective relaxation effects on the NMR spectra of OmpX and DHPC from a water-soluble relaxation agent and from nitroxide spin-labels attached to lipophilic molecules, confirmed data obtained previously with more complex NMR studies of the diamagnetic OmpX/DHPC system, and yielded additional novel insights into the protein–detergent interactions in the mixed micelles.

10.3
Nucleic Acids

10.3.1
Introduction

The general principles of the application of nitroxides for solving problems of structure and dynamics of DNA and RNA are similar to those for the synthetic polymers and proteins (see Section 8.1). Nevertheless, several essential specific features might be pointed out. Among them are the most important site-directed and specific nitroxide-labeled oligonucleotides, identification of a single genome, nucleic acids complexation with proteins and membranes, antitumor drugs (say Pt^{II}-complexes), spin trapping DNA radicals. Three trends, general for both DNA and RNA research, may be pointed out: (1) ues of new methods of nuclear acid labeling; (2) studies of the landscape and dynamics of local environments of interest in the vicinity of a label position; and (3) measurement of the distance between two sites.

An important step in the investigation of any nucleic acid is its chemical modification by nitroxide spin labels. A number of methods have been utilized ([6, 8, 10, 136–138] and references therein). These methods include covalent attachment of nitroxide to the nucleotide bases, ribose glucosyl rings, alkylation SH groups, intercalation, enzymatic syntheses with the use of a labeled base, and complexation with paramagnetic ions. More recently, new specific nitroxide derivatives suitable for modification DNA and RNA have been synthesized. A series of nitroxides that attach to the 4 or 5 positions of pyrimidines via alkylation was reported [139]. In [140] the rigid label 2,2,5,5-tetramethyl-pyrrolin-1-yl-oxyl-3-acetylene (TPA) was introduced during solid-phase synthesis through a Sonogashira cross-coupling with 5-iodo-uridine. Reaction of the 2′-amino group of 2′-amino uridine in RNA with a nitroxide containing an aliphatic isocyanate yielding spin-labeled RNA has been reported [141]. A guanine derivative with a covalently linked TEMPO for probing oligodeoxynucleotides has been devised in [136]. The synthesis of a new spin-labeled DNA base, 5-(2,2,6,6-tetramethyl-4-ethynylpiperidyl-3-ene-1-oxyl)-uridine was reported [137]. Spin-labeled 2′-deoxyuridine with π-conjugated nitroxide forming a one-dimensional ferromagnetic chain has been recently synthesized [138].

Site-direction spin-labeling opens new horizons for deeper insight into the microstructure and dynamics of nuclear acids. The methane thiosulfonate spin-label (1-oxyl-2,2,5,5-tetramethylpyrroline-3-methyl) (MTSL) is most often used in SDSL studies due to its sulfhydryl specificity and its small molecular volume (Figure 10.4) ([142] and references therein). Nevertheless, the high flexibility of this label causes some difficulties in interpreting ESR spectra of objects of interest. As examples of recent modifications, six uridine sites within TLR RNA were singly substituted with 4-thiouridine and modified with MTSL [143] and this label was attached to the SDR1su7 RNA that contains a 4-thio-U-modified base [144].

Three modes of motion influence the observed EPR spectra: (1) tumbling of the entire molecule; (2) torsional oscillations about bonds that connect the nitroxide moiety to the macromolecule; and (3) local macromolecule structural fluctuations at the labeling site. In addition to commonly used X-band ESR, utilization of multifrequency (high frequency, in particular) ESR spectroscopy and Fourier deconvolution methods essentially widen the ability to characterize quantitatively the dynamic behavior of spin labels. The stochastic Liouville equation and molecular dynamic simulations are especially effective for describing the effect of the nitroxide motion on the line-shape of ESR spectra.

Examination of effect of spin–spin exchange and dipole–dipole interactions on the line-shapes of the nitroxide ESR spectra makes it possible to measure interspin distances in the range 0.4–2.5 nm while analysis of the CW ESR saturation curves and pulse methods such as PELDOR allow us to evaluate the distances up to 70 Å (see Chapter 4).

10.3.2
DNA

Work done during the 1970s, 1980s, and 1990s on chemical modification of DNA with nitroxide derivatives and ESR examination of structure and dynamics of the labeled biopolymer, has laid the basis for further detailed investigations in this extremely important area of molecular biology [140, 144–150]. In this section we describe recent developments in DNA spin-labeling and SDSL in particular.

Eight distances between spin-labels attached to selected positions on a dodecamer DNA duplex with a known NMR structure, ranging from 2.0 to 4.0 nm were measured using double electron–electron resonance (DEER) [151]. Arbitrary nucleic acid sequences were chemically substituted by phosphorothioates followed by labeling with nitroxide R5. The distances measured correlated with the values predicted in calculation based on a search of sterically allowable R5 conformations in the known NMR structure. The flexibility of a series of linear duplex DNAs containing 14–100 base pairs on the submicrosecond timescale was studied using a site-specific spin probe that rigidly locked to the DNA [152]. The EPR spectra of the spin-labeled duplex were acquired and simulated by the stochastic Liouville equation for anisotropic rotational diffusion suggesting the diffusion tensor for a right circular cylinder. The weakly bending rod model, which was modified to take into account the finite relaxation times of the internal modes, was considered. The ESR data were in agreement with the model. From the length and position dependence of the internal flexibility of the DNA, a submicrosecond dynamic bending persistence length of around 1500 to 1700 Å was found. The modified weakly bending rod model was used for interpreting experimental data on the short-time (submicrosecond) bending dynamics of duplex DNA modified by a single, rigidly tethered spin probe 2AP-Q (Figure 10.10) [153]. It was found that purine–purine steps (which are the same as pyrimidine–pyrimidine steps) were near average in flexibility, but that pyrimidine–purine steps (5′ to 3′) were nearly twice as flexible, whereas purine–pyrimidine steps were only half as flexible as average DNA.

ESR data on DNA dynamics were analyzed using the slowly relaxing local structure model [153]. The model suggested global tumbling of the DNA lattice, and at a faster level, internal dynamics. The analysis revealed two spectra representing different components, one a highly restricted site yielding a large order parameter (0.61) and slower internal motions, and the other a much less restricted site (with order parameter of 0.18) and faster internal motions. It was concluded that that the spin labels trapped in the highly restricted/slow motional site have a stronger interaction with the base than those in the other site and the longer the tether the faster are the correlation times for internal dynamics. For all tethers it was found that the correlation time for internal motion perpendicular to the internal symmetry axis systematically becomes slower as the size of the oligomer increases. A manifestation of collective modes of motion of the DNA was suggested. An alternative model of spin-labeled DNA dynamics based on monitoring the highly ordered/slow motional site was discussed in [154].

The submicrosecond bending dynamics at a single site as a function of the sequence of the nucleotides constituting the duplex DNA was investigated in [155]. ESR results indicated that the dominant feature of the dynamics is best explained in terms of purine- and pyrimidine-type steps, although distinction is made among all 10 unique steps. It was found that purine–purine steps and pyrimidine–pyrimidine steps exhibit similar flexibility, but the pyrimidine–purine steps (5′ to 3′) were nearly twice as flexible, whereas purine–pyrimidine steps were more than half as flexible as average DNA. The authors suggested that these findings

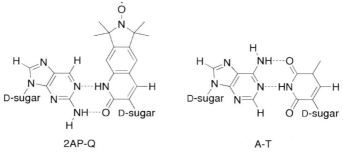

Figure 10.10 2AP-Q spin-probe base pair and the naturally occurring adenosine–thymidine (A-T) base pair shown for comparison [153]. (Reproduced with permission).

provide a quantitative basis for interpreting the dynamics and kinetics of DNA-sequence-dependent biological processes, including protein recognition and chromatin packaging.

Oligonucleotides labeled with flexible tethered nitroxides were proposed as molecular size sensors for the identification and quantitation of specific DNA by ESR spectroscopy [156]. Site-directed spin-labeling (SDSL) was used for the preparation of different probes specific to the detection of *E. hellem* DNA under polymerase chain reaction (PCR) amplification conditions. The possibility of the identification of *E. hellem* genomes and DNA isolated from *E. hellem* spores by the examination of changes in nitroxide dynamics caused by target-specific degradation of the probe was demonstrated.

Nanometer distances in site-directed spin-labeled oligonucleotides were measured by 4-PELDOR [157]. Five double-stranded oligonucleotides were doubly spin-labeled during solid-phase synthesis of the oligonucleotides utilizing a palladium-catalyzed cross-coupling reaction between 5-iodo-2′-deoxyuridine and the rigid spin-label TPA. PELDOR technique yielded spin–spin distances of 1.92–4.48, and 5.25 nm for different oligonucleotides. MD simulations on the same spin-labeled oligonucleotides gave a mean value of the spin–spin distances of 1.96–5.25 nm, respectively, in very good agreement with the measured distances.

Data on molecular mechanisms of gene regulation studied by site-directed spin-labeling were reviewed in [142]. The conformational changes of Tet repressor (TetR) and the human immunodeficiency virus type 1 reverse transcriptase (RT) on the interaction with nucleic acid substrates or inhibitors in solution were investigated. A twisting motion of the DNA reading heads of TetR on induction by tetracycline (tc) was observed in solution by changes of the interspin distances between interacting nitroxides at positions 22/22′ or 47/47′. Conformations of the tc- and DNA-complexed repressor at the core of the protein were found to be different from the conformation at the tc-binding pocket or at position 202.

The application of molecular dynamic simulations to spin-labeled oligonucleotides with the goal of examining the effect of spin label binding on the polymers was reported [158]. Molecular dynamic (MD) simulation on the double-stranded spin-labeled oligonucleotides (ONs) bearing a five- (5sp) or six- (6sp) membered ring nitroxides was performed. It was found that attaching the spin labels may cause widening of the major groove, decreasing of the helical twist, and a more negative X-displacement of the base pairs.

To study backbone dynamics along the GCN4-58 bZip sequence in a solution of a protein in a free-state and bound to DNA, a site-directed spin method was utilized [159]. ESR spectra revealed a mobility gradient similar to that observed in NMR relaxation, indicating that side chain motions mirror backbone motions. It was shown that the backbone motions are damped in the DNA-bound state, although a gradient of motion persists with residues at the DNA-binding site being the most highly ordered, similar to those of helixes on globular proteins.

10.3.3
RNA

The technique of site-directed spin-labeling (SDSL) has been used to study RNAs with complex three-dimensional structures [160–167]. Similar to protein SDSL, RNA SDSL utilizes a site-specific attached nitroxide moiety to obtain local structural information by analyzing the electron paramagnetic resonance (EPR) spectrum of the nitroxide. It is capable of monitoring both solution structure and conformational changes at specific sites of RNA molecules [161–166].

A dodecamer RNA duplex, 7.6 kDa SDR, with a known structure was utilized as a model system generating the first RNA SDSL library for a nitroxide (Figure 10.11) [168]. The thiol-reactive nitroxide derivative, 1-oxyl-3-methanesulfonylthiomethyl-2,5-dihydro-2,2,5,5-tetramethyl-1H-pyrrole was attached to the SDR1su7 RNA that contains a 4-thio-U modified base followed by examination of the ESR spectra of the labeled RNA. To minimize spectral effects due to molecular tumbling as a whole, SDR duplex was tethered to an avidin molecule (~60 kDa) via a biotin moiety incorporated at the 5' terminus of one of the SDR strands. The $(CH_2)_6$ linker allows the biotin to reach into the deep binding pocket in avidin without generating excess relative motion between RNA and avidin. Different line-shapes were obtained for different secondary structures, including single-strand, stacked A/U pair, and U/U mismatch structures. Similarity in the ESR spectra of SDR and stacked A/U pair that are observed in different RNA molecules was revealed.

The use of EPR spectroscopy to study changes in the internal structure and dynamics of a hammerhead ribozyme containing a nitroxide spin-label at position U7 under a variety of experimental conditions including temperature, metal ion identity, ionic strength, and the presence of ribozyme inhibitors was reported [169]. A 4-isocyanato TEMPO spin-label was attached to nucleotide U7 of the HH16 catalytic core by reaction of 2'-amino modified RNA. At 40 °C in the absence of Mg^{2+}, U7 of the hammerhead ribozyme exhibited motion in the fast regime and had an apparent correlation time τ_c of 1.1 ns, whereas U7 of the hammerhead ribozyme-non-cleavable substrate complex had τ_c of 1.79 ns. The spin-labeled ribozyme retained catalytic activity. U7 was shown by EPR spectroscopy to be more mobile in the ribozyme-product complex than in either the unfolded ribozyme or the ribozyme-substrate complex. EPR results indicated that the first RNA folding occurs around 0.25 mM Mg^{2+}, while the second folding event is completed at 10 mM Mg^{2+}.

Changes in HIV-1 TAR RNA internal dynamics induced by each of 10 metal ions were detected by changes in EPR spectral width for TAR RNAs containing spin-labeled nucleotides (U23, U25, U38, and U40) [170]. Experiments indicated different groups according to effect on TAR RNA internal dynamics: Li^+ and K^+ had little effect; Na^+ had a dynamic signature that was similar to Ca^{2+} and Sr^{2+}, with a decrease in mobility at U23 and U38, little or no change at U25, and an increase in mobility at U40; Mg^{2+}, Co^{2+}, Ni^{2+}, Zn^{2+}, and Ba^{2+} had similar effects on U23, U38, and U40, but the mobility of U25 was markedly increased.

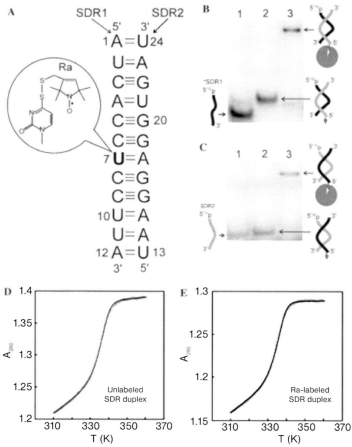

Figure 10.11 The SDR model RNA. (A) Sequence and numbering scheme for the SDR duplex. (B) Tethering of the SDR duplex to avidin as detected by native gel, with radio-labeled SDR1 (*SDR1) as the probe. (C) Tethering of the SDR duplex to avidin as detected using radio-labeled SDR2 (*SDR2) as the probe. (D) Thermal denaturation of the unlabeled SDR duplex.. (E) Thermal denaturation of the Ra-labeled SDR duplex. [168] (Reproduced from Qin, P.Z., Iseri, J. and Oki, A. (2006) A model system for investigating lineshape/structure correlations in RNA site-directed spin labeling, Biochem. Bioph. Res. Commun., 343, 117–124, with permission of Elsevier).

The 2′-amino group of 2′-amino uridine in RNA was labeled with a nitroxide containing an aliphatic isocyanate active group [171]. Thermal denaturation experiments indicated that the 2′-urea linked nitroxide had a minor effect on RNA duplex stability while experiments showed dependence of ESR spectra on the label nitroxide motion. In contrast, a label attached by fusing a nitroxide to the base moiety of 2′-deoxycytidine using phosphoramidite moved independently from the nucleic acid tumbling. An RNA spin-labeling technique, based on incorporating guanosine monophosphorothioate (GMPS) at the 5' end of the Rev response element (RRE) using T7 RNA polymerase and covalently attaching a thiol-specific nitroxide spin label, was developed. [164] Examination of the effect of spin–spin dipole–dipole interaction on the line-shape of ESR spectra of the labeled RNA allowed us to measure distances of up to 25 Å between the labeled sites.

Identification of amino acids that promote specific and rigid TAR RNA–Tat protein complex formation was reported in [172]. A goal of the work was to determine which amino acids in the basic region of the Tat protein are responsible for the observed decreases in U23 and U38 mobility in the wild-type TAR RNA–Tat peptide complex. Using ESR spectroscopy, it was shown that the changes in mobility of nitroxide spin-label attached to four RNA sites, correlate with differences in nucleotide mobility in the presence of a series of Tat-derived mutant peptides. Results indicate that mutations in the C-terminal end of the Tat basic region, but not the N-terminal end, interfere with wild-type RNA dynamics in the TAR–Tat complex, and that, specifically, arginine 56 contributes most prominently to U38 immobilization.

Site-directed spin-labeled 10-mer RNA duplexes and HIV-1 TAR (transactivating responsive region) RNA motifs with various interspin distances (R) were examined and distances measured in RNA molecules by continuous wave (CW) EPR spectroscopy [165]. The interspin distances R between spin labels attached to the $2'\text{-NH}_2$ positions of different uridines in the duplexes were measured by both molecular dynamic simulations (MD) and Fourier deconvolution methods. The experimental R values were found to be ranging from 1.0 nm to 2.0 nm, while MD predicted a range of 1.0–3.1 nm. The conformational changes in TAR RNA in the presence and in the absence of different divalent metal ions were monitored by measuring distances between two nucleotides in the bulge region.

10.4
Polysacchrides and Dextrins

10.4.1
Cotton and Cellulose

Polysaccharides are included in immunoglobulins and other glycoproteins such as cotton fibers, celluloses, dextrins, and so forth. Data obtained during the 1970s and 1980s clearly demonstrated the potential of the spin-labeling method for studies of microstructure, molecular dynamics, and conformational transitions in polysuccharides. The methods for polysaccharide modification by nitroxide labels are similar to those for the ribose and glycosyl rings of nuclear acids [6, 173–178]. Recently, a new solid-phase method of mechanochemical incorporation of a spin-label into cellulose has been reported [177].

After the first publication on the labeling of linear polymers such as cotton, silk, and wool by trichlorotriazine-based nitroxide [179] a series of works on the investigation of cotton fibers and cellulose were reported by the Marupov and Likhtenshein groups. The spin-labeling method has been used for the investigation of the dependence of ESR parameters of spin-labeled cotton fibers in liquid medium on viscosity and temperature [180], the thermal transitions of plasticized cellulose [181], the supermolecular structure of cotton cellulose [182], the effect of UV-radiation on cotton microstructure [183], and changes in the structure of cotton cellulose during biosynthesis [184].

High-frequency 2-mm band ESR spectroscopy was first used for the study of spin-labeled cotton fiber [185]. Raising the temperature from 150 to 280 K did not markedly change the ESR spectrum of spin-labeled cotton fibers. A further increase in temperature from 280 to 335 K was accompanied by a shift of x and y components of the anisotropic g-tensor to low fields, probably ensuing from gradual conformation changes of the fiber. The width of the z-components increased only slightly as they moved toward the center of the spectra. Such a combination indicated anisoptropic motion of the label in the nano-second time range. Recent investigation of the effect of humidity and temperature on the supramolecular structure of cotton, studied by spin-labeling, has confirmed the main results of the works cited above, and added important details [186]. It was shown that three different components contribute differently to the experimental EPR spectra, corresponding to (a) mobile radicals absorbed in the bulk amorphous region, (b) slow moving radicals adsorbed on the crystallite surfaces in cotton, and (c) aggregated radicals. For all water loadings and temperatures, the local polarity and mobility of the label environment increased when the water content increased up to 3 wt.-%.

A study of the effect of a polysaccharide capsule of the microalgae *Staurastrum iversentii var. americanum* on diffusion of charged and uncharged molecules, using the spin-probe technique was reported [187]. The results regarding the diffusion of charged spin-probes suggested that the polysaccharide capsule made a barrier for probe penetration into the cells and the interaction of cell capsule occurs more strongly with negatively charged molecules than with positively charged ones.

Nitroxide-mediated oxidation of cellulose in water at pH 10–11 using TEMPO derivatives has been studied [188]. The process was monitored by means of high-performance size-exclusion chromatography and NMR. The spin probes 1,1,3,3-tetramethylisoindolin-2-yloxyl and the sodium salt of its sulfonate, 1,1,3,3-tetramethylisoindolin-2-yloxyl-5-sulfonate were used to monitor the microviscosity changes of water during starch gelatinization [189]. In cereal starch, the amylopectin and amylose regions underwent a transition at about 55 °C and a large increase in the microviscosity on cooling. Pea starch containing the two above-mentioned domains also exhibited the 55 °C transition. Spin-labeled (SL-) polysaccharides (SL-pullulan, SL-xylan, and SL-maltoheptaose) were employed to monitor the sorption of SL-polysaccharides to natural sediment surfaces and to measure polysaccharide size due to enzymic hydrolysis [190]. The properties of spin-labeled polysaccharides grafted on nanoparticles were investigated in [178]. It was concluded that the mobility of the polysaccharide depended on the capacity of the polysaccharide chains to fold, making possible hydrophobic interactions between the label and the nanoparticle core and the transition between the unfolded–folded regimen depended on the nature of the polysuccharide.

The use of the spin-labeling for study of polysaccharides to natural sediment surfaces and montmorillonite was reported [191].

10.4.2
Cyclodextrins

Application of spin-labeling methods to cyclodextrins have three main goals: (1) studies of the formation of inclusion complexes between nitroxides of different structures and various dextrins; (2) investigation of the probes and substrate dynamics within the complexes; and (3) the use of dextrins for catching and stabilizing spin probes and free radical–nitron adducts. The latter issue was described in Chapter 7.

An inclusion complex of 1,1-dimethylethyl 2-methyl-1-phenylpropyl nitroxide was covalently bound to a permethylated-β-cyclodextrin and properties of an inclusion complex were investigated by ESR spectroscopy [192]. The experimental activation parameters of the Eyring equation for the forward and backward steps of complex formation were consistent with an equilibrium between a non-associated form and a weakly associated form, with activation free enthalpies for each reaction of approximately $34\,kJ\,mol^{-1}$. The free radical trapping properties of eight 5-alkoxycarbonyl-5-methyl-1-pyrroline N-oxide and 5,5-dimethyl-1-pyrroline N-oxide were evaluated for trapping of superoxide anion radicals in the presence of 2,6-di-O-methyl-β-cyclodextrin [193].

Inclusion complexes of spin-labeled pyrrolidine- (**1**) and piperidine-containing (**2**) indole derivatives with β-cyclodextrin and γ-cyclodextrin (CD) were prepared in the solid phase and studied by Saturation Transfer (ST) ESR in a wide temperature interval [194]. It was shown that the p-orbit axis of the NO group of both guest molecules in γ-CD undergo rapid librations the amplitude of which reaches about 16° at 333 K. Line-shape analysis of the spectra suggested that they consist of two components, one corresponding to strong spin–spin interaction between guest molecules and the other corresponding to almost an absence of this interaction. In the temperature range 238–333 K, probes **1** and **2** move by a mechanism of rotational jumps and the rotational frequencies of **1** and **2** are in intervals 1.8×10^7–$6 \times 10^7\,s^{-1}$ and 4×10^7–$3 \times 10^8\,s^{-1}$, respectively. In both cyclodextrins the rotational mobility of molecules **2** is higher than that of **1** owing to intramolecular conformational transitions in the piperidine ring of **2** and steric hindrances produced by the methyl group in **1**.

In work [195] the synthesis of mono-functionalized spin-labeled β-cyclodextrins (Figure 10.12) was reported. The formation of inclusion complexes of the cyclodextrins with large guest molecules such as phenolphthalein and 1-adamantyl amine was monitored by ESR spectroscopy.

Thus the nitroxide spin-labeling method has proved its effectiveness in studies of the dynamic and structural properties of proteins, enzymes, biomembranes,

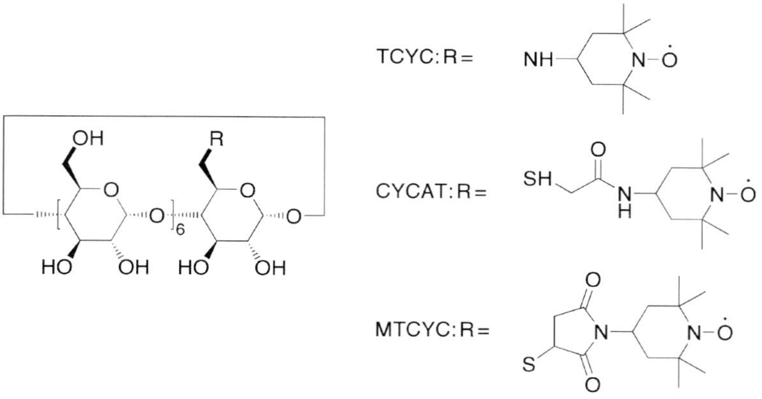

Figure 10.12 Schematic representation of spin-labeled dextrins [195]. (Reproduced with permission).

nucleic acids, lipoproteins, glycoproteins, and polyssacharides. Recent new directions such as site-directed spin-labeling and the employment of advanced magnetic resonance technique have added "fresh blood" to the field.

References

1 McConnell, H.M. and McFarland, H.F. (1970) *Quarterly Review of Biophysics*, 3, 91–136.
2 Berliner, L. (ed.) (1979) *Spin Labeling. Theory and Applications*, Vol. 2, Academic Press, New York.
3 Berliner, L. (ed.) (1998) *Spin Labeling. The Next Millennium*, Vol. 14, Academic Press, New York.
4 Hemminga, M.A. and Berliner, L.J. (eds) (2007) *ESR Spectroscopy in Membrane Biophysics*, Vol. 27, Biological Magnetic Resonance, Springer Verlag.
5 Berliner, L.J., Eaton, S.S. and Eaton, G.R. (eds) (2000) *Biological Magnetic Resonance*, Vol. 19, Kluwer Academic/ Plenum Publisher, New York.
6 Likhtenshtein, G.I. (1976) *Spin Labeling Method in Molecular Biology*, Wiley Interscience, New York.
7 Likhtenshtein, G.I. (2005) Labeling, biophysical, in *Encyclopedia of Molecular Biology and Molecular Medecine*, Vol. 7 (ed. R. Meyers), VCH, New York, pp. 157–78.
8 Likhtenshtein, G.I. (1993) *Biophysical Labeling Methods in Molecular Biology*, Cambridge University Press, Cambridge, NY.
9 Likhtenshtein, G.I., Febrario, F. and Nucci, R. (2000) *Spectrochimica Acta, Part A: Molecular and Biomolecular Spectroscopy*, 56, 2011–31.
10 Kocherginsky, N. and Swartz, H.M. (1995) *Nitroxide Spin Labels*, CRC Press, Boca Raton.
11 Lumry, R. and Eyring, H. (1954) *Journal of Physical Chemistry*, 58, 110–20.
12 Linderstrom-Lang, K.H. and Schellmann, X.X. (1959) *The Enzyme*, Vol. 1 (eds P.D. Boyer, M. Lardy and X. Myrback), Academic Press, NY, pp. 443–510.
13 Koshland, D.J. (1959) *The Enzyme*, Vol. 1 (eds V. Boyer, P.D. Lardy and M. Myrback), Academic Press, NY, pp. 305–46.
14 Likhtenshtein, G.I., Troshkina, T.V., Akhmedov, Yu.D. and Shuvalov, V.F. (1969) *Molecular Biology (Moscow)*, 3, 413–20.

15 Frolov, E.N., Mokrushin, A.N., Likhtenshtein, G.I., Trukhtanov, V.A. and Goldansky, V.I. (1973) *Doklady Akademii Nauk SSSR*, 212, 165–8.

16 Likhtenshtein, G.I. (1976) *L'eau et les Systemes Biologoque* (eds A. Alfsen and A.J. Bertran), CNR, Paris, pp. 45–3.

17 Likhtenshtein, G. (1979) *Special Collogue Amper on Dynamic Processes in Molecular Systems* (ed. A. Losche), Karl-Marx University, Leipzig, pp. 100–7.

18 Berg, A.I., Kononenko, A.F., Noks, P.P., Khrymova, I.N., Frolov, E.N., Rubin, Likhtenshtein, G.I., Uspenskaya, N. and Hideg, K. (1979) *Molecular Biology*, 13, 469–74.

19 Berg, A.I., Kononenko, A.F., Noks, P.P., Khymova, I.N., Frolov, E.N., Rubin, A.B., Likhtenshtein, G.I., Goldansky, V.I., Parak, F., Bukl, M. and Mossbauer, R.L. (1979) *Molecular Biology*, 13, 81–9.

20 Fogel, V.R., Rubtsova, E.T., Likhtenshtein, G.I. and Hideg, K. (1994) *Journal of Photochemistry and Photobiology A: Chemistry*, 83, 229–36.

21 Likhtenshtein, G.I., Bogatyrenko, V.R., Kulikov, A.V., Hideg, K., Khankovskaya, G.O., (Hankovsky, H.O.), Lukoyanov, N.V., Kotel'nikov, A.I. and Tanaseichuk, B.S. (1980) *Doklady Akademii Nauk SSSR*, 253, 481–4.

22 Likhtenshtein, G.I., Bogatyrenko, V.R. and Kulikov, A.V. (1993) *Applied Magnetic Resonance*, 4, 513–21.

23 Frolov, E.N., Kharakhonicheva, N.V. and Likhtenshtin, G.I. (1974) *Molecular Biology (Moscow)*, 8, 886–93.

24 Likhtenshtein, G.I. (2003) *New Trends in Enzyme Catalysis and Mimicking Chemical Reactions*, Kluwer Academic/ Plenum Publishers, New York.

25 Parak, F., Knapp, E.W. and Kucheida, D. (1982) *Journal of Molecular Biology*, 161, 177–94.

26 Krinichnyi, V.I., Grinberg, O.Y., Bogatirenko, V.R., Likhtenshtein, G.I. and Lebedev, Y.S. (1985) *Biofizika*, 30, 216–19.

27 Vitkup, D., Ringe, D., Petsko, G.A. and Karplus, M. (2000) *Nature Structural Biology*, 7, 34–5.

28 Tarek, M., Martina, G.J. and Tobias, D. (2000) *Journal of the American Chemical Society*, 122, 10459–1.

29 Frolov, E.N., Belongova, O.V. and Likhtenshtein, G.I. (1977) Investigation of the mobility of spin and Mössbauer labels bound to macromolecules, in *Equilibrium Dynamics of the Native Structure of Protein* (ed. E.A. Burshtein), Izdatelstvo Akademii Nauk SSSR, Pushchino, pp. 99–142.

30 Victor, K., Van-Quynh, A. and Bryant, R.G. (2005) *Biophysical Journal*, 88, 443–54.

31 Frauenfelder, H. and McMahon, B.H. (2001) *Springer Series in Chemical Physics (Single Molecule Spectroscopy)*, 67, 257–76.

32 Frolov, E.N., Likhtenshtein, G.I., Khurgin, Yu.I. and Belonogova, O.V. (1973) *Izvestiya Akademii Nauk SSSR, Seriya Khimicheskaya*, 1, 231–2.

33 Likhtenshtein, G.I. (1996) *Journal of Photochemistry and Photobiology A: Chemistry*, 96, 79–92.

34 Ogava, S. and McConnell, H.M. (1967) *Proceedings of the National Academy of Sciences of the United States of America*, 58, 19–26.

35 Tabak, M., de Sousa Neto, D. and Salmon, C.E.G. (2006) *Brazilian Journal of Physics*, 36, 3–89.

36 Fajer, P.G. (2005) *Journal of Physics: Condensed Matter*, 17, S1459–69.

37 Delannoy, S.U., Tombline, G., Senior, A.E. and Vogel, P.D. (2005) *Biochemistry*, 44, 14010–19.

38 Lozinsky, E., Febbraio, F., Shames, A.I., Likhtenshtein, G.I., Bismuto, E. and Nucci, R. (2002) *Protein Science: A Publication of the Protein Society*, 11, 2535–44.

39 Kirby, T.L., Karim, C.B. and Thomas, D.D. (2004) *Biochemistry*, 43, 5842–52.

40 Lozinsky, E., Iametti, S., Barbiroli, A., Likhtenshtein, G.I., Kalai, T., Hideg, K. and Bonomi, F. (2006) *The Protein Journal*, 25, 1–15.

41 Alfimova, E.Ya. and Likhtenshtein, G.I. (1979) *Advances in Molecular Relaxation and Intraction Processes*, 14, 47–9.

42 Privalov, P.L. and Gill, J.G. (1998) *Advances in Protein Chemistry*, 39, 191–234.

43 Battistel, E., Attanasio, F. and Rialdi, G. (2000) *Journal of Thermal Analysis and Calorimetry*, 61, 513–25.

44 Lumry, R. (1995) *The new paradigm for protein*, in Gregery R.B. (ed), Protein-solvent interaction, Marcel Dekker, New York.

45 Zavodsky, S., Kardos, J., Svingor, A. and Petsko, G.A. (1995) *Proceedings of the National Academy of Sciences of the United States of America*, 95, 7406.

46 Brown, L.J., Klonis, N., Sawyer, W.H., Fajer, P.G. and Hambly, B.D. (2001) *Biochemistry*, 40, 8283–91.

47 Karplus, M. and McCammon, M. (1986) *Scientific American*, 251, 4–12.

48 Zhou, H.X., Wlodek, S.T. and McCommon, J.A. (1998) *Proceedings of the National Academy of Sciences of the United States of America*, 95, 9280–3.

49 Murzyn, K., Rog, T., Blicharski, W., Dutka, M., Pyka, J., Szytula, S. and Froncisz, W. (2006) *Proteins: Structure, Function, and Bioinformatics*, 62, 1088–100.

50 Berliner, L.J. and McConnell, H.M. (1966) *Proceedings of the National Academy of Sciences of the United States of America*, 55, 708–12.

51 Makinen, M.W. (1998) *Spectrochimica Acta, Part A: Molecular and Biomolecular Spectroscopy*, 54A, 2269–81.

52 Mustafi, D., Hofer, J.E., Huang, W., Palzkill, T. and Makinen, M.W. (2004) *Spectrochimica Acta, Part A: Molecular and Biomolecular Spectroscopy*, 60A, 1279–89.

53 Mustafi, D., Sosa-Peinado, A. and Makinen, M.W. (2001) *Biochemistry*, 40, 2397–409.

54 Park, J.H. and Trommer, W.E. (1989) *Biological Magnetic Resonance*, 8, 547–95.

55 Palm, T., Coan, C. and Trommer, W.E. (2001) *Biological Chemistry*, 382, 417–23.

56 Delannoy, S., Urbatsch, I.L., Tombline, G., Senior, A.E. and Vogel, P.D. (2005) *Biochemistry*, 44, 14010–19.

57 Keyes, R.S., Cao, Y.Y., Bobst, E.V., Rosenberg, J.M. and Bobst, A. (1996) *Journal of Biomolecular Structure and Dynamics*, 14, 163–72.

58 Likhtenshtein, G.I. (2000) *Biological Magnetic Resonance*, Vol. 19 (eds L.J. Berliner, S.S. Eaton and G.R. Eaton), Kluwer Academic/Plenum Publisher, New York.

59 Moebius, K., Savitsky, A., Wegener, C., Plato, M., Fuchs, M., Schnegg, A., Dubinskii, A.A., Grishin, Y.A., Grigor'ev, I.A., Kuehn, M., Duche, D., Zimmermann, H. and Steinhoff, H.-J. (2005) *Magnetic Resonance in Chemistry: MRC*, 43 (Spec. Issue), S4–19.

60 Hubbell, W.L., Gross, R.L. and Lietzow, M.A. (1988) *Current Opinion in Structural Biology*, 8, 649–56.

61 Steinhoff, H.-J. (2004) *Biological Chemistry*, 385, 913–20.

62 Fanucci, G.E. and Cafiso, D.S. (2006) *Current Opinion in Structural Biology*, 16, 644–53.

63 Sammalkorpi, M. and Lazaridis, T. (2007) *Biophysical Journal*, 92, 10–22.

64 Guo, Z., Cascio, D., Hideg, K., Kalai, T., Hubbell, W.L. (2007) *Protein Science*, 16, 1069–86.

65 Berliner, L.J. (1971) *Journal of Molecular Biology*, 61, 189–94.

66 Akhmedov, Iu.D., Likhtenshtein, G.I., Ivanov, L.V. and Kokhanov, Iu.V. (1972) *Doklady Akademii Nauk SSSR*, 205, 372–4.

67 Likhtenshein, G.I., Akhmedov, Yu.D., Ivanov, L.V., Krinitskaya, L.A. and Kokhanov, Yu.V. (1974) *Molecular Biology (Moscow)*, 8, 40–8.

68 Shimanovskii, N.L., Stepaniants, A.U., Lexina, V.P., Ivanov, L.V. and Likhtenshtein, G.I. (1977) *Biofizika*, 22, 811–15.

69 Tombolato, F., Ferrarini, A. and Freed, J.H. (2006) *Journal of Physical Chemistry. B*, 110, 26260–71.

70 Borbat, P.P., Mchaourab, H.S. and Freed, J.H. (2002) *Journal of the American Chemical Society*, 124, 5304–14.

71 Columbus, L., Kámás, T., Jekö, J., Hideg, K. and Hubbell, W.L. (2001) *Biochemistry*, 40, 3828–34.

72 Jacobsen, K., Oga, S., Hubbell, W.L. and Risse, T. (2005) *Biophysical Journal*, 88, 4351–65.

73 Jacobsen, K., Hubbell, W.L., Ernst, O.P. and Risse, T. (2006) *Angewandte Chemie (International Ed. in English)*, 45, 3874–7.

74 Hustedt, E.J., Stein, R.A., Sethaphong, L., Brandon, S., Zhou, Z. and DeSensi, S.C. (2006) *Biophysical Journal*, 90, 340–56.

75 Barnes, J.P., Liang, Z., Mchaourab, H.S., Freed, J.H. and Hubbell, W.L. (1999) *Biophysical Journal*, 76, 3298–306.

76 Mchaourab, H.S., Oh, K.J., Fang, C.J. and Hubbell, W.L. (1997) *Biochemistry*, 36, 307–16.

77 Columbus, L., Kalai, T., Jekoe, J., Hideg, K. and Hubbell, W.L. (2001) *Biochemistry*, 40, 3828–46.

78 Borbat, P.P., Mchaourab, H. and Freed, J.H. (2002) *Journal of the American Chemical Society*, 124, 5304–14.

79 Pyka, J., Osyczka, A., Turyna, B., Blicharski, W. and Froncisz, W. (2001) *European Biophysics Journal: EBJ*, 30, 367–73.

80 Steinhoff, H.-J., Savitsky, A., Wegener, C., Pfeiffer, M., Plato, M. and Möbius, K. (2000) *Biochimica et Biophysica Acta*, 1457, 253.

81 Rink, T., Pfeiffer, M., Oesterhelt, D., Gerwert, K. and Steinhoff, H.J. (2000) *Biophysical Journal*, 78, 1519–30.

82 Bordignon, E., Klare, J.P., Doebber, M., Wegener, A.A., Martell, S., Engelhard, M. and Steinhoff, H.-J. (2005) *The Journal of Biological Chemistry*, 280, 38767–75.

83 Radzwill, N., Gerwert, K. and Steinhoff, H.J. (2001) *Biophysical Journal*, 80, 2856–66.

84 Scoville, D., Stamm, J.D., Toledo-Warshaviak, D., Altenbach, C., Phillips, M., Shvetsov, A., Rubenstein, P.A., Hubbell, W.L. and Reisler, E. (2006) *Biochemistry*, 45, 13576–84.

85 Gingras, A.R., Klaus-Peter Vogel, K.-P., Steinhoff, H.-J., Ziegler, W.H., Patel, B., Jonas Emsley, J., Critchley, D.R., Roberts, G.C.K. and Barsukov, I.L. (2006) *Biochemistry*, 45, 1805–17.

86 Brown, L.J., Sale, K.L., Hills, R.R., Rouviere, C., Song, L., Zhang, X. and Fajer, P.G. (2002) *Proceedings of the National Academy of Sciences of the United States of America*, 99, 12765–70.

87 Steigmiller, S., Boersch, M., Graeber, P. and Huber, M. (2005) *Biochimica et Biophysica Acta, Bioenergetics*, 1708, 143–53.

88 Stopar, D., Strancar, J., Spruijt, R.B. and Hemminga, M.A. (2006) *Biophysical Journal*, 91, 3341–8.

89 Vos, W.L., Vermeer, L.S. and Hemminga, M.A. (2007) *Biophysical Journal*, 92, 138–46.

90 Chelius, D., Loeb-Hennard, C., Fleischer, S., McIntyre, J.O., Marks, A.R., De, S., Hahn, S., Jehl, M.M., Moeller, J.P.R., Wise, J.G. and Trommer, W.E. (2000) *Biochemistry*, 39, 9687–97.

91 Motz, C., Hornung, T., Kersten, M., McLachlin, D.T., Dunn, S.D., Wise, J.G. and Vogel, P.D. (2004) *The Journal of Biological Chemistry*, 279, 49074–81.

92 Hornung, T., McLachlin, D.T., Dunn, S.D., Hustedt, E.J., Wise, J.G. and Vogel, P.D. (2004) *Biophysical Journal*, 86, 337A.

93 Fanucci, G.E. and Cafiso, D.S. (2006) *Current Opinion in Structural Biology*, 16, 644–53.

94 Savitsky, A., Kühn, M., Duché, D., Möbius, K. and Steinhoff, H.J. (2004) *Journal of Physical Chemistry. B*, 108, 9541.

95 Möbius, K., Savitsky, A., Schnegg, A., Plato, M. and Fuchs, M. (2005) *Physical Chemistry Chemical Physics: PCCP*, 7, 19–42.

96 Marsh, D. and Pali, T. (2004) *Biochimica et Biophysica Acta, Biomembranes*, 1666 (1–2), 118–41.

97 Borbat, P.P., Costa-Filho, A.J., Earle, K.A., Moscicki, J.K. and Freed, J.H. (2001) *Science (Washington, DC, United States)*, 291, 266–9.

98 Goloshchapov, A.N. and Burlakova, E.B. (2005) *Essential Results in Chemical Physics and Physical Chemistry* (eds A.N. Goloshchapov, G.E. Zaikov and V.V. Ivanov), Nova Science Publishers, Hauppauge, pp. 1–18.

99 Bartucci, R., Erilov, D.A., Guzzi, R., Sportelli, L., Dzuba, S.A. and Marsh, D. (2006) *Chemistry and Physics of Lipids*, 141, 142–57.

100 Karp, E.S., Inbaraj, J.J., Laryukhin, M. and Lorigan, G.A. (2006) *Journal of the American Chemical Society*, 128, 12070–1.

101 Cherepanova, E.S., Kulikov, A.V. and Likhtenshtein, G.I. (1990) *Biologicheskie Membrany*, 7, 51–6.

102 Strashnikova, N.V., Medvedeva, N. and Likhtenshtein, G.I. (2001) *Journal of Biochemical and Biophysical Methods*, 48, 43–60.

103 Peric, M., Alves, M. and Bales, B.L. (2006) *Chemistry and Physics of Lipids*, 142, 1–13.

104 Marsh, D., Dzikovski, B.G. and Livshits, V.A. (2006) *Biophysical Journal*, 90, L49–51.

105 Bartucci, R., Guzzi, R., Marsh, D. and Sportelli, L. (2003) *Biophysical Journal*, 84, 1025–30.

106 Erilov, D.A., Bartucci, R., Guzzi, R., Shubin, A.A., Maryasov, A.G., Marsh, D., Dzuba, S.A. and Sportelli, L. (2005) *Journal of Physical Chemistry. B*, 109, 12003–13.

107 Noethig-Laslo, V., Cevc, P., Arcon, D. and Sentjurc, M. (2004) *Origins of Life and Evolution of the Biosphere*, 34, 237–42.

108 Alves, M. and Peric, M. (2006) *Biophysical Chemistry*, 122, 66–73.

109 Chiang, Y.-W, Zhao, J., Wu, J., Shimoyama, Y., Freed, J.H. and Feigenson, G.W. (2005) *Biochimica et Biophysica Acta, Biomembranes*, 1668, 99–105.

110 Swamy, M.J., Ciani, L., Ge, M., Smith, A.K., Holowka, D., Baird, B. and Freed, J.H. (2006) *Biophysical Journal*, 90, 4452–65.

111 Cardon, T.B., Dave, P.C. and Lorigan, G.A. (2005) *Langmuir: The ACS Journal of Surfaces and Colloids*, 21, 4291–8.

112 Rappolt, M., Amenitsch, H., Strancar, J., Teixeira, C.V., Kriechbaum, M., Pabst, G., Majerowicz, M. and Laggner, P. (2004) *Advances in Colloid and Interface Science*, 111, 63–77.

113 Krinichnyi, V.I. (1994) *2-mm Wave Band ESR Spectroscopy of Condensed Systems*, CRC Press, Boca Raton.

114 Krinichnyi, V.I., Grinberg, O.Ya., Judanova, E.I., Borin, M.L., Lebedev, Ya.S. and Likhtenshtein, G.I. (1988) *Biofizika*, 32, 59–65.

115 Hyde, J.S. and Feix, J.B. (1989) Theory and application, in *Biological Magnetic Resonance*, Vol. 8 (eds L.J. Berliner and J. Reubin), Plenum Press, New York, pp. 305–39.

116 Erilov, D.A., Bartucci, R., Guzzi, R., Marsh, D., Dzuba, S.A. and Sportelli, L. (2004) *Biophysical Journal*, 87, 3873–8.

117 Medvedeva, N., Papper, V. and Likhtenshtein, G.I. (2005) *Physical Chemistry Chemical Physics: PCCP*, 7, 3368–74.

118 Iwamoto, S., Yoshioka, D. and Sueishi, Y. (2006) *Chemical Physics Letters*, 430, 314–18.

119 Mravljak, J., Konc, J., Hodoscek, M., Solmajer, T. and Pecar, S. (2006) *Journal of Physical Chemistry. B*, 110, 25559–2556.

120 McConnell, H.M. and Radhakrishnan, A. (2006) *Abstracts of Papers*, 232nd ACS National Meeting, San Francisco, CA, USA, September 10–14, 2006, PHYS-012.

121 Korb, J.P., Ahadi, M. and McConnell, H.M. (1987) *Journal of Physical Chemistry*, 91, 255–9.

122 Stopar, D., Strancar, J., Spruijt, B. and Hemminga, M.A. (2006) *Biophysical Journal*, 91, 3341–8.

123 MirAfzali, Z., Leipprandt, J.R., McCracken, J.L. and DeWitt, D.L. (2006) *The Journal of Biological Chemistry*, 281, 28354–64.

124 Xu, Q., Ellena, J.F., Kim, M. and Cafiso, D.S. (2006) *Biochemistry*, 45, 10847–54.

125 Ozalp, C., Szczesna-Skorupa, E. and Kemper, B. (2006) *Biochemistry*, 45, 4629–37.

126 Liang, B., Bushweller, J.H. and Tamm, L.K. (2006) *Journal of the American Chemical Society*, 128, 4389–97.

127 Herrick, D.Z., Sterbling, S., Rasch, K.A., Hinderliter, A. and Cafiso, D.S. (2006) *Biochemistry*, 45, 9668–74.

128 Hedin, E.M.K., Hoyrup, P., Patkar, S.A., Vind, J., Svendsen, A. and Hult, K. (2005) *Biochemistry*, 44, 16658–71.

129 Nielsen, R.D., Che, K., Gelb, M.H. and Robinson, B.H. (2005) *Journal of the American Chemical Society*, 127, 6430–42.

130 Salnikov, E.S., Erilov, D.A., Milov, A.D., Tsvetkov, Yu.D., Peggion, C., Formaggio, F., Toniolo, C., Raap, J. and Dzuba, S.A. (2006) *Biophysical Journal*, 91, 1532–40.

131 Milov, A.D., Samoilova, R.I., Tsvetkov, Yu.D., Formaggio, F., Toniolo, C. and Raap, J. (2005) *Applied Magnetic Resonance*, 29, 703–16.

132 Milov, A.D., Erilov, D.A., Salnikov, E.S., Tsvetkov, Yu.D., Formaggio, F., Toniolo, C. and Raap, J. (2005) *Physical Chemistry Chemical Physics: PCCP*, 7, 1794–9.

133 Dzikovski, B., Earle, K., Pachtchenko, S. and Freed, J. (2006) *Journal of Magnetic Resonance*, 179, 273–9.

134 Carmieli, R., Papo, N., Zimmermann, H., Potapov, A., Shai, Y. and Goldfarb, D. (2006) *Biophysical Journal*, 90, 492–505.

135 Hilty, C., Wider, G., Fernández, C. and Wüthrich, K. (2003) *Journal of Biomolecular NMR*, 27, 377–82.
136 Okamoto, A., Inasaki, T. and Saito, I. (2004) *Bioorganic and Medicinal Chemistry Letters*, 14, 3415–18.
137 Gannett, P.M., Darian, E., Powell, J., Johnson, I.E.M., Mundoma, C., Greenbaum, N.L., Ramsey, C.M., Dalal, N.S. and Budil, D.E. (2002) *Nucleic Acids Research*, 30, 5328–37.
138 Das, K., Pink, M., Rajca, S. and Rajca, A. (2006) *Journal of the American Chemical Society*, 128, 5334–5.
139 Robinson, B.H., Mailer, C. and Drobny, G. (1997) *Annual Review of Biophysics and Biomolecular Structure*, 26, 629.
140 Piton, N., Schiemann, O., Mu, Y., Stock, G., Prisner, T. and Engels, J.W. (2005) *Nucleosides, Nucleotides and Nucleic Acids*, 24, 771–5.
141 Cekan, P. and Sigurdsson, S.Th. (2005) Collection Symposium Series 7 (Chemistry of Nucleic Acid Components), pp. 225–8.
142 Steinhoff, H.J. and Suess, B. (2003) *Methods*, 29, 188–95.
143 Qin, P.Z., Hideg, K., Feigon, J., Hubbell, W.H. (2003) *Biochemistry*, 42, 6772–83.
144 Qin, P.Z., Iseri, J. and Oki, A. (2006) *Biochemical and Biophysical Research Communications*, 343, 117–24.
145 Strobel, O.K., Keyes, R.S. and Bobst, A.M. (1990) *Biochemical and Biophysical Research Communications*, 166, 1435–40.
146 Artiukh, R.I., Postnikova, G.B., Sukhorukov, B.I. and Kamzolova, S.G. (1972) *Biokhimiia (Moscow, Russia)*, 37, 902–6.
147 Bobst, A.M. (1972) *Biopolymers*, 11, 1421–33.
148 Mil', E.M., Zavriev, S.K., Grigorian, G.L. and Krugliakova, K.E. (1973) *Doklady Akademii Nauk SSSR*, 209, 217–20.
149 Strobel, O.K., Kryak, D.D., Bobst, E.V. and Bobs, A.M. (1991) *Bioconjugate Chemistry*, 2, 89–95.
150 Robinson, B.H., Mailer, C. and Drobny, G. (1997) *Annual Review of Biophysics and Biomolecular Structure*, 26, 629.
151 Cai, Q., Kusnetzow, A.K., Hubbell, W.L., Haworth, I.S., Gacho, G.P.C., Van Eps, N., Hideg, K., Chambers, E.J. and Qin, P.Z. (2006) *Nucleic Acids Research*, 34, 4722–30.
152 Okonogi, T.M., Reese, A.W., Alley, S.C., Hopkins, P.B. and Robinson, B.H. (1999) *Biophysical Journal*, 77, 3256–76.
153 Okonogi, T.M., Alley, S.C., Reese, A.W., Hopkins, P.B. and Robinson, B.H. (2002) *Biophysical Journal*, 83, 3446–59.
154 Liang, Z., Freed, J.H., Keyes, R.S. and Bobst, A.M. (2000) *Journal of Physical Chemistry. B*, 104, 5372–81.
155 Robinson, B.H., Mailer, C. and Drobny, G. (1997) *Annual Review of Biophysics and Biomolecular Structure*, 26, 629–35.
156 Bobst, A.M., Hester, J.D. and Bobst, E.V. (2004) *Abstracts of Papers*, 227th ACS National Meeting, Anaheim, CA, United States, March 28–April 1.
157 Schiemann, O., Piton, N., Mu, Y., Stock, G., Engels, J.W. and Prisner, T.F. (2004) *Journal of the American Chemical Society*, 126, 5722–9.
158 Darian, E. and Gannett, P.M. (2005) *Journal of Biomolecular Structure and Dynamics*, 22, 579–93.
159 Columbus, L. and Hubbell, W.L. (2004) *Biochemistry*, 43, 7273–87.
160 Qin, P.Z. and Dieckmann, T. (2004) *Current Opinion in Structural Biology*, 14, 350–9.
161 Edwards, T.E. and Sigurdsson, S.T. (2002) *Biochemistry*, 41, 14843–7.
162 Qin, P.Z., Feigon, J. and Hubbell, W.L. (2005) *Journal of Molecular Biology*, 351, 1–8.
163 Kim, N.K., Murali, A. and DeRose, V.J. (2005) *Journal of the American Chemical Society*, 127, 14134–5.
164 Macosko, J.C., Pio, M.S. and Shin, Y.-K. (1999) *RNA*, 5, 1158–66.
165 Kim, N., Murali, A. and DeRose, V.J. (2004) *Chemistry and Biology*, 11, 939–48.
166 Schiemann, O., Weber, A., Edwards, T.E., Prisner, T.F. and Sigurdsson, S.T. (2003) *Journal of the American Chemical Society*, 125, 3334–9.
167 DeRose, V.J. (2003) *Current Opinion in Structural Biology*, 13, 317–24.
168 Qin, P.Z., Iseri, J. and Oki, A. (2006) *Biochemical and Biophysical Research Communications*, 343, 117–24.

169 Edwards, T.E. and Sigurdsson, S.Th. (2005) *Biochemistry*, 44, 12870–8.
170 Edwards, T.E. and Sigurdsson, S.Th. (2003) *Biochemical and Biophysical Research Communications*, 303, 721–5.
171 Cekan, P. and Sigurdsson, S.Th. (2005) *Collection and Symposium Series (Chemistry of Nucleic Acid Components)*, 7, 225–8.
172 Edwards, T.E., Robinson, B.H. and Sigurdsson, S.Th. (2004) *Bioorganic and Medicinal Chemistry Letters*, 14, 3415–18.
173 Sykulev, Y.K. and Nezlin, R.S. (1990) *Glycoconjugate Journal*, 7, 163–82.
174 Nezlin, R.S., Pankratova, E.V. and Timofeev, V.P. (1988) *Biologicheskie Membrany*, 5, 258–62.
175 Nezlin, R.S. and Sykulev, Y.K. (1982) *Molecular Immunology*, 19, 347–56.
176 Timofeev, V.P., Nikolsky, D.O., Lapuk, V.A. and Alyoshkin, V.A. (2002) *Journal of Biomolecular Structure and Dynamics*, 20, 389–95.
177 Dushkin, A.V., Troitskaya, I.B., Boldyrev, V.V. and Grigor'ev, I.A. (2005) *Russian Chemical Bulletin*, 54, 1155–9.
178 Bertholon, I., Hommel, H., Labarre, D. and Vauthier, Ch. (2006) *Langmuir: The ACS Journal of Surfaces and Colloids*, 22, 5485–90.
179 Bobodzhanov, P.Kh. and Likhtenshtein, G.I. (1974) *Dokl Akad Nauk Tadzhikskoi SSR*, 17, 34–7.
180 Yusupov, I.Kh., Bobodzhanov, P.Kh., Marupov, R., Islomov, S., Antsiferova, L.I., Kol'tover, V.K. and Likhtenshtein, G.I. (1984) *Vysokomolekulyarnye Soedineniya, Seriya A*, 26, 369–73.
181 Islomov, S., Marupov, R. and Likhtenshtein, G.I. (1989) *Cellulose Chemistry and Technology*, 23, 13–21.
182 Kulikov, A.V., Yusupov, I.Kh., Bobodzhanov, P.Kh., Marupov, R. and Likhtenshtein, G.I. (1991) *Zhurnal Prikladnoi Spektroskopii*, 55, 961–5.
183 Kostina, N.V., Marupov, R.M., Anzifereva, L.I. and Likhtenshtein, G.I. (1987) *Bifizika*, 32, 736–42.
184 Islomov, S., Bobodzhanov, P.Kh., Marupov, R.M., Likhtenshtein, G.I. and Zhbankov, R.G. (1986) *Cellulose Chemistry and Technology*, 20, 277–87.
185 Krinichnyi, V.I., Grinberg, O.Ya., Yusupov, I.Kh., Marupov, R.M., Bobodzhanov, P.Kh., Likhtenshtein, G.I. and Lebedev, Ya.S. (1986) *Biofizika*, 31, 482–5.
186 Frantz, S., Huebner, G.A., Wendland, O., Roduner, E., Mariani, C., Ottaviani, M.F. and Batchelor, S.N. (2005) *Journal of Physical Chemistry. B*, 109 (23), 11572–9.
187 Freire-Nordi, S., Vieira, A.A.H., Nakaie, C.R. and Nascimento, O.R. (2006) *Brazilian Journal of Physics*, 36, 75–82.
188 Saito, T., Shibata, I., Isogai, A., Suguri, N. and Sumikawa, N. (2005) *Carbohydrate Polymers*, 61, 414–19.
189 Robertson, J.A. and Sutcliffe, L.H. (2005) *Magnetic Resonance in Chemistry: MRC*, 43, 457–62.
190 Steen, A.D., Arnosti, C., Ness, L. and Blough, N.V. (2006) *Journal of Marine Chemistry*, 101, 266–76.
191 Steen, A.D., Arnosti, C., Ness, L. and Blough, N.V. (2006) *Journal of Marine Chemistry*, 101, 266–76.
192 Bardelang, D., Rockenbauer, A., Jicsinszky, L., Finet, J.-P., Karoui, H., Lambert, S., Marque, S.R.A. and Tordo, P. (2006) *Journal of Organic Chemistry*, 71, 7657–67.
193 Bardelang, D., Rockenbauer, A., Karoui, H., Finet, J.-P., Biskupska, I., Banaszak, K. and Tordo, P. (2006) *Journal of Organic Chemistry B*, 4, 2874–82.
194 Livshits, V.A., Dzikovskii1, B.G., Samardak, E.A. and Alfimov, M.V. (2006) *Russian Chemical Bulletin*, 55, 238–46.
195 Ionita, G. and Chechik, V. (2005) *Organic Biomolecular Chemistry*, 3, 3096–8.

11
Biomedical and Medical Applications of Nitroxides
Gertz I. Likhtenshtein

11.1
Cells and Tissues. Biomedical Aspects

The nitroxide spin-labeling method applied to cells and tissues is a rich source of information on proteins, membranes, nuclear acids, and so forth, in the same way that it gives information about individual biomolecules. In addition, detailed data on a number of specific problems associated with the intrinsic specific organization, inhomogeneity, redox processes, and special distribution of spin-labeled species in organs and living organism can readily be obtained. Several reviews of such studies carried out during the 1970s, 1980s, and 1990s are available in the literature [1–15]. In this section we focus on recent developments in this area, which include the following topics of interest: fluidity of cell membranes, chemical reduction and metabolism of nitroxides, nitroxide as an inhibitor of oxidation, detection of nitric oxide (NO), superoxide (SO) and other ROS, mapping redox status, oxygen concentration, and acidity of tissues by magnetic resonance imaging (MRI).

11.1.1
Cell Membrane Fluidity

Molecular dynamics is the determining factor for several chemical and physical processes in cell biomembranes that are of fundamental importance in determining chemical and physical processes in cell biomembranes.

Continuous-wave ESR and ESR tomography were applied to investigate nitroxide spin-label space distribution and motion in human skin [16]. Rates of rotational diffusion and Heisenberg spin exchange of the label were measured as a function of the nitroxide's concentration, temperature, and time. A comparative spin-labeling study of the fluidity of bovine erythrocytes and types of cell-cultured fibroblasts, baby hamster kidney (BHK), and hamster lung (V-79) was reported [17]. The membrane lipid phases were labeled with a lipophilic nitroxide spin-probe, a Me ester of 5-doxyl palmitate (MeFASL, 10.3). Temperature, pH, calcium concentration, and osmolality were varied and followed by examination of the labeled

Nitroxides: Applications in Chemistry, Biomedicine, and Materials Science
Gertz I. Likhtenshtein, Jun Yamauchi, Shin'ichi Nakatsuji, Alex I. Smirnov, and Rui Tamura
Copyright © 2008 WILEY-VCH Verlag GmbH & Co. KGaA, Weinheim
ISBN: 978-3-527-31889-6

membrane's ESR spectra line-shapes. The experiments indicated that the plasma membrane of fibroblasts exhibits a lower ordering and motional restrictions in the lipid phase than the erythrocyte membrane. The latter is characterized by a high cholesterol content. The imposed variations of pH, calcium concentration, and osmolality promoted different perturbations of the investigated cell membrane characteristics, though the most pronounced was the effect on the population of the two classes of lipid domains.

11.1.2
Cells Redox Status

Among the experimental problems to be solved in spin-labeling studies of the local properties of active biological systems, such as molecular dynamics, polarity, pH, oxygen concentration, and imaging, one of the remaining is the stability of nitroxides to redox and other intercell chemical reactions. Several approaches to this problem have been described in Chapter 7. Recently, the synthesis and biological testing of aminoxyls designed for long-term retention by living cells has been reported [18]. Specifically, it was shown that the spin label (2,2,5,5-tetramethylpyrrolidin-1-oxyl-3-ylmethyl)amine-N,N-diacetic acid has been accumulated by cells at high intracellular concentrations, with an intracellular exponential lifetime of 114 minutes. The authors suggested that it should be feasible to use EPR imaging to perform *in vivo* tracking of populations of cells which have accumulated high intracellular levels of aminoxyls.

The reduction kinetics of the dual chromophore–nitroxide probe, 5-dimethylaminonaphthalene-1-sulfonyl-4-amino-2,2,6,6-tetramethyl-1-piperidine-oxyl (R*), by human blood and its components were studied using the EPR technique [19]. It was found that this probe was adsorbed on the outer surface of the erythrocyte membrane, that it does not penetrate into the erythrocytes, and that it is reduced only by ascorbic acid. The apparent first-order rate of the nitroxide reduction rate constant, k, was found to be in the following sequence: $k_{blood} = k_{eryth} = k_{plasma} > k_{eryth} > k_{plasma}$. The experiments indicated two important findings: (i) the erythrocytes catalyze the reduction of R* by ascorbate and (ii) there is an efficient electron transfer route through the cell membrane.

A reduction of the level of glutathione in human red blood cells in the presence of nitroxide was demonstrated [20]. TEMPOL, TEMPamine, and (3-carbamido-2,2,5,5-tetramethylpyrrolidine-1-oxyl in concentration of $2\,mmol\,L^{-1}$) decreased GSH levels in RBCs by more than 90%, while five-membered ring pyrroline radicals (3-carbamido- and 3-carboxy-2,2,5,5-tetramethylpyrroline-1-oxyl, pyrrolin and carboxypyrrolin) were less effective. The mechanisms for nitroxide effects on the glutathione levels in the presence of oxygen are discussed in Chapter 7. One of the effects of the TEMPOL reduction is an increase of oxidant stress in cultured EA.hy926 endothelial cells and tissues and this was investigated [21]. Cells loaded with ascorbic acid showed increased rates of TEMPOL-dependent ferricyanide reduction. and a more rapid loss of the TEMPOL EPR signal than the cells without ascorbate. The intracellular ascorbate prevented dihydrofluorescein oxidation and spared GSH from oxidation by the radical. The authors made an important conclusion: "whereas TEMPOL may scavenge other more toxic radicals, care must be taken to ensure that it does not itself induce an oxidant stress, especially with regard to depletion of ascorbic acid".

Reduction of piperidine nitroxide spin-labels was investigated in human keratinocytes of the cell line HaCaT by GC and GC-MS techniques combined with S-band ESR [22]. Besides reduction these studies indicated further conversion of hydroxylamines to the corresponding secondary amines. Both reactions were inhibited by the thiol blocking agent N-ethylmaleimide and also by the strong inhibitors of thioredoxin reductase (TR) 2-chloro-2,4-nitrobenzene and 2,6-dichloroindophenol. Furthermore, a correlation between the lipid solubility of the nitroxides and inhibition has been found. The authors concluded that the hydroxylamines and the secondary amines represent one of the major metabolites of nitroxides besides the hydroxylamine inside keratinocytes formed via the flavoenzyme thioredoxin reductase.

Another potential complexity in the reduction of nitroxide spin labels *in vivo* has been analyzed in reference [23]. The spin-probe 2,2,5,5-tetramethyl-4-phenylimidazolin-3-oxide-1-oxyl (TPI) was exposed to both native and partially oxidized human low-density lipoprotein (LDLn and LDLpox, resectively). According to ESR experiments, reduction of TPI in the presence of LDLpox occurred via a complex mechanism involving the consumption of tocopherol, the vitamin E analog, Trolox C, and cholesteryl linoleate hydroperoxides.

11.1.3
Nitroxides as Cell Protectors

In ischemia–reperfusion injuries, elevated calcium and reactive oxygen species (ROS) induce mitochondrial permeability transition (mPT), which plays a pivotal role in mediating damage and cell death. It was shown that the SOD (superoxide dismutase)-mimetic mPT inhibitor (HO-3538), based on the amiodarone structure, inhibited mPT and the release of proapoptotic mitochondrial proteins [24]. It was shown that the inhibitor can eliminate ROS in the microenvironment of the permeability pore in isolated mitochondria and had a ROS scavenging and antiapoptotic effect in a cardiomyocyte line. The antioxidant effects of several nitroxides, including TEMPOL, CAT-1, and N-acetyl-L-cysteine (L-NAC), on radiation-induced apoptosis pathways in human lymphoblastoid TK6 cells was investigated [25]. The cells were irradiated with the same dose (6 Gy) with and without antioxidants, followed by assessment of survival (clonogenic assay) and the apoptotic index. It was shown that TEMPOL at 10 mM partially radioprotected TK6 cells against clonogenic killing, but had no effect on radiation-induced apoptotic parameters.

Pyrene-nitronyl was used for real-time monitoring of nitric oxide outflux from pig trachea epithelia on the nanomolar concentration scale [26]. Conversion of pyrene-nitronyl (PN) to pyrene-imino nitroxide radical (PI) and NO_2 upon reacting with NO is accompanied by changes in the electron paramagnetic resonance (EPR) spectrum and a drastic increase in pyrene fluorescence. The method was applied to the monitoring of NO flux of 0.9 nmol/g × min from tissue stimulated by extracellular adenosine 5′-triphosphate. The dual fluorophore–nitroxide compound, acridine–TEMPO (Ac–TEMPO) was used to characterize interactions between the nitroxide and the glutathionyl radical (GS·) generatd through phenoxyl radical recycling by peroxidase [27].

Cytotoxicity of a nitroxide derivative of 4-ferrocenecarboxyl-2,2,6,6 tetramethyl-piperidine-1-oxyl (FC-TEMPO) in high metastatic lung carcinoma cells (95-D) was studied [28]. FC-TEMPO showed a strong inhibitor effect on the viability of cancer cells, while it was less toxic to a normal human cell line. Further studies found that FC-TEMPO suppressed the growth of tumor cells by induced apoptosis through activating caspase-3 and by cell cycle arrest at the G1 phase. In the presence of the radical superoxide dismutase (SOD) and catalase (CAT) activities were increased. Alkyl phospholipid analogs of perifosine and miltefosine bearing a nitroxide moiety at different positions on an alkyl chain caused strong inhibitor effects on the growth of MT1, MT3, and MCF7 breast cancer cell lines [29]. The inhibitor was delivered to the cell by large unilamellar liposomes containing cholesterol and dicetyl phosphate or directly through aqueous solution.

Factors influencing TEMPOL reduction and cytotoxicity in wild-type and glucose-6-phosphate dehydrogenase (G6PD)-deficient Chinese hamster ovary cells were evaluated [30]. TEMPOL was reduced at a faster rate when cells were under hypoxic compared with aerobic conditions. TEMPOL-induced cytotoxicity was markedly less for G6PD-deficient cells compared with the wild-type cells. The results indicated the bioredaction of TEMPOL can be influenced by a number of factors, such as glutathione depletion and inhibition of 6-phosphogluconate dehydrogenase, and so forth.

11.2
Nitroxides *In Vivo*

11.2.1
Introduction

An increasing volume of literature reports confirm that *in vivo* nitroxide spin-labeling and nitron trapping methods provide unique and useful information in animal studies. Recently, the most significant progress has been made in applying nitroxides and related nitrons to biochemical and biomedical problems *in vivo*. This progress is attributed to the unique feature of these methods which provide data on several local properties of the label's environment simultaneously: (1) molecular dynamics and fluidity; (2) polarity; (3) redox status and anioxidant reactivity, including reactive thiol groups; (4) redox-sensitive metabolism; (5) charges and electrostatic potential; (6) nature and distribution of reactive radicals; (7) nitric oxide and superoxide dynamics; (8) metal ion concentrations; (9) pH; and (10) oxygen concentration.

Practically all the power of modern magnetic resonance methods, including time-domain ESR spectroscopy, pulsed double electron electron resonance (pulse

DEER, PELDOR), Overhauser-enhanced magnetic resonance imaging (MRI), low-field RF-EPR, longituidinal EPR (LODEPR), field circle dynamic polarization (FC-DNP), multifrequency ESR, and multiquantum ESR (Chapter 4), has been applied, at least in some degree, to *in vivo* studies. In principle, all these measurements can be used in combination imaging space mapping techniques. This information has proved to be of great importance for biomedical research, pharmokinetics, and the therapeutic and clinical application of nitroxides and nitrons. Advantages for the application of nitroxides *in vivo* have been reviewed ([31–40] and references therein). Here we describe recent progress in this field which appears to be of great basic and applied importance for both fundamental and applied biomedical research, pharmacology, and clinical chemistry.

11.2.2
Nitroxide *In Vivo* Biochemistry. Biomedical Aspects

11.2.2.1 Antioxidant Activity of Nitroxides

The protective antitumor activity of nitroxide in animals was first demonstrated in the pioneering work of Konovalova *et al.* [41]. Later other groups reported a series of investigations in this field ([42] and references therein). It was shown that nitroxyl derivatives of 5-fluorouracil [43], daunorubicin [44], rubomycin, and TEMPOL [45] exibit significant antitumor activity. As an example, inhibition by ruboxyl, a nitroxyl derivative of daunorubicin, and 5-fluorouracil was compared in metastases of experimental colorectal carcinoma in murine liver. The indexes of metastasis inhibition were 84, 43, and 70%, respectively. In rats receiving the drugs by continuous intravenous infusion for 7 days, the numbers of metastases was measurably reduced (ruboxyl -1.0 ± 1.4; 5-fluorouracil 3.2 ± 1.3) [44]. Recently, $cis,trans$-PtIV(RNH$_2$)(NH$_3$)(OAc)2Cl$_2$, where R is 2,2,6,6-tetramethyl-1-oxylpiperidin-4-yl (1b) or 2,2,5,5-tetramethyl-1-oxylpyrrolidin-3-yl (2b) have been synthetized [46]. Complex 1b exhibited high antitumor activity comparable with that of Cisplatin against leukemia P388 used as the experimental tumor. Simultaneous administration of low doses of 1b and Cisplatin (1/20 of LD50 each) results in synergism of the antitumor activity and 100% cure of animals.

Evaluation of the hydroxylamineTEMPOL-H as an *in vivo* radioprotector of C3H mice was reported [47]. TEMPOL-H was administered in increasing doses via an intraperitonial route to the mice and the whole-blood pharmacology of the nitroxide was investigated with ESR spectroscopy. The results demonstrated the appearance of TEMPOL in whole blood immediately after the injection. TEMPOL-H also provided protection against the lethality of whole-body radiation in C3H mice similar to the results obtained with TEMPOL. Hemodynamic measurements in C3H mice after the injection showed that that TEMPOL-H produced little effect on blood pressure or pulse compared with TEMPOL. The authors concluded that the systemic administration of TEMPOL-H *in vivo* leads to an equilibration between TEMPOL and TEMPOLl-H that limits the toxicity of the nitroxide and provides *in vivo* radioprotection. The kinetics of free radical reactions in rat liver as a result of exposure to low-dose β-radiation was evaluated by monitoring the reduction of the nitroxyl spin probe after *in vivo* administration [48]. The EPR signal intensity of a nitroxyl probe in bile flow was monitored by cannulating the bile duct through the cavity of an X-band EPR spectrometer. The results *in vivo* and *in vitro* showed that the rate of nitroxide signal loss was higher in rats whose livers were exposed to β-rays compared to unexposed rats.

The spin probes TEMPOL, 3-carbamoyl-2,2,5,5-tetramethylpyrrolidine-1-oxyl (carbamoyl-PROXYL) and the cell impermeable probe CAT-1 were utilized for two methods: X-band ESR (9.4 GHz) bile flow monitoring (BFM) and 300 MHz *in vivo* EPR measurement [49]. ESR signal decay of a nitroxyl spin probe in the bile flow and in the liver region (upper abdomen) of several rat groups with different selenium status were measured by both methods. The experiments indicated that the EPR signal intensity of CAT-1 in the bile was markedly weaker compared with that of either carbamoyl-PROXYL or TEMPOL and the *in vivo* decay rate of the spin probe in the normal rat may reflect its kinetics in the liver. The authors stressed that the nitroxide clearance measured with the *in vivo* EPR method may be affected not only by the redox status in the liver but also by information from other tissues in the measured region of the rat.

With the aim of evaluating the location where *in vivo* free radical and redox reactions were enhanced in adjuvant arthritis (AA) model rats, the nitroxide spin probe technique was utilized [50]. It was shown that the signal decay after intravenous injection of a nitroxide spin probe was enhanced in AA compared to a control and was suppressed by pre-treatment with dexamethasone (DXT) and by a simultaneous injection of free radical scavengers. The effect of nitroxides, including TEMPOL, 4-amino-TEMPO, 4-oxo-tempo, CAT-1,3-carbamoyl-proxyl, or 3-carboxy-proxyl, on blood pressure and heart rate in the spontaneously hypertensive rat was investigated [51]. The authors' attention was concentrated on the nitroxide SOD mimetic activity or lipophilicity. The results indicated that pyrrolidine nitroxides are ineffective antihypertensive agents.

11.2.2.2 Detection of Reactive Radicals: Spin-Trapping

Electron spin resonance and spin-trapping techniques provide a unique opportunity to follow radical processes *in vitro* and *in vivo* (Chapter 7). In this section this opportunity is illustrated by several typical examples. *In vivo* detection of reactive radicals and change of redox status induced by diethylnitrosamine (DEN) in rat liver tissue was observed using the *in vivo* spin probe technique [52–54].

The protection from myocardial ischemia–reperfusion injury given by trimetazidine, the well-known anti-anginal and anti-ischemic drug, modified with nitroxides and their precursors, was investigated [55]. Trimetazidine was modified by pyrroline and tetrahydropyridine nitroxides and their hydroxylamine and sterically hindered secondary amine precursors. Two of the investigated compounds, containing 2,2,5,5-tetramethyl-2,5-dihydro-1H-pyrrole and 4-phenyl-2,2,5,5-tetramethyl-2,5-dihydro-1H-pyrrole substituents on the piperazine ring, provided significant protection from the cardiac dysfunction caused by ischemia/reperfusion in Langendorff-perfused rat hearts. The authors suggested that this protection could be attributed to a combination of anti-ischemic and antioxidant effects.

With the aim of evaluating the effect of different doses and times of administration of TEMPOL, the heads of C3H mice were exposed to a single irradiation dose of 15 Gy [56]. Different routes of the nitroxide effect were analyzed. Results showed that TEMPOL treatment (137.5 or 275 mg kg^{-1}) significantly (by approximately 50–60%) reduced irradiation-induced salivary hypofunction and that TEMPOL is a promising candidate for clinical application to protect salivary glands in patients undergoing radiotherapy for head and neck cancers. The protection is likely to occur via a reactive radical trapping mechanism. The protective effect of the coadministration of NCX-4016 [2-(acetyloxy)benzoic acid 3-(nitrooxymethyl)phenyl ester] (an NO donor) with antioxidants TEMPOL, superoxide dismutase (SOD), or urate on postischemic myocardial reperfusion injury in isolated rat hearts was also investigated [57]. Treatment of hearts with NCX-4016 and TEMPOL showed a significantly enhanced recovery of heart function compared with NCX-4016 alone, enhanced NO generation, and decreased ROS and dityrosine (a marker of peroxynitrite) formation.

The efficacy of TEMPOL in cell culture and animal models of the central and peripheral dysfunction associated with Parkinson's disease, a disorder in pathogenesis, was examined [58]. *In vivo*, intraperitoneal TEMPOL protected mice from intrastriatal 6-OHDA-induced cell, from ptosis, activity level decrement, and mortality induced by intraperitoneal and by intrastriatal administration of d-amphetamine. The authors suggested that the observed neuroprotective effects of TEMPOL are due to its reaction with ROS. TEMPOL protection from human cancer prone syndrome in a mouse model, Atm-deficient mice, was reported [59]. Continuous TEMPOL administration reduced ROS, restored mitochondrial membrane potential, reduced tissue oxidative damage and oxidative stress, lowered weight gain of tumor-prone mice, and exhibited an anti-proliferative effect *in vitro*. As a result, the mice lifespan was prolonged.

Local ROS levels in frozen tissue samples taken from different organs of animals were measured using spin probe (1-hydroxy-3-carboxy-pyrrolidine CP), and a low temperature [60]. It was found that ROS formation in tissue of control animals increased in the following order: liver < heart < brain < cerebellum < lung < muscle < blood < ileum < kidney < duodenum < jejunum. Endotoxin, which commonly causes septic shock, increased the formation of ROS in rat liver, heart, lung, and blood, but not in gut, brain, or skeletal muscles.

Nitrons prevent acute ischemia–reperfusion injury of the liver by limiting free radical-mediated tissue damage [61]. Liver-derived radical production in mice was assessed in bile samples by measuring α-(4-pyridyl-1-oxide)-N-tert-butylnitrone (POBN) radical adducts using ESR [62]. The administration of 4-chloro-6-(2,3-xylidino)-2-pyrimidinylthioacetic acid (WY-14,643) caused a sustained increase in POBN radical adducts in mouse liver. Experiments with NADPH oxidase-deficient (p47 phox-null) and PPARα-null mice indicated that that PPARα, not NADPH oxidase, is critical for a sustained increase in POBN radical production in Kupffer cells caused by peroxisome proliferators in rodent liver.

Mason's group developed a technique to detect the generation of *in vivo* free radicals in rat bile using *in vivo* spin-trapping/*ex vivo* detection [62–65]. ESR investigation of the oxidative damage in lungs caused by asbestos and air pollution particles has also been reported [63]. The protective effect of a novel free radical scavenger, 3-methyl-1-phenyl-2-pyrazolin-5-one (edaravone) against acute ischemic stroke injury in several animal models was investigated using the spin-trapping technique [64]. When edaravone was administered prior to ischemia and at the time of initiation of the reperfusion, liver injury was markedly reduced. In addition, an increase in free radical adducts caused by ischemia/reperfusion caused an increase in free radical adducts; pro-inflammatory cytokines, infiltration of leukocytes, and lipid peroxidation in the liver were markedly blocked. These results demonstrated that edaravone is an effective blocker of free radicals *in vivo* in the liver after ischemia/reperfusion. Free radical formation has been investigated in experimental animal models of LPS-induced inflammation. [65]. The ESR spectra of α-(4-pyridyl-1-oxide)-N-tert-butylnitrone radical adducts were consistent with the trapping of lipid-derived radical adducts. A secondary radical-trapping technique using DMSO demonstrated Me radical formation, revealing the production of hydroxyl radical. These results suggested that both inducible nitric oxide synthase (iNOS) and xanthine oxidase (XO) play a role in free radical formation. Free radical formation in the hemolymph of insects (intact *Galleria mellonella* and *G. mellonella* infected with the fungus *Metarhizium anisopliae*) was detected by ESR spin-trapping methods [66]. The formation of the spin adduct DOPA-semiquinone in hemolymph was demonstrated using spin stabilization of o-semiquinones by Mg^{2+}.

With the aim of clarifying the relationship between *in vivo* reactive oxygen species (ROS) generation and an induced gastric lesion formation, an *in vivo* 300-MHz ESR (ESR) spectroscopy/nitroxyl probe technique to detect ·OH generation in the stomach of rats was used [67]. A nitroxide probe was orally administered to water-immersion-restraint treated rats, and the spectra in the gastric region were obtained by *in vivo* ESR spectroscopy. Based on the finding that the decay was suppressed by mannitol, desferrioxamine, and catalase, the authors concluded that it was caused by ·OH generation which induces mucosal lesion formation via the hypoxanthine/xanthine oxidase system. *In vivo* oxygen radical generation in the skin of the protoporphyria model mouse with visible light exposure was reported [68]. A nitroxide was administered followed by detection of the nitroxide ESR spectra in the mouse skin using an ESR spectrometer equipped with a surface-coil-type resonator that could detect radicals within about 0.5 mm of the skin surface. Light irradiation enhanced the decay of the nitroxyl signal in griseofulvin-treated mice which was suppressed by administration of hydroxyl radical scavengers, superoxide dismutase or catalase, or desferrioxamine.

11.2.2.3 Spin Farmokinetics *In Vivo*

The pharmacokinetics of a spin-labeled analog of rubomycin (ruboxyl) (I) were studied by ESR spectroscopy [45]. Differences were found in ruboxyl pharmacokinetics in normal and tumor-bearing animals and in the pharmacokinetics of ruboxyl and TEMPOL. Most of the drug was excreted within 6 hours. The differences in the pharmacokinetics of ruboxyl and nitroxyl radical were established.

The pharmacokinetics of the nitroxyl radicals (TEMPO, TEMPONE (oxo-TEMPO), and amino-TEMPO) were investigated by bile flow monitoring (BFM) and blood circulating monitoring (BCM) methods using X-band EPR [69]. In the presence of both nitroxides, additional .EPR signals were detected in the bile. These signals were attributed to metabolites formed during transport from blood to bile through the liver. It was shown that carboxy-TEMPO and carboxy-PROXYL can be transported via an anion transporter into hepatic cells. Delivery of the nitroxide 3-acetoxy-methoxycarbonyl-2,2,5,5-tetramethyl-1-pyrrolidinyloxyl across the blood–brain barrier (BBB) was examined by low-frequency ESR (EPR) spectroscopy [70]. Two nitroxides, 3-methoxycarbonyl-2,2,5,5-tetramethyl-1-pyrrolidinyloxyl and 3-acetoxy-methoxycarbonyl-2,2,5,5-tetramethyl-1-pyrrolidinyloxyl, used as pro-imaging agents to deliver 3-carboxy-2,2,5,5-tetramethyl-1-pyrrolidinyloxyl across the BBB were examined for their potential for *in vivo* measurement of tissue oxygenation in the mouse BBB. Figure 11.1 depicts the diffusion of nitroxide into a neuronal cell, where esterase hydrolysis liberates nitroxide. *In vivo* pharmacokinetic and pharmacodynamic studies in mice suggested that esterase-labile nitroxide crossed the BBB, and was esterificated and retained.

The pharmacokinetics of several nitroxyl spin-probes in the circulating blood of a living mouse using an X-band (9.4 GHz) ESR spectrometer were examined [71]. The studies indicated that the six-member piperidine nitroxyls are suitable for estimating. redox status in the circulation, whereas the five-member pyrrolidine nitroxyl radicals are suited for tissue redox status detecton. The result also showed that for carbamoyl-PROXYL, the hydroxylamine form of TEMPOL might give radioprotection *in vivo*. Co-administration of amifostine with the carbamoyl-proxy spin probe (CP) via injection and oral administration of the amifostine in C3H

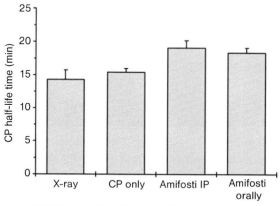

Figure 11.1 Comparision of values of the first order rate of decay of the CP EPR signal after radiation in different condition (see text) [72]. (Reproduced from Elas, M., Parasca, A. Grdina, D.J. and Halpern, H.J. (2003) Oral administration is as effective as intraperitoneal administration of amifostine in decreasing nitroxide EPR signal decay *in vivo*. *Biochimica et Biophysica Acta. Molecular Basis of Disease*, 1637, 151–155, with permission of Elsevier).

mice was performed [72]. These decreased the first-order rate of decay of the CP EPR signal after radiation as compare with the rate of decay of the CP EPR signal without amifostine (Figure 11.1).

Theoretical aspects of pharmokinetics *in vivo* were considered in detail in [73].

11.2.2.4 Spin pH Probing

Spatially and temporarily addressed spin pH measurements in animals *in vivo* and potentially in the human body are of considerable biomedical relevance [13]. This includes local acidosis induced by ischemia, infection, or inflammation, extracellular acidosis in tumors, and depth-specific tissue pH variations during skin treatments or wound healing. Upon therapeutic intervention, the delivery, absorption, and pharmacological effectiveness of drugs can be altered by changing the pH of their local environment. Approaches based on EPR spectroscopy have the advantage for *in vivo* applications in animals and humans

The spin pH applications of *in vitro* probes include studies of water-in-oil (w/o) ointments, proteins and proteineous matrix, biodegradable polymers, and phospholipid membranes. The influence of treatment with therapeutically used acids, such as salicylic acid and azelaic acid, on microacidity in rat and human skin, and *in vitro* ESR imaging study was investigated using spin pH probes for noninvasive direct and depth-specific measurement of pH [73]. Spectral-spatial ESR imaging (ss-ESRI) and pH-sensitive nitroxides were used to obtain a pH map of rat and human skin *in vitro*. These works, together with advantages of synthetic chemistry [74–76] and recently developed low-field EPR-based techniques, such as low-field RF-EPR, LODEPR, FC-DNP [77–82], offer a unique opportunity for non-invasive measurements of functional EPR probes, including pH probes *in vivo*.

Figure 11.2 demonstrates the efficiency of the spin pH approach for monitoring kinetics of pH after ischemia-induced miocardium acidosis in the isolated rat heart

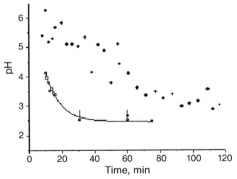

Figure 11.2 Time dependencies of pH changes in the stomach measured by LODEPR after giving 3 ml of gavage containing 5 mM pH sensitive nitroxide alone (□) or with 50mM bicarbonate (●) [83] (Reproduced from Potapenko, I., Foster, M.A., Lurie, D.J. Kirilyuk, I.A., Hutchison, J.M.S., Grigor'ev, I.A., Bagryanskaya, E.G. and Khramtsov, V. (2006) Real-time monitoring of drug-induced changes in the stomach acidity of living rats using improved pH-sensitive nitroxides and low-field EPR techniques. *Journal of Magnetic Resonance*, 182, 1–11, with permission of Elsevier).

and in the stomach using L-band EPR spectroscopy. Real-time monitoring of drug-induced changes in the stomach acidity of living rats using improved pH-sensitive nitroxides and low-field EPR techniques was reported [83]. To improve pH sensitivity in the range from pH 1.8 to 6 and stability toward reduction of nitroxides *in vivo* the nitroxide 4-[bis(2-hydroxyethyl)amino]-2-pyridine-4-yl-2,5,5-triethyl-2,5-dihydro-1H-imidazol-oxyl, bearing the bulky ethyl groups and two ionizable groups was synthesized. 290 MHz radiofrequency (RF) and X-band ESR, longitudinally detected EPR (LODEPR), proton–electron double resonance imaging (PEDRI), and field-cycled dynamic nuclear polarization (FC-DNP) techniques were utilized to detect nitroxide localization and acidity in the rat stomach. The authors suggested the applicability of the techniques for monitoring drug pharmacology and disease in living animals.

11.2.2.5 Spin Imaging

ESR imaging (EPRI) enables noninvasive spatial mapping of free radical metabolism to provide *in vivo* physiological information regarding alterations in the redox state of tumors and neoplastic tissues. The EPRI technique has been successfully utilized to clarify the location of nitroxide probes and therefore provide real-time monitoring of free radical reactions [84–87]. Progress in nitroxide applications in research *in vivo* are due to a great extent to the development of sophisticated imaging techniques such as multimodality magnetic resonance ([RF CW-ESR, LODESR, PEDRI, and combination of these modalities to allow sequential PEDRI and CW-ESR, sequential LODESR and proton NMR imaging, and simultaneous LODESR and CW-ESR [76, 77], a new experimental apparatus for multimodal resonance imaging: [88] and references therein), Overhauser-enhanced MRI for simultaneous molecular imaging [89], EPRI/MRI technique [90], 3D spatial and 3D or 4D spectral-spatial [86, 91], and spin echo ESR imaging [92].

Using the EPRI technique, reactive oxygen radicals were detected and their kinetics were monitored in a number of disease models, such as ionizing radiation-damaged animals [93], gastric lesion [94], streptozotocin-induced diabetes [95], ischemia–reperfusion injury [96], nitroxide dynamics in human skin [97], and hypoxia [98]. An example of the successive application of ESRI for the study of space distribution of a nitroxide probe in human skin is represented in Figure 11.3 [98].

Proton electron double resonance imaging instrumentation (PEDRI) at a magnetic field of 0.02 T was applied to imaging the *in vivo* distribution, clearance, and metabolism of nitroxide radicals in living mice using 2,2,5,5-tetramethyl-3-carboxylpyrrolidine-N-oxyl (I) and 3-carbamoyl-proxyl (PCA) and 2,2,6,6-tetramethyl-4-oxopiperidine-N-oxyl [99]. Coronal images of nitroxide-infused mice enabled visualization of the kinetics of spin-probe uptake and clearance in different organs including the great vessels, heart, lungs, kidneys, and bladder with an in-plane spatial resolution of 0.6 mm. It was found that PCA, due to its intravascular compartmentalization, provided the sharpest contrast for the vascular system and highest enhancement values in the PEDRI images among the three nitroxides. High-resolution mapping of tumor redox status by magnetic resonance imaging

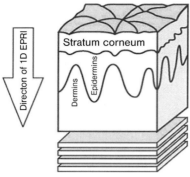

Figure 11.3 Human skin ESR imaging. The nitroxide concentration was measured invasively [98]. (Reproduced with permission).

using nitroxide 3-carbamoyl-PROXYL (3CP) as redox-sensitive contrast agents was performed [100]. The kinetics of the reduction of 3CP were examined by EPR imaging and magnetic resonance imaging instrumentation in muscle and squamous cell carcinoma implanted in the hind leg of C3H mice. Experiments indicated that the tumor regions exhibited a faster decay rate of the nitroxide compared to muscle (0.097 min^{-1} versus. 0.067 min^{-1}, respectively).

Simultaneous molecular imaging of redox reactions monitored by Overhauser-enhanced MRI with ^{14}N- and ^{15}N-labeled carbamoyl-PROXYL was reported [101]. To examine the imaging capabilities where both nitroxide spin probes can be simultaneously imaged, a phantom containing both probes in several different tubes was tested. Kinetics of oxidation and reduction of nitroxides in phantom objects, containing encapsulated ascorbic acid were monitored at nanometer scale by labeling membrane-permeable and -impermeable nitroxyl radicals with ^{14}N and ^{15}N nuclei. The result demonstrated that with judicious choice of the spin probe and isotopic substitution, the simultaneous molecular imaging of redox reactions monitored by Overhauser allows us to perceive *changes at nanometer scale* referred to positioning on nitroxide with respect to the lipid bilayer. To assess the sites of ROS generation in rats treated by indomethacin (a nonsteroidal anti-inflammatory drug), the noninvasive measurement of ROS was undertaken using *in vivo* 300-MHz electron spin resonance (ESR) spectroscopy [54].

EPRI was used for early detection and visualization of gastric carcinoma in a rat gastric cancer model induced by 1-methyl-3-nitro-1-nitrosoguanidine [102]. *In vivo* three-dimensional EPRI experiments at 750 MHz on the stomach of rats with continuous intravenous administration of nitroxide 3-carboxamido-2,2,5,5-tetramethylpyrrolidine-N-oxyl (3-carbamoyl-proxyl) [3CP] were performed (Figure 11.4). Pharmacokinetic studies indicated that 3CP in the tumor region was rapidly reduced to an undetectable level, whereas the 3CP levels in normal stomach tissue persisted (Figure 11.5).

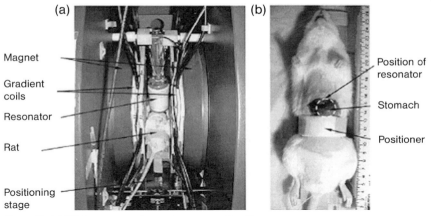

Figure 11.4 ESRI instrumentation and rat preparation for measurement in the stomach of a rat bearing tumor [102]. (Reproduced with permission).

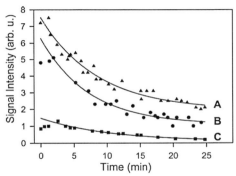

Figure 11.5 Pharmokinetics of reduction of the CP3 nitroxide probe in the stomach of tumor-bearing rats: in control stomach (A), in stomach with two small adjacent tumors of 2 mm diameter (B), and large tumor of 8 mm in diameter (C) [102]. (Reproduced with permission).

11.2.2.6 Nitroxide Spin Probe Oximetry

There are many reasons for the development and use of *in vivo* spin-probe oximetry using EPR techniques for experimental applications in animals and human body in clinics. Molecular oxygen plays a key role in many biochemical, physiological, and pathophysiological processes such as mitochondrial respiration, synthetic and degradative reactions, oxidative damage, cell signaling, and so forth. ESR oximetry is a technique that can make noninvasive sensitive and localized measurements of oxygen. Over the last few years, significant progress has been achieved in EPR oximetry through increased availability of instrumentation and paramagnetic materials capable of measuring pO_2 in tissues with an accuracy and sensitivity greater than that available by any other method. A principle advantage of spin-probe oximetry is its remarkable ability to noninvasively monitor oxygenation of different parts of a whole animal and in some specific cases of human body. ESR spectroscopy-based oximetry with a highly sensitive probe such as nitroxides could be very useful also for detection of oxygen consumption.

The EPR-probe oximetry technique is based on the magnetic interaction of a probe with molecular oxygen (Section 6.3). Using a small amount of sample (10 μl) or less if HF ESR or loop resonators are used, it allows us to determine molecular oxygen from the oxygen-induced EPR line-broadening of a suitable paramagnetic probe. EPR-probe oximetry can provide real-time, noninvasive measurements of oxygen.

The entire arsenal of modern ESR and NMR techniques (Chapter 4) has been intensively used in spin oximetry biomedical research and medical applications; for recent reviews see [33, 40, 103–106].

It is now customary to use several types of spin oximetry based on molecular probes: nitroxides or stable aromatic radicals (derivatives of triarylmethyl radical, Indian ink), phthalocyanine, and micrococrystals (fusinite and carbohydrate chars) (Section 6.3). Each type of probe has its own specific advantages and limitations. The special advantage of nitroxide probes is that the ESR spectra exhibit a hyperfine structure and the possibility of specific incorporation of such a probe into every part of a system of interest. The analysis of nitroxide ESR spectra hyperfine structure provides research information not only about oxygen concentration but also about local viscosity, polarity, electrostatic potential, pH, and so forth.

We should also consider that experimentally detected broadening of the ESR linewidth and spin electron relaxation rates and connected parameters are related to $k_{ex} [O_2]$ where k_{ex} is the rate constant of the dynamic exchange interaction and $[O_2]$ is the oxygen concentration (Section 6.3). The k_{ex} value can be different in a control and the systems of interest. Examination of nitroxide ESR spectra allows us to quantitatively characterize the local viscosity effect after estimation of k_{ex} using the Stokes–Einstein equation. One of the problems in oximetry with nitroxides is their relatively high chemical reactivity to redox processes in living systems. In some cases the initial linewith of the nitroxide ESR spectra is relatively broad and this decreases the method's sensitivity. This limitation can be overcome by direct measurement of the spin electron lattice relaxation rate $(1/T_1)$ or related parameters (that is, CW ESR saturation curves, for example), which are essentially more sensitive to exchange interaction. Spin oximetry probes, aromatic radicals, and micromaterials are more stable and more suitable for oximetry in reactive biological and living systems, including implantation in the whole body.

In this section we focus on nitroxide spin oximetry (NSO) *in vivo*. Examples concerning spin oximetry in medicine will be given in Section 11.2.2.6. Because of the above-mentioned limitations of nitroxide spin oximetry this method will not find widespread application. Nevertheless, recent progress in the synthesis of nitroxides more stable to redox processes (Section 7.1) and the development of new kinetic approaches [12] open the way to overcoming this limitation.

An ability of nitroxides to monitor redox and oxygenation state in *vitro* and *in vivo* was clearly demonstrated in a series of works. With the aims of characterizing the alterations of *in vivo* tissue redox status, oxygenation, formation of ROS, and their effects on the postischemic heart of a mouse, a combined approach, including nitroxide spin-labeling, fluorometry, and high-performance liquid chromatography, was utilized [106]. Experiments demonstrated that the reduction rate of nitroxide in the heart tissue was increased 100% during ischemia and decreased 33% after reperfusion compared to the nonischemic tissue. After reperfusion the tissue oxygenation and formation of ROS drastically increased. Tissue GSH/GSSG level showed a 48% increase during ischemia and 29% decrease after reperfusion The authors concluded that the hypoxia during ischemia limited mitochondrial respiration and caused a shift of tissue redox status to a more reduced state.

To test the use of nitroxides as potential O_2-sensitive probes *in vivo*, the EPR spectral linewidths of nitroxides at different O_2 concentrations in PBS at pH 7.4 were measured [107, 108]. Experiments showed that the EPR linewidths of 3-acetoxymethoxycarbonyl-2,2,5,5-tetramethyl-1-pyrrolidinyloxyl were more O_2-sensitive than that of the commonly used oximetry probe 4-oxo-2,2,6,6-tetramethylpiperidine-d16-1-15N-oxyl. The probe retention occurred in brain tissue and not in the extensive vasculature. With a goal of searching for nitroxide probes which would meet the requirement as EPR imaging agents for mapping O_2 distribution in the brain following stroke, a series of pharmacokinetic and pharmacodynamic experiments in Sprague-Dawley rats were performed [107, 108]. From these experiments, the nitroxides 3-acetoxymethoxycarbonyl-2,2,5,5-tetramethyl-1-pyrrolidinyloxyl and *trans*-3,4-di(acetoxymethoxycarbonyl)-2,2,5,5-tetramethyl-1-pyrrolidinyloxyl, but not 2,2,5,5-tetramethylpyrrolidin-1-oxyl-3-methyl)amine-N,N-diacetic acid diacetoxymethyl ester, exhibited favorable pharmacokinetic, pharmacodynamic, and oximetry profiles. Figure 11.6 shows a schematic depicting the diffusion of nitroxide into a neuronal cell where esterase hydrolysis liberates nitroxide [107].

11.3
Medical Application of Nitroxides

11.3.1
Nitroxides and Nitrons as Drugs

Stable nitroxide free radicals and nitrons are used in animal models and human diseases to protect processes of reactive radical formation, reactive oxygen species (ROS) O_2^-, H_2O_2, ·OH in particular, involving oxidative stress. ROS are the normal product of native oxidative metabolism. Enhanced production of these free radicals is implicated in many troublesome processes such as Alzheimer and Parkinson diseases, stroke, multiple sclerosis, adult respiratory distress syndrome, ischemia–reperfusion injury, congestive heart failure, cardiovascular disease, wound angiogenesis, thrombosis, Down syndrome, skin diseases, cancer, hemorrhagic stroke,

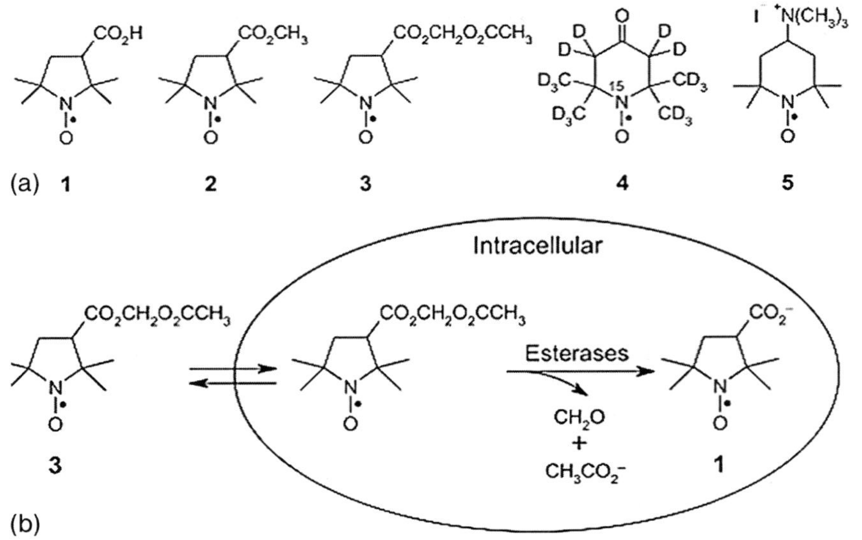

Figure 11.6 Schematic depicting the diffusion of nitroxide into a neuronal cell, where esterase hydrolysis liberates nitroxide, which is anionic at physiologic pH. The anionic nitroxide is membrane-impermeant and therefore is retained intracellularly [107]. (Reproduced with permission).

hearing loss, retinal light damage, neuroinflammatory diseases, aging, and so forth. Nitroxides at nontoxic concentration are effective as *in vitro* and *in vivo* antioxidants when oxidation is induced by superoxide, hydrogen peroxide, organic hydroperoxides, ionizing radiation, or specific DNA-damaging anticancer agents are therefore very promising as medicines for human therapeutic and clinical purposes. Nevertheless, preclinical studies on animal model diseases are absolutely necessary.

The special advantages of nitroxides lay in the dual use of this compound, firstly, as drugs and, secondly, as spin-probes that provide facilities for investigation of properties such as viscosity, pH, electrostatic potential and oxygenation, and pharmokinetics, in different parts of the object of interest using modern ESR techniques. For general information on the advantages and limitations in this important field the reader is referred to the following reviews [40, 109–116]. The antioxidant activity of nitroxides has been reviewed in Chapter 7 and is described in references [41–46, 51, 56–59] cited above. In the next section we describe recent data on nitroxides as drugs in animal models of human disease.

11.3.2
Protection in Animal Model Diseases

Neuroprotection by the stable nitroxide 3-carbamoyl-proxyl (3CP) during reperfusion in the brain of a rat model of transient focal ischemia was reported [115].

Reversible ischemia was induced by a thread placed intraluminally in the middle cerebral artery of rats and different amounts of 3CP (1, 10, and 100 mg kg^{-1}) was given during reperfusion. A statistically significant reduction in infarct size was achieved in the 10- and 100-mg kg^{-1} 3CP-treated groups. No effects of 3CP on blood pressure or brain temperature were observed.

The ability of nitroxide cap to protect cells, tissues, organs, and whole organisms from oxidative stress and radiation injury was shown in [117]. A series of nitroxide derivatives of spirocyclopentane, spirocyclohexane, spirocycloheptane, spirocyclooctane, 5-cholestane, or norbornane showed significant protecting effect. The abilities of polynitroxyl human serum albumin (PNA), TEMPOL, and the combination of PNA and TEMPOL to prevent lung microvascular injury secondary to prolonged gut ischemia and reperfusion in the rat was tested [118]. TEMPOL readily accessed the intracellular compartment while PNA acted only in the extracellular compartment, or in concert with TEMPOL, to provide additional antioxidant protection within cells. It was shown that the combination of PNA and TEMPOL but not PNA alone or TEMPOL alone prevented lung microvascular injury. Anti-inflammatory therapeutic activity was correlated with blood TEMPOL levels in the presence of PNA. The antitumor and antioxidant activity of spin-labeled derivatives of podophyllotoxin (GP-1) and congeners was demonstrated in [119]. The spin-labeled derivative of podophyllotoxin, podophyllic acid-[4-(2,2,6,6,-tetramethyl-1-piperidyloxy)] hydrazone (GP-1,2) and its congeners (GP-1-OH, 3, GP-1-H, 4), were synthesized, and its inhibition activity to the transplanted tumor S180 and HepA and the LD50 values of these compounds in mice were tested. The results showed that the anticancer activity of these compounds followed the order GP-1 > GP-1-OH > GP-1-H and could be attributed to the influence of their partition coefficients and ionization constants on the compound's properties. The authors concluded that the redox potentials of nitroxides play a key role to the antioxidant activity.

With the aim of developing methods of protection from radiation-induced alopecia (hair loss) with the application of nitroxides, a series of experiments on a guinea-pig model was performed [120]. A solution containing TEMPO or TEMPOL in ethanol was topically applied to the skin surface of one side and ethanol was applied to the control side 10 minutes before irradiation. After radiation treatment, over weeks 4 to 11 post-irradiation hair loss was much more pronounced in control animals when compared with nitroxide-treated animals.

It is well-documented that nitrons have an ability to trap reactive radicals like nitroxides and therefore could be used as drugs [121–125]. This ability can be illustrated by the following examples.

Nitronaphthal-NU showed antitumor activity in mice [126]. Several compounds from a family of glycolipidic nitrones were used as drugs for neurodegenerative disorders [127]. Disodium 2,4-disulfophenyl-N-tert-butylnitrone (NXY-059) was shown to be effective for treating acute ischemic stroke in rats and neuroprotection in hemorrhagic stroke [124].

A series of nitron syntheses and applications of these compounds as potential drugs were recently patented: methods for treating multiple sclerosis by acryl nitrone compositions [128]; nitrone compounds, prodrugs and pharmaceutical compositions containing them for the treatment of human disorders [129]; preparation of adamantyl nitroxides and nitrones as antioxidants and NMDA antagonists [130]; methods of making and using of azulenyl nitrone spin-trapping agents. [131].

11.3.3
Human Diseases. Therapeutic Aspects

Numerous biochemical and biophysical studies involving nitroxides as well as results of preclinical studies have formed the basis for the therapeutic applications of nitroxides.

Nowadays, potential clinical applications have been realized step by step. Here, we briefly outline the current status of this field, which could be found in several patents. For reviews readers are addressed to references [105, 109, 113, 114, 123, 132–135].

A series of methods for treating and preventing a number of human diseases has been patented: essential hypertension and oxidative stress [136]; wound and photodamage oxidative stress [137]; dermatological diseases using cosmetic compositions which include aromatic nitroxides [138]; neurological diseases and disorders using propargyl nitroxides and indanyl nitroxides [139]; inflammation [140]; diabetes [141]; cardiovascular disease [142]; neoplastic disease and amyloid-related diseases such as Alzheimer disease [143]; immunological diseases [144]; diseases involving cell proliferation, migration, or apoptosis of myeloma cells, or angiogenesis [145]; ocular diseases [146]; cancer [147]; radiation damage, ischemia, and other applications [148]; apoptosis of myeloma cells of patients undergoing radiotherapy [145]; and alleviation of radical toxicity [148]. As an illustration, several typical examples will be discussed below.

Several compositions of a pharmaceutically acceptable carrier containing effective therapeutic or prophylactic amounts of TEMPOL that alters the expression of one or more genes related to the cardiovascular disease have been patented [142]. For a 70-kg patient found to be at risk for myocardial infarction a dose of 1500 mg of TEMPOL per day for 180 days has been administered. The nitroxide was administered in a single dose, or may be administered as a number of smaller doses over a 24-hour period; for example, three 500-mg doses at eight-hour intervals. Following treatment, the protein level of hepatocyte growth factor and adiponectin in the plasma was increased, and the protein level of caspase 3 was decreased. Propargyl nitroxides and indanyl nitroxides with alkyl, alkynylamino, NH substituents and its enantiomers were used for alleviating symptoms of neurolological, autoimmune, and inflammatory disorders caused by the presence of reactive oxygen species.

A charged nitroxide preparation was administered to a patient in a therapeutically effective amount [137]. The nitroxide interacted with reactive oxygen species in the patient for a longer period of time and modulated adverse effects of those reactive oxygen species and gave protection against UV skin damage. The effect of another nitroxide, disodium 2,4-disulfophenyl-N-tert-butylnitrone (PMX-DHP), on cytokines upon treatment in 25 patients who underwent emergency abdominal surgery and were immediately started on a postoperative regimen of continuous hemodiafiltration (CHDF) and nitroxide PMX-DHP, was reported [149]: after treatment with nitroxide 80% of patients survived for more than one month. The authors concluded that PMX-DHP treatment may be limited in clinical applications for its ability to remove inflammatory cytokines and humoral mediators but is useful for hemodynamic stabilization.

A report of nitroxide skin toxicity was documented in [123]. The nitroxides TEMPO, 2,2,5,5-tetramethyl-3-oxazolidinoxyl (Doxo), 2,2,5,5-tetramethyl-1-dihydropyrrolinoxyl (Proxo), 2,2,3,4,5,5-hexamethyl imidazoline-1-yloxyl (Imidazo) and the nitrones 5,5-dimethyl-1-pyrroline-N-oxide (DMPO) and N-tert-butyl-phenylnitrone (PBN) in concentrations of 100 mM were used to assess cutaneous tolerance to nitroxides in human skin. The order of nitroxide irritation potency (TEMPO > Doxo > Imidazo = Proxo) was found to be inverse to the order of nitroxide biostability in murine and human skin (Imidazo = Proxo > Doxo > TEMPO).

According to [150] no neuroprotective drug including disodium 4-[(tert-butylimino) methyl] benzene-1, 3-disulfonate N-oxide has yet been shown to be effective in treating acute ischemic stroke in the clinic, despite evidence of efficacy in animal models. The neuronal excitotoxicity of the blood–brain barrier permeable nitroxide radical, 3-methoxycarbonyl-2,2,5,5-tetramethylpyrrolidine-1-oxyl on the central nervous system of the hippocampus of conscious rats was revealed in [134]. The authors warned that more detailed studies on the possible toxicity of nitroxide radicals will be needed for evaluating prospects of moving nitroxide from the experimental to the clinical arena where nitroxide radicals would be used for treating CNS disease in future.

11.3.4
Nitroxides in Clinics

At present several commonly employed applications of nitroxides in clinics can be pointed out [151–157]: including (1) an analysis of redox status and ROS of patients blood, wound, ibroblasts, skin, and so forth [151]; (2) assessing human membrane fluidity [155]; (3) spin oximetry in 4-day chick embryo and human blood [156], clinically used as drug carriers [133]; and (5) immunoassay using nitroxides (spin-immunoassay) [157].

A highly sensitive enzyme immunoassay (EIA) that detects serum hepatitis B surface antigen (HBsAg) by measuring nitroxide radicals using an ESR technique was developed [157]. To reveal hepatitis B virus (HBV) infection in patients, serum samples from 30 patients with acute or fulminant hepatitis have been investigated. Serum HBV DNA by amplification of the HBV S gene, using the polymerase chain reaction (PCR) technique, was also examined. The authors suggested that this method is useful for screening and diagnosing HBV infection in patients with liver diseases who are negative for conventional HBV-related serological markers.

Direct detection of ROS in human arterial and venous blood during open heart surgery using ESR with a spin-trapping technique was reported [158]. ESR spectroscopy has been applied to measure the generation of free radicals before, during, and after hyperoxic extracorporeal circulation in 12 patients (6 men, 6 women) with a mean age of 69.5 ± 6.2 years undergoing elective cardiopulmonary bypass for myocardial revascularization. Mean ischemic time during extracorporeal circulation was 45.5 ± 13.3 min. Blood samples were gained before, during, and after extracorporeal circulation and measured for free radicals in arterial and venous blood samples by means of ESR spectroscopy using N-tert-butyl-α-phenylnitrone (PBN) as a spin trap. These data provided direct clinical evidence for enhanced generation of oxygen-derived free radicals in arterial and venous human blood during open heart surgery using hyperoxic ($PaO_2 \geq 150$ mmHg) extracorporeal circulation. Reactive oxygen species generation in gingival fibroblasts (DS-GF) of Down syndrome patients were detected by an ESR spin-trapping technique with 5,5-dimethyl-1-pyrolline-N-oxide (DMPO) as the spin trap [159]. The formation of the DMPO-OH spin adduct, indicating HO· generation from cultured DS-GF and non-DS-GF was observed. The HO· generation in cultured DS-GF was strongly decreased in the presence of catalase, or the iron chelator, desferal. The authors suggested that this effect may be due to the enzymic ability of overexpressed CuZn-superoxide dismutase in Down syndrome. An association between plasma asymmetric dimethylarginine and membrane fluidity of erythrocytes in hypertensive and normotensive men was investigated by a nitroxide spin-probe method [160]. It was shown that increased levels of asymmetric dimethylarginine (ADMA), an endogenous inhibitor of nitric oxide synthase that is associated with increased risk of vascular dysfunction, also affected membrane fluidity of erythrocytes. A parameter of fluidity, the order parameter (S) for the spin-label agent (5-nitroxide stearate) in the EPR spectra of erythrocyte membranes was measured in the erythrocyte membranes from hypertensive ($n = 38$) and normotensive ($n = 35$) men. The order parameter (S) for the spin-label agent 5-nitroxide stearate in the EPR spectra of erythrocyte membranes was significantly higher in hypertensive men than in normotensive men.

The nitroxide spin probe method was used for studying the effect of a selective estrogen receptor modulator, tamoxifen, on membrane fluidity of erythrocytes in normotensive and hypertensive postmenopausal women [160]. Tamoxifen significantly decreased the order parameter (S) for 5-nitroxide stearate (5-NS) and it also decreased the peak height ratio (h_o/h_{-1}) for 16-NS obtained from EPR spectra of erythrocyte membranes in normotensive postmenopausal women and therefore increased the membrane fluidity of erythrocytes and improved the rigidity of cell membranes. The effect of tamoxifen was significantly potentiated by the nitric oxide donors, L-arginine and S-nitroso-N-acetylpenicillamine, and a cGMP analog 8-bromo-cGMP

The inhibitory effect of selen, GSH, and vitamin E on ROS in human brain olfactory bulb tissues (OBT) taken from Alzheimer and normal patients was investigated using a spin trap, 1-hydroxyl-3-methoxycarbonyl-2,2,5,5,-tetramethyl pyrrolidin [161]. OBT were obtained postmortem about 3 hours after death of patients. Results showed increase of ROS and pronounced inhibitory effect in OBT from Alzheimer disease patients. The dual fluorophore–nitroxide probe, 5-Dimethylaminonaphthalene-1-sulfonyl-4-amino 2,2,6,6-tetramethyl-1-piperidine-oxyl was used as a model of a persistent radical that can mimic moderate, prolonged oxidative stress [O. Saphir and G.I. Likhtenshtein, private communication]. The method was applied to blood samples of healthy pregnant women in the third trimester of pregnancy. It was found that in the third trimester of pregnancy, there is a significant decline in the rate constant of reduction of the radical in blood as compared with healthy pregnant women. In pre-eclampsia, this effect is more pronounced. However, there is no decrease in the antioxidant capacity in both cases. Biochemical reactions in blood and its components taking into account oxidative stress processes have been discussed.

11.4
Areas Related to Future Development of Nitroxide Applications in Biomedicine

Among promising areas of nitroxide synthesis and applications the following directions stand apart: (1) spin-labeled physiologically active compounds; (2) biological assays and analyses; (3) new ESR instumentation; (4) new systems for specific delivery of drugs to various organ parts.

From the 1970s to the 1990s numerous nitroxide analogs of molecule of biological and biomedical importance have been synthesized and been investigated, including amino acids and peptides, steroids, phosphororganic compounds, alkoloids, terpenes, nucleotides and their fragments, coenzymes, haptens, morphine, local anesthetics, sterine protease substrates and inhibitor analogs, and so forth ([8, 10, 109, 162–165] and references therein). Recently, syntheses of new spin-labeled compounds were reported: alkyl phospholipid analogs of perifosine and miltefosine [166], chromons [167], and flavones and flavanones [168]. For recent results in chemistry and biology of nitroxides readers are referred to Chapters 5 and 7 and reviews [169–171]. A variety of these compounds can be utilized as spin-probes, pharmacokinetic agents in biomedical research *in vivo* and as potential medicines in clinics.

The significant potential of nitroxide chemistry in analytical and bioanalytical chemistry has not yet been realized. Determination of metal ions, (Ca^{2+}, Ni^{2+}, Cu^{2+}, and Co^{2+}) in solution by chelate formation with spin-labeled and intramolecular luminescence-quenched spin-labeled reagents has proved to be a very sensitive analytical technique [172, 173]. Though immunological analysis nitroxides are already used in medical, veterinary, and biological research ([174, 175] and references therein), this highly specific, sensitive and fast method is not commonly employed. Two main trends may be pointed out in this field: (1) substitution of labeled haptens from the antibody active centers for natural haptens [176, 177] and (2) monitoring of the changes in the interaction between nitroxide spin probes after destroying liposome sacks loaded by the probes upon complexation between haptens and antibodies [178]. Spin immunoassay procedures have been developed for analysis using thyroxine, dinitro phenol, morphine, extol, catecholamines, and other compounds ([8, 174, 175] and references therein).

Figure 11.7 Professor Harold Swartz making *in vivo* oximetry measurements in his own foot (personal communication).

Spin oximetry, spin redox probing, spin trapping, and spin imaging of the human body are promising techniques in future medical and clinical research and practice. The possibility of direct monitoring of the oxygen dynamic in a human body was demonstrated by Swartz's group (Figure 11.7).

Recently, promising data on the development of new systems of drug delivery using microcapsulation techniques were reported. With the aim being to obtain stable microcapsules for delivering drugs *in vivo*, two new systems were established. [179]. A spin-probe was incorporated inside the microcapsules (Figure 11.8) and the resistance of the probe to reduction with ascorbic acid was measured. The experiment showed that the probe inside the new system was protected to a higher degree than the one incorporated in other known microcapsules (Figure 11.9).

A new method, electron spin resonance microscopy (ESRM) which is an extension of the conventional millimeter-scale ESR imaging technique, was applied to the study of controlled drug release [180]. This method may enable us to obtain 3D spatially resolved information about the drug concentration, its self-diffusion tensor, rotational correlation time, and the pH in the release matrix, with a resolution of approximately $3 \times 3 \times 8\,\mu m$.

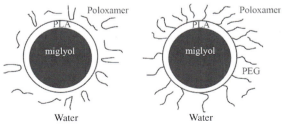

Figure 11.8 Schematic representation of PLA and BEG-PLA microcapsules [179]. (Reproduced with permission).

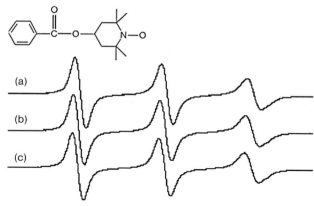

Figure 11.9 ESR spectra of TEPPOL benzoate located in different nanocapsules: (a) PLA, (b) BEG-PLA 10%, (c) BEG-PLA 25% [179]. (Reproduced with permission).

This chapter briefly reviews recent progress in examining molecular structure and the biological implication for cells, active organs, and animal models of the use of nitroxides and nitron spin-traps. The biomedical aspects of the problem have been stressed. Nowadays we witness a real burst of developing new ESR techniques for biomedical and medical research and for spin-imaging, spin oximetry, and spin-pharmokinetics, in particular (Chapter 4). This trend will, without doubt, be continued.

References

1 McConnell, H.M. and McFarland, B.G. (1970) *Quarterly Reviews of Biophysics*, **3**, 91–136.
2 Likhtenshtein, G.I. (1974) *Spin Labeling Method in Molecular Biology*, Nauka, Moscow (in Russian).
3 Hemminga, M.A. and Berliner, L.J. (eds) (2007) *ESR Spectroscopy in membrane biophysics*, in *Biological Magnetic Resonance*, Vol. 27, Springer Verlag.
4 Berliner, L. (ed.) (1998) *Spin Labeling. The Next Millennium*, Vol. 14, Academic Press, New York.
5 Berliner, L.J., Eaton, S.S. and Eaton, G.R. (eds) (2000) *Biological Magnetic Resonance*, Vol. 19, Kluwer Academic/ Plenum Publisher, New York.
6 Likhtenshtein, G.I. (1976) *Spin Labeling Method in Molecular Biology*, Wiley Interscience, New York.
7 Likhtenshtein, G.I. (2005) Labeling, biophysical, in *Encyclopedia of Molecular Biology and Molecular Medecine*, Vol. 7 (ed. R. Meyers), VCH, New York, pp. 157–78.
8 Likhtenshtein, G.I. (1993) *Biophysical Labeling Methods in Molecular Biology*, Cambridge University Press, Cambridge, NY.
9 Likhtenshtein, G.I., Febrario, F. and Nucci, R. (2000) *Spectrochimica Acta, Part A: Molecular and Biomolecular Spectroscopy*, **56**, 2011–31.
10 Kocherginsky, N. and Swartz, H.M. (1995) *Nitroxide Spin Labels*, CRC Press, Boca Raton.
11 Swartz, H.M. *Biochemical Society Transaction* (2002), **30**, 248–52.
12 Samouilov, A., Roubaud, V., Kuppusamy, P. and Zweier, J.L. (2004) *Analytical Biochemistry*, **334**, 145–54.

13 Khramtsov, V.V. (2005) *Current Organic Chemistry*, **9**, 909–23.

14 Goloshchapov, A.N. and Burlakova, E.B. (2005) *Essential Results in Chemical Physics and Physical Chemistry* (eds A.N. Goloshchapov, G.E. Zaikov and V.V. Ivanov), Nova Science Publishers, Hauppauge, New York, pp. 1–18.

15 Harrison, D.G. and Dikalov, S. (2006) *Molecular Mechanisms in Hypertension* (ed. R.N. Re), Taylor and Francis, Abingdon, UK, pp. 297–320.

16 Herrmann, W., Stosser, R., Moll, K.P. and Borchert, H.H. (2005) *Applied Magnetic Resonance*, **28**, 85–106.

17 Zuvic-Butorac, M., Batista, U. and Schara, M. (2001) *Periodicum Biologorum*, **103**, 235–40.

18 Gallez, B. and Swartz, H.M. (2004) *NMR in Biomedicine*, **17**, 223–5.

19 Saphier, O., Silberstein, T., Shames, A.I., Likhtenshtein, G.I., Maimon, E., Mankuta, D., Mazor, M., Katz, M., Meyerstein, D. and Meyerstein, N. (2003) *Free Radical Research*, **37**, 301–8.

20 Bujak, S. and Gwozdzinski, K. (2003) *Meeting of the Society for Free Radical Research--European Section: Free Radicals and Oxidative Stress: Chemistry, Biochemistry and Pathophysiological Implications* (ed. D. Galaris), Ioannina, Greece, pp. 105–8.

21 James, M.Q., Zhi-C., Juliao, S. and Cobb, C.E. (2005) *Free Radical Research*, **39**, 195–202.

22 Kroll, C., Langner, A.L., Borchert, A. and Radical, H.-H. (1999) *Free Radical Biology and Medicine*, **26**, 850–7.

23 Witting, P.K. and Stocker, R. (1997) *Magnetic Resonance in Chemistry*, **35**, 100–10.

24 Bognar, Z., Kalai, T., Palfi, A., Hanto, K., Balazs Bognar, L.M., Zoltan, S.Z., Tapodi, A., Radnai, B., Sarszegi, Z., Szanto, A., Gallyas, F., Jr., Hideg, K., Balazs, S. and Varbiro, G. (2006) *Free Radical Biology and Medicine*, **41**, 835–48.

25 Samuni, A.M., DeGraff, W., Cook, J.A., Krishna, M.C., Russo, A. and Mitchell, J.B. (2004) *Free Radical Biology and Medicine*, **37**, 1648–55.

26 Lozinsky, E.M., Martina, L.V., Shames, A.I., Uzlaner, N., Masarwa, A., Likhtenshtein, G.I., Meyerstein, D., Martin, V.V. and Priel, Z. (2004) *Analytical Biochemistry*, **326**, 139–45.

27 Borisenko, G.G., Martin, I., Zhao, Q., Amoscato, A.A. and Kagan, V.E. (2004) *Journal of the American Chemical Society*, **126**, 9221–32.

28 Wu, Y., Tang, W., Li, C.L., Liu, J.W., Miao, L.D., Han, J. and Lan, M.B. (2006) *Pharmazie*, **61**, 1028–33.

29 Mravljak, J., Zeisig, R. and Pecar, S. (2005) *Journal of Medicinal Chemistry*, **48**, 6393–9.

30 Samuni, Y., Gamson, J., Samuni, A., Yamada, K., Russo, A., Krishna, M.C. and Mitchell, J.B. (2004) *Antioxidants and Redox Signaling*, **6**, 587–95.

31 Halpern, H.J., Chandramouli, G.V.R., Barth, E.D., Yu, C., Peric, M., Grdina, D.J. and Teicher, B.A. (1999) *Cancer Research*, **59**, 5836–41.

32 Fujii, H. and Berliner, L.J. (2004) *NMR in Biomedicine*, **17**, 311–18.

33 Swartz, H.M. and Berliner, L.J. (2003) *Biological Magnetic Resonance*, **18** (In Vivo EPR (ESR)), 1–20.

34 Gallez, B. and Swartz, H.M. (2004) *NMR in Biomedicine*, **17**, 223–5.

35 Subramanian, S., Mitchell, J.B. and Krishna, M.C. (2003) *Biological Magnetic Resonance*, **18** (In Vivo EPR (ESR)), 23–40.

36 Zweier, J.L. and Kuppusamy, P. (1988) *Proceedings of the National Academy of Sciences of the United States of America*, **85**, 5703–7.

37 Kuppusamy, P. (ed.) (2005) ESR, A joint conference of 11th in vivo EPR spectroscopy and imaging and 8th International EPR spin trapping. Abstracts. Columbus, Ohio, September 2005, 4–8.

38 Zweier, J.L. and Hassan Talukder, H. (2006) *Cardiovascular Research*, **70**, 181–90.

39 Zweier, J.L. and Villamena, F.A. (2003) *Oxidative Stress and Cardiac Failure* (eds M.L. Kukin and V. Fuster), Futura Publishing Company, Inc., Armonk, NY, pp. 67–95.

40 Swartz, H.M. (2002) *Biochemical Society Transactions*, **30**, 248–52.

41 Konovalova, N.P., Bogdanov, G.N., Mille, V.B., Neiman, M.B., Rozanzev, E., Emanuel, G. and Doklady, N.M. (1964) *Doklady Akademii nauk SSSR*, **157**, 707–9.

42 Emanuel, N.M. and Konovalova, N.P. (1992) *Bioactive Spin Labels* (ed. R.I. Zhdanov), Springer, Berlin, Germany, pp. 439–60.

43 Emanuel, N.M., Rozenberg, A.N., Golubev, V.A., Bogdanov, G.N., Vasil'eva, L.S. and Konovalova, N.P. (1984) *PCT International Patent Application*, 24 pp.

44 Konovalova, N.P., Volkova, L.M., Codaci-Pisanelli, D., Seminara, P. and Franchi, F. (2000) *Voprosy Onkologii*, **46**, 438–41.

45 Konovalova, N.P., Diatchkovskaya, R.F., Kukushkina, G.V., Volkova, L.M., Varfolomeev, V.N., Dombrovsky, L.S. and Shapiro, A.B. (1988) *Neoplasma*, **35**, 185–90.

46 Sen', V.D., Golubev, V.A., Lugovskaya, N.Yu., Sashenkova, T.E., Konovalova, N.P. (2006) *Russian Chemical Bulletin*, **55**, 62–65

47 Hahn, S.M., Krishna, M.C., DeLuca, A.M., Coffin, D. and Mitchell, J.B. (2000) *Free Radical Biology and Medicine*, **28**, 953–8.

48 Matsumoto, K.-I., Okajo, A., Kobayashi, T., Mitchell, J.B., Krishna, M.C. and Endo, K. (2005) *Journal of Biochemical and Biophysical Methods*, **63**, 79–90.

49 Ui, I., Okajo, A., Endo, K., Utsumi, H. and Matsumoto, K.-I. (2006) *Journal of Magnetic Resonance*, **81**, 107–12.

50 Yamada, K.-I., Nakamura, T. and Utsumi, H. (2006) *Free Radical Research*, **40**, 455–60.

51 Patel, K., Chen, Y., Dennehy, K., Blau, J., Connors, S., Mendonca, M., Tarpey, M., Krishna, M., Mitchell, J.B., Welch, W.J. and Wilcox, C.S. (2006) *American Journal of Physiology*, **290** (1, Pt. 2), R37–43.

52 Hahn, S.M., Krishna, M.C., DeLuca, A.M., Coffin, D. and Mitchell, J.B. (2000) *Free Radical Biology and Medicine*, **28**, 953–8.

53 Yamada, K.-I., Yamamiya, I., Utsumi, H. (2006) *Free Radical Biology and Medicine*, **40**, 2040–6.

54 Utsumi, H., Yasukawa, K., Soeda, T., Yamada, K.-I., Shigemi, R., Yao, T. and Tsuneyoshi, M. (2006) *The Journal of Pharmacology and Experimental Therapeutics*, **317**, 228–35.

55 Kálai, T., Khan, M., Balog, M., Kutala, V.-K., Kuppusamy, P. and Hideg, K. (2006) *Bioorganic and Medicinal Chemistry*, **14**, 5510–16.

56 Cotrim, A.P., Sowers, A.L., Lodde, B.M., Vitolo, J.M., Kingman, A., Russo, A., Mitchell, J.B. and Baum, B. (2005) *Clinical Cancer Research*, 7564–8.11

57 Kutala, V.K., Khan, M., Mandal, R., Potaraju, V., Colantuono, G., Kumbala, D. and Kuppusamy, P. (2006) *Journal of Cardiovascular Pharmacology*, **48**, 79–87.

58 Liang, Q., Amanda, D., Pan, S.S., Tyurin, V.A., Kagan, V.E., Hastings, T.G. and Schor, N.F. (2005) *Biochemical Pharmacology*, **70**, 1371–81.

59 Schubert, R., Erker, L., Barlow, C., Yakushiji, H., Larson, D., Russo, A., Mitchell, B. and Wynshaw-Boris, A. (2004) *Human Molecular Genetics*, **13**, 1793–802.

60 Kozlov, A.V., Szalay, L., Umar, F., Fink, B., Kropik, K., Nohl, H., Redl, H. and Bahrami, S. (2003) *Free Radical Biology and Medicine*, **34**, 1555–62.

61 Kono, H., Woods, C.G., Maki, A., Connor, H.D., Mason, R.P., Rusyn, I. and Fujii, H. (2006) *Free Radical Research*, **40**, 579–88.

62 Woods, C.G., Burns, A.M., Maki, A., Bradford, B.U., Cunningham, M.L., Connor, H.D., Kadiiska, M.B., Mason, R.P., Peters, J.M. and Rusyn, I. (2007) *Free Radical Biology and Medicine*, **42**, 335–42.

63 Kadiiska, M.B., Ghio, A.J. and Mason, R.P. (2004) *Spectrochimica Acta, Part A: Molecular and Biomolecular Spectroscopy*, **60A**, 1371–7.

64 Kono, H., Woods, C.G., Maki, A., Connor, H.D., Mason, R.P., Rusyn, I. and Fujii, H. (2006) *Free Radical Research*, **40**, 579–88.

65 Nakai, K., Kadiiska, M.B., Jiang, J.-J., Stadler, K. and Mason, R.P. (2006) *Proceedings of the National Academy of Sciences of the United States of America*, **103**, 4616–21.

66 Komarov, D.A., Slepneva, I.A., Glupov, V.V. and Khramtsov, V.V. (2005) *Applied Magnetic Resonance*, **28**, 411–19.

67 Yasukawa, K., Kasazaki, K., Hyodo, F. and Utsumi, H. (2004) *Free Radical Research*, **38**, 147–55.

68 Takeshita, K., Takajo, T., Hirata, H., Ono, M. and Utsumi, H. (2004) *The Journal of Investigative Dermatology*, **122**, 1463–70.

69 Okajo, A., Matsumoto, K.-I., Mitchell, J.B., Krishna, C. and Endo, K. (2006) *Magnetic Resonance in Medicine*, **56**, 422–31.

70 Shen, J., Liu, S., Miyake, M., Liu, W., Pritchard, A., Kao, J.P.Y., Rosen, G.M., Tong, Y. and Liu, K.J. (2006) *Magnetic Resonance in Medicine*, **55**, 1433–40.

71 Matsumoto, K., Krishna, M.C. and Mitchell, J.B. (2004) *The Journal of Pharmacology and Experimental Therapeutics*, **310**, 1076–108.

72 Elas, M., Parasca, A., Grdina, D.J. and Halpern, H.J. (2003) *Biochimica et Biophysica Acta. Molecular Basis of Disease*, **1637**, 151–5.

73 Kamatari, M., Yasui, H., Nakamura, M., Ogata, T. and Sakurai, H. (2006) *Chemistry Letters*, **35**, 1170–1.

74 Kroll, C., Herrmann, W., Stosser, R., Borchert, H.-H. and Mader, K. (2001) *Pharmaceutical Research*, **18**, 525–30.

75 Kirilyuk, A., Bobko, A.A., Khramtsov, V.V. and Grigor'ev, I.A. (2005) *Organic and Biomolecular Chemistry*, **3**, 1269–74.

76 Khramtsov, V.V., Grigor'ev, I.A., Foster and, M.A. and Lurie, D.J. (2004) *Antioxidants and Redox Signaling*, **6**, 667–76.

77 Mccallum, S.J., Nicholson, I. and Lurie, D.J. (1998) *Physics in Medicine and Biology*, **43**, 1857–61.

78 Hyodo, F., Yasukawa, K., Yamada, K.-I. and Utsumi, H. (2006) *Magnetic Resonance in Medicine*, **56**, 938–43.

79 Mailer, C., Sundramoorthy, S.V., Pelizzari, C.A. and Halpern, H.J. (2006) *Magnetic Resonance in Medicine*, **55**, 904–12.

80 Halpern, H.J. (2003) *Biological Magnetic Resonance*, **18** (In Vivo EPR (ESR)), 469–82.

81 Gallez, B. and Swartz, H.M. (2004) *NMR in Biomedicine*, **17**, 223–5.

82 Klug, C.S., Camenisch, T.G., Hubbell, W.L. and Hyde, J.S. (2005) *Biophysical Journal*, **88**, 3641–7.

83 Potapenko, I., Foster, M.A., Lurie, D.J., Kirilyuk, I.A., Hutchison, J.M.S., Grigor'ev, I.A., Bagryanskaya, E.G. and Khramtsov, V. (2006) *Journal of Magnetic Resonance*, **182**, 1–11.

84 Takeshita, K., Utsumi, H. and Hamada, A. (1991) *Biochemical and Biophysical Research Communications*, **177**, 874–80.

85 Kuppusamy, P., Li, H., Ilangovan, G., Cardounel, A.J., Zweier, J.L., Yamada, K., Krishna, M.C. and Mitchell, J.B. (2002) *Cancer Research*, **62**, 307–12.

86 Elas, M., Ahn, K.-H., Parasca, A., Barth, E.D., Lee, D., Haney, C. and Halpern, H.J. (2006) *Clinical Cancer Research*, **12**, 4209–17.

87 Sano, H., Naruse, M., Matsumoto, K., Oi, T. and Utsumi, H. (2000) *Free Radical Biology and Medicine*, **28**, 959–69.

88 Lurie, D.J. (2001) *British Journal of Radiology*, **74**, 782–4.

89 Benial, A., Milton, F., Ichikawa, K., Murugesan, R., Yamada, Ken.-I. and Utsumi, H. (2006) *Journal of Magnetic Resonance*, **182**, 273–82.

90 He, G., Deng, Y., Li, H., Kuppusamy, P. and Zweier, J.L. (2002) *Magnetic Resonance in Medicine*, **47**, 571–8.

91 Zweier, J.L., Samouilov, A. and Kuppusamy, P. (2003) *Biological Magnetic Resonance*, **18** (In Vivo EPR (ESR)), 441–68.

92 Mailer, C., Sundramoorthy, S.V., Pelizzari, C.A. and Halpern, H.J. (2006) *Magnetic Resonance in Medicine*, **55**, 904–12.

93 Hahn, S.M., Krishna, M.C., DeLuca, A.M., Coffin, D. and Mitchell, J.B. (2000) *Free Radical Biology and Medicine*, **28**, 953–8.

94 He, G., Petryakov, S., Fallouh, M.M., Deng, Y., Ishihara, R., Kuppusamy, P., Tatsuta, M. and Zweier, J.L. (2004) *Cancer Research*, **64**, 6495–502.

95 Sonta, T., Inoguchi, T., Matsumoto, S., Yasukawa, K., Inuo, M., Tsubouchi, H., Sonoda, N., Kobayashi, K., Utsumi, H. and Nawata, H. (2005) *Biochemical and Biophysical Research Communications*, **330**, 415–22.

96 Kuppusamy, P., Wang, P., Zweier, J.L., Krishna, M.C., Mitchell, J.B., Ma, L., Trimble, C.E. and Hsia, C.J. (1996) *Biochemistry*, **35**, 7051–7.

97 Herrmann, W., Stosser, R., Moll, K.-P. and Borchert, H.-H. (2005) *Applied Magnetic Resonance*, **28**, 85–106.

98 Hochkirch, U., Herrmann, W., Stoesser, R., Borchert, H.-H. and Linscheid, M.W. (2006) *Spectroscopy (Amsterdam, Netherlands)*, **20**, 1–17.

99 Li, H., He, G., Deng, Y., Kuppusamy, P. and Zweier, J.L. (2006) *Magnetic Resonance in Medicine*, **55**, 669–75.

100 Matsumoto, A., Koretsky, A.P., Sowers, A.L., Mitchell, J.B. and Krishna, M.C. (2006) *Clinical Cancer Research*, **12**, 2455–62.

101 Utsumi, H., Yamada, K.I., Ichika, K., Sakai, K., Yuichi, K., Matsumoto, S. and Nagai, M. (2006) *Proceedings of the National Academy of Sciences of the United States of America*, **103**, 1463–8.

102 Mikuni, T., He, G., Petryakov, S., Fallouh, M.M., Deng, Y., Ishihara, R., Kuppusamy, P., Tatsuta, M. and Zweier, J.L. (2004) *Cancer Research*, **64**, 6495–502.

103 Gallez, B. and Mader, K. (2000) *Free Radical Biology and Medicine*, **29**, 1078–84.

104 Ilangovan, G., Zweier, J.L. and Kuppusamy, P. (2004) *Methods in Enzymol*, **381**, 747–62 (Oxygen Sensing).

105 Swartz, H. (2005) *Advanced Drug Delivery Reviews*, **57**, 1085–6.

106 Zhu, X., Zuo, X., Li, X., Cardounel, X., Zweier, A.J. and He, J.L. (2007) *Antioxidants and Redox Signaling*, **9**, 447–55.

107 Rosen, G.M., Scott, R., Burks, M., Kohr, M.J. and Kao, J.P.Y. (2005) *Organic and Biomolecular Chemistry*, **3**, 645–8.

108 Miyake, M., Shen, J., Liu, S., Shi, H., Liu, W., Yuan, Z., Pritchard, A., Kao, J.P.Y., Liu, K.J. and Rosen, G.M. (2006) *The Journal of Pharmacology and Experimental Therapeutics*, **318**, 1187–93.

109 Zhdanov, R.I. (ed.) (1992) *Bioactive Spin Labels*, Springer, Berlin, Germany.

110 Henke, S.L. (1999) *Expert Opinion on Therapeutic Patents*, **9**, 169–80.

111 Mitchell, J.B., Krishna, M.C., Kuppusamy, P., Cook, J.A., Russo, A. (2001) *Experimental Biology and Medicine (Maywood, NJ, United States)*, **226**, 620–1.

112 Maeder, K. and Gallez, B. (2003) *Biological Magnetic Resonance*, **18** (In Vivo EPR (ESR)), 515–45.

113 Mitchell, B.J. (2005) A joint conference of 11th in vivo EPR spectroscopy and imaging and 8th International EPR spin trapping (ed. P. Kuppusamy), Abstracts. Columbus, Ohio, p. 56, 2005, September 4–8.

114 Wink, D.A. (2005) A joint conference of 11th in vivo EPR spectroscopy and imaging & 8th International EPR spin trapping (ed. P. Kuppusamy), Abstracts. Columbus, Ohio, p. 38, 2005, September 4–8.

115 Hu, G., Lyeth, B.G., Zhao, X., Mitchell, J.B. and Watson, J.C. (2003) *Journal of Neurosurgery*, **98**, 393–6.

116 Kálai, T., Khan, M., Balog, M., Kutala, V.K., Kuppusamy, P. and Hideg, K., (2006) *Bioorganic and Medicinal Chemistry*, **14**, 5510–16.

117 Mitchell, J.B., Samuni, A., Degraff, W.G. and Hahn, S. (2003) (The United States of America, The Secretary of the Department of Health and Human Services, USA), 21 pp.

118 Zhang, S. Li, H. Ma, L., Trimble, C.E., Kuppusamy, P., Hsia, C.J.C. and Carden, D.L. (2000) *Free Radical Biology and Medicine*, **29**, 42–50.

119 Tian, X., Zhang, F.M. and Li, W.-G. (2002) *Life Sciences*, **70**, 2433–43.

120 Cuscela, D., Coffin, D., Lupton, G.P., Cook, J.A., Krishna, M.C., Bonner, R.F. and Mitchell, J.B. (1996) *The Cancer Journal From Scientific American*, **2**, 273–8.

121 Kulkarni, A.P., Kellaway, L.A., Lahiri, D.K. and Kotwal, G.J. (2004) *Annals of the New York Academy of Sciences*, **1035**, 147–64.

122 Floyd, R.A. (2006) *Aging Cell*, **5**, 51–7.

123 Fuchs, J., Groth, N. and Herrling, T. (2003) *Biological Magnetic Resonance*, **18** (In Vivo EPR (ESR)), 483–513.

124 Maples, K.R., Green, A.R. and Floyd, R.A. (2004) *Drugs*, **18**, 1071–84.

125 Bulut, G., Oktav, M. and Ulgen, M. (2004) *European Journal of Drug Metabolism and Pharmacokinetics*, **29**, 237–48.

126 Pain, A., Dutta, S., Saxena, A., Shanmugavel, M., Pandita, R.M., Qazi, G.N. and Sanyal, U. (2004) *Journal of Experimental Therapeutics and Oncology*, **5**, 15–22.

127 Durand, G., Polididori, A., Salles, J.-P. and Pucci, B. (2003) *Bioorganic and Medicinal Chemistry Letters*, **13**, 859–62.

128 Kelly, M.G., Serafini, T. and Chen, H. (2005) *PCT International Patent Application*, 84 pp.

129 Kelly, M.G., Upasani, R.B. and Janagani, S. (2005) *PCT International Patent Application*, 105 pp.

130 Wang, Y. and Larrick, J.W. (2002) *PCT International Patent Application*, 140 pp.

131 Becker, D. and Ley, J.J. (2006) *PCT International Patent Application*, 63 pp.

132 Ahmad, N., Misra, M., Husain, M.M. and Srivastava, R.C. (1996) *Ecotoxicology and Environmental Safety*, **34**, 141–4.

133 Maeder, K. and Gallez, B. (2003) *Biological Magnetic Resonance*, **18** (In Vivo EPR (ESR)), 515–45.

134 Ueda, Y., Yokoyama, H., Tokumaru, J., Doi, T. and Nakajima, A. (2004) *Biological Magnetic Resonance*, **29**, 1695–701.

135 Tsuda, K. and Nishio, I. (2005) *American Journal of Hypertension: Journal of the American Society of Hypertension*, **18**, 1243–8.

136 Wilcox, C.S. (1999) *PCT International Patent Application*, 51 pp.

137 Hsia, J.-C. and Ma, L. (2003) *(USA). U.S. Pat. Appl. Publ.*, 33 pp.

138 Greci, L. and Damiani, E. (2003) *U.S. Pat. Appl. Publ.*, 12 pp.

139 Sterling, J., Sklarz, B., Herzig, Y., Lerner, D., Falb, E. and Ovadia, H. (2006) *PCT International Patent Application*, 25 pp.

140 Garvey, D.S. (2006) *PCT International Patent Application*, 61 pp.

141 Habash, L. and Jones, C. (2006) *PCT International Patent Application*, 14 pp.

142 Habash, L. and Jones, C. (2006) *PCT International Patent Application*, 15 pp.

143 Habash, L. and Jones, C. (2006) *PCT International Patent Application*, 25 pp.

144 Habash, L. and Jones, C. (2006) *PCT International Patent Application*, 24 pp.

145 Hilberg, F., Solca, F., Stefanic, M.F., Baum, A., Munzert, G., Van, M. and Jacobus, C.A. (2004) *PCT International Patent Application*, 101 pp.

146 Bernstein, E.F. (2004) *PCT International Patent Application*, 38 pp.

147 Mitchell, J.B., Russo, A., Deluca, A.M. and Cherukuri, M.K. (1998) *PCT International Patent Application*, 31 pp.

148 Habash, L. (2006) *PCT International Patent Application*, pp. 18.

149 Matsuno, N., Ikeda, T., Ikeda, K., Hama, K., Iwamoto, H., Uchiyama, M., Kozaki, K., Narumi, Y., Kikuchi, K., Degawa, H. and Nagao, T. (2001) *Therapeutic Apheresis*, **5**, 36–9.

150 Green, A.R., AstraZeneca, R., Charnwood, D. and Loughborough, X. (2002) *Clinical and Experimental Pharmacology and Physiology*, **29**, 1030–4.

151 Saphier, O., Silberstein, T., Shames, A.I., Likhtenshtein, G.I., Maimon, E., Mankuta, D., Mazor, M., Katz, M., Meyerstein, D. and Meyerstein, N. (2003) *Free Radical Research*, **37**, 301–8.

152 Tanigawa, T. (2005) *Science and Engineering Review of Doshisha University*, **45** (4, Suppl.), 89–94.

153 Hochkirch, U., Herrmann, W., Stoesser, R., Borchert, H.-H. and Linscheid, M.W. (2006) *Spectroscopy (Amsterdam, Netherlands)*, **20**, 1–17.

154 Swartz, H.M., Khan, N., Buckey, J., Comi, R., Gould, L., Grinberg, O., Hartford, A., Hopf, H., Hou, H., Hug, E., Iwasaki, A., Lesniewski, P., Salikhov, I. and Walczak, T. (2004) *NMR in Biomedicine*, **17**, 335–51.

155 Kazushi, T. and Ichiro, N. (2005) *American Journal of Hypertension: Journal of the American Society of Hypertension*, **18**, 1243–8.

156 Rajala, G.M., Lai, C.S., Kolesari, G.L. and Cameron, R.H. (1985) *Life Sciences*, **36**, 291–7.

157 Aoki, M., Saito, T., Watanabe, H., Taku, M., Saito, K., Togashi, H., Kawata, S., Ishikawa, K., Aoyama, M., Kamada, H. and Shinzawa, H. (2002) *Journal of Medical Virology*, **66**, 166–70.

158 Etz, C., Soeparwata, R., Scheld, H.H. and Steinhoff, H.-J. (2005) Abstracts of

International Conference, SPIN 2005, Novosibirsk.
159 Komatsu, T., Lee, M.-C.-I., Miyagi, A., Shoji, H., Yoshino, F., Maehata, Y., Maetani, T., Kawamura, Y., Ikeda, M. and Kubota, E. (2006) *Redox Report: Communications in Free Radical Research*, **11**, 71–7.
160 Tsuda, K. and Nishio, I. (2005) *American Journal of Hypertension: Journal of the American Society of Hypertension*, **18**, 1243–8.
161 Dufault, R., Le Blanc, B., Kumar, S., Romanyukha, A., Romanyukha, L. and Chad, M. (2005) A joint conference of 11th in vivo EPR spectroscopy and imaging and 8th International EPR spin trapping (ed. P. Kuppusamy), Abstracts. Columbus, Ohio, p. 23.
162 Keana, J.F.W. (1981) *Spin Labeling in Pharmacology* (ed. J.L. Holtsman), Academic Press, New York, pp. 2–86.
163 Park, J.H. and Trommer, W.E. (1989) *Biological Magnetic Resonance*, **8**, 547–95.
164 Hideg, K. (1990) *Pure and Applied Chemistry. Chimie Pure et Appliquee*, **62**, 207–12.
165 Rozantsev, E.G. (1990) *Pure and Applied Chemistry. Chimie Pure et Appliquee*, **62**, 311–16.
166 Mravljak, J., Zeisig, R. and Pecar, S. (2005) *Journal of Medicinal Chemistry*, **48**, 6393–9.
167 Muller, E., Kalai, T., Jeko, J. and Hideg, K. (2000) *Synthesis*, **10**, 1415–20.
168 Kalai, T., Kulcsar, G., Osz, E., Jeko, J., Suemegi, B. and Hideg, K. (2004) *ARKIVOC* (Gainesville, FL, USA), **7**, 266–76.
169 Sar, C.P., Osz, E., Jeko, J. and Hideg, K. (2005) *Synthesis*, **2**, 255–9.
170 Hideg, K., Kalai, T. and Sar, C.P. (2005) *Journal of Heterocyclic Chemistry*, **42**, 437–50.
171 Palm, T., Coan, C. and Trommer, W.E. (2001) *Biological Chemistry*, **382**, 417–23.
172 Nagy, V.Yu. (1988) *Imidazoline Nitroxides*, Vol. 2 (ed. L.B. Volodarskii), CRC Press, Boca Raton, pp. 115–35.
173 Nagii, V.Yu., Bystryak, I.M., Kotel'nikov, A.I., Likhtenshtein, G.I., Petrukhin, O.M., Zolotov, Yu.A. and Volodarskii, L.B. (1990) *Analyst (Cambridge, United Kingdom)*, **115**, 839–41.
174 Likhtenshtein, G.I. (1996) Spin and fluorescence immunoassays in solution, in *Immunology Methods Manual* (eds I. Levkovits and R. Nezlin), Pergamon Press, London, pp. 540–50.
175 Likhtenshtein, G.I. (2005) Biophysical labeling, in *Encyclopedia of Molecular Biology and Molecular Medecine*, Vol. 7 (ed. R. Meyers), VCH, New-York, pp. 157–78.
176 Montgomery, M.R., Holtzman, J.L. and Leute, R.K. (1975) *Clinical Chemistry*, **21**, 221–6.
177 Ashirov, P.M., Likhtenshtein, G.I., Smotrov, S.P. (1980) Patent SSSR N, 789752.
178 Chan, S.W., Tan, C.T. and Hsia, J.C. (1978) *Journal of Immunological Methods*, **21**, 185–95.
179 Mader, K. and Rube, A. (2005) *Biomed Nanotechnol*, **1**, 1–6.
180 Blank, A., Freed, J.H., Kumar, N.P. and Wang, C.-H. (2006) *Journal of Controlled Release: Official Journal of the Controlled Release Society*, **111**, 174–84.

12
Conclusion

It is known that there are delicate links and fine parallels between art and science. Both these spheres of human endeavor involve a unique combination of professional skill and creative search. Sometimes an intuitive line that a great poet or philosopher pursues may be likened to the opening of a new horizon in science. Thus, the composer Maurice Ravel in his famous "Bolero" allegorically depicts the process of birth and development of an epochal discovery that gives rise to many advantages. At first, a musical tune arises whose sound is so weak, so feeble, that it can be easily drowned out by the surroundings. In the second movement, the music is repeated with the same melody, but now with an additional hue. The process repeats itself, again and again, until eventually the most powerful strains of bold, majestic music are then performed by the symphony orchestra.

Like that opening musical movement, the initial publications on the chemistry and application of a novel class of stable radicals, nitroxides, were first met with skepticism, and even strong criticism, from qualified and very professional members of the scientific community. But, later, more and more young enthusiasts joined the ranks of scientists applying this new tool in their research, and ever-increasing numbers of reports of nitroxides were published in the various fields of chemistry, physics, biology, material science, and biomedicine. The theoretical and experimental data presented in this book clearly demonstrate both the current progress and prospects for future developments within the nitroxide "empire".

This progress resulted to a great extent from interdisciplinary cooperation. Experts in natural science formulate up-to-date structural, dynamic, and functional problems to be solved. Modern synthetic chemistry provides researchers with a wide assortment of nitroxide labels and probes, and paves the way for specific modification of certain portions of objects under interest. Twists to traditional ESR techniques and advanced magnetic resonance methods, such as pulse-ESR, ESR imaging, double electron–electron, and electron–nuclear resonances, in particular, and theoretical approaches to analysis of experimental data ensure profound investigation of structure and functional activity of labeled systems on a molecular level.

Let us summarize briefly the main possibilities, advantages, and limitations of various fields of nitroxide application.

Thousands of new nitroxide radicals, related nitrons, and chromophore–nitroxide compounds of different chemical reactivity, redox potential, hydrophobic and hydrophilic properties, electrostatic charge, and so forth have been synthesized and characterized. The incorporation of nitroxides in various biological and non-biological objects followed by the use of ESR and other physical methods allows us to investigate molecular dynamics within a wide range of correlation times ($\tau_c = 10^{-3} - 10^{-11}$ s) and amplitudes (0.002–0.2 nm), to measure the distance between labeled groups up to 6 nm apart and depth of radical immersion up to 4 nm. Special nitroxides intended for the measurement of local pH, polarity, electrostatic potential, redox status, oxygen concentration, antioxidants, and functional groups have been synthesized and successfully employed.

The molecular mechanism of polymerization, nitroxide-mediated leaving polymerization in particular, of photochemical and photophysical processes, quantomechanical interactions in complexes of nitroxides with paramagnetic metals, and multi-spin systems, the structures and dynamics of organic and inorganic materials are subjects of interest for researchers employing nitroxides.

Nitroxides and their complexes with paramagnetic metals appear to be promising materials possessing unique magnetic properties. The construction of utterly organic ferromagnetic materials on the basis of nitroxides is a real challenge to chemists and physicists.

The spin-labeling method has proved its effectiveness in studies of molecular dynamics and the structure of proteins, enzymes, nucleic acids, polysaccharides, biological and model membranes, and its combinations *in vitro* and *in vivo*. Recent new directions such as site-directed spin-labeling and the employment of advanced magnetic resonance techniques added "fresh blood" to the field.

Nitroxides as protectors against radiation and reactive-oxygen-species damage and as therapeutic drags have massive applications. Spin redox probing, measurements of local pH, and spin-oximetry in functioning organs, living animals, and even in the human body, using the ESR imaging technique have arisen as new promising areas and are a matter of growing interest, especially from researchers who attend directly to human well-being.

Approaches based on the utilization of nitroxides have, along with their advantages, a number limitations and drawbacks. An attentive and qualified reader may find in every particular case specific "risks" that arise, for example from the danger of destroying the native form of a biological structure. The information obtained with the aid of these approaches sometimes is of indirect nature. In spite of successful application of nitroxides as therapeutic preparations, these compounds appear not to be "alchemical philosophical stones", a remedy against all diseases. At present several researchers have pointed out the side-effects of nitroxides in animal models.

Nevertheless, the data cited in the present book indicate that the use of nitroxides offers unique information on the structure and properties of many objects and forms the basis for new magnetic materials and applications in biology and medicine. One of the authors of this monograph wrote in the first book on spin-labeling [Likhtenshtein, G.I. (1974) *Spin Labeling Method in Molecular Biology*, Nauka, Moscow (in Russian)]: "It is thus our hope that spin labeling will continue to be an effective tool for solving various complicated problems in molecular biology". Now, after 34 years, it is evident that present-day reality has surpassed those optimistic expectations.

Index

a
A-anisotropy 96f.
A-value 93, 96f., 146f., 206f., 220, 288
absorption line 78ff., 81, 83
acceptor carrying NR 172f.
adjuvant arthritis 376
alkoxyamine 270ff.
– bond energy of dissociation (BDE) 272f.
– TEMPO-based 273
allosteric transition 336
aminoxylamine, chemical structure 271
Ampére's circuital law 1, 3
angular constraint 134
angular selectivity 134
animal model disease 386ff.
anisotropic parameter 94f., 99
– ESR powder pattern 99
anodic aluminium oxide (AAO) membrane 139f.
– DMPC deposited 140
– EPR spectrum 140
anthracene derivative 189f.
antibody 253
antiferromagnetic material 6f., 61
– magnetic susceptibility 29, 289
– molecular orbital 290
– one-dimensional organic antiferromagnets 61
antiferromagnetic resonance (AFMR) 30, 116f.
antiferromagnetism 6f., 25f., 28f., 117, 289, 291, 294, 317, 319
– canting 7
antioxidant activity 375, 387
antioxidant mechanism 242
antioxidant status 240ff., 245, 251, 259, 294
– analysis 257ff., 260
– characterization 242, 259
antitumor activity 375, 387
antitumor drug 194

ascorbic acid 257ff.
– analysis 258
– fluorescence enhancement 258
Au nanoparticle 183
– nitroxide spin-labeled 295
– preparation 296
Aufbau principle 15
axial anisotropy 100
axial case 101

b
backbone dynamics 257
bacteriorhodopsin 344f.
– structure 344
bicelle alignment 138f.
bifunctional stilbene-nitroxide label 260, 294
bilayer alignment method 136ff.
bile flow monitoring (BFM) 376, 379
bio-functional nitroxide radical 191
biomembrane, see also membrane 346ff.
Bloch equation 77f., 81ff.
– modified 82f.
– solution 77f.
blood circulating monitoring (BCM) 379
blood-brain barrier (BBB) 379, 389
Bohr magneton 10, 14f., 24
bond energy of dissociation (BDE) 272f.
Brillouin function 22f., 26f.
Brownian dynamics (BD) model 145
tert-butylnitrone 248

c
C–O bond 270
– bond energy of dissociation (BDE) 272
– reversible thermal homolysis 270
calix[n]arene carrying NR 175ff.
– *t*-butylnitroxide group carrying 176
– spin-spin interaction 176

Nitroxides: Applications in Chemistry, Biomedicine, and Materials Science
Gertz I. Likhtenshtein, Jun Yamauchi, Shin'ichi Nakatsuji, Alex I. Smirnov, and Rui Tamura
Copyright © 2008 WILEY-VCH Verlag GmbH & Co. KGaA, Weinheim
ISBN: 978-3-527-31889-6

cancer cell   374
Car-Parrinello molecular dynamics   207
cascade method   225f.
cascade reactant   225
cathode-active material   325f.
cell membrane   371f., 390
– electron transfer   372
– fluidity   371f.
cell protection   373f., 386f.
cellulose   360ff.
– nitroxide-mediated oxidation   361
cgs unit   74
charge distribution   213
charge recombination   284
charge-transfer (CT) complex   50, 168ff., 171ff., 174
– conductivity   169ff.
– NN radical-based formation   169
– NR-based   168ff., 171ff., 174
– TEMPO radical-based formation   168
chemical exchange   94
chiral multispin system   316f.
chiral nitroxide radical   303ff.
– bicyclic   309
– five-membered cyclic   304ff.
– liquid crystalline state properties   318
– magnetic properties   315
– Mn^{2+} complex magnet   315
– α-nitronyl nitroxide (α-NN)   304ff., 316
– properties   303
– PROXYL-type   309
– six-membered cyclic   314
– solid state   315
– structure   304ff., 309
– synthesis   304ff., 309, 315
chiral organic liquid crystalline radical compound   131
cholesterol   137ff., 348
chromophore   218f., 225, 256f., 280
– quencher   219
cisplatin   194
coercivity   28
continuous-wave electron spin resonance (CW ESR)   72, 112, 125ff., 205, 352, 355
– nitroxide-nitroxide pair   122ff.
coordinate bond   49
coordinate compound   50ff., 57
correlation time   91f.
– rotational   92f.
cotton   360ff.
– spin-labeled   360f.
– supramolecular structure   361
Coulomb integral   30f.
covalent bond   49

critical field   29
crown compound   175, 293
– NR carrying   175
– spin-labeled ether   175f., 293
– TEMPO-substituted   175
cryoprotection   333
cryptand carrying NR   175f.
crystal-to-liquid crystal-to-liquid phase transition   318f.
CTMN (complex of transition metal with nitroxide ligand)   287ff., 290ff.
– energy level   291
– ESR properties   289
– exchange coupling   291
– magnetic properties   287
– optical properties   287
– quantomechanical effect   287
– structure   291f.
Cu^{2+} complex   65f., 130, 283, 288
– coordination   66
– structure   65f., 288
Curie law   3, 21ff., 24f., 36f., 45
Curie temperature   25f.
Curie-Weiss behavior   320
Curie-Weiss law   3, 24ff., 38, 57, 64
Curie-Weiss model   189f.
cyclic hydroxylamine   245
– superoxide anion radical detection   246
cyclic nitron   248, 255
cyclodextrin   362f.
– spin-labeling   362f.
cyclotriphosphazene   316f.
cyclotron motion   9f.
cytochrome c   224, 343f.
cytotoxicity   374

d
D-anisotropy   99
D-value   99
DANO (di-p-anisyl nitroxide)   63, 99, 115
– crystal structure   63
– magnetic phase transition temperature   64
– magnetic susceptibility   63
Davis method   113
Debye equation   223
demagnetization   115
denaturation   336
dendrimer   180, 182
– carrying NR   182
– PROXYL radical-functionalized   183
density functional theory (DFT)   255
dextrin   360ff., 363
– spin-labeled   363

DHPC, see 1,2-dihexanoyl-*sn*-glycero-
  3-phosphocholine
diagonalization   95ff.
diamagnetic material   2, 6
diamagnetism   5f., 10, 49
– origins   8ff.
diastereomer method   304f.
dibenzene chromium iodide   215
diffusion-controlled rate constant   227
diffusion-controlled two-dimensional
  reaction theory   226
1,2-dihexanoyl-*sn*-glycero-3-phosphocholine
  (DHPC)   138
dimer model   37ff., 40, 55
5,5-dimethylpyrroline-*N*-oxide (DMPO)
  196, 255f., 286, 389
1,2-dimyristoyl-*sn*-glycero-3-phosphocholine
  (DMPC)   138
– bilayer   138
dioxygen   215
– affecting NR spin-lattice and transverse
  relaxation rate   215
diphenylanthracene   185f.
dipolar broadening effect   124
– computer simulation   124
dipolar interaction   31ff., 34f., 52, 98, 114,
  123, 125, 128f., 217
– relaxation rate   221
– static   131f.
discoidal bilayered micelle   138
– magnetic force alignment   138
distance constraint   122ff., 128ff.,
  131, 133f.
distance distribution   132
disulfide derivative carrying NR   183
DMPC, see 1,2-dimyristoyl-*sn*-glycero-
  3-phosphocholine
DMPC/DHPC bicelle   138
DMPO, see 5,5-dimethylpyrroline-*N*-oxide
DNA   129, 131ff., 251
– complex   129
– DEER measurement   133
– dynamics   354f., 357
– flexibility   355f.
– free radical detection   253
– identification   357
– kinetics   357
– long-range distance constraint   131
– nitroxide spin-labeled   129, 354ff.
– quantitation   357
– structure   354
donor carrying NR   172ff.
double electron-electron resonance (DEER)
  133, 279, 351, 355, 375

double quantum coherence (DQC)
  experiment   133, 342
double resonance   108ff., 113
– pulsed methods   113f.
double superoxide trapping   252
drug, spin-labeled   191, 193f.
drug-biomaterial interaction   191
DTBN (di-*tert* Bu nitroxide)   276, 347
dual fluorophore-nitroxide compound
  227ff., 230, 256ff., 263, 279ff., 283f., 286,
  334, 374, 390
– Dansyl derivative   229, 281
– fluorescence   228
– phosphorescence   228
– properties   227f.
– pyrene derivative   229
– structure   228f.
– synthesis   228
dual molecule   228ff., 256ff., 263, 279ff.,
  283
– chromophore-imidazoline compound
  283
– chromophore-nitroxide probe   372
– paramagnetic nitroxide-naphthalene
  257
– pyren-nitronyl probe   260
– reduction kinetics   372
dual spin-trapping   251
dynamic exchange interaction   218
dynamic model   124

e

echo-detected EPR (ED-EPR)   205, 211f.,
  349
– dynamical transition detection   212
echo-detected ESR (ED-ESR)   106f., 205,
  210, 349
edaravone   378
electron magnetic moment   72, 127
electron magnetic resonance (EMR)   72
electron paramagnetic resonance (EPR)   71,
  185, 224, 231f., 248, 258, 263, 276, 278,
  295, 345f., 372, 376, 383
– doxyl-stearic acid   148
– imidazolidine radical   230
– *in vivo*   376
– low-frequency   216
– oximetry   217, 383f.
– pH effect   230
– pyrene-imino-nitroxide   247
– spectrum   140, 145, 148, 278
– spin-marked PEVP   278
– thiol group detection   231f.
– tomography   279

electron spin echo (ESE) 104, 131
– two-pulse method 104
electron spin echo envelope modulation (ESEEM) 108, 352
electron spin resonance (ESR) 71ff., 74ff., 175, 178, 225, 227, 239, 258f., 262, 269, 274f., 283, 285, 292ff., 296, 338, 340, 346, 356, 361f., 374
– additive formation process monitoring 272
– fundamental 71ff.
– historical background 71f.
– *in vivo* 378
– magnetic material 114f.
– microscopy (ESRM) 392
– oximetry 217
– parameter 206, 361
– pH-dependent 191
– powder pattern, *see* ESR powder pattern
– resonance condition equation 73, 78
– resonance interpretation 74
– rotational diffusion parameter 332
– segmental dynamics 276
– sensitivity 225
– slow motion, *see* slow motion ESR
– spectrum, *see* ESR spectrum
– spin-labeling, *see* spin-labeling ESR
– time-resolved, *see* time-resolved ESR
– titration curve 151
electron transfer 219, 279f., 283, 372
– photo-induced 283
electron-electron double resonance (ELDOR) 111ff., 114
– ESR sublevels 112
– three-pulsed echo-detected 114
electron-exchange induced intersystem crossing 219
electron-nuclear double resonance (ENDOR) 108ff., 113, , 148f., 214, 338
– absorption 109
– HF pulsed 148f.
– pattern 109
– resolution 110
– spectrum 109ff.
– superimposed Mims-type HF 149
electronic relaxation 126f.
– spin pair enhancement 130
– time 126f.
electrostatic effect 221
– charge effect on dipolar interaction 221ff.
– quantitative characterization 221
electrostatic field 222f.
– influence factor 222

electrostatic interaction 221
– characterization 221
electrostatic potential 221, 223f., 331
– DNA surface 224
Elliot mechanism 106
enantioselective oxidation of chiral secondary alcohol 314, 326
enzyme 331ff., 336ff., 339f.
– active center structure 338ff.
– catalysis mechanism 338
– conformational transition 332ff., 336f.
– dynamic properties 331
– electron transfer 332
– intramolecular dynamics 332ff., 336
– low-temperature molecular dynamics 332f.
– structure 331
epimerization mechanism 309, 311
erythrocyte 372
ESR imaging (ESRI) 279, 282, 381
ESR microscopy (ESRM) 392
ESR powder pattern 99ff.
– fine structure 101
ESR spectrum 73f., 79, 86, 91, 110, 207, 216, 249f., 257, 277, 288, 293, 342f., 349f., 355, 393
– concentration dependence 94
– DEPMPO-O_2H 249f.
– DTBNO 93f., 97f.
– EMPO-O_2H 249f.
– hyperfine structure 86
– nitronyl 247
– nitroxide motion effect 355
– nitroxide radical 94
– pH-sensitive parameter 296
– superoxy adduct 250f.
– temperature-dependent 93
– VO^{2+} 92f.
exchange integral 30f., 287, 289
exchange interaction 30f., 34, 40, 51f., 55, 213f., 224
– charge impact 223
– equilibrium 240
– radical-paramagnetic complex 222
exchange relaxation 213ff.
– rate constant 213

f

Faraday 1f., 43
– balance 2
– method 43
fatty acid 127, 139f.

Fermi interaction   85
ferrimagnetic resonance   72, 116, 118
ferrimagnetism   7, 51, 67, 116, 118, 315
ferrocene   172
ferroelectric liquid crystalline (FLC) state   320
ferroelectricity   322
ferromagnetic material   6f., 28
– molecular orbital   290
ferromagnetic resonance (FMR)   72, 115f.
ferromagnetism   6f., 25ff., 289, 291, 293, 316
fibroblast   312
field-cycled dynamic nuclear polarization (FC-DNP)   375, 381
fine structure   90f., 101
– magnetic field orientation dependence   91
– splitting   101
– triplet state   90
Fischer equation   273
flopping   29
fluidity   206
fluorescence   218f., 224f., 227f., 256ff., 259ff., 262f., 279ff., 296
– intensity   258
– kinetics   266f.
– label mobility   338
fluorescence quenching   219, 224f., 227f., 256f., 279ff., 296, 352
– limitation   225
– mechanism   219, 282
fluorescence spectroscopy   227, 263
flux density   6
folding   336
forbidden absorption band   288
forbidden transition   91, 291
Forrester-Hepburn mechanism   253
F_5PNN (pentafluorophenyl nitronyl nitroxide)   59f.
– crystal structure   59
– magnetic properties   59f.
– magnetic susceptibility   59
– molecular network   59
free induction decay (FID)   77, 102ff.
– quadrature-detected   103
free radical detection   253
– DNA   253
free radical formation *in vivo*   378
Freed model   208
Fremy's salt (dipotassium nitrosodisulfonate)   55, 102f., 161, 210, 212
– FID   102f.

– FT-ESR   102
– molecular reorientation   210
– pulsed MF EPR   210
– relaxation mechanism   210
FT-ESR   102f.
– fundamental concept   102f.
– quadrature-detected   103
– time-evolution   102
fullerene   283
functional nitrone   196ff.
– dual functionality   197
– NR detection   196
– phosphorous-containing   196
– synthesis   196
functional nitroxide radical (FNR)   161ff., 191ff., 194ff.
– biomedical application   191ff.
– historical survey   161ff.
– preparation   161ff.

g
g-anisotropy   94, 100
g-value (gyromagnetic ratio)   14, 20f., 35, 52, 74, 94ff., 99f., 116, 146f., 207, 220, 230, 288
– effective   95, 97
gastric carcinoma visualization   382f.
Gaussian distribution   81
Gaussian line shape   80, 114
Gd^{3+} ion   127
– electronic properties   127
Gd-DOTAP   127f.
gene expression   320
gene regulation   357
general TRIPLE (GTR)   110
– spectrum   111
Gouy balance   2
Gouy method   43

h
Hahn's echo   72
Hahn's echo method (two-pulse method)   104
Haldane gap   42
half-field resonance   91
Hamiltonian   16, 19, 30ff., 33ff., 36ff., 56, 84, 96, 141f.
– hyperfine   84f.
– time-dependent   142f.
Heisenberg antiferromagnet   294
Heisenberg Hamiltonian   31, 52
Heisenberg model (1-D)   189
helical tilt   139, 141
heme protein   334f., 337

hemoglobin 334, 336
Henderson-Hasselbalch titration curve 151
heterospin complex 291f.
hexafluoroacetyl acetonate (hfac) 65
HF CW ESR 134
HF ELDOR 135
HF ESR 134, 141ff., 147, 150, 349, 353
– advantages 144
– hydrogen bond sensitivity 150
– nitroxide microenvironment 146ff.
– nitroxide spectra changes 141
– pH-dependent 150f.
– recent development 142
– spin-labeling 141ff.
hfac, see hexafluoroacetyl acetonate
high-resolution NMR 122, 131
highest occupied molecular orbital (HOMO) 49
host-guest complex 178
HSQC, see ^{15}N heteronuclear single quantum coherence spectrum
human disease 388ff.
human serum albumin (HSA) 127f., 333, 335
– intramolecular mobility 333
– nanosecond intramolecular mobility 334f.
– polynitroxyl (PNA) 387
human skin ESR imaging 382
Hund's rule 15
hydrogen atom 87ff.
– ESR spectrum 88
hydrogen bond formation 147ff., 150, 220
– geometrical parameter 150
hyperfine coupling 94, 207
hyperfine interaction 84ff.
hyperfine splitting (hfs) 85f., 88ff., 92, 96, 100, 175, 206, 220, 230f., 288
– anisotropy 100, 206
– effective 96
– isotropic 88
– pH dependence 231
hyperfine structure 84, 86
hysteresis curve 28, 65

i
imidazolidine NR 150, 191, 230, 232
imidazoline NR 191, 230
imino nitroxide 246, 262
immersion depth 217f., 347, 353
– chromophore 217ff., 347
– determination 217ff.
– fluorescent center 217ff., 347
– luminescent chromophore 218f.

– membrane 219, 347
– paramagnetic center 218
– radical center 217f.
immuno-spin-trapping technique 253
IMTSL (methanethiosulfonic acid S-(1-oxyl-2,2,3,5,5-pentamethyl-imidazolidin-4-ylmethyl)ester 151
in vivo spin-trapping/ex vivo detection 377f.
inducible nitric oxide synthase (iNOS) 378
interchain interaction 60ff.
intermolecular magnetic interaction tuning 190
interspin distance determination 124
intersystem crossing (IC) 281, 283
– reverse 283
inversion recovery method 105f.
ion-relaxater 218
ionomer 276
ionophore 175
– spin-labeled 175
IPNN, see 2-isopropyl nitronyl nitroxide
ischemia-induced myocardium acidosis pH kinetics 380
ischemia-reperfusion injury 373, 376ff., 381, 385ff.
Ising model 42f., 52, 60
isopotential spin-dry ultracentrifugation (ISDU) 137f.
2-isopropyl nitronyl nitroxide (IPNN) 66

k
Kittel's equation 115f.
Kittel's mode 116f.
Kornblum reaction 304
Kramer transfer 209
Kubo-Tomita theory 114

l
Langevin function 21ff.
Langmuir-Blodgett (LB) film 292f.
– ionic channel 293
– molecular surface motion 293
– preparation 293
Larmor precession 74f.
laser flash photolysis (LFP) 286
Lentz law 9
life-time broadening 92
ligand 65f.
line shape 78ff., 81, 83, 114, 128, 143f., 215, 342, 360
– analysis 215
line-width 79ff., 84, 127, 214, 349

– dipolar broadening   81
– exchange narrowing   81
– oxygen molecule   94
linear combination of atomic orbitals-MO (LCAO-MO)   56
Liouville-von Neumann equation   143f.
lipid bilayer   136ff.
– macroscopic alignment   136ff.
– mechanical alignment   136f.
– nanotubular   139
– oxygen profile   347
– preparation   138
– spin-labeling ESR   136ff.
– substrate-supported   139
– water concentration profile   348
– water penetration depth   348
lipid nanotube array   139, 141
– advantages   141
liquid crystal (LC)   131, 318ff.
– DOXYL   318f.
– paramagnetic rod-like   318
– phase-transition behavior   320ff.
– PROXYL, see also PROXYL liquid crystal   320ff.
– spin probe   318
– TEMPO   318f.
local charge   223f.
– calculation   223
– determination   224
– electrostatical   224, 339
local pH   150, 230, 296, 331
– inside sorbent pore   297
local pKa   150
local polarity   146f., 152, 281, 331
local proticity   146ff.
long-range distance constraint   128ff., 131, 135
longitudinal EPR (LODEPR)   375, 381
Lorentzian line shape   79ff., 83, 114
– homogeneous broadening   128
low-dimensional magnetic material   114f.
low-field RF-EPR   375
low-temperature dynamic behavior model   210
lowest unoccupied molecular orbital (LUMO)   49
luminescent chromophore   218, 315

m

macromolecular distance constraint   122ff.
macroscopic alignment   136ff.
magic angle spinning (MAS)   130

magnetic cluster   38
magnetic conductor   170
magnetic dipole-electric dipole interaction (magnetoelectric interaction)   320
magnetic energy   3
magnetic field   2f.
– alignment   138
magnetic flux   6
magnetic induction   3f., 6
– method   43
magnetic interaction   30ff.
– network   40f.
magnetic linear chain   40f., 59
magnetic material
– three-dimensional   290
– two-dimensional   290
magnetic measurement   2
magnetic moment   3f., 13ff., 17ff., 20, 44
– $3d^n$ ions   20
– disordered state   7
– evaluation   44
– $4f^n$ ions   20
– general cases   17f
– ordered state   7, 26f.
magnetic permeability   5f.
magnetic quantum number   11ff.
magnetic resonance   71ff.
– classification   72
– historical background   71f.
magnetic susceptibility   4ff., 9, 21ff., 24ff., 29, 35f., 38f., 41, 43f., 52, 59, 283, 289, 317f.
– calculating method   35
– diamagnetic substances   44
– measurement   43f.
– temperature dependence   21ff., 318
– van Vleck formula   35ff., 38f.
magnetic transition temperature   62
– pressure dependence   62
magnetism   1ff.
– atom-based   47
– historical background   1ff., 47f.
– molecular   47ff.
– one-dimensional   40f., 59
– origins   8ff., 47f.
– two-dimensional   63
magnetization   4, 21ff., 25ff., 28, 76, 102, 105f., 115
– residual   28
– rotation   102
– spontaneous   26ff.
magneto-chiral dichroism (MChD)   315f.
Marcus equation   242
Marcus region   282

Marcus theory of electron transfer 281
McConnell's formulation 56, 89f.
Meissner effect 6
membrane 136ff., 146, 350ff.
– depth parameter 352
– dynamics 349f.
– fluidity 371f.
– label location 347f.
– liquid-(dis-)ordered lipid domain 348
– microstructure 348f
– orientation 353
– oxygen location 347f.
– peptide aggregation 352
– peptide position 352f.
– protein 345f.
– protein location 352
– spin-labeled 143f.
– water association 347
– water location 347f.
merocyanine 186f.
metal ion determination 390
metal ion-nitroxide distance measurement 130
metalloenzyme 224, 339
metalloprotein 131, 224, 339
metamagnetism 30, 315
metastasis inhibition 375
methane thiosulfonate spin-label 354f.
methyl nitronyl nitroxide (MeNN) 60
micelle 347
micro-relief 213
microcapsulation technique 392
micropolarity 280
microviscosity 206
Mims method 113
mitochondrial permeability transition (mPT) 373
Mn^{2+} complex 65f., 315
– magnet 315
$Mn(hfac)_2$ 66
$Mn(hfac)_2$-IPNN 66f.
– ferrimagnetism 67
– interacting network 67
– magnetic susceptibility 67
Mn^{2+}-IPNN 66
– magnetic phase transition 66
– magnetic susceptibility 66
Mössbauer atom 333
Mössbauer label 333f., 338
Mössbauer spectroscopy 333f.
molecular breathing 331
molecular dynamics (MD) 142ff., 205ff., 225, 227f., 230, 274, 280f., 357
– cascade method 225
– method 145f., 225
– multi-nanosecond simulation 343
– nitroxide spin-label modeling 146
– simulation 145f., 343
– surrounding molecule 206f.
molecular motion 206ff.
– high-amplitude low-frequency 208
– low-amplitude high-frequency wobbling 208, 210
molecular tumbling 208
MTSL (1-oxyl-2,2,5,5-tetramethylpyrroline-3-methyl) 340f., 355
multifrequency ESR 142ff., 146, 206, 375
– advantages 144
multilayer structure, planar-supported 137
multiple-quantum experiment 133
multiple-spin cluster case 38f.
multiquantum (MQ) ESR 135f., 375
– advantages 135
– nitroxide-labeled protein 135
muscle protein 344f.
myocardial ischemia-reperfusion injury 376f.
myoglobin 224

n

^{15}N heteronuclear single quantum coherence (HSQC) spectrum 129
N–C-bond 53
N–O-bond 53f., 57
– spin density distribution 57
– stability 53
– structural resonance 54
nanocomposite 278
nanoparticle 295f., 393
– metal 295
– nitroxide spin-labeled 295
nanopore-confined cylindrical bilayer 139f.
nanoporous channel 139ff.
nanotubular bilayer 141
near-edge X-ray absorbtion fine structure (NEXAFS) spectroscopy 293
Nèel temperature 26
nitric oxide 246
– analysis 261f.
– analysis by pyren-nitronyl 261
– dynamics real-time monitoring 262
– quantitative analysis 247
– trapping 254f.
nitrone, see also functional nitrone
– addition reaction 249, 252, 254ff.
– chemical structure 248ff.
– drug 385, 387f.

- inhibitor activity   251
- kinetics   254ff.
- nitric oxide addition   254
- NMLP regulation agent   212
- non-radical reaction   253
- nucleophilic addition   253f.
- radical reaction   248ff.
- spin-trapping   247ff., 252, 256, 393
- synthesis   252
- thermodynamics   254ff.

nitronyl complex   290
- binuclear   290
- quantomechanical effect   290

nitronyl nitroxide (NN)   56f., 66, 163, 246, 291
- complex   291
- cytosine-substituted   178
- general preparative scheme   163
- phenyl-substituted, deoxygenation reaction   171
- uracil-substituted   178

p-nitrophenyl nitronyl nitroxide, *see p*-NPNN

nitroxide
- A-tensor   96, 206
- antioxidant activity   375, 386f.
- antioxidant status   240ff., 373
- antitumor activity   375
- biochemistry   375ff.
- biomedical application   371ff., 390f.
- clinical application   389
- coordination effect   288
- CW ESR spectrum   216
- diffusion into neuronal cell, schematic depicting   386
- dissociation energy   240
- disulfide-modified spin-label   295
- drug   385f., 388
- dynamic theory   206
- exchange reaction equilibrium   240
- future development   391f.
- g-tensor   96
- hydrogen bond formation   147, 220
- *in vivo* nitron trapping   374
- *in vivo* spin-labeling   374, 386
- inhibitor activity   251
- inorganic chemistry   292ff.
- ligand   287f.
- low-dimensional properties   57ff.
- magnetic interaction   55f.
- magnetic phase transition   55
- magnetism   55, 57f.
- MD   205ff.
- MD applied to ESR   145
- medical application   371ff., 385ff.
- metal ion coordination   65f.
- microenvironment   146ff.
- molecular interaction   55f.
- motion   206
- multispin system   284f., 316
- neuroprotection   386
- one electron redox process   239
- orientational distribution   279
- pH probe   230f.
- polarity probe   220
- polarized resonance structure   220
- polymer investigation   274
- polyradical   67
- power saturation curve   216
- protecting effect   386f.
- radical polymerization inhibitor   245, 269
- redox kinetics   382
- redox potential   240, 242, 387
- reduction   240f.
- SH-group probe   231
- skin toxicity   389
- solvent effect   147
- spatial distribution   279
- spin oximetry (NSO) *in vivo*   383f., 392
- therapeutic application   388ff.
- toxicity   389

nitroxide mediated living polymerization (NMLP)   269ff., 272ff.
- chemistry   270ff.
- kinetics   272f.
- performance improvement   274
- principle reaction   270
- technology   273f.
- thermodynamics   272f.

nitroxide mediated polymerization (NMP)   272

nitroxide radical (NR)   50, 60, 67, 151
- building block   53, 56, 178
- chiral, *see* chiral nitroxide radical
- clearance   318
- diaryl   161f.
- ESR powder pattern   100f.
- ESR spectrum   220
- exchange interaction with paramagnetic species   213
- functional, *see* functional nitroxide radical
- g-value   52, 100
- hyperfine splitting   93
- *in vivo* distribution   381
- intramolecular fluorescence quencher   227f., 280f.
- isotropic hyperfine splitting   90
- ligation   57
- metabolism   381
- pH-sensitive   296f., 380f.

- quantitative characterization of redox processes 239
- quencher 219, 227
- redox properties 323ff.
- site switching 57
- spin polarization 90
- synthesis 162
- TEMPO-based 323
nitroxide spin labeling 128f., 130, 134, 373, 390
- *in vivo* reduction 373
nitroxide-metal ion pair 125ff., 128
- CW-ESR spectrum 125ff.
nitroxide-nitroxide pair 122ff., 125, 128
- dipolar interaction 123ff., 128
- distance determination 133
- spin exchange 123
- spin-spin dipolar interaction 123
NMLP, *see* nitroxide mediated living polymerization
non-Kramer transfer 209
nonlinear mesoscopical-ferromagnetic interaction 323
p-NPNN (*p*-nitrophenyl nitronyl nitroxide) 64f.
- ac susceptibility 64
- crystal structure 64f.
- heat capacity 64
- hysteresis curve 65
- transition temperature 64
nuclear magnetic moment 14
nuclear magnetic resonance (NMR) 71, 128, 217
nuclear Overhauser effect (NOE) 122
nuclear spin relaxation parameter 217
- paramagnetic center effect 217
nucleic acid 354ff.
- nitroxide spin label modification method 354
nucleobase 178f.
- NR spin-labeled 179
nucleoside 178f.
- NN spin-labeled 178
- NR spin-labeled 178f.
nucleotide 338f.
- spin-labeled mobility 338f.
nutation frequency 108
nutation spectroscopy 107f.

o

one-ion anisotropy 35
orbital angular momentum 12ff., 16ff., 19f.
- vector model 12
orbital multiplicity 17

orbital quantum number 16
orbital quenching 19f., 23f., 34
organic biradical 180
organic ferromagnet 165ff.
- benzylideneamino-TEMPO radical-based 166
- ferromagnetic behavior 166f.
- NN radical-based 166
- NR-based 165ff.
- phase transition temperature 166
- substituted-phenyl NN radical-based 165
- synthesis 167
- TEMPO radical-based 166
organic photofunctional spin system 184
organic radical 48f., 52f., 161ff., 163ff.
- crystal 171
- magnetic properties 52f., 55
organoferromagnetic conductor 172
orientation distribution function (ODF) 279f.
orthorhombic case 101
outer membrane protein A (OmpA) 129
Overhauser-enhanced magnetic resonance imaging (MRI) 375, 382
oxazolidine-1-oxyl (DOXYL) 304, 313f.
- chiral 313f.
- spin label 313
- synthesis 313f.
oxidation 239f., 314, 323, 325f.
- alcohol 314, 323, 325f.
- catalyst 323f.
- enantioselective 314, 325f.
4-oxo-TEMPO 163f., 379
oxoammonium cation 242ff.
oxygen sensor 216

p

Pake doublet 114
Pake model 124
paramagnetic complex 168
paramagnetic material 2, 6f., 214ff., 217
- rod-like liquid crystal 318
paramagnetic relaxation enhancement (PRE) 128ff.
- protein topology 129
paramagnetism 5, 7, 10ff., 49f., 214f.
- molecular 49, 51f.
- origins 10ff.
- temperature-independent 39f.
Parkinson's disease 377
Pauli exclusion principle 15
PELDOR 114, 352, 355, 357, 375
pentafluorophenyl nitronyl nitroxide, *see* F$_5$PNN

persistent radical effect (PRE)   272f.
pH marker   230f.
phase memory time   105
phase transition   25f., 29, 55, 60
phonon dynamics   228
phonon process   210
phosphatidylcholine lipid   137
phospholipid   136, 140f.
– nitroxide spin-label modified   351
phospholipid bilayer   136, 348, 350
– phase behavior   146
– polarity gradient   147
– self-assembly   140
phosphorescent chromophore   218
photo-dimerization   189
photo-switching   185
photochemistry   279ff., 285f.
– reaction mechanism   285
photochrome   226f.
photochromic reaction   184, 187, 189f.
– anthracene   189f.
– norbornadiene   189f.
photochromic spin system carrying NR   184ff.
– intramolecular magnetic interaction switching   185
photoelectron transfer   279ff.
photoisomerization   188f., 225ff., 260, 294
– azobenzene   188
– kinetics   294
photolysis   286
photonucleation   286
photophysics   279ff.
photoreduction   279ff.
– mechanism   282
– rate constant dependence on reduction power   282
picosecond fluorescent time-resolved technique   334f.
platinum complex   194
polar relaxation   228
polymer   269ff.
– destruction   279
– dynamic properties   274f.
– MD   274ff., 277
– microstructure   274ff.
– mineral filler   277
– nitroxide modification   274f.
– segmental dynamics   275ff.
– self-associating   276
– spin-labeling   275f.
– thermal degradation   279
polymeric alkoxyamine   270ff.
polymerization   164, 180, 183, 269ff.

– alkoxyamine-mediated   271
– nitroxide-mediated   270ff.
– radical   180, 270ff., 273ff.
– styrene   180, 271f.
– TEMPO-substituted biradical   164f.
– thermal   180
polysaccharide   360ff.
– spin-labeled   360f.
porphyrexide   161f.
– synthesis   162
potassium ferricyanide   215
power saturation curve   216ff.
– CW ESR analysis   217f.
precession mode   116
predenaturation phenomena   336
Predici simulation   273
principle axis transformation   95
probe mobility   227, 333ff.
protein   125, 146, 331ff., 350ff.
– binding site mapping   129
– conformational change   336f.
– dynamic properties   331f., 334f.
– electron transfer   332
– free radical detection   253
– functional properties   335
– helix packing   127
– intramolecular dynamics mechanism   336
– intramolecular mobility   333ff.
– long-range distance constraint   128ff.
– MD   332
– nanosecond intramolecular mobility   334f.
– orientation   137
– soluble   341ff.
– spin-labeled   125, 135, 143f.
– structure   331
– structure determination   128, 137
protein-lipid interaction   350f., 353
proton-electron double resonance imaging (PEDRI)   381
PROXYL derivative, *see also* pyrrolidine-1-oxyl   180, 183, 188f., 386f.
PROXYL liquid crystal   320ff.
– ferroelectric properties   322
– magnetic properties   324
– optical response time   322
– phase-transition behavior   320ff.
– phase-transition temperature   321
– spontaneous polarization   322
pulse sequence   105, 113
– 2+1   131
pulsed double-electron electron resonance (pulsed DEER)   114, 131ff., 180, 374

– four-pulse experiment   131f., 180
– pulse pattern   132
– pulse-train   132
pulsed electron spin resonance
   (pulsed ESR)   72, 102ff., 130, 205, 342
– distance measurement   342
– principle   77
pumped transition   109
purine-purine step   356
pyrimidine-pyrimidine step   356
pyrrolidine-1-oxyl (PROXYL)   304ff., 307ff.
– chiral   305ff., 308
– conglomerate   308ff.
– enantiomeric resolution   311f.
– epimerization   309, 311
– liquid crystal, see PROXYL liquid crrystal
– racemization   308ff., 311
– racemization mechanism   311f.
– synthesis   305ff., 308f.
– X-ray structure   309f.

q

quencher   219, 227, 257, 262, 280
quenching   19f., 23f., 34, 219, 224ff., 257, 279ff., 347, 351
– excited-triplet state   226
– mechanism   284

r

radiation-induced alopecia   387
radical battery   325f.
radical ion salt, NR-based   168ff., 171ff., 174
– anion salt, TEMPO-based   171
– cation salt, NR-based   170
– conductivity   171f., 174
– ferromagnetic   170
– IN-substituted alkylpyridinium   170
– magnetic behavior   172, 174
– nitroxide-substituted sulfonate   171
– semiconductivity   171f., 174
radical location determination   218
radical pair (RP)   209
– EPR spectrum   209
radical polymerization   269ff., 272ff.
– reversible radical inhibition   272
radical scavenger   194
radioprotection   375, 379
– in vivo   379
Raman three-dimensional (non-)Kramer transfer   209
reactive oxygen species (ROS)   373, 377f., 381f., 385
– detection   381

– in vivo   378, 381, 385
rechargeable lithium battery   325
red blood cell   372
– glutathione level reduction   372
redox potential   240, 242, 255, 258
redox reaction, molecular imaging   382
redox sensor   256f.
redox status   372f., 376, 379, 381, 385
relaxation   77, 210
– enhancement effect   130, 351
– orientation-dependent   210
– spin-lattice   81f.
– spin-spin   82
– transverse   210
relaxation time   81, 105ff.
– longitudinal   81
– spin-lattice   81f., 105f.
– spin-spin   81f., 105
– temperature-dependent   106f.
– transverse   82
residual dipolar coupling (RDC) NMR   122
resonance position   74
resorcinarene   178
restricted open-shell Hartree-Fock (ROHF) method   147
reversible intramolecular electron transfer   281f.
reversible thermal C–O bond homolysis   270
rhodium complex   283
rhodopsin   344f.
– structure   344
RNA   129, 354f., 358ff.
– complex   129
– conformational change   358, 360
– dynamics   354
– folding   358
– HIV-1 TAR RNA internal dynamics   358, 360
– long-range distance constraint   131
– SDR model   359
– spin-labeled   133, 354, 359f.
– structure   354, 358
– thermal denaturation   359
rotational diffusion   126, 331
Russell-Saunders coupling   16ff., 21

s

salivary hypofunction   377
saturation transfer (ST) ESR   362
Schrödinger equation   11, 13, 20
secular broadening   92
segmental dynamics   207f.
self-assembly   140, 276, 293
sensitizer   225ff.

SI units 8, 74
silica layer formation 294
silicon phthalocyanine (SiPc) 284f.
– electronic state 284
– NR covalently linked 284f.
single crystal study 94, 99
single-frequency technique for refocusing (SIFTER) 189
singlet-triplet (ST) model 38, 102, 189
singly occupied molecular orbital (SOMO) 49
site-directed nitroxide spin-labeling (SDSL) method 121, 128, 131, 152, 340f., 344f., 350f., 354f., 357f.
– pH-sensitive 150ff.
slow anisotropic rotation model 208
slow motion ESR 143ff.
– line shape determination 144
– line shape theory 143
– spectra simulation 143ff.
slowly relaxing local structure (SRLS) model 144
smectic phase 318
snapshot effect 144f.
solid-state nuclear magnetic resonance (ssNMR) 130, 138, 141
solvent effect 147
special TRIPLE 110
spectral-spatial ESR imaging (ss-ESRI) 380
spin adduct 248
spin alignment 185f.
– diphenylanthracene 186
– photo-induced 185f.
spin angular momentum 12, 14, 16f.
– vector model 12
spin cluster case 37
spin density 54, 56
spin diffusion model 114
spin echo 77
– principle image 104
spin imaging technique 381, 392
spin label 205, 213, 390
– mobility 338
– structure 340
spin label-spin probe method 212ff.
spin long-range ordering 60ff.
spin multiplicity 17, 30f.
spin only case 18ff.
spin oximetry 215ff., 383f., 392
– *in vivo* measurement 216, 383f., 392
spin packet 80
spin pH probing *in vivo* 380f.
spin pharmacokinetics *in vivo* 378f.

spin polarization 88ff.
– mechanism 88f.
spin probe 191f., 205ff., 209, 213, 215, 223, 277, 280, 294, 318, 349, 390, 392
– adsorption 294
– aqueous solution 215
– degradation 295
– different charged 223
– diffusion 361
– distribution 294
– glassy system 210f.
– hydrophobic 274, 294
– *in vivo* 376
– low-amplitude high-frequency and phonon dynamics sensitive parameters 209
– mobility 333ff.
– paramagnetic 216
– pH-sensitive 191f., 380f.
– photochrome 225, 227
– pulsed MF EPR 210
– synthesis 192
– triplet 225
spin quantum number 12f., 16
spin relaxation rate 221f.
spin state 48f.
spin-exchange interaction 38
spin-flopped state 29f.
spin-labeled drug 191, 193, 388
spin-labeled spin-trap 194f.
– synthesis 194f.
spin-labeling ESR 121, 141, 277f.
– high magnetic field 141ff.
spin-labeling magnetic resonance 122ff.
spin-labeling method 205, 212, 232, 269, 276ff., 293, 331ff.
– macromolecule 207f., 269
spin-lattice relaxation rate 222, 287
– spin probe concentraton dependence 222
spin-orbit coupling 16f., 19, 34f.
spin-probe-partitioning electron paramagnetic resonance (SPPEPR) 347
spin-redox probing 227, 239, 256
– requirement 239
spin-relaxation parameter 223
spin-spin exchange 287, 293, 355
spin-spin dipole-dipole interaction 287f., 297, 360
spin-spin interaction 217, 287, 293
– mesoscopical ferromagnetic 320, 323
spin-trap 194f., 227, 244, 247ff., 252ff., 255ff., 263f., 285f., 376, 378, 392f.

- characterization 248
- dual molecule 263
- half-life 255f.
- hydroxyl radical 255, 263
- photochemical reaction 285f.
- photochromic 197

spin-triplet-fluorescence-photochrome method 224ff.

spiropyran 186f.
- nitrone-substituted 197
- photochromic reaction 187
- TEMPO radical-carrying 186f.

stimulated echo 105
stochastic Liouville equation (SLE) 143, 206, 341f., 355
stochastic Liouville theory 143
Stokes-Einstein equation 208, 216, 384
Stokes-Einstein model 93
stomach acidity 380f.
super exchange 287
super slow rotation 207
superconducting quantum interference device (SQUID) 43
superconductive material 6
superoxide 245, 255
- adduct 250ff., 255f.
- analysis by pyren-nitronyl 261
- anion radical 245
- detection 245f., 248, 256, 263
- nitronyl reaction 246
- production rate 261

superoxide dismutase (SOD) 242, 377
- mimetic activity mechanism 242

supramolecular rod incorporating NR 180f.
supramolecular spin system carrying NR 175ff., 178
supramolecular wire incorporating NR 180f.
surface electrostatic potential (SEP) 297
surface microstructure 294

t

T4 lysozyme (T4L) mutant 126f., 137, 342f.
- conformational transition 343
- doubly labeled 342
- ESR spectrum 126
- nitroxide-labeled 127
- orientation 342
- ribbon structure diagram 126, 342
- secondary structure 342
- spin-labeled 126f., 133, 137

- tertiary structure 342

TANOL, see TEMPOL

TEMPO (2,2,6,6-tetramethyl-1-piperidinyl-oxy) 163ff., 170ff., 178, 180, 186ff., 189f., 228, 257, 263, 273, 276, 283, 294f., 374, 379, 387, 389
- Dansyl-TEMPO 228, 257, 259, 281, 335
- indol-TEMPO 228
- 4-oxo-TEMPO preparation 163

TEMPOL (2,2,6,6-tetramethyl-4-piperidinol-1-oxyl) 57f., 61, 64f., 164f., 220, 271, 280, 339, 372ff., 375ff., 387f.
- crystal structure 58, 65
- Cu^{2+} complex 65f.
- cytotoxicity 374
- magnetic susceptibility 58
- magnetism 58f., 61f.
- Mn^{2+} complex 65f.
- molecular network 58
- neuroprotective effect 377
- ODF 280
- phase transition temperature 62
- pressure-dependent magnetic interaction 62
- reduction 374
- structural formula 164

tether-in-a-cone model 125, 132, 342
Teuber reaction 161
thiol group detection 231
thiol group labeling 152
thiol-disulfide exchange 232
three-axis anisotropy 100
three-pulse method 105f.
through-bond interaction 51ff.
through-space molecular interaction 51, 53, 58, 316
time domain magnetic resonance 128ff.
time-resolved chemically induced dynamic nuclear polarization (TR-CIDNP) 286
time-resolved ESR (TRESR) 185, 284
time-resolved fluorescence quenching (TRFQ) 347
tissue redox status in vivo 385
total angular momentum 13
transactivating responsive region (TAR) 358, 360
transition metal complex, see also CTMN 287ff., 291ff.
TRIPLE 110f.
triplet-triplet energy transfer 225ff.
TTF 170ff., 173f.
- ESR 172f.
- 4-hydroxy-TEMPO-substituted 174

– magnetic behavior   172
– radical cation   172f.
– TEMPO-substituted   174
– tetrakis-TEMPO-substituted   172
TTP (tetrathiapentalene)   174
tumor redox status   381
two-pulse method   104f., 108

u
ultra-slow motion   209

v
valence-bond theory   54
van Leeuwen theorem   10
van Vleck formula   35ff., 38f.
vibrating sample magnetometer (VSM)   43
viscosity   260

w
Walker mode   116
Watson-Crick-type molecular complex   178
weakly bending rod model   355f.
wobbling   208, 276
– low-amplitude high-frequency   208, 210

x
xanthine oxidase   378

z
Zeeman anisotropy   141
Zeeman effect   13, 18f., 73
Zeeman energy   19, 34
Zeeman Hamiltonian   19, 35, 37, 40
Zeeman splitting   73
Zeeman term   19, 36, 141
zero-field splitting   94